T0296504

Fungi in Biogeochemical Cycles

Fungi have important roles in the cycling of elements in the biosphere but are frequently neglected within microbiological and geochemical research spheres. Symbiotic mycorrhizal fungi are responsible for major transformations and redistribution of inorganic nutrients, while free-living fungi have major roles in the decomposition of organic materials, including xenobiotics. Fungi are also major biodeterioration agents of stone, wood, plaster, cement and other building materials, and are important components of rock-inhabiting microbial communities. The aim of this book is to promote further understanding of the key roles that free-living and symbiotic fungi (in mycorrhizas and lichens) play in the biogeochemical cycling of elements, the chemical and biological mechanisms that are involved, and their environmental and biotechnological significance. Where appropriate, relationships with bacteria are also discussed to highlight the dynamic interactions that can exist between these major microbial groups and their integrated function in several kinds of habitat.

The British Mycological Society promotes all aspects of fungal science. It is an international society with members throughout the world. Further details regarding membership, activities, and publications can be obtained at *www.britmycolsoc.org.uk.*

GEOFF GADD is Professor of Microbiology, Head of the Division of Environmental and Applied Biology, and Deputy Research Director in the School of Life Sciences at the University of Dundee.

Fungi in Biogeochemical Cycles

EDITED BY

GEOFFREY MICHAEL GADD

School of Life Sciences
University of Dundee

CAMBRIDGE
UNIVERSITY PRESS

CAMBRIDGE UNIVERSITY PRESS
Cambridge, New York, Melbourne, Madrid, Cape Town,
Singapore, São Paulo, Delhi, Tokyo, Mexico City

Cambridge University Press
The Edinburgh Building, Cambridge CB2 8RU, UK

Published in the United States of America by Cambridge University Press, New York

www.cambridge.org
Information on this title: www.cambridge.org/9781107403215

First published 2006
First paperback edition 2011

A catalogue record for this publication is available from the British Library

ISBN 978-0-521-84579-3 Hardback
ISBN 978-1-107-40321-5 Paperback

In loving memory of my mother, Mary (Sheila) Gadd, who encouraged my love of the natural world from an early age

Contents

Contributors

Michael Allen
Center for Conservation Biology
Departments of Plant Pathology and Biology
University of California
Riverside
CA 92521–0334
USA

Sarah R. Barratt
Faculty of Life Sciences
1.800 Stopford Building
University of Manchester
Manchester M13 9PT
UK

Dan Bebber
Department of Plant Sciences
University of Oxford
South Parks Road
Oxford OX1 3RB
UK

Lynne Boddy
Cardiff School of Biosciences
Cardiff University
Main Building
Park Place
Cardiff CF10 3TL
UK

Olivier Braissant
Institut de Géologie
Université de Neuchâtel
11, rue Emile Argand
CH-2007 Neuchâtel 7
Switzerland

Euan P. Burford
Division of Environmental and Applied Biology
Biological Sciences Institute
School of Life Sciences
University of Dundee
Dundee DD1 4HN
Scotland UK

Guillaume Cailleau
Institut de Géologie
Université de Neuchâtel
11, rue Emile Argand
CH-2007 Neuchâtel 7
Switzerland

Carl Cerniglia
National Center for Toxicological Research
HFT250
US Food and Drug Administration
Division of Microbiology
3900 NCTR Road
Jefferson
AR 72079
USA

Nicholas Clipson
Department of Industrial Microbiology
University College Dublin
Belfield
Dublin 4
Ireland

Peter Darrah
Department of Plant Sciences
University of Oxford
South Parks Road
Oxford OX1 3RB
UK

Henry L. Ehrlich
Biology Department
Rensselaer Polytechnic Institute (RPI)
110 8th St
Troy
NY 12180
USA

Roger Finlay
Department of Forest Mycology and Pathology
Swedish University of Agricultural Sciences
SE-75007 Uppsala
Sweden

Marina Fomina
Division of Environmental and Applied Biology
Biological Sciences Institute
School of Life Sciences
University of Dundee
Dundee DD1 4HN
Scotland UK

Mark Fricker
Department of Plant Sciences
University of Oxford
South Parks Road
Oxford OX1 3RB
UK

Geoffrey M. Gadd
Division of Environmental and Applied Biology
Biological Sciences Institute
School of Life Sciences
University of Dundee
Dundee DD1 4HN
Scotland UK

Anna A. Gorbushina
Geomicrobiology
ICBM
Carl Von Ossietzky Universitat
Oldenburg
POB 2503
D-26111 Oldenburg
Germany

Malcolm Greenhalgh
Arch Biocides
Hexagon House
Blackley
Manchester
UK

Dr Vladislav Gulis
Department of Biological Sciences
University of Alabama
Tuscaloosa
AL 35487
USA

Johnson R. Haas
Department of Geosciences/Department of Environmental Studies
Western Michigan University
Kalamazoo
MI 49008
USA

Pauline S. Handley
Faculty of Life Sciences
1.800 Stopford Building
University of Manchester
Manchester M13 9PT
UK

Erik A. Hobbie
Complex Systems Research Center
University of New Hampshire
Durham
03824
USA

David Johnson
Department of Plant and Soil Science
University of Aberdeen
Cruickshank Building
St Machar Drive
Aberdeen AB24 3UU
UK

Kevin Kuehn
Department of Biology
Eastern Michigan University
Ypsilanti
MI
USA

Eleanor Landy
School of Biomedical and Molecular Sciences
University of Surrey
Guildford GU2 7XH
UK

Jonathan R. Leake
Department of Animal and Plant Science
University of Sheffield
Alfred Denny Building
Western Bank
Sheffield S10 2TN
UK

Marinus Otte
Department of Botany
University College Dublin
Belfield
Dublin 4
Ireland

Karsten Pedersen
Göteborg University
Department of Cell and Molecular Biology
Box 462
SE-405 30 Göteborg
Sweden

O. William Purvis
Department of Botany
The Natural History Museum
Cromwell Road
London SW7 5BD
UK

David J. Read
Department of Animal and Plant Science
University of Sheffield
Alfred Denny Building
Western Bank
Sheffield S10 2TN
UK

Joachim Reitner
Göttinger Zentrum Geowissenschaften CZG
Abt Geobiologie
Universität Göttingen
Goldschmidtstr. 3
D-37077 Göttingen
Germany

Karl Ritz
National Soil Resources Institute
Cranfield University
Silsoe MK45 4DT
UK

Geoffrey D. Robson
School of Biological Sciences
1.800 Stopford Building
University of Manchester
Oxford Road
Manchester M13 9PT
UK

Anna Rosling
Department of Earth & Planetary Science
University of California
Berkeley
CA 94720-4767
USA

Hristo A. Sabev
Faculty of Life Sciences
1.800 Stopford Building
University of Manchester
Manchester M13 9PT
UK

Gabriela Schumann
Göttinger Zentrum Geowissenschaften CZG
Abt Geobiologie
Universität Göttingen
Goldschmidtstr. 3
D-37077 Göttingen
Germany

Mark Smits
Laboratory of Soil Science and Geology
Wageningen University
POB 37
6700 AA Wageningen
The Netherlands

Keller Suberkropp
Department of Biological Sciences
University of Alabama
Tuscaloosa
AL 35487
USA

John B. Sutherland
National Center for Toxicological Research
HFT250
US Food and Drug Administration
Division of Microbiology
3900 NCTR Road
Jefferson
AR 72079
USA

Monica Tlalka
Department of Plant Sciences
University of Oxford
South Parks Road
Oxford OX1 3RB
UK

Eric P. Verrecchia
Institut de Géologie
Université de Neuchâtel
11, rue Emile Argand
CH-2007 Neuchâtel 7
Switzerland

Håkan Wallander
Department of Microbial Ecology
Ecology Building
Lund University
223 62 Lund
Sweden

Sarah Watkinson
Department of Plant Sciences
University of Oxford
South Parks Road
Oxford OX1 3RB
UK

Preface

In praise of geomycology

Interactions between the microbially dominated biosphere and the geosphere have and are profoundly affecting our planet and all life on it. Geomicrobiology can be defined as the study of the role that microbes have played and are playing in processes of fundamental importance to geology, and within the diffuse boundaries enclosed by this definition, fungi are important components. Some of the major geological processes affected by microbial activities include mineral formation, mineral degradation (including weathering, bioleaching, soil and sediment formation), element cycling and fossil fuel genesis and degradation. The cycling of component elements from organic and inorganic substrates as a result of these processes can be termed biogeochemical cycling, which again emphasizes the interplay between physicochemical and biological mechanisms. The study of the roles and importance of fungi as agents of geological change can be termed *geomycology* and fungi are ideally suited for this purpose. The branching, filamentous mode of growth allows efficient colonization and exploration of solid substrates while extracellular release of enzymes and other metabolites mediates many organic and inorganic transformations. Considerable physical force can arise from hyphal penetration while translocation of resources through the mycelium enables exploitation of environments where nutrients have an irregular distribution. Fungi can attack silicates, carbonates, phosphates and other minerals while their carbonaceous predilections are well-known, extending to recalcitrant organic molecules of natural origin, e.g. lignin and chitin, or from anthropogenic activity, e.g. pesticides and other xenobiotics.

While considerable attention has been directed towards bacteria in geomicrobiology because of their incredible environmental and metabolic diversity, which includes major transformations in anaerobic environments such as the deep subsurface, fungi have received less attention or are rather taken for granted, biogeochemical activities being submerged in more general perspectives. For example, fungi are more commonly associated with organic matter decomposition and therefore carbon (and nitrogen) cycling, although it should be appreciated that every stable element can be associated with living organisms: any decomposition process involves cycling of all constituent elements present in dead organic matter including those accumulated from the environment. Fungi are often associated with phosphate mobilization from inorganic sources yet little attention is given to the fate of associated metals and other components in minerals. Symbiotic fungi such as mycorrhizas, associated with the vast majority of plant species, have an enormous influence on plant growth and nutrient cycling while lichens are primary rock colonizers and therefore involved in the very earliest stages of rock and mineral transformation, and soil development. As well as being of major importance in the plant root-soil aerobic zone, our knowledge of the boundaries of the biosphere are being extended continually and fungi are now known to be components of such habitats as the deep subsurface and aquatic sediments, and 'extreme' locations such as those of high salinity or containing high levels of toxicants. Biogeochemical activities of fungi are also relevant to the natural attenuation of polluted habitats and the bioremediation of organic and inorganic pollution, but, on the negative side, are also involved in the biodeterioration of stone, wood, plaster, cement and other building materials.

The prime objective of this book is to highlight the roles and importance of fungi in biogeochemical cycles and give an account of the latest understanding of the physical, chemical and biological mechanisms involved and their significance in the biosphere and for other organisms. Where appropriate, relationships with bacteria are also discussed to highlight the dynamic interactions that can exist between these major microbial groups and their integrated function in several kinds of habitat. The chapters are written by leading international authorities and represent a unique synthesis of this subject area, hopefully with broad appeal not only to mycologists and microbiologists but to environmental scientists, geologists, earth scientists, ecologists and environmental biotechnologists.

I would like to thank all the authors who have contributed to this work in an enthusiastic manner, and all at Cambridge University Press who have facilitated progress. In Dundee, special thanks go to Diane Purves who greatly assisted communication, collation, editing and formatting of chapters. Finally, I would like to thank the British Mycological Society and the Society for General Microbiology whose wholehearted initial joint support for a Symposium in this area provided the impetus for subsequent development and production of this book, and my family, Julia, Katie and Richard.

Geoffrey Michael Gadd

1

Geomicrobiology: relative roles of bacteria and fungi as geomicrobial agents

HENRY L. EHRLICH

Introduction

The following definition of geomicrobiology will provide a proper context for the discussion in this essay. *Geomicrobiology* is a study of the role that microbes have played in the geologic past from the time of their first appearance on the planet Earth about 4 eons ago to the present, and the role they are playing today and are likely to play in the future in some of the processes that are of fundamental importance to geology. The discussion will be restricted to current geomicrobial activities because being able to observe them directly, we know most about them. Geomicrobial activities in the geologic past have been deduced from the detection in the geologic record of (1) microbial fossils that morphologically resemble present-day microorganisms of geologic significance and (2) relevant biomarkers. Past geomicrobial activities have also been inferred from present-day geomicrobial activities that occur under conditions similar to those presumed to have existed in the geologic past. Molecular phylogeny is providing information that supports inferences about ancient geomicrobial activity.

Geomicrobial agents

Phylogenetic distribution

Although geomicrobial agents that are presently recognized include members of the domains Bacteria (Eubacteria) and Archaea in the Prokaryota and members of Algae, Protozoa and Fungi in the Eukaryota, the following discussion will emphasize mainly geomicrobial activities of members of the Bacteria, Archaea and Fungi.

Fungi in Biogeochemical Cycles, ed. G. M. Gadd. Published by Cambridge University Press. © British Mycological Society 2006.

Geomicrobial activities

Types of geomicrobial activities
Geomicrobial activities play a role in (1) mineral formation, (2) mineral degradation, (3) the cycling of organic and inorganic matter, (4) chemical and isotopic fractionation and (5) fossil-fuel genesis and degradation. Microbial mineral degradation includes phenomena such as weathering, bioleaching, and soil and sediment formation and transformation (diagenesis). Microbes contribute, to varying extents, to the genesis and degradation of fossil fuels, including methane, peat, coal and petroleum. Some geomicrobial activities can be commercially exploited in processes such as metal extraction from ores, biogas genesis, commercial tertiary petroleum recovery and environmental bioremediation.

Physiological processes involved in geomicrobial activity
The physiological basis for different forms of geomicrobial activity depends on the type of activity, the substance being transformed and the organism(s) involved. Some geomicrobial activity involves enzymatic oxidations or reductions of inorganic substances. Such reactions are promoted mostly by prokaryotic organisms and may contribute to mineral formation, mineral diagenesis and mineral degradation. Other geomicrobial activity involves enzymatic synthesis or degradation of naturally occurring organic carbon compounds in which both prokaryotes and eukaryotes participate extensively. Such organic transformations involve many other types of enzymatic reactions besides oxidations and/or reductions. In microbial physiology, microbial degradation of organic carbon to CO_2 is sometimes called *mineralization*, but in this book that term is strictly reserved for the process of mineral formation.

Some geomicrobial activity may involve non-enzymatic reactions in which inorganic or organic products of microbial metabolism serve as chemical reagents in reactions such as heavy metal precipitation, mineral weathering and dissolution, or mobilization of viscous petroleum hydrocarbons. Thus, heavy metals can be precipitated by H_2S formed by sulphate-reducing bacteria. Carbonates, silicate and aluminosilicate minerals and phosphate minerals may be weathered by microbially formed inorganic acids such as H_2SO_4, HNO_3, H_2CO_3, and organic acids such as acetic, oxalic, lactic, propionic, butyric and citric acids, or by microbially formed bases such as ammonia and amines. Metal constituents in some minerals may also be mobilized by microbially synthesized ligands, as for instance the mobilization of ferric iron by siderophores. Water-insoluble components of petroleum may be emulsified and thereby mobilized through the action of microbially formed surface-active agents.

Some geomicrobial activity is attributable to physical effects exerted on the environment by growing microbes. Thus, growing microbes may transform an aerobic environment into an anaerobic one by consuming oxygen in their respiration faster than it can be replaced by contact with air. Conversely, oxygenically photosynthesizing microbes (cyanobacteria, algae) can transform a quasi-anaerobic environment into an aerobic one by generating O_2 faster than it is consumed by respiring organisms accompanying them. Microbes can raise or lower the pH of their environment, thereby rendering it more or less fit for other organisms present in the same environment. Microbes growing in rock fissures may contribute to the break-up of the rock by the pressure that their increasing biomass exerts on the rock fissures, causing the fissures to enlarge. Finally, some geomicrobial activity may be the result of a combination of several of the activities mentioned above.

Conditions that determine whether a geomicrobial attack of a mineral is enzymatic or non-enzymatic

Direct enzymatic attack Direct enzymatic attack of a mineral is either oxidative or reductive and can occur if three conditions are met. The first condition is the presence of one or more oxidizable or reducible mineral constituents. The second condition is that the cells involved in the oxidation or reduction of an appropriate mineral constituent attach to the mineral surface. The third condition is that the enzyme capable of catalysing the oxidation or reduction of a mineral constituent resides at the cell surface. Besides being in contact with the mineral surface, this enzyme must also be in contact with other enzymes and electron carriers residing below the cell surface. In a mineral oxidation by a Gram-negative bacterium, enzymes and electron carriers below the cell surface in the periplasm and plasma membrane convey electrons removed from an oxidizable mineral constituent by the oxidase at the cell surface (outer membrane) to a terminal electron acceptor, which is oxygen in an aerobic process (Fig. 1.1a). In a mineral reduction by a Gram-negative bacterium, the enzymes and electron carriers below the cell surface in the plasma membrane and the periplasm convey electrons from an electron donor within the cell to the reductase at the cell surface (outer membrane) in contact with the mineral, an appropriate constituent of which will be reduced in serving as terminal electron acceptor (Fig. 1.1b). Only prokaryotic microbes have the capacity for direct enzymatic attack of minerals because only they include representatives with oxidases or reductases at their cell surface capable of interacting with an oxidizable or reducible mineral. So far, such enzymes with a cell-surface location have only been identified in Gram-negative bacteria, i.e. in the aerobe *Acidithiobacillus ferrooxidans* (Yarzábal *et al.*, 2002), and in anaerobically growing *Shewanella oneidensis* MR-1, a facultative organism (Myers & Myers, 1992), and in the strict anaerobe *Geobacter sulfurreducens* (Lovley, 2000). Such enzymes probably also exist in Gram-negative marine isolates strains BIII 32, BIII 41 and BIII 88 (Ehrlich, 1980, 1993a, b). Circumstantial evidence suggests that Gram-positive *Bacillus* 29 and *Bacillus* GJ33 are capable of MnO_2 reduction by a direct mechanism similar to that proposed for marine strain BIII 88 when it reduces MnO_2 aerobically (Ghiorse & Ehrlich, 1976; Ehrlich, 1993a, b; 2002a, p. 451). Although *Sulfolobus* spp. and *Acidianus brierleyi*, which

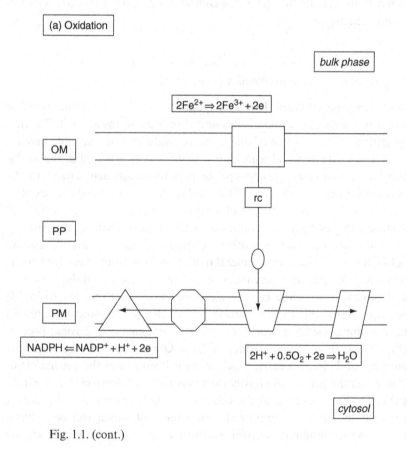

Fig. 1.1. (cont.)

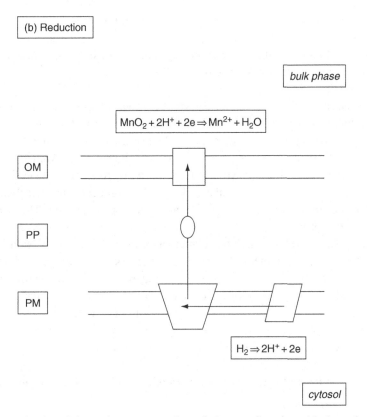

Fig. 1.1. Schematic representation of electron flow in oxidation of an inorganic electron donor (a) and reduction of a terminal inorganic electron acceptor (b) at the cell surface of respective Gram-negative bacteria. The inorganic electron donors or acceptors do not penetrate the outer membrane in these bacteria. The diagram in (a) summarizes electron flow in Fe^{2+} oxidation by *Acidithiobacillus ferrooxidans* (Ehrlich *et al.*, 1991; Yarzábal *et al.*, 2002). The diagram in (b) summarizes electron flow in reduction of MnO_2 by the anaerobe *Geobacter sulfurreducens* (Lovley, 2000). The general reduction scheme for *G. sulfurreducens* also applies to anaerobically grown *Shewanella oneidensis* MR-1 when reducing MnO_2 (Myers & Myers, 1992). OM, outer membrane; PP, periplasm; PM, plasma membrane. The arrows indicate direction of electron flow. In (a), the rectangle (OM), oval (PP) and trapezoid (PM) represent different c-type cytochromes, the parallelogram (PM) represents cytochrome a_1, the octagon (PM) represents a bc1 complex and the triangle represents NADPH dehydrogenase complex. The framed rc in the periplasm represents rusticyanin. In (b), the parallelogram (PM) represents a NADH dehydrogenase complex, and the trapezoid (PM), oval (PP) and rectangle (OM) represent different cytochromes of the c-type.

belong to the domain Archaea, are known to attack various sulphide minerals such as pyrite (FeS_2), chalcopyrite ($FeCuS_2$), arsenopyrite (FeAsS) and nickel sulphide (NiS) (see, for instance, summary by Ehrlich, 2002a, p. 632), the mechanism of attack employed by them is not known.

Non-enzymatic attack In non-enzymatic attack of minerals by microbes, reactive products of microbial metabolism come into play. The microbial enzymes responsible for metabolic product formation are located below the cell envelope, in the cytoplasm of prokaryotes (Bacteria and Archaea) and in cell organelles and/or the cytoplasm of eukaryotes (e.g. fungi, algae, lichens). In these instances of microbial attack, physical contact of the microbial cells with the surface of a mineral being attacked is not essential. The reactive metabolic products are formed intracellularly and are then excreted into the bulk phase where they are able to interact chemically, i.e. non-enzymatically, with a susceptible mineral. Depending on the type of metabolic product and mineral, the interaction with the mineral may result in mineral dissolution or mineral diagenesis by oxidation or reduction or acid or base attack. Mineral dissolution or diagenesis may also be the result of complexation by a microbial metabolic product with that capacity. In some instances mineral attack may involve a combination of some of these reactions.

Enzymatically catalysed inorganic geomicrobial transformations

Oxidations

Aerobic oxidation of dissolved inorganic substances resulting in end-product immobilization and mineral formation Aerobic bacterial oxidation of dissolved Fe^{2+} to a Fe(III) oxide or oxyhydroxide and of Mn^{2+} to Mn(IV) oxide are examples of end-product immobilization by mineral formation (Ehrlich, 1999).

Aerobic mineral oxidation resulting in mineral degradation and product mobilization Aerobic bacterial oxidation of elemental sulphur (S°), of various mineral sulphides such as pyrite (FeS_2), chalcopyrite ($CuFeS_2$), arsenopyrite (FeAsS), sphalerite (ZnS), cobalt sulphide (CoS) and nickel sulphide (NiS) to corresponding metal sulphates, and of uraninite (UO_2) to UO_2^{2+} are examples in which oxidizable minerals undergo dissolution of one or more of their constituents, which are thus mobilized (see Ehrlich, 2002a).

Anaerobic oxidation of dissolved Fe(II) to Fe(III) oxide Ferrous iron has been shown to be anaerobically oxidized to Fe(III) by some bacteria using nitrate as terminal electron acceptor (Straub *et al.*, 1996). Ferrous iron has also been shown to be anaerobically oxidized to Fe(III) in anoxygenic bacterial photosynthesis in which the oxidation of Fe(II) is the source of reducing power for assimilation of CO_2, i.e. carbon fixation (Widdel *et al.*, 1993).

Aerobic and anaerobic oxidation of arsenite to arsenate Aerobic oxidation of arsenite to arsenate by bacteria has been known for a long time (see Ehrlich, 2002a, pp. 305–8). Recently, anaerobic oxidation of arsenite by a chemoautotrophic bacterium, strain MLHE-1, using nitrate as terminal electron acceptor was discovered in the monimoliminion of meromictic Mono Lake, CA, USA by Oremland *et al.* (2002a). This organism together with heterotrophic, anaerobically arsenate-respiring organisms (see later, p. 9) from the sediment in Mono Lake make possible a complete arsenic cycle that is promoted entirely by bacteria in the anaerobic region of the lake (Oremland *et al.*, 2004).

Reductions

Bacterial reduction in air of MnO_2 to Mn^{2+} and of $CrO_4{}^{2-}$ to Cr^{3+} Several examples of enzymatic reduction of MnO_2 to Mn^{2+} in air have been reported (see Ehrlich, 2002a, pp. 449–55). Glucose and acetate have been shown to be effective electron donors in Mn(IV) reduction. In at least one instance of Mn(IV) reduction, some energy appeared to be conserved in the process (Ehrlich, 1993a, b).

A number of examples of enzymatic reduction of $CrO_4{}^{2-}$ to Cr(III) in air have also been reported (see Ehrlich 2002a, pp. 531–4). Glucose and citrate were found to be effective electron donors. Involvement of the electron transport system in the plasma membrane of *Pseudomonas fluorescens* LB300 suggested that some energy may be conserved in this process.

Anaerobic bacterial reduction of MnO_2 to Mn^{2+} A number of different bacteria have been shown to reduce MnO_2 to Mn^{2+} only anaerobically (see Lovley, 2000; Ehrlich, 2002a). Some of these, like *S. oneidensis*, are facultative, whereas others, like *G. metallireducens*, are obligately anaerobic. Depending on the specific organism, effective electron donors include H_2, acetate, lactate, glucose and others. Regardless of the electron donor, all these reductions of MnO_2 are a form of anaerobic respiration from which energy required by the organism to function is conserved.

Anaerobic bacterial reduction of Fe(III) oxide to Fe_3O_4 or Fe^{2+} A variety of different bacteria have been shown to use Fe(III) as a terminal electron acceptor in anaerobic respiration (see Lovley, 2000; Ehrlich, 2002a). In some cases (e.g. *S. oneidensis* and *S. putrefaciens*) these bacteria are facultative, whereas in many other instances they are strict anaerobes (e.g. *Geobacter* spp., *Geospirillum barnesii; Geothrix fermentans, Geovibrio ferrireducens; Pyrobaculum islandicum, Desulfobulbus propionicus; Desulfovibrio desulfuricans; Desulfuromonas acetoxidans; Desulfuromusa* spp.; *Ferribacterium limneticum*). Depending on the organism, H_2, acetate and a variety of other organic compounds, including aromatic compounds, can serve as electron donors. Reduction of Fe(III) is a very important respiratory process in anaerobic carbon decomposition in soil and sediment environments.

Anaerobic bacterial reduction of SO_4^{2-} to HS^- In anaerobic marine environments, especially in estuaries and coastal nearshore regions, bacterial sulphate reduction is a very important respiratory process for degradation of organic carbon because of the ready availability of sulphate in seawater. The importance of sulphate reduction in anaerobic carbon decay was not recognized until a seminal observation by Widdel and Pfennig (1977). They found that their newly discovered *Desulfotomaculum acetoxidans* was able to oxidize acetate to CO_2 and H_2O with sulphate as terminal electron acceptor. Before that time it was thought that the sulphate reducers known up to then (*Desulfovibrio* spp., *Desulfotomaculum nigrificans*, and *Desulfomonas pigra*) were only able to degrade a limited number of carbon compounds, in particular lactate, pyruvate, fumarate, malate and ethanol, to acetate but not to CO_2 and H_2O. The discovery of *Desulfotomaculum acetoxidans* led to the isolation of many other sulphate reducers that collectively are able to degrade a wide range of organic compounds. In addition, some were isolated that could reduce sulphate autotrophically using H_2 or formate as electron donor. Until the early 1980s all known sulphate reducers were eubacterial, but since then some archaeal sulphate reducers, e.g. *Archeoglobus fulgidus* (Stetter *et al.*, 1987) and *A. profundus* (Burggraf *et al.*, 1990), have been identified as well.

Anaerobic reduction of SeO_4^{2-} and SeO_3^{2-} to Se^o and methylated selenides Selenate and selenite reduction by microorganisms has been noted for some time. Organisms with this capacity include bacteria and fungi (Trudinger *et al.*, 1979; Stolz & Oremland, 1999). In bacteria the reductions are mostly a form of anaerobic respiration and result in selenium immobilization with the formation of water-insoluble Se^o. In fungi the reductions involve the formation of methylated selenium, e.g. $(CH_3)_2Se$ and $(CH_3)_2Se_2$. The formation of the methylated selenium compounds is a form of selenium detoxification because these compounds are volatile and can thus be vented from soil, sediment and water into the atmosphere. In trace amounts, selenium meets a nutritional requirement of diverse organisms and is assimilated by them.

Anaerobic reduction of AsO_4^{3-} to AsO_2^- and methylated arsenides Although the ability of bacteria and fungi to reduce arsenate has been known for many decades (see Stolz & Oremland, 1999; Ehrlich, 2002a), clear recognition that bacterial reduction of arsenate to arsenite is mostly a form of anaerobic respiration has come only recently. Glucose, lactate and acetate have been found to be effective electron donors in this process (Oremland *et al.*, 2002b). *Chrysiogenes arsenatis* was the first organism found to be able to use acetate as reductant (Macy *et al.*, 1996). At least one of the organisms that can respire with arsenate as terminal electron acceptor is an archeon, *Pyrobaculum arsenaticum* (Huber *et al.*, 2000). Fungi as well as bacteria can form methylated arsenides from arsenite, bacteria forming $(CH_3)_2As$ and fungi forming $(CH_3)_3As$ in a process that is a form of arsenic detoxification.

Anaerobic reduction of CO_2 to CH_4 Carbon dioxide can serve as terminal electron acceptor in anaerobic hydrogen respiration. The product of this reduction is methane when certain members of the domain Archaea, the methanogens, are involved:

$$4H_2 + CO_2 \rightarrow CH_4 + 2H_2O \qquad (1.1)$$

The product is acetate when certain members of the domain Bacteria, homoacetogens, are involved:

$$4H_2 + 2CO_2 \rightarrow CH_3COOH + 2H_2O \qquad (1.2)$$

These transformations are an important phase of the anaerobic part of the carbon cycle.

Non-enzymatic inorganic transformations of geomicrobial significance

Mineral formation

Metal sulphides Sulphide produced in bacterial sulphate respiration can precipitate heavy metal ions from solution when the sulphide concentration is in excess of that demanded by the solubility product of the corresponding metal sulphide. Although most microbial metal sulphide precipitation is attributable to the bacterial reduction of sulphate, at least one instance of CuS precipitation by fungal reduction of S° has been observed in the laboratory (Ehrlich & Fox, 1967).

Carbonates Biogenic carbonate can precipitate calcium ions as calcite or aragonite, ferrous ions as siderite and manganous ions as rhodochrosite when the carbonate ion concentration is in excess of that demanded by the solubility product of the respective carbonate precipitates and if some other factors do not interfere with the reaction. The carbonate ions may be formed from CO_2 generated in microbial respiration or fermentation of organic carbon (see below, p. 14), or it may be the result of photosynthesis in an aqueous environment in which the assimilation of dissolved CO_2 causes an increase in the concentration of carbonate ion due to the removal of CO_2 from HCO_3^--containing solution ($2HCO_3^- \leftrightarrow CO_3^{2-} + CO_2\uparrow + H_2O$). Although the formation of the carbonate anions in these instances depends on enzymatic processes, the subsequent carbonate precipitations are not enzymatically catalysed. Ehrlich (2002a, pp. 184–211) reviewed this topic in more detail.

Phosphates Microbially mobilized phosphate resulting from the enzymatic breakdown of phosphate esters may precipitate as calcium phosphate (e.g. apatite) when encountering critical environmental concentrations of Ca^{2+}. Such precipitation may occur in the bulk phase (see Ehrlich, 2002a, pp. 275–7) or at a cell surface (Macaskie *et al.*, 1987, 1992).

Mineral weathering

Carbonate, silicate and phosphate minerals may be subject to weathering by corrosive products of bacterial and fungal metabolism. Corrosive metabolic products produced by bacteria and/or fungi include the mineral acids H_2CO_3, H_2SO_4 and HNO_3, and organic acids such as formic, acetic, oxalic, propionic, pyruvic, lactic, succinic, butyric, gluconic, 2-ketogluconic, citric and others. An organic acid may corrode a mineral by promoting acid dissolution or by withdrawing cationic constituents like Ca^{2+} from the mineral surface through complexation, thereby promoting eventual collapse of the crystal lattice. Thus, the calcium of limestone may be mobilized as a result of dissolution of its $CaCO_3$ by H_2CO_3 formed from respiratory CO_2 produced by bacteria and/or fungi growing on the limestone surface. Similarly, the calcium of limestone may be mobilized by HNO_3 formed in the oxidation of NH_4^+ by nitrifying bacteria or by H_2SO_4 formed in the oxidation of H_2S by sulphide-oxidizing bacteria when these organisms are growing on the limestone surface. Some fungi are capable of mobilizing Ca^{2+} and Fe^{3+} by forming the ligand citrate from sugars like glucose or sucrose, which can remove these ions from the exposed crystal structure of minerals containing them. Some bacteria are capable of mobilizing Ca^{2+} in the crystal structure of calcium-containing minerals by forming the ligand 2-ketogluconate in their breakdown of sugars derived from polysaccharide hydrolysis, which can then remove the Ca^{2+} from the exposed crystal lattice of calcium-containing minerals. The raw materials used by the fungi and bacteria to produce the corresponding ligands commonly occur in the litter zone of soils.

Silicate, aluminosilicate and phosphate minerals may be weathered via attack by microbially produced corrosive inorganic and organic acids or by ligands such as fungally produced citric acid and bacterially produced 2-ketogluconic acid. Unlike $CaCO_3$ minerals, these minerals are not generally susceptible to attack by H_2CO_3, which is too weak an acid. Silicates and aluminosilicates may also be attacked by a base such as NH_4OH, which may be formed in the hydrolysis of urea.

Microbially produced exopolysaccharides have been directly implicated in weathering of silicates and aluminosilicates through either complexation of cationic constituents of these minerals or, in the case of acid mucopolysaccharides, through acidolysis (Barker & Banfield, 1996). In some instances, however, exopolysaccharides may inhibit weathering, as in the case of plagioclase by gluconate at circumneutral pH (Welch & Vandevivere, 1995).

Geomicrobial transformations by a single microbial species and by microbial consortia

Transformation by a single microbial species

Some geomicrobial transformations in nature involve a single microbial species. An example of such a transformation is the anaerobic reduction of a Mn(IV) oxide to Mn^{2+} by *S. oneidensis* or *G. metallireducens* in an environment with a plentiful supply of an appropriate electron donor like lactate for *S. oneidensis* or acetate for *G. metallireducens*. Because each of these two microbial species can perform the reduction by themselves, and because electron donors like lactate and acetate are formed as major end-products in the energy metabolism of a variety of microbes present in the same environment that harbours *S. oneidensis* or *G. metallireducens*, the latter do not need to form specific microbial associations to bring about Mn(IV) oxide reduction.

Transformation by microbial collaborations (consortia)

Deposition of elemental sulphur formed from sulphate Essential collaboration of at least two different microbial species occurs in the transformation of sulphate to $S°$ in salt domes or similar sedimentary formations (see Ivanov, 1968). This transformation is dependent on the interaction of a sulphate reducer like *Desulfovibrio desulfuricans*, which transforms sulphate to H_2S in its anaerobic respiratory metabolism, and an H_2S oxidizer like *Thiobacillus thioparus*, which, under conditions of limited O_2 availability, transforms H_2S to $S°$ in its respiratory metabolism (van den Ende & van Gemerden, 1993). The collaboration of these two physiological types of bacteria is obligatory in forming $S°$ from sulphate because sulphate reducers cannot form $S°$ from sulphate, even as a metabolic intermediate. It should be noted, however, that the sulphate reducers and H_2S oxidizers are able to live completely independent of each other as long as the overall formation of $S°$ from sulphate is not a requirement.

Growth enhancement of the chemolithoautotroph Acidithiobacillus ferrooxidans *by an organotrophic associate* Some microbial associations in nature are very intimate. One example in acidic aerobic environments is the association of the Fe(II) oxidizer *Acidithiobacillus ferrooxidans* with *Thiobacillus acidophilus* (renamed *Acidiphilium acidophilum*) (Zavarzin, 1972; Guay & Silver, 1975; Harrison, 1981). *Acidiphilium acidophilum* is an acidophilic, facultative autotroph, incapable of oxidizing Fe^{2+} but capable of oxidizing $S°$ as energy source. Zavarzin (1972) called this type of organism a satellite of *At. ferrooxidans*. Its role in this microbial association can also be filled by certain acidophilic organotrophs found in acid mine-water. In these microbial associations, the chemolithotrophic *At. ferrooxidans* lives by fixing CO_2 at the expense of the energy it gains from the oxidation of Fe^{2+}. It fixes more CO_2 than it needs for assimilation during growth and excretes some or all of the excess of the fixed carbon into the bulk phase in the form of pyruvate and some amino acids. If these compounds accumulate in sufficient quantities in the bulk phase surrounding *At. ferrooxidans*, they inhibit its growth. Association of *At. ferrooxidans* with an acidophilic organism like *A. acidophilum*, which can grow organotrophically at the expense of the excreted carbon, leads to the depletion of this carbon by consumption and thereby measurably improves growth of *At. ferrooxidans* (Harrison, 1984). The association of the chemolithotrophic iron oxidizer with the organotroph is so intimate that the separation of *At. ferrooxidans* from its satellite(s) is very difficult in mineral salts medium; indeed, until Zavarzin first reported it in 1972, the satellite organisms went undetected.

Methanogenesis from ethanol by a methanogen and an ethanol fermenter Another example of a very intimate association of two bacteria is *Methanobacterium bryantii* and the S organism, in this instance under anaerobic conditions. The two organisms cannot be separated from each other on an ethanol-containing medium. When first isolated by Barker (1940) on such a medium from anaerobic sediment from San Francisco Bay, CA, the culture he obtained was deemed by him to be pure by the criteria applicable at that time. It became known as *Methanobacterium omelianskii*. In the laboratory, the culture transformed ethanol anaerobically to methane and acetate in the presence of CO_2. It was not until 1967 when a group at the University of Illinois, Urbana, USA (Bryant *et al.*, 1967), demonstrated that when *M. omelianskii* was grown anaerobically with H_2 as the sole energy source and CO_2 the sole carbon source, the culture lost its ability to grow on ethanol. Further study showed that *M. omelianskii* was a mixed culture consisting of two organisms, an ethanol fermenter and a methanogen. The reason for the inability of *M. omelianskii* to grow on ethanol after cultivation on H_2 was shown to have been due to the loss of the ethanol fermenter in the culture, which was unable to grow on H_2 as an energy source and CO_2 as a carbon source. The ethanol fermenter, which became known as the S organism, was able to grow on ethanol as carbon and energy source by transforming it to acetate and H_2, but only if the accumulating H_2 it produced from ethanol was removed. Accumulation of H_2 was found to inhibit its growth. In the *M. omelianskii* culture, the methanogen was the agent that removed the H_2 that the S organism produced in its ethanol fermentation, thereby making the continued growth of the S organism possible. The methanogen was named *Methanobacterium bryantii*. The S organism could be grown in the absence of *M. bryantii* in a modified medium in which H_2 did not acccumulate.

Lichens Lichens represent a third example of an intimate, geomicrobially important consortium (Ahmadjian, 1967; Hale, 1967; see Haas & Purvis, Chapter 15, this volume). Some lichens of geomicrobial interest grow on the surface of rocks, some others within certain types of rock. In either case, their growth occurs in extremely oligotrophic environments. Although to the naked eye, lichens appear to be a single plant-like organism, they are really an intimate association of a fungus with either a cyanobacterium or an alga. When growing on the surface of rocks, the photosynthetic partner of a lichen is usually surrounded by the mycelium of the fungal partner, shielding it from excess sunlight that may be inhibitory to optimal photosynthesis. The fungus forms protoplasmic extensions (haustoria) penetrating the cells of the algal or cyanobacterial partner of the lichen for sharing photosynthate produced by the latter in return for nutritionally required minerals mobilized from the rock substratum by the fungal partner. When growing inside fractured rock, the photosynthetic partner of the lichen grows close to the rock surface where light penetrates whereas the fungal partner grows in fractures deeper in the rock where light does not penetrate. The fungal partner communicates with the photosynthesizing partner by protoplasmic extensions (Friedmann 1982; Friedmann & Ocampo, 1984).

Mechanisms of biogenic mineral formation: examples

Biogenic carbonates ($CaCO_3$: calcite, aragonite)

Extracellular deposition Some bacteria and fungi contribute to calcium carbonate precipitation in the bulk phase with CO_2 generated in their respiration on organic energy sources. This precipitation is dependent on a circumneutral pH range and an excess of Ca^{2+} in the reactive zone in the bulk phase. The formation of extracellular calcium carbonate from respiratory CO_2 formed by heterotrophic bacteria and fungi involves the following sequence of equilibrium reactions or variants thereof:

$$(CH_2O) + O_2 \rightarrow CO_2 + H_2O \tag{1.3a}$$

$$(CH_2O) + 0.5SO_4{}^{2-} + 0.5H^+ \rightarrow CO_2 + 0.5\,HS^- + H_2O \tag{1.3b}$$

$$CO_2 + H_2O \leftrightarrow HCO_3{}^- + H^+ \tag{1.4}$$

$$2HCO_3{}^- \leftrightarrow CO_3{}^{2-} + CO_2\uparrow + H_2O \tag{1.5}$$

$$Ca^{2+} + CO_3{}^{2-} \leftrightarrow CaCO_3\downarrow \tag{1.6}$$

Equations (1.3a) and (1.3b) illustrate the overall reactions of CO_2 generation in aerobic and anaerobic respiration, respectively. Although not shown here, some types of bacterial fermentations can also be a source of CO_2 for carbonate generation. Equation (1.5) explains the origin of $CO_3{}^{2-}$. This reaction is forced in the direction of $CO_3{}^{2-}$ by loss of CO_2 to the atmosphere in an open system, and by precipitation of calcium carbonate ($CaCO_3$), illustrated in Eq. (1.6), or by the formation of other carbonates not shown here.

Extracellular calcium carbonate formation by some cyanobacteria and algae, which are photosynthetic autotrophs that obtain their carbon from CO_2, can be explained by Eq. (1.5) above. In that instance, the uptake and fixation of CO_2 by the cyanobacteria and algae promotes CO_3^{2-} formation needed for precipitation of extracellular $CaCO_3$. In the cyanobacterium *Synechococcus*, the Ca^{2+} precipitated by CO_3^{2-} from photosynthesis is derived from that bound to the cell surface of *Synechococcus* (Thompson *et al.*, 1990).

Structural deposition Some algae and protozoa form structural $CaCO_3$ that is usually integrated into structures just beyond the plasma membrane. In some algae it becomes an integral part of the cell wall and in some protozoa, like calcareous foraminifera, it becomes a shell-like stucture around the cell, i.e. a test. The calcium carbonate crystallites for these structures are assembled in special membrane vesicles inside the algal or protozoan cells. In at least some instances these vesicles are derived from the Golgi apparatus. Once formed, the $CaCO_3$ crystallites are exported to the cell surface where they are integrated into the wall structure of the algae (e.g. coccolithophores) or tests of protozoa (e.g. calcareous foraminifera) (see de Vrind-de Jong & de Vrind, 1997). The carbonate may form as a by-product of photosynthesis (Eq. (1.5)) or from CO_2 generated in respiration (Eqs. (1.3a) and (1.4)) followed by Eq. (1.5). In the calcareous alga *Chara corallina*, $CaCO_3$ deposition at the cell surface may involve an ATP-driven H^+/Ca^{2+} exchange (McConnaughey, 1991).

Biogenic metal oxides

Oxidative processes Various members of the domains Bacteria and Archaea, have a capacity to oxidize Fe^{2+} to Fe(III) aerobically. Most of these organisms are chemolithoautotrophic (Blake & Johnson, 2000), but at least one heterotrophic (mixotrophic?) organism in the Bacteria has been described (Johnson *et al.*, 1993). The large majority of Fe^{2+} oxidizers are acidophilic, but some are neutrophilic (Emerson, 2000) of which the first to have been recognized is *Gallionella ferruginea*. It was discovered in 1836 by Ehrenberg, but its physiology was not fully understood until well into the twentieth century. The iron minerals the iron oxidizers, especially acidophiles, produce are in most cases deposited in the bulk phase, but in the case of *G. ferruginea*, the iron oxide is deposited on its lateral stalk, and in the case of sheathed bacteria like *Leptothrix ochracea* it is deposited on the sheath.

In nature, the acidophile *At. ferrooxidans* oxidizes ferrous iron that is derived mainly from ferrous sulphide (FeS_2, FeAsS, $CuFeS_2$, Cu_5FeS_4, etc.)-containing rock formations. The Fe(III) minerals resulting from this biooxidation of Fe(II) are chiefly jarosites ($AFe_3(SO_4)_2(OH)_6$, where A may represent Na^+, K^+, NH_4^+, H_3O^+ and some other cations) (Lazaroff *et al.*, 1985). The acid pH on which *At. ferrooxidans* depends results from the H_2SO_4 formed in the biooxidation of mineral sulphides like pyrite, chalcopyrite, and arsenopyrite. In abiotic iron oxidation under the same conditions, amorphous hydrated iron sulphates and goethite (α-FeOOH) are formed.

The iron minerals generated by *G. ferruginea* and by *L. ochracea* in their oxidation of Fe(II) to Fe(III) have not been well characterized but have been assumed to be a form of FeOOH. The exact role that these organisms play in transforming the Fe(III) they produce into FeOOH remains to be elucidated.

Several members of the domain Bacteria have a capacity to form Mn(IV) oxides by oxidizing Mn(II), which some deposit on their cells, sheaths or appendages, or in the case of marine *Bacillus* sp. strain SG-1 on the surface of the spores this organism forms (see Ehrlich, 1999). In most cases the mineral type of the manganese oxide formed has not been well characterized. In the case of the free spores of *Bacillus* sp. strain SG-1, the nature of the mineral formed varies depending on culture conditions, such as Mn(II) concentration, ionic strength and temperature, under which the mineral was formed, and on mineral aging (Mandernack *et al.*, 1995). The initial product of the Mn(II) oxidation by the spores is poorly crystallized.

Reductive processes Under certain growth conditions, amorphous ferric oxide has been shown to be anaerobically reduced by certain bacteria such as *G. metallireducens* GS-15 to magnetite (Fe_3O_4) that is accumulated extracellularly (Lovley *et al.*, 1987; Lovley & Phillips, 1988; Lovley 1991). The electron donor in experimental demonstrations of this phenomenon was acetate. The ferric iron from ferric citrate, on the other hand, was found to be completely reduced to Fe^{2+} by *G. metallireducens* GS-15 with acetate as electron donor (Lovley & Phillips, 1988).

Magnetotactic bacteria form magnetite and/or greigite (Fe_3S_4) or pyrrhotite crystals intracellularly (Bazylinski & Frankel, 2000). These crystals are contained in membrane-bound structures called magneto-somes. The magnetite crystals behave as single-domain magnets, which these bacteria employ as compasses in locating their preferred partially reduced growth environment. The mechanism by which these organisms form magnetite is still not fully understood but appears to involve reduction of ferric to ferrous iron and subsequent partial re-oxidation of the ferrous iron to ferric iron. Magnetite consists of 1 part ferrous and 2 parts ferric iron ($FeO \cdot Fe_2O_3$). Less is known about the mechanism of greigite and pyrrhotite formation.

Siderite ($FeCO_3$) has been found to accumulate extracellularly in some instances of anaerobic bacterial Fe(III) reduction (Pye, 1984; Pye *et al.*, 1990; Coleman *et al.*, 1993; Ehrlich & Wickert, 1997). The exact conditions under which the biogenic siderite forms in nature remain to be elucidated. A major or exclusive source of the carbonate in biogenic siderite is assumed to be the CO_2 formed from organic electron donors used in the anaerobic respiration on Fe(III).

Mechanisms of microbial mineral degradation: examples

Aerobic examples – metal sulphides

Acidithiobacillus ferrooxidans is able to degrade chalcocite (Cu_2S) by oxidizing the Cu(I) in the mineral to Cu^{2+} and the sulphide-S to SO_4^{2-}. Because the end-products of this oxidation are soluble, the oxidation results in the mobilization of the copper and sulphur in the chalcocite. Evidence exists for two distinct mechanisms by which *At. ferrooxidans* can perform the oxidation of chalcocite. One mechanism involves a direct attack of the crystal lattice of chalcocite by cells attached to the surface of chalcocite. The other mechanism involves an indirect attack of the crystal lattice of chalcocite by the chemical oxidant Fe^{3+} generated from Fe^{2+} in the bulk phase by planktonic cells (unattached) of *At. ferrooxidans*.

In direct attack, electrons removed by *At. ferrooxidans* attached to the surface of chalcocite are transferred by the electron transport system of the organism to O_2 as terminal electron acceptor. The organism conserves energy in this electron transfer, which is a respiratory process. The following equation summarizes the initial steps in chalcocite oxidation:

$$Cu_2S + 0.5O_2 + 2H^+ \rightarrow Cu^{2+} + CuS + H_2O \qquad (1.7)$$

Digenite, Cu_9S_5, containing Cu(I) and Cu(II), appears to be an intermediate in this process (Fox, 1967; Nielsen & Beck, 1972).

Equation (1.8) summarizes the subsequent steps in the oxidation:

$$CuS + 2O_2 \rightarrow Cu^{2+} + SO_4^{2-} \qquad (1.8)$$

In indirect attack, planktonic *At. ferrooxidans* forms ferric iron in the bulk phase by oxidizing dissolved ferrous iron. The resultant ferric iron then acts as a chemical oxidant that attacks the chalcocite, forming Cu^{2+}, S° and Fe^{2+} as end-products. A form of cupric sulphide, CuS, appears also to be an intermediate in this chemical oxidation (Sullivan, 1930). Planktonic *At. ferrooxidans* regenerates Fe^{3+} from the Fe^{2+} produced in the chemical oxidation of chalcocite and H_2SO_4 from the S° produced in the same oxidation and so prolongs the chemical oxidation of chalcocite. In the field, original dissolved Fe^{3+} in the bulk phase derives mainly from the oxidation of iron-containing sulphides such as pyrite and chalcopyrite that accompany chalcocite. This oxidation is also promoted by *At. ferrooxidans*. For further discussion of direct and indirect attack of chalcocite by *At. ferrooxidans*, the reader is referred to Ehrlich (2002a, pp. 632–9; 2002b).

Other acidophilic Fe(II)-oxidizing bacteria may also promote oxidation of chalcocite, but whether by a direct or indirect mechanism remains to be elucidated.

Some metal sulphide minerals can be degraded by fungi. The metal mobilization mechanism in this instance involves complexation of the metal constituent in the sulphide mineral by a ligand such as gluconate or citrate produced in the metabolism of a sugar like sucrose by some fungi (Wenberg *et al.*, 1971; Hartmannova & Kuhr, 1974; Burgstaller & Schinner, 1993). *Penicillium* sp. and *Aspergillus* sp. are examples of fungi with this potential.

Aerobic examples – metal oxide

Certain fungi have the capacity to reduce MnO_2 non-enzymatically with oxalate (e.g. Stone, 1987) that they produce from glucose, for instance. Under acidic conditions, the oxalate reduces MnO_2 chemically to Mn^{2+}:

$$MnO_2 + HOOCCOO^- + 3H^+ \rightarrow Mn^{2+} + 2CO_2 + 2H_2O \quad (1.9)$$

Anaerobic examples – Mn(IV) oxide

As previously mentioned (p. 7), bacteria like *S. oneidensis* and *G. metallireducens* have the capacity to reduce insoluble MnO_2 to soluble Mn^{2+} enzymatically by anaerobic respiration with a suitable electron donor in a direct process in which the respective organisms attach to the surface of the oxide. In the case of *S. oneidensis*, the electron donor may be lactate, pyruvate, formate or H_2, but not acetate. The lactate and pyruvate are oxidized to acetate as end-product. *Geobacter metallireducens* can use butyrate, propionate, lactate and acetate as electron donors, but not H_2 or formate, and oxidizes the organic electron donors completely to CO_2 and H_2O. *Geobacter sulfurreducens* can use H_2 as electron donor in MnO_2 reduction (see Lovley, 2000).

Fossil-fuel genesis, recovery and biodegradation

Fossil-fuel genesis

Methane Much of the methane found on Earth is of biogenic origin. It has been and continues to be formed by methanogens under strictly anaerobic conditions. These organisms are members of the domain Archaea. The majority of these organisms are able to form methane chemoautotrophically by reducing CO_2 with H_2:

$$CO_2 + 4H_2 \rightarrow CH_4 + 2H_2O \tag{1.10}$$

and some heterotrophically by fermenting acetate to methane and CO_2:

$$CH_3COOH \rightarrow CH_4 + CO_2 \tag{1.11}$$

A few lack the ability to reduce CO_2 with H_2.

Peat and coal Peat and coal have been formed by partial aerobic decomposition of accumulated plant debris. Peat formation has been viewed as an intermediate step in coal formation. Some peat is forming at the present time. Fungi have been identified as chiefly responsible for peat formation, but bacteria have been implicated in the attack of labile components of the plant debris and organic residue from the fungal attack. Burial by sediment of the decomposing plant debris gradually changes conditions from aerobic to anaerobic and ultimately terminates the microbial decomposition process by limiting access to moisture. The organic residue plus the microbial biomass from these transformations constitute peat.

If extensive additional burial of the peat by sediment occurs after this time, the peat may undergo physicochemical changes from the effects of pressure and heat and is gradually transformed to coal. The extent of transformation of the peat determines the rank of coal, the lowest rank being lignite, the highest being bituminous and anthracite coal. The rank of coal is determined chiefly by its carbon, hydrogen and moisture content and its heat value.

Petroleum As in the formation of coal, bacteria and fungi play a role only in the intial stages of petroleum formation. The source of the biomass from which petroleum originates is aquatic (algae and cyanobacteria), chiefly of marine origin, rather than terrestrial (non-vascular and vascular plants). Fungi and bacteria partially decompose this biomass on the sea floor until it becomes sufficiently buried by accumulating sediment to stop biological activity. Thereafter, pressure and geothermal sources of heat gradually transform the buried biomass physicochemically to petroleum. The petroleum at this point is highly dispersed in porous strata of the sediment.

Natural formation of petroleum reservoirs

Water migrating through permeable petroleum-containing sediment strata transports the dispersed petroleum to traps such as the apex of anticlinal folds in the strata where the petroleum collects. Since petroleum is not readily miscible with water, it forms a discontinuous phase in an aqueous matrix. Some living bacteria present in the porous sediment may form sufficient quantities of gases (e.g. CH_4, CO_2, H_2), such that the combined gas pressure helps to facilitate the migration of the water with the petroleum. Some of the bacteria may also form surface-active agents that facilitate the mixing of petroleum constituents with the water and thereby assist petroleum migration. Methane, being miscible with petroleum to some degree, may reduce its viscosity. It is from traps to which the dispersed petroleum is transported that it is harvested commercially.

Fossil fuel degradation

Methane Methane is readily degraded aerobically by methanotrophs, which are members of the domain Bacteria that form CO_2 and H_2O from the methane, using it as a source of energy and carbon. Methane may also be anaerobically degraded in some marine environments by a microbial consortium consisting of an archaeon (e.g. a member of the Methanosarcinales) and a eubacterium (e.g. a member of the Desulfosarcinales) (e.g. Orphan *et al.*, 2001). The archaeon forms H_2 and CO_2 from the methane, and the eubacterium uses the H_2 to reduce seawater sulphate to H_2S. The overall reaction brought about by this consortium is:

$$CH_4 + SO_4^{2-} + 2H^+ \rightarrow CO_2 + H_2S + 2H_2O \qquad (1.12)$$

Peat and coal Peat and most forms of coal are not readily degraded microbiologically. However, fungi and some bacteria that produce manganese peroxidases and lignin peroxidase have been shown to grow on lignite and some sub-bituminous coals. Some dibenzothiophene-degrading bacteria have been found to break down partially the carbon framework of liquefied bituminous coal and remove bound sulphur from it (see Ehrlich, 2002a, pp. 704–6).

Petroleum A number of different bacteria and fungi have been found to break down aliphatic, aromatic and heterocyclic components of petroleum. Some of these processes are aerobic and some anaerobic. Some yield useful energy to the degrader and others do not. The latter processes are the result of co-oxidation in which an organic compound that may not be related to the hydrocarbon being degraded serves as carbon and energy source and enables the organism to degrade the hydrocarbon. Fungi and especially bacteria have been found useful in removing organic sulphur from petroleum without significantly altering its calorific value (Atlas, 1984; Baker & Herson, 1994; Grossman *et al.*, 2001).

Element cycling

Both bacteria and fungi play important roles in elemental cycling, often specializing in the promotion of particular reactions. Fungi often, though not always, contribute to some of the degradative steps in cycles. In the carbon cycle, for instance, fungi play a primary role in initial degradation of polymers like cellulose and lignin in plant residues, which the large majority of bacteria are unable to attack because they are not genetically endowed to produce the necessary depolymerases. The bacteria, however, participate in the further degradation of the depolymerization products that the fungi produce in far greater amounts than are needed for their own growth. In the sulphur cycle, some fungi as well as some bacteria can play important roles in the removal of divalent sulphur from organic compounds like dibenzothiophenes while at the same time breaking some carbon–carbon bonds. In the phosphorus cycle, fungi can play important roles in mobilizing phosphate contained in phosphate minerals like apatite by producing organic acids that can solubilize the phosphate. In the nitrogen cycle, fungi play a significant although not exclusive role in heterotrophic nitrification, whereas bacteria play an exclusive role in autotrophic nitrification. As mentioned earlier, fungi as well as bacteria can affect the arsenic cycle in soil and sediment, fungi forming dimethyl arsenide and bacteria forming trimethyl arsenide. Both of the arsenides are volatile and can escape into the atmosphere. Fungi and bacteria can also affect the selenium cycle by forming methylated selenides from selenate and/or selenite (see earlier, pp. 8–9). A more detailed discussion of the roles of bacteria and fungi in elemental cycles is beyond the scope of this chapter and the reader is referred to other reviews (e.g. Ehrlich, 2002a).

Chemical and isotopic fractionation

Chemical fractionation by microbes

Bacteria can promote differential solubilization of the constituents of rock, ore and minerals. One such example is the differential reductive solubilization of Mn(IV) and other metallic constituents in marine ferromanganese concretions by Mn(IV)-reducing bacteria under aerobic conditions (e.g. Ehrlich *et al.*, 1973). Ferromanganese concretions contain significant quantities of Mn(IV) and Fe(III) oxides as well as varying quantities of some other transition metals, especially copper, nickel and cobalt. In laboratory experiments the bacteria did not mobilize significant amounts of the iron in the nodules. The residue from bacterial leaching of nodules thus becomes enriched in Fe(III) oxide.

Bacteria associated with bauxite ore have been shown to reductively mobilize significant amounts of the Fe(III) oxides in the bauxite under anaerobic conditions without mobilizing significant amounts of the Al(III) oxides like gibbsite or boehmite as long as the pH of the reaction system was kept above 4.5. The bauxite residue was thus enriched in aluminium (Ehrlich *et al.*, 1995; Ehrlich & Wickert, 1997).

Acidophilic iron-oxidizing bacteria like *At. ferrooxidans* are being used on a commercial scale for oxidative removal of pyrite and arsenopyrite from pyritic gold ores in which the pyrites encapsulate the small grains of gold in the ore. The bacterial oxidation of the pyrites leads to sufficient exposure of the grains of gold to make them directly accessible to chemical extractants like cyanide- or thiourea-solutions (e.g. Livesey-Goldblatt *et al.*, 1983; Brierley *et al.*, 1995).

Isotopic fractionation by microbes

Stable isotopic discrimination by some microbes occurs in nature and has been experimentally demonstrated in the laboratory. The microbes generally prefer to interact metabolically with the lighter stable isotope of an element. For instance, in attacking $CaSO_4$, in which the sulphur contains both ^{32}S and ^{34}S, sulphate-reducing bacteria such as *Desulfovibrio desulfuricans* can be shown to form H_2S in which the sulphur is significantly depleted in ^{34}S, measured in parts per thousand (‰), compared to the sulphate-sulphur in the original $CaSO_4$. Some other bacteria can discriminate between D/H, $^{13}C/^{12}C$, $^{15}N/^{14}N$, $^{18}O/^{16}O$ and even $^{56}Fe/^{54}Fe$ (see Ehrlich 2002a).

Conclusions

It is now well recognized that some bacteria and fungi serve as geomicrobial agents in nature, playing important roles in formation of some minerals, in weathering of rocks and minerals, in genesis and degradation of fossil fuels, in elemental cycling and in chemical and isotopic fractionation. Although some of these geomicrobial activities are not restricted to bacteria and fungi, but also involve certain algae and proto-zoa, in general bacteria and fungi have played and are playing a central role in the promotion of many geomicrobial processes. Although bacteria are more versatile than fungi with respect to the types of physiological activity they can put to work geomicrobiologically, fungi nevertheless play a central role in some types of geomicrobial transformation, as, for example, in the weathering of some rocks, in the initial stages of the degradation of complex natural polymers like cellulose and lignin that are part of the organic matter in the litter zone of soil, and in some bioremedia-tion processes involving xenobiotics in soil and water. Lichens play an important role in the weathering of some rocks. In sum, bacteria, archaea and fungi include geomicrobial agents that control some important present-day geological processes and have done so in the geologic past.

Acknowledgements

I am very much indebted to Geoff Gadd for inviting me to write this chapter for his book. His invitation was a stimulus for me to take a new look at some of the various roles that bacteria and fungi play in geomicrobiology.

References

Ahmadjian, V. (1967). *The Lichen Symbiosis*. Waltham, MA: Blaidsell.

Atlas, R. M. (1984). *Petroleum Microbiology*. New York: McGraw-Hill.

Baker, K. H. & Herson, D. S. (1994). Microbiology and biodegradation. In *Bioremediation*, ed. K. H. Baker & D. S. Herson. New York: McGraw-Hill, pp. 9–60.

Barker, H. A. (1940). Studies upon the methane fermentation. IV. The isolation and culture of *Methanobacterium omelianskii*. *Antonie van Leeuwenhoek*, **6**, 201–20.

Barker, W. W. & Banfield, J. F. (1996). Biologically versus inorganically mediated weathering reactions: relationships between minerals and extracellular microbial polymers in lithobiontic communities and minerals. *Chemical Geology*, **132**, 55–69.

Bazylinski, D. A. & Frankel, R. B. (2000). Biologically controlled mineralization of magneto-tactic iron minerals by magnetotactic bacteria. In *Environmental Microbe-Metal Interactions*, ed. D. R. Lovley. Washington, DC: ASM Press, pp. 109–44.

Blake, R. II & Johnson, D. B. (2000). Phylogenetic and biochemical diversity among acido-philic bacteria that respire on iron. In *Environmental Microbe-Metal Interactions*, ed. D. R. Lovley. Washington, DC: ASM Press, pp. 53–78.

Brierley, J. A., Wan, R. Y., Hill, D. L. & Logan, T. C. (1995). Biooxidation-heap pretreatment technology for processing lower grade refractory gold ores. In *Biohydrometallurgical Processing*, Vol. 1, ed. T. Vargas, C. A. Jerez, J. V. Wiertz & H. Toledo. Santiago: University of Chile, pp. 253–62.

Bryant, M. P., Wolin, E. A., Wolin, M. J. & Wolfe, R. S. (1967). *Methanobacillus omelianskii*, a symbiotic association of two species of bacteria. *Archiv für Mikrobiologie*, **59**, 20–31.

Burggraf, S., Jannasch, H. W., Nicolaus, B. & Stetter, K. O. (1990). *Archeoglobus profundus* sp. nov. represents a new species within the sulfate-reducing archaebacteria. *Systematic and Applied Microbiology*, **13**, 24–8.

Burgstaller, W. & Schinner, F. (1993). Leaching of metals with fungi. *Journal of Biotechnology*, **27**, 91–116.

Coleman, M. L., Hedrick, D. B., Lovley, D. R., White, D. C. & Pye, K. (1993). Reduction of Fe(III) in sediments by sulfate reducing bacteria. *Nature*, **361**, 436–8.

de Vrind-de Jong, E. W. & de Vrind, J. P. M. (1997). Algal deposition of carbonates and silicates. In *Reviews in Mineralogy*, Vol. 35. *Geomicrobiology: Interactions Between Microbes and Minerals*, ed. J. F. Banfield & K. H. Nealson. Washington, DC: Mineralogical Society of America, pp. 267–307.

Ehrlich, H. L. (1980). Bacterial leaching of manganese ores. In *Biogeochemistry of Ancient and Modern Environments*, ed. P. A. Trudinger, M. R. Walter & B. J. Ralph. Canberra: Australian Academy of Science, pp. 609–14.

Ehrlich, H. L. (1993a). Electron transfer from acetate to the surface of MnO_2 particles by a marine bacterium. *Journal of Industrial Microbiology*, **12**, 121–8.

Ehrlich, H. L. (1993b). A possible mechanism for the transfer of reducing power to insoluble mineral oxide in bacterial respiration. In *Biohydrometallurgical Technologies*, Vol. II, ed. A. E. Torma, M. L. Apel & C. L. Brierley. Warrendale, PA: The Minerals, Metals & Materials Society, pp. 415–22.

Ehrlich, H. L. (1999). Microbes as geologic agents: their role in mineral formation. *Geomicrobiology Journal*, **16**, 135–53.

Ehrlich, H. L. (2002a). *Geomicrobiology*, 4th edn. New York: Marcel Dekker.

Ehrlich, H. L. (2002b). How microbes mobilize metals in ores: a review of current understandings and proposals for further research. *Minerals and Metallurgical Processing*, **19**, 220–4.

Ehrlich, H. L. & Fox, S. I. (1967). Copper sulfide precipitation by yeasts from acid mine-waters. *Applied Microbiology*, **15**, 135–9.

Ehrlich, H. L. & Wickert, L. M. (1997). Bacterial action on bauxites in columns fed with full-strength and dilute sucrose-mineral salts medium. In *Biotechnology and the Mining Environment*, ed. L. Lortie, P. Bédard & W. D. Gould. Proceedings of 13th Annual General Meeting Biominet SP 97–1, CANMET, Natural Resources, Canada, Ottawa, Canada, pp. 73–89.

Ehrlich, H. L., Yang, S. H. & Mainwaring, J. D. Jr. (1973). Bacteriology of manganese nodules. VI. Fate of copper, nickel, cobalt and iron during bacterial and chemical reduction of the manganese(IV). *Zeitschrift für Allgemeine Mikrobiologie*, **13**, 39–48.

Ehrlich, H. L., Ingle, J. L. & Salerno, J. C. (1991). Iron and manganese oxidizing bacteria. In *Variations in Autotrophic Life*, ed. J. M. Shively & L. L. Barton. London: Academic Press, pp. 147–70.

Ehrlich, H. L., Wickert, L. M., Noteboom, D. & Doucet, J. (1995). Weathering of bauxite by heterotrophic bacteria. In *Biohydrometallurgical Processing*, Vol. 1, ed. T. Vargas, C. A. Jerez, J. V. Wiertz & H. Toledo, Santiago: University of Chile, pp. 395–403.

Emerson, D. (2000). Microbial oxidation of Fe(II) and Mn(II) at circumneutral pH. In *Environmental Microbe-Metal Interactions*, ed. D. R. Lovley. Washington, DC: ASM Press, pp. 31–52.

Fox, S. I. (1967). Bacterial oxidation of simple copper sulfides. Unpublished Ph.D. thesis, Rensselaer Polytechnic Institute, Troy, NY.

Friedmann, E. I. (1982). Endolithic microorganisms in the Antarctic cold desert. *Science*, **215**, 1045–53.

Friedmann, E. I. & Ocampo, R. (1984). Endolithic microorganisms in extreme dry environments: Analysis of a lithobiontic microbial habitat. In *Current Perspectives in Microbial Ecology*, ed. M. J. Klug & C. A. Reddy. Proceedings of the

Third International Symposium on Microbial Ecology. Washington, DC: American Society of Microbiology, pp. 177–85.

Ghiorse, W. C. & Ehrlich, H. L. (1976). Electron transport components of the MnO_2 reductase system and the location of the terminal reductase in a marine bacillus. *Applied and Environmental Microbiology*, **31**, 977–85.

Grossman, M. J., Lee, M. K., Prince, R. C. *et al.* (2001). Deep desulfurization of extensively hydrodesulfurized middle distillate oil by *Rhodococcus* sp. strain ECRD-1. *Applied and Environmental Microbiology*, **67**, 1949–52.

Guay, R. & Silver, M. (1975). *Thiobacillus acidophilus* sp. nov. Isolation and some physiological characteristics. *Canadian Journal of Microbiology*, **21**, 281–8.

Hale, M. E. Jr. (1967). *The Biology of Lichens*. London, UK: Edward Arnold.

Harrison, A. P. Jr. (1981). *Acidiphilium cryptum* gen. nov., heterotrophic bacterium from acidic mineral environments. *International Journal of Systematic Bacteriology*, **31**, 327–32.

Harrison, A. P. Jr. (1984). The acidophilic thiobacilli and other acidophilic bacteria that share their habitat. *Annual Review of Microbiology*, **38**, 265–92.

Hartmannova, V. & Kuhr, I. (1974). Copper leaching by lower fungi. *Rudy*, **22**, 234–8.

Huber, R., Sacher, M., Vollman, A., Huber, H. & Rose, D. (2000). Respiration of arsenate and selenate by hyperthermophilic Archaea. *Systematic and Applied Microbiology*, **23**, 305–14.

Ivanov, M. V. (1968). *Microbiological Processes in the Formation of Sulfur Deposits*. Israel Program for Scientific Translations. US Department of Agriculture and National Science Foundation, Washington, DC.

Johnson, D. B., Ghauri, M. A. & Said, M. F. (1993). Isolation and characterization of an acidophilic, heterotrophic bacterium capable of oxidizing ferrous iron. *Applied and Environmental Microbiology*, **58**, 1423–8.

Lazaroff, N., Melanson, L., Lewis, E., Santoro, N. & Pueschel, C. (1985). Scanning electron microscopy and infrared spectroscopy of iron sediment formed by *Thiobacillus ferrooxidans*. *Geomicrobiology Journal*, **4**, 231–68.

Livesey-Goldblatt, E., Norman, P. & Livesey-Goldblatt, D. R. (1983). Gold recovery from arsenopyrite/pyrite ore by bacterial leaching and cyanidation. In *Recent Progress in Biohydrometallurgy*, ed. G. Rossi & A. E. Torma. Iglesias, Italy: Associazione Mineraria Sarda, pp. 627–41.

Lovley, D. R. (1991). Dissimilatory Fe(III) and Mn(IV) reduction. *Microbiological Reviews*, **55**, 259–87.

Lovley, D. R. (2000). Fe(III) and Mn(IV) reduction. In *Environmental Microbe-Metal Interactions*, ed. D. R. Lovley. Washington, DC: ASM Press, pp. 3–30.

Lovley, D. R. & Phillips, E. J. P. (1988). Novel mode of microbial energy metabolism: organic carbon oxidation coupled to dissimilatory reduction of iron or manganese. *Applied and Environmental Microbiology*, **54**, 1472–80.

Lovley, D. R., Stolz, J. F., Nord, G. L. & Phillips E. J. P. (1987). Anaerobic production of magnetite by dissimilatory iron-reducing microorganisms. *Nature*, **330**, 252–4.

Macaskie, L. E., Dean, A. C. R., Cheetham, A. K., Jakema, R. J. B. & Skarnulis, A. J. (1987). Cadmium accumulation by a *Citrobacter* sp.: the chemical nature of the accumulated metal precipitate and its location on the bacterial cells. *Journal of General Microbiology*, **133**, 539–44.

Macaskie, L. E., Empson, R. M., Cheetham, A. K., Grey, C. P. & Skarnulis, A. J. (1992). Uranium bioaccumulation by a *Citrobacter* sp. as a result of enzymatically mediated growth of polycrystalline HUO_2PO_4. *Science*, **257**, 782–4.

McConnaughey, T. (1991). Calcification in *Chara carollina*: CO_2 hydroxylation generates protons for bicarbonate assimilation. *Limnology and Oceanography*, **36**, 619–28.

Macy, J. M., Nunan, K., Hagen, K. D. *et al.* (1996). *Chrysogenes arsenatis* gen. nov, spec. nov., a new arsenate-respiring bacterium isolated from gold mine wastewater. *International Journal of Systematic Bacteriology*, **46**, 1153–7.

Mandernack, K. W., Post, J. & Tebo, B. M. (1995). Manganese mineral formation by bacterial spores of a marine *Bacillus* sp., strain SG-1: evidence for the direct oxidation of Mn(II) to Mn(IV). *Geochimica et Cosmochimica Acta*, **59**, 4393–408.

Myers, C. R. & Myers, J. M. (1992). Localization of cytochromes in the outer membrane of anaerobically grown *Shewanella putrefaciens* MR-1. *Journal of Bacteriology*, **174**, 3429–38.

Nielsen, A. M. & Beck, J. V. (1972). Chalcocite oxidation and coupled carbon dioxide fixation by *Thiobacillus ferrooxidans*. *Science*, **175**, 1124–6.

Oremland, R. S., Hoeft, S. E., Santini, J. M. *et al.* (2002a). Anaerobic oxidation of arsenite in Mono Lake water by a facultative arsenite oxidizing chemoautotroph, strain MLHE-1. *Applied and Environmental Microbiology*, **68**, 4795–802.

Oremland, R. S., Newman, D. K., Kail, B. W. & Stolz, J. F. (2002b). Bacterial respiration of arsenate and its significance in the environment. In *Environmental Chemistry of Arsenic*, ed. W. T. Frankenberger. New York: Marcel Dekker, pp. 273–95.

Oremland, R. S., Stolz, J. F. & Hollibaugh, J. T. (2004). The microbial arsenic cycle in Mono Lake, California. *FEMS Microbiology Ecology*, **48**, 15–27.

Orphan, V. J., Hinrichs, K.-U., Ussler, W. III *et al.* (2001). Comparative analysis of methane oxidizing Archaea and sulfate-reducing bacteria in anoxic marine sediments. *Applied and Environmental Microbiology*, **67**, 1922–34.

Pye, K. (1984). SEM analysis of siderite cements in interstitial marsh sediments, Norfolk, England. *Marine Geology*, **56**, 1–12.

Pye, K., Dickson, A. D., Schiavon, N., Coleman, M. L. & Cox, M. (1990). Formation of siderite-Mg-calcite-iron sulfide concentrations in intertidal marsh and sandflat sediments, north Norfolk, England. *Sedimentology*, **37**, 325–43.

Stetter, K. O., Laurer, G., Thomm, M. & Neuner, A. (1987). Isolation of extremely thermophilic sulfate reducers: evidence for a novel branch of archaebacteria. *Science*, **236**, 822–4.

Stolz, J. F. & Oremland, R. S. (1999). Bacterial respiration of arsenic and selenium. *FEMS Microbiology Reviews*, **23**, 615–27.

Stone, A. T. (1987). Microbial metabolites and the reductive dissolution of manganese oxides: Oxalate and pyruvate. *Geochimica et Cosmochimica Acta*, **51**, 153–7.

Straub, K. L., Benz, M., Schink, B. & Widdel, F. (1996). Anaerobic, nitrate-dependent microbial oxidation of ferrous iron. *Applied and Environmental Microbiology*, **62**, 1458–60.

Sullivan, J. D. (1930). *Chemistry of leaching chalcocite*. Tech. Paper 473. US Department of Commerce, Bureau of Mines, Washington, DC.

Thompson, J. B., Ferris, F. G. & Smith, D. A. (1990). Geomicrobiology and sedimentology of the mixolimnion and chemocline of Fayetteville Green Lake, New York. *Palaios*, **5**, 52–75.

Trudinger, P. A., Swaine, D. J. & Skyring, G. W. (1979). Biogeochemical cycling of elements – general considerations. In *Biogeochemical Cycling of Mineral-Forming Elements*, eds. P. A. Trudinger & D. J. Swaine. Amsterdam: Elsevier, pp. 1–27.

van den Ende, F. P. & van Gemerden, H. (1993). Sulfide oxidation under oxygen limitation by a *Thiobacillus thioparus* isolated from a marine microbial mat. *FEMS Microbiology Ecology*, **13**, 69–78.

Welch, S. A. & Vandevivere, P. (1995). Effect of microbial and other naturally occurring polymers on mineral dissolution. *Geomicrobiology Journal*, **12**, 227–38.

Wenberg, G. M., Erbisch, F. H. & Violon, M. (1971). Leaching of copper by fungi. *Transactions of the Society of Minerals Engineering* AIME, **250**, 207–12.

Widdel, F. & Pfennig, N. (1977). A new anaerobic, sporing, acetate-oxidizing sulfate-reducing bacterium, *Desulfotomaculum* (emend.) *acetoxidans*. *Archives of Microbiology*, **112**, 119–22.

Widdel, F., Schnell, S., Heising, S. *et al.* (1993). Ferrous oxidation by anoxygenic phototrophic bacteria. *Nature*, **362**, 834–6.

Yarzábal, A., Brasseur, G., Ratouchniak, J. *et al.* (2002). The high-molecular-weight cytochrome c Cyc2 of *Acidithiobacillus ferrooxidans* is an outer membrane protein. *Journal of Bacteriology*, **184**, 313–17.

Zavarzin, G. A. (1972). Heterotrophic satellite of *Thiobacillus ferrooxidans*. *Mikrobiologiya*, **41**, 369–70.

2

Integrated nutrient cycles in boreal forest ecosystems – the role of mycorrhizal fungi

ROGER D. FINLAY AND ANNA ROSLING*

Introduction

Mycorrhizal fungi play a central role in biogeochemical cycles since they obtain carbon from their photosynthetic plant hosts and allocate this via their mycelia to the soil ecosystem. The mycelia interact with a range of organic and inorganic substrates, as well as with different organisms such as bacteria, fungi, soil micro- and meso-fauna and the roots of secondary hosts or non-host plants. Some of the carbon allocated to the mycelium is used to make compounds such as enzymes, organic acids, siderophores or antibiotics, which influence biotic or abiotic substrates through processes such as decomposition, weathering or antibiosis. Organic and inorganic nutrients mobilized from these substrates can be taken up by the mycorrhizal mycelia and translocated to their plant hosts, influencing plant growth, community structure and vegetation dynamics. Ultimately these changes have further impacts on biogeochemical cycles. Different types of mycorrhizal symbiosis have evolved as adaptations to different suites of edaphic parameters, resulting in the characteristic vegetation types that dominate different terrestrial biomes. Other chapters in this book consider specific contributions of ectomycorrhizal fungi to mineral dissolution (see Wallander, Chapter 14, this volume), carbon and nitrogen cycling (see Hobbie & Wallander, Chapter 5, this volume) and mineral tunnelling (see Smits, Chapter 13, this volume). In this chapter we concentrate on how these activities are integrated and on ways in which ectomycorrhizal hyphae may interact with other microorganisms to influence biogeochemical cycles. The main emphasis will be placed on ectomycorrhizal fungi and

* Present address: *Department of Earth & Planetary Science, University of California, Berkeley, CA 94720-4767, USA.*

Fungi in Biogeochemical Cycles, ed. G. M. Gadd. Published by Cambridge University Press. © British Mycological Society 2006.

boreal forests but the influence of other types of mycorrhizal fungi is also considered where similar types of interaction may take place in other types of ecosystem (see Johnson *et al.*, Chapter 6, this volume).

Boreal forest ecosystems are dominated by tree species that typically form ectomycorrhizal associations with basidiomycete or ascomycete fungi. The majority of fine tree roots, commonly more than 95%, are colonized by ectomycorrhizal fungi (Taylor, 2002) and these fungi play a central role in integrating the cycling of nutrients and carbon within forest ecosystems. Mycorrhizal mycelia are the major organs for nutrient uptake by trees and mycelial colonization of the soil substrate has a quantitative effect on nutrient uptake since the mycelium increases the effective absorptive surface area of the root system. The enzymatic potential of the fungi to hydrolyse organic substrates and mobilize organic polymers also has a qualitative effect on nutrient uptake through affecting the availability of different organic forms of nitrogen and phosphorus to roots (Read & Perez-Moreno, 2003). In addition to colonizing organic substrates, fungal hyphae are also able to colonize mineral substrates and may be able to modify them through chemical interactions that release mineral nutrients. This is made possible through the supply of energy-rich carbon compounds to the fungal mycelia from their plant hosts. A considerable fraction of the carbon fixed through photosynthesis may initially be allocated to ectomycorrhizal mycelia (Finlay & Söderström, 1992; Leake *et al.*, 2001; Rosling *et al.*, 2004a) and the consequences of this for carbon and nitrogen cycling are discussed elsewhere (see Hobbie & Wallander, Chapter 5, this volume). Field experiments involving disruption of the carbon supply to mycorrhizal mycelia in both forest (Högberg *et al.*, 2001) and grassland ecosystems (Johnson *et al.*, 2002) have demonstrated a close coupling between current photosynthesis and respiratory loss of carbon from roots and mycorrhizal structures, underlining the importance of mycorrhizal fungi in carbon cycling. In soil ecosystems different mineral and organic substrates are distributed as discrete resources within a three-dimensional matrix. Filamentous fungi promote spatial integration by growing as interconnected functional units, mycelia, translocating resources within a network of hyphae and rhizomorphs. The ability of mycelia to connect different mineral and carbon sources enables translocation of heterogeneously distributed nutrients and moisture through the mycelium, makes resource utilization more effective and increases stress tolerance in filamentous fungi compared to single-celled organisms (Hirsch *et al.*, 1995). Interactions of mycorrhizal fungi with different types of substrate including dissolved mineral nutrients, solid mineral

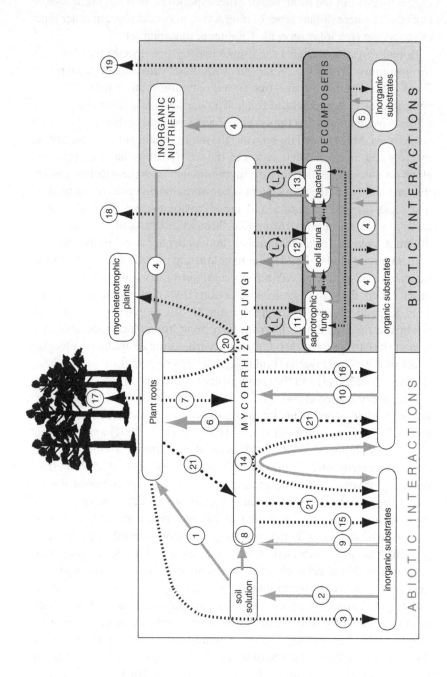

Fig. 2.1. Diagram of possible abiotic and biotic interactions between mycorrhizal fungal mycelia and other components of forest soil ecosystems. Carbon flow is represented by dashed black arrows, flow of nutrients is represented by solid grey arrows and flow of water by dotted arrows. Conventional theories of nutrient uptake involve direct uptake by roots from dissolved inorganic nutrient sources (1) or from mineral substrates via physicochemical weathering (2) or through dissolution/weathering induced by roots (3). Conventional theories of nutrient cycling are also based on mobilization of nutrients by decomposers from organic and inorganic substrates via cycles of immobilization and mobilization ('mineralization') (4), (5). More modern theories of nutrient cycling acknowledge the intermediate position of symbiotic mycorrhizal hyphae in nutrient acquisition (6) and carbon allocation (7), their uptake of inorganic nutrients from soil solution (8) and from mineral substrates via weathering (9) and their possible intervention in the decomposition pathway through direct mobilization of organic nutrients (10) and interactions with saprotrophic fungi (11), soil fauna (12) and bacteria (13). The filamentous nature of the mycorrhizal mycelium permits interchange of carbon and nutrients between different substrates (14) and allocation of photoassimilate-derived compounds to microsites within inorganic (15) and organic (16) substrates, which may be inaccessible to plant roots and isolated from the bulk soil solution. Nutrients mobilized by the mycorrhizal mycelium in these microsites (9), (10) or through interaction with decomposers (11, 12, 13) can be translocated directly to the plant hosts without entering the soil solution. Respiratory loss of carbon from roots (17), mycorrhizal fungi (18) and other soil organisms (19) may be accompanied by transfer of carbon from mycorrhizal hosts via mycelial connections to mycoheterotrophic hosts (20). Transfer of water via hydraulic lift from roots to mycorrhizal fungi (21) may help to maintain mycelial integrity in dry soil and to condition the environment round hyphal tips, allowing more stable conditions for microbial interactions at this interface. L represents nutrient microloops involving exchange of carbon and nutrients (see p. 43).

substrates, soluble organic compounds, dead organic matter and other organisms are shown in Fig. 2.1, together with their likely influences on different biogeochemical cycles. These different types of interactions are discussed in more detail in the sections below in an attempt to summarize current knowledge and gaps in our knowledge where further research is required. Bold numbers in parenthesis refer to Fig. 2.1.

Mycelial interactions with inorganic substrates

Soil chemistry and calculations of weathering budgets are largely based on the concept of the soil solution as the interface controlling processes such as weathering and plant nutrient uptake (**1, 2**) (Sverdrup & Warfvinge, 1995). Typically, the release of low-molecular-weight organic acids from plant roots is assumed to play a role in nutrient availability through interaction with both dissolved and solid-phase cations (**3**) (Sandnes et al., 2005). The capacity of the mycelia to absorb and transport a range of inorganic ions from soil solution (**8**) is also well documented (Finlay, 1992; Smith & Read, 1997; Jentschke et al., 2000). Translocation of phosphorus, nitrogen, calcium, magnesium, sodium and potassium (^{86}Rb) through the mycelium over distances of up to 40 cm has been demonstrated using tracer experiments (see Finlay, 1992). The surface area of the fungal mantle is greatly increased compared with that of non-mycorrhizal short roots and the extensive mycelial structures emanating from the fungal mantle increase the surface area available for nutrient absorption still further. The development of rhizomorphs in some ectomycorrhizal species facilitates the translocation of nutrients over long distances and finer undifferentiated hyphae at the mycelial margin are able to penetrate micro-sites and take up nutrients that are inaccessible to thicker plant roots. Microcosm experiments by Jentschke et al. (2001) have also demonstrated stimulation of hyphal density in response to localized sources of phosphorus under otherwise nutrient-limited conditions. As well as improving the uptake of essential plant nutrients, ectomycorrhizal fungi are also able to decrease the loss of base cations through leaching in acid soils (Ahonen-Jonnarth et al., 2003). Ectomycorrhizal mycelia are also able to modify the effects of acidification stress on plants in a number of other ways including chelation of toxic metals (Finlay, 1995; Ahonen-Jonnarth et al., 2000).

In addition to their interactions with dissolved mineral nutrients, mycorrhizal mycelia may also interact with solid mineral substrates (**9, 15**). Fungi are well-known biogeochemical agents (Sterflinger, 2000) and the capacity of lichens to weather rock surfaces is possibly the most striking of biological weathering processes and relatively well studied (Barker et al., 1997).

In forest soils, some ectomycorrhizal fungi, such as *Hysterangium crassum* [Tul. & Tul.] Fischer, *Hysterangium setchellii* (Fischer) and *Gautieria monticola* (Harkness), are known to form dense mycelial mats, strongly affecting the nutrient availability and weathering rate of the colonized mineral (Graustein *et al.*, 1977; Cromack *et al.*, 1979; Entry *et al.*, 1991; Griffiths *et al.*, 1994). Landeweert *et al.* (2000) reviewed the evidence for weathering of minerals by mycorrhizal fungi and concluded that it could be of significance for replacing nutrients lost through increased biomass harvesting or leaching. On the basis of calcium:strontium ratios in soil water, minerals in the soil and different mycorrhizal and non-mycorrhizal trees, Blum *et al.* (2002) concluded that direct calcium uptake by ectomycorrhizal fungi weathering apatite in the parental material could compensate for calcium loss in base-poor ecosystems. Data on element ratios should, however, be interpreted with care, because of high variation of Ca/Sr ratios in different plant tissues and limited understanding of the cycling of these elements in plants (Watmough & Dillon, 2003). Ectomycorrhizal fungi that colonize root tips and produce extramatrical mycelia in mineral soils will allocate carbon to this habitat. Allocated carbon is used to build up mycelial biomass and to mediate mycelial activities, such as nutrient uptake mediated through exudation of organic compounds that condition the environment surrounding the hyphae **(15/21)**. Consequences of this conditioning for biotic interactions with fungi **(11)** and bacteria **(13)** are considered later in this chapter (see pp. 41 and 43).

The high fine-root density in the organic and upper mineral soil has resulted in most studies of ectomycorrhizal fungal communities being restricted to sampling in the upper, organic part of the soil profile (Horton & Bruns, 2001). Tree roots are also found at greater depths (Jackson *et al.*, 1996) and a recent study of the ectomycorrhizal community in a podzol soil profile in northern Sweden demonstrated that two-thirds of all root tips were located in the mineral soil (Rosling *et al.*, 2003). In this study the composition of the ectomycorrhizal community colonizing ectomycorrhizal root tips was shown to change depending on the soil horizon, with over half of the ectomycorrhizal fungal taxa occurring exclusively in the mineral soil horizons. A complementary study at the same site demonstrated a similar relationship for the extramatrical mycelium of the same fungi (Landeweert *et al.*, 2003). Identification of ectomycorrhizal fungi through terminal restriction fragment length polymorphism (T–RFLP) analysis of DNA extracted from soil mycelium has also been used to demonstrate differences in ectomycorrhizal species composition between different components of the forest floor (the litter, fermentation and humus layers) and the B horizon of the mineral soil in a North American *Pinus resinosa* Ait. stand (Dickie *et al.*, 2002). These results suggest that there may be a functionally specific relationship between certain ectomycorrhizal species and the mineral substrate they colonize. In our laboratory recent studies of ectomycorrhizal mycelia colonizing different mineral substrates in vivo, or in plant microcosms, also suggest that patterns of mycelial growth, carbon allocation and substrate acidification may be influenced by mineral composition (Rosling *et al.*, 2004a, b).

Weathering as a result of organic acid exudation by fungi and bacteria (**5, 15**), was recognized early in studies of biogeochemical processes (e.g. Duff *et al.*, 1963; Henderson & Duff, 1963). The ability to produce and exude low-molecular-weight organic acids is widespread in the fungal kingdom (Gadd, 1999). Accumulation of calcium oxalate crystals on ectomycorrhizal hyphae in the field, indicates that oxalic acid is an important agent in biological weathering, resulting in increased phosphorus availability to plants (**6, 9**) (Graustein *et al.*, 1977). Field estimates of organic acid concentrations in the soil solution are, however, generally too low to cause weathering of minerals, such as feldspar (Drever & Stillings, 1997). This is likely to be a result of current methods used to measure organic acid concentrations in field samples, which systematically underestimate the real concentration by not taking into account the probable large spatial variation and microbial decomposition (**11, 13**) (Jones *et al.*, 2003; van Hees *et al.*, 2000, 2003, 2005a).

Although organic acids, such as oxalic acid, play an important role in biogeochemical weathering, the many underlying mechanisms for their production in fungi are not fully characterized. In soil systems, increased exudation of organic acids, by plants, fungi and bacteria (**3, 5, 15**), has been demonstrated in response to high concentrations of toxic metals, such as aluminium (Gadd, 1993; Ma *et al.*, 1997; Hamel *et al.*, 1999). Ahonen-Jonnarth *et al.* (2000) demonstrated increased production of oxalic acid by the ectomycorrhizal fungi *Suillus variegatus* and *Rhizopogon roseolus* colonizing seedlings in response to increased concentrations of aluminium and copper. Fungal regulation of organic acid production in response to cation deficiencies varies, depending on experimental conditions and the fungal species tested. Phosphorus deficiency in plants results in increased production and exudation of organic acids (Ryan *et al.*, 2001) and mycorrhiza-induced apatite dissolution is increased under conditions of phosphorus deficiency in forest soils (Hagerberg *et al.*, 2003). In pure cultures, six tested ectomycorrhizal isolates did not regulate oxalic acid production in response to phosphorus limitations but instead increased production when $CaCO_3$

was present in the medium (Arvieu *et al.*, 2003). In experiments by Paris *et al.* (1995, 1996) magnesium and potassium deficiency significantly increased oxalic acid exudation in the external mycelium of *Paxillus involutus* (Batsch.: Fr) Fr. and *Pisolithus tinctorius* (Pers.) Coker & Couch compared to non-deficient conditions. Oxalic acid production in *P. tinctorius* was increased regardless of the nitrogen source supplied, whereas *P. involutus* increased oxalic acid production when nitrogen was supplied as NH_4^+ but not when the nitrogen source was NO_3^- (Paris *et al.*, 1996). Element deficiency does not always induce increased organic acid production in fungi and deficiency responses may also be difficult to separate from general stress responses. The production and exudation of oxalic acid may fulfil a range of different functions in the physiology and ecology in different groups of fungi. For instance the form and availability of carbon and nitrogen sources influence the production of oxalic acid (Dutton & Evans, 1996). More oxalate is generally produced when nitrogen is supplied as nitrate, compared to ammonium (Gharieb *et al.*, 1998; Gadd, 1999). This pattern, however, varies for different fungi (Casarin *et al.*, 2003). Increased production has been found in response to excess carbon compared to other elements in the fungal growth substrate (Gadd, 1999) and suggested to be a result of incomplete oxidation of sugars (Richards, 1987). Jacobs *et al.* (2002) investigated the fungal carbon requirements of *Rhizoctonia solani* Kühn for producing and exuding protons or organic acid needed for the dissolution of tri-calcium phosphate (TCP) in arrays of discrete agar droplets. Dissolution of TCP in drops not containing a carbon source only occurred when hyphae spanning different droplets were able to translocate carbon to the TCP-containing droplets and dissolution did not occur unless the glucose concentration was greater than 2% w/v in the droplets containing the carbon source. If nutrient deficiency in the plant increases the carbon allocated to the root system, this carbon may be allocated to mycorrhizal fungi and result in increased exudation of organic compounds by the hyphae. Under field conditions, however, poor potassium and phosphorus status of the trees does not appear to result in increased production of ectomycorrhizal mycelium (Hagerberg *et al.*, 2003). Production of organic acids may have other effects on biotic actions. Oxalic acid production is also involved in the degradation of plant-derived organic matter by saprotrophic fungi (4) (Connolly & Jellison, 1995; Palfreyman *et al.*, 1996; Ruijter *et al.*, 1999) and in the plant colonizing activity of pathogenic fungi (Dutton & Evans, 1996; Clausen *et al.*, 2000). Through ion-complex formation, exuded oxalic acid results in the formation of biominerals on the surface of many fungal hyphae. It has been suggested that this may protect hyphae from dehydration, as well as providing a physical barrier against grazing microfauna (Arocena *et al.*, 2001).

Siderophores, organic polymers released by plants (3), fungi and bacteria (5) in response to iron deficiency, may also play a role in mycorrhizal weathering of minerals (15). Strong complexes are formed with Fe^{3+} and these are then taken up through specific transporters in the plasma membrane in some plants, fungi and bacteria (Shenker *et al.*, 1995; Marschner, 1998). High etching rates of amorphous and crystalline silicates were observed when these were colonized by the fungi *Penicillium notàtum* and *Aspergillus amstellodami*. The intense etching was suggested to be a result of the presence of siderophores in the cell walls of the fungi (Callot *et al.*, 1987). Strong complex formation with elements in the mineral structure, such as binding of siderophores to iron, reduces the stability of the mineral structure thereby enhancing weathering (Ehrlich, 1998). The hydroxamate siderophores ferrichrome and ferricrocin were found by Holmström *et al.* (2004) in forest soil solution associated with ectomycorrhizal hyphae and production of ferricrocin by the ectomycorrhizal fungus *Cenococcum geophilum* has also been shown by Haselwandter and Winkelmann (2002). It is possible that siderophores may play a role in the formation of the tunnels in mineral grains found by Jongmans *et al.* (1997). These authors reported that minerals in podzol surface soils and shallow granitic rock under European coniferous forests were commonly criss-crossed by networks of tubular pores 3–10 µm in diameter. The aetiology of these has not yet been established although the authors speculated that they might have been caused by hyphae of ectomycorrhizal fungi exuding organic acids at their tips (van Breemen *et al.*, 2000a). Further studies by Smits *et al.* (2005) have now shown, however, that the contribution of these tunnels to total feldspar weathering is less than 1%, suggesting that the weathering of mineral surfaces is quantitatively more important (see also Smits, Chapter 13, this volume).

Weathering does not take place unless the mineral surface is in contact with a solution. Extracellular mucilage produced by fungi and bacteria may provide this through improved moisture retention on colonized mineral particles (Hirsch *et al.*, 1995; Barker *et al.*, 1998). Many fungal hyphae are extensively coated in rich extracellular mucilage enabling adhesion of the fungi to surfaces as well as to other hyphae (Jones, 1994). The mucilage consists of organic polymers, such as carbohydrates, proteins and lipids exuded by the fungi, but the composition varies between different fungi (Jones, 1994; Cooper *et al.*, 2000). Less variation is found in mucilage of bacteria and algae, where polysaccharides are the major component

(Jones, 1994). Fungal mucilage has mainly been studied in the context of spore germination and hyphal adhesion of plant pathogens and wood decay fungi (e.g. Chaubal *et al.*, 1991; Abu *et al.*, 1999). The biogeochemical importance of extracellular mucilage has, however, been studied in bacteria, predominantly in aquatic biofilm systems (Little & Wagner, 1997). While strong attachment of some polymers to a mineral surface may inhibit dissolution, other polymers form complexes with components of the mineral surface, resulting in reduced stability and thus increased dissolution of the mineral (Ullman *et al.*, 1996). Polysaccharides can change the weathering rate of minerals by a factor of three, either enhancing or suppressing the process (Banfield *et al.*, 1999). The water-holding capacity of extracellular mucilage may be one of the major weathering effects resulting from microbial attachment to mineral surfaces (Barker *et al.*, 1998). Formation of biominerals, such as calcium oxalate, is commonly observed in fungal mucilage and has been suggested to be a method by which fungi regulate external calcium concentrations (Connolly *et al.*, 1999). Extracellular mucilage may provide a habitat for associated bacteria (Fig. 2.2) (**13**).

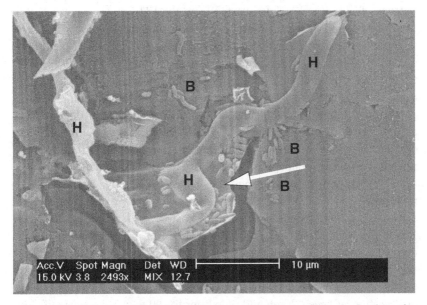

Fig. 2.2. Mycelium of *Hebeloma crustuliniforme* after colonization of a potassium feldspar surface for seven months. The sample was prepared by fixation and critical point drying followed by gold coating and analysis by scanning electron microscopy. Hyphae (H) and bacteria (B) are visible. Scale bar = 10 μm. The hyphal surface contact is mediated by a film of extracellular mucilage (arrow) and bacteria are seen in the mucilage.

Interactions with organic substrates

During recent years it has become increasingly apparent that ectomycorrhizal fungi are able to supply their host plants with nitrogen and phosphorus (6) that they sequester from a range of organic polymers that are otherwise unavailable to plant roots (10) (Read & Perez-Moreno, 2003). The enzymatic activities of ectomycorrhizal mycelia (16) have been reviewed by Lindahl *et al.* (2005) who point out that so far only very few species have been investigated and that very little is known about the potential roles of other decomposer groups in the aquisition (4, 11, 12, 13). Abuzinadah *et al.* (1986) first suggested that the ability of ectomycorrhizal fungi to use organic nutrient sources would be of significance to nutrient cycling in forests since it would restrict immobilization of nutrients in tissues of decomposer organisms and enable tighter nutrient cycling than the conventionally proposed immobilization–mobilization pathways (4), facilitating the direct return of organic nitrogen to the host trees. Much of the evidence for mobilization of nitrogen and phosphorus from organic substrates has come from microcosm experiments by Read and co-workers. Read (1991) demonstrated proliferation of ectomycorrhizal hyphae in patches of organic material collected from the FH horizon (humic layer) of forest soil and introduced into peat-containing microcosms with *Larix* seedlings colonized by *Boletinus cavipes*, suggesting that the mycelium is able to forage for nutrients contained in organic residues. Further support for this idea was obtained in studies by Bending and Read (1995a) showing depletion of nitrogen, phosphorus and potassium from similar organic patches. Although the role played by saprotrophs in this nutrient mobilization was not clear, Bending and Read (1995b) were able to measure elevated activities of nutrient-mobilizing enzymes such as protease, polyphenol oxidase and phosphomonoesterase in substrates intensively colonized by ectomycorrhizal fungi. In these experiments newly formed mycelial patches of *Suillus bovinus*, extending from a colonized *Pinus sylvestris* seedling, were strong sinks of host-derived carbon (Bending & Read, 1995a) and ectomycorrhizal fungi are clearly able to use host-derived assimilates to manufacture enzymes and support hyphal colonization of organic substrates (15). In experiments by Perez-Moreno and Read (2000) on *Betula pendula* Roth seedlings colonized by *P. involutus*, mycelial exploitation of beech, birch and pine litter for 90 days led to a significant decrease in the phosphorus content of all litter types and to enhanced growth of the seedlings compared to non-litter controls. As in the experiments by Bending and Read (1995a), the possible role of other associated microorganisms in nutrient mobilization was not clear in these experiments. There was a clear loss of phosphorus from the litter but the loss of nitrogen was smaller than the nitrogen gained by the plants and it is possible this could be accounted for through N_2-fixation by bacteria associated with the mycorrhizal mycelium (**13**) (Sen, 2000). These types of biotic interaction and their possible significance for nutrient cycling are discussed below. Recent experiments by Hodge *et al.* (2001) demonstrated that the arbuscular mycorrhizal fungus *Glomus hoi* was able to enhance decomposition and increase plant nitrogen capture from grass leaves. However, further research is still needed to distinguish between the direct capacity of arbuscular mycorrhizal fungi to mobilize organic substrates and their possible indirect effects on decomposition and plant nutrient uptake, caused by stimulation of decomposers and subsequent uptake of their decomposition products by mycorrhizal hyphae.

Direct utilization of a range of organic substrates, including nematodes (Perez-Moreno & Read, 2001a) and pollen (Perez-Moreno & Read, 2001b), has been demonstrated and it seems likely that in many cases ectomycorrhizal fungi are able to intervene (**10, 11, 12, 13**) in conventional decomposition cycles, bypassing the longer pathways, which involve immobilization of nutrients in decomposer organisms (**4**). New models of nutrient cycling based on antagonistic interactions involving nutrient transfer between decomposer and mycorrhizal fungi have been proposed (Lindahl *et al.*, 2002) and these are discussed in the following section. While the nutrient mobilizing potential of some ectomycorrhizal fungi has been established, only a restricted range of fungi have been examined, and there is a need to investigate the extent to which these processes occur in a wider range of fungi under field conditions. Although uptake of simple amino compounds has been shown under field conditions (Näsholm *et al.*, 1998), further experiments need to be conducted in which mycelial uptake and root uptake can be partitioned. Further experiments are still needed to determine the relative importance of different organic nutrient sources, and to examine interspecific variation in the efficiency of nutrient transfer across the fungal–host interface. Analyses of the true enzymatic potential of mycorrhizal fungi, using *in situ* measurements in the field, will require the continued development of molecular tools, as well as their application under ecologically realistic conditions (Lindahl *et al.*, 2005).

Interactions with other organisms

Experiments involving mycorrhizal interactions with plants are frequently carried out without regard to the existence and possible influence of other soil organisms. It is increasingly apparent, however, that important interactions may take place with other fungi (11), soil animals (12), bacteria (13) and secondary plant hosts (Leake et al., 2004) (20). The boreal forest ecosystems in which ectomycorrhizal fungi predominate are nutrient poor and competitive interactions between nutrient-conservative saprotrophic and mycorrhizal fungi might result in rapid movement of resources from one biological pool to another. In these types of interaction, with intense competition for nutrients, the concept of mineral nutrient uptake from a soil solution continually replenished by decomposition processes (2) may not be realistic (Lindahl et al., 1999, 2002; Schimel & Bennett, 2004).

Saprotrophic and mycorrhizal fungi are not separate groups from an evolutionary perspective. Symbiotic association with plants is a life strategy that has evolved repeatedly from ancestral saprotrophic precursors during evolutionary history, although there appear to have been many reversals to the free-living condition (Hibbett et al., 2000). The mycorrhizal fungal strategy of deriving carbon from a symbiotic plant host should release these fungi from competition with other soil fungi for carbon from sources of dead organic materials. In soil, mycelia of both mycorrhizal and saprotrophic fungi interact and compete for nutrients (11). Direct transfer of labelled ^{32}P from the saprotrophic mycelium of Hypholoma fasciculare to mycorrhizal plants colonized by Suillus variegatus or P. involutus has been demonstrated by Lindahl et al. (1999), suggesting that direct interactions may alter the nutrient status of mycorrhizal plants. The balance of competition between ectomycorrhizal and saprotrophic fungi, and the direction of nutrient transfer resulting from antagonistic interactions can be strongly influenced by resource availability (Lindahl et al., 2001). In microcosms where the mycelial vigour of S. bovinus is decreased as a result of the presence of the wood-decomposing fungus Phanerochaete velutina (DC.: Fr.) P. Karts., less carbon was allocated to the mycelium and allocation was slower than in non-disturbed mycelia (Leake et al., 2001). The carbon pools available to the different mycelia largely determine the outcome of competitive interactions. As one mycelium achieves competitive superiority over another, it may utilize the other mycelium as a source of nutrients (Lindahl et al., 2001). Clearly, the existence of such direct interactions has important implications for models of nutrient uptake, nutrient cycling, carbon flow and forest ecology.

Interactions also occur with soil animals and Klironomos and Hart (2001) have recently demonstrated that the ectomycorrhizal fungus Laccaria bicolor can kill the soil micro-arthropod Folsomia candida, take up nitrogen from the animals and transfer the derived nitrogen to *Pinus strobus* host plants. Perez-Moreno and Read (2001a) demonstrated similar uptake from dead nematodes and Bonkowski (2004) recently showed that mycorrhizal fungi and protozoa competed for plant-derived carbon, maximizing plant uptake of nitrogen and phosphorus. Bacterial communities in soil are influenced by a number of factors, such as moisture, nutrients and carbon availability. Release of photosynthetically derived carbon and input of organic matter by roots and mycorrhizal hyphae alters the local soil environment to allow microbial proliferation. The soil–root interface, the rhizosphere (e.g. Marschner, 1998), and the soil–hyphal interface, the mycorrhizosphere (Rambelli, 1973), commonly differ from the bulk soil with higher carbon availability, lower pH, changed redox potential and increased concentrations of low-molecular-weight organic compounds (Marschner, 1998). Microbial activity may further be stimulated by stabilized moisture conditions in the mycorrhizosphere compared to the bulk soil. Under drought conditions, nocturnal water transfer from deep host roots to external mycorrhizal mycelia has been demonstrated (Querejeta *et al.*, 2003) and drops are commonly observed on aerial hydrophobic hyphae of ectomycorrhizal fungi (Unestam & Sun, 1995). In studies of mycorrhizal fungi inoculated with different bacteria, the mycorrhizosphere environment has been reported to have both synergistic and antagonistic effects on microbial activity (Leyval & Berthelin, 1989). Olsson *et al.* (1996) showed a reduction of bacterial activity (thymidine and leucine incorporation) by the mycelium of *P. involutus*, a result that was consistent with those of Nurmiaho-Lassila *et al.* (1997). Patterns of fungal enzyme expression in different parts of the mycelium have been shown to vary (Timonen & Sen, 1998) and the composition and substrate utilization patterns of bacteria associated with ectomycorrhizal mycelium have been shown to vary according to both species and location within the mycorrhizosphere (Timonen *et al.* 1998) as well as locations within the soil profile (Heinonsalo *et al.*, 2001). There are still few functionally based studies of mycorrhiza-associated bacteria but recent experiments by Frey-Klett *et al.* (2005) suggest that the ectomycorrhizal fungus *Laccaria bicolor* may significantly influence the functional diversity of bacteria in the mycorrhizosphere of *Pseudotsuga menziesii* seedlings in ways that are potentially beneficial to the symbiosis and the plant host. Endosymbiotic bacteria have been detected within arbuscular mycorrhizal fungi (Bianciotto *et al.*, 2000) and more recently within ectomycorrhizal fungi (Bertaux *et al.*, 2003) and further research is needed to determine the range of functional roles these bacteria play in biogeochemical cycles.

Carbon cycling

Carbon resources in the forest ecosystem originate from the photosynthetic activity of trees. The synthesized carbon compounds are allocated to growth, respiration and exudation in the plant (Marschner, 1998). A substantial proportion, 20%–25%, of the photosynthate allocated to tree roots is required for the growth and maintenance of the mycorrhizal fungi (Smith & Read, 1997). Current photosynthate is allocated mainly to sites of active growth (Erland et al., 1991). Together, roots, mycorrhizal fungi and the associated microbial community respire a large part of the carbon derived from photosynthesis in trees, accounting for approximately half of the soil respiration in a boreal pine forest in the north of Sweden (Högberg et al., 2001). Energy-rich assimilates from host plants may be used to produce a range of compounds such as organic acids, antibiotics, enzymes and siderophores. Experiments by Bending and Read (1995a), Leake et al. (2001), Mahmood et al. (2001) and Rosling et al. (2004a) suggest that these can be allocated in a selective manner to exploit soil heterogeneity. Soil respiration appears to be tightly coupled to current photosynthesis (Högberg et al., 2001) and carbon from simple organic compounds such as amino acids is rapidly lost through respiration during nitrogen assimilation (Finlay et al., 1996; Näsholm et al., 1998). In laboratory experiments by Leake et al. (2001), close to 60% of the host-derived carbon allocated to mycorrhizal mycelia of P. involutus colonizing P. sylvestris seedlings was allocated within 48 h to mycelial patches proliferating in discrete sources of organic matter. Organic acids involved in weathering interactions may also be used as respiratory substrates by bacteria and van Hees et al. (2005b) have argued that this carbon source may account for a substantial proportion of soil respiration. A further potential fate of carbon allocated via mycorrhizal mycelia may be in production of chemicals promoting the formation of soil aggregates. Tisdall and Oades (1979) pointed out that, in addition to penetration and exploitation of soil microsites, mycorrhizal hyphae also play a role in stabilizing soil aggregates. Production of glycoproteins such as glomalin may play a role in stabilizing soil aggregates (Piotrowski et al., 2004) and the potential role of mycorrhizal fungi in soil carbon storage has been discussed (Treseder & Allen, 2000). Carbon allocation may take place between different plants joined by a common mycorrhizal mycelium. The ecological significance of this is questioned by some scientists (Robinson & Fitter, 1999) but common mycorrhizal fungi interconnecting plants, forming different types of mycorrhizal associations (Villarréal-Ruiz et al., 2004), provide the potential for movement of carbon and the process may be of significance where achlorophyllous mycoheterotrophs are connected to plants which can supply photosynthetically derived carbon (**20**) (Leake *et al.*, 2004). Although the flux of carbon involved in this process may not be significant in biogeochemical terms it may be important for maintaining floristic diversity or populations of locally distributed plants.

Conclusions

Mycorrhizal mycelia play a number of integrating roles in nutrient cycling in forest ecosystems by connecting organic and inorganic substrates and allowing interchange of carbon and nutrients (**14**). In addition to connecting substrates with different C/N ratios, mycelia may link discrete mineral substrates, enriched with different nutrient elements. The mycelial growth form, in conjunction with the direct supply of host assimilates, permits selective allocation of carbon for production of enzymes or organic acids (**15, 16**) (Bending & Read, 1995a; Rosling *et al.*, 2004a, b) resulting in exploitation of soil heterogeneity. Vertical translocation of elements from mineral soil to organic top layers by saprotrophic (**4, 5**) (Connolly *et al.*, 1999) and mycorrhizal (**14**) (van Breemen *et al.*, 2000b) mycelia has also been suggested to play a major role in forest nutrient cycling and soil formation. Seasonal patterns of storage of phosphorus and nitrogen within the mycorrhizal mantles of short roots (Lussenhop & Fogel, 1999; Genet *et al.*, 2000) also suggest that mycorrhizal fungi may integrate nutrient supply over time, influencing storage and mobilization of nutrients in systems where there is a seasonal fluctuation in supply of and demand for nitrogen and phosphorus. The apparent ability of mycorrhizal mycelia to allocate water derived from hydraulic lift by roots to fungal hyphae colonizing dry substrates (**21**) (Sun *et al.* 1999; Querejeta *et al.*, 2003) maintains hyphal integrity and conditions the immediate environment surrounding the hyphae, enabling interactions between the hyphae and other microorganisms at the soil–mycelial interface. These interactions may take place in protected microsites within mineral or organic substrates, which are effectively isolated from the bulk soil solution and involve nutrient microloops involving exchange of carbon and nutrients (L in Fig. 2.1). The real significance of mycorrhizal fungi is that they connect the primary producers of ecosystems, plants, to the heterogeneously distributed nutrients required for their growth, enabling the flow of energy-rich compounds required for nutrient mobilization whilst simultaneously providing conduits for the translocation of mobilized products back to their hosts. Elucidating the diversity of mechanisms involved, the range of interactions with other organisms and the ways in which these are regulated remains the ultimate challenge in understanding the role of these fungi in biogeochemical cycles.

Acknowledgements

We thank Torgny Unestam, Andy Taylor, Björn Lindahl, Les Paul, Petra Fransson, Audrius Menkis and Jonas Johansson for stimulating discussions and valuable comments. Financial support from the Swedish Natural Sciences Research Council (VR), the Swedish Research Council for Environment, Agricultural Sciences and Spatial Planning (FORMAS), the European Commission, the Swedish Environmental Protection Agency (SNV) and the Swedish National Energy Administration (STEM) is gratefully acknowledged.

References

Abu, A. R., Murphy, R. J. & Dickinson, D. J. (1999). Investigation of the extracellular mucilaginous materials produced by some wood decay fungi. *Mycological Research*, 103, 1453–61.

Abuzinadah, R. A., Finlay, R. D. & Read, D. J. (1986). The role of proteins in the nitrogen nutrition of ectomycorrhizal plants. II. Utilisation of protein by mycorrhizal plants of *Pinus contorta*. *New Phytologist*, 103, 495–506.

Ahonen-Jonnarth, U., van Hees, P. A. W., Lundström, U. & Finlay, R. D. (2000). Production of organic acids by mycorrhizal and non-mycorrhizal *Pinus sylvestris* seedlings exposed to elevated concentrations of aluminium and heavy metals. *New Phytologist*, 146, 557–67.

Ahonen-Jonnarth, U., Göransson, A. & Finlay, R. D. (2003). Growth and nutrient uptake of ectomycorrhizal *Pinus sylvestris* seedlings treated with elevated Al concentrations. *Tree Physiology*, 23, 157–67.

Arocena, J. M., Glowa, K. R. & Massicotte, H. B. (2001). Calcium-rich hypha encrustations on *Piloderma*. *Mycorrhiza*, 10, 209–15.

Arvieu, J-C., Leprince, F. & Plassard, C. (2003). Release of oxalate and protons by ectomycorrhizal fungi in response to P-deficiency and calcium carbonate in nutrient solution. *Annals of Forest Science*, 60, 815–21.

Banfield, J. F., Barker, W. W., Welch, S. A. & Taunton, A. (1999). Biological impact on mineral dissolution: application of the lichen model to understanding mineral weathering in the rhizosphere. *Proceedings of the National Academy of Sciences of the United States of America*, 96, 3404–11.

Barker, W. W., Welch, S. A. & Banfield, J. F. (1997). Biogeochemical weathering of silicate minerals. *Reviews in Mineralogy*, 35, 391–428.

Barker, W. W., Welch, S. A., Chu, S. & Banfield, J. F. (1998). Experimental observations of the effects of bacteria on aluminosilicate weathering. *American Mineralogist*, 83, 1551–63.

Bending, G. D. & Read, D. J. (1995a). The structure and function of the vegetative mycelium of ectomycorrhizal plants: V. Foraging behaviour and translocation of nutrients from exploited litter. *New Phytologist*, 130, 401–9.

Bending, G. D. & Read, D. J. (1995b). The structure and function of the vegetative mycelium of ectomycorrhizal plants. VI. Activities of nutrient mobilizing enzymes in birch litter colonized by *Paxillus involutus* (Fr.) Fr. *New Phytologist*, **130**, 411–17.

Bertaux, J., Schmid, M., Chemidlin Prevost-Boure, N. *et al.* (2003). *In situ* identification of intracellular bacteria related to *Paenibacillus* spp. in the mycelium of the ectomycorrhizal fungus *Laccaria bicolor* S238N. *Applied and Environmental Microbiology*, **69**, 4243–8.

Bianciotto, V., Lumini, E., Lanfranco, L. *et al.* (2000). Detection and identification of bacterial endosymbionts in arbuscular mycorrhizal fungi belonging to Gigasporaceae. *Applied and Environmental Microbiology*, **66**, 4503–9.

Blum, J. D., Klaue, A., Nezat, C. A. *et al.* (2002). Mycorrhizal weathering of apatite as an important calcium source in base-poor forest ecosystems. *Nature*, **417**, 729–31.

Bonkowski, M. (2004). Protozoa and plant growth: the microbial loop revisited. *New Phytologist*, **162**, 617–31.

Callot, G., Maurette, M., Pottier, L. & Dubois, A. (1987). Biogenic etching of microfeatures in amorphous and crystalline silicates. *Nature*, **328**, 147–9.

Casarin, V., Plassard, C., Souche, G. & Arvieu, J.-C. (2003). Quantification of oxalate ions and protons released by ectomycorrhizal fungi in rhizosphere soil. *Agronomie*, **23**, 461–9.

Chaubal, R., Wilmot, V. A. & Wynn, W. K. (1991). Visualization, adhesiveness, and cytochemistry of the extracellular matrix produced by urediniospore germ tubes of *Puccinia sorghi*. *Canadian Journal of Botany*, **69**, 2044–54.

Clausen, C. A., Green, F. III, Woodward, B. M., Evans, J. W. & DeGroot, R. C. (2000). Correlation between oxalic acid production and copper tolerance in *Wolfiporia cocos*. *International Biodeterioration and Biodegradation*, **46**, 69–76.

Connolly, J. H. & Jellison, J. (1995). Calcium translocation, calcium oxalate accumulation, and hyphal sheath morphology in the white-rot fungus *Resinicium bicolor*. *Canadian Journal of Botany*, **73**, 927–36.

Connolly, J. H., Shortle, W. C. & Jellison J. (1999). Translocation and incorporation of strontium carbonate derived strontium into calcium oxalate crystals by the wood decay fungus *Resinicium bicolor*. *Canadian Journal of Botany*, **77**, 179–97.

Cooper, L. L. D., Oliver, J. E., De Vilbiss, E. D. & Doss, R. P. (2000). Lipid composition of the extracellular matrix of *Botrytis cinerea* germlings. *Phytochemistry*, **53**, 293–8.

Cromack, K., Sollins, P., Graustein, W. *et al.* (1979). Calcium oxalate accumulation and soil weathering in mats of the hypogeous fungus *Hysterangium crassum*. *Soil Biology and Biochemistry*, **11**, 463–8.

Dickie, I. A., Xu, B. & Koide, R. T. (2002). Vertical niche differentiation of ectomycorrhizal hyphae in soil as shown by T-RFLP analysis. *New Phytologist*, **156**, 527–35.

Drever, J. I. & Stillings, L. L. (1997). The role of organic acids in mineral weathering. *Colloids and Surfaces*, **120**, 167–81.

Duff, R. B., Webley, D. M. & Scott, R. O. (1963). Solubilization of minerals and related materials by 2-ketoglutonic acid-producing bacteria. *Soil Science*, **95**, 105–14.

Dutton, M. V. & Evans, C. S. (1996). Oxalate production by fungi: its role in pathogenicity and ecology in the soil environment. *Canadian Journal of Microbiology*, **42**, 881–95.

Ehrlich, H. L. (1998). Geomicrobiology: its significance for geology. *Earth-Science Reviews*, **45**, 45–60.

Entry, J. A., Donnelly, P. K. & Cromack, K. (1991). Influence of ectomycorrhizal mat soils on lignin and cellulose degradation. *Biology and Fertility of Soils*, **11**, 75–8.

Erland, S. E., Finlay, R. D. & Söderström, B. (1991). The influence of substrate pH on carbon translocation in ectomycorrhizal and non-mycorrhizal pine seedlings. *New Phytologist*, **119**, 235–42.

46   *R. D. Finlay and A. Rosling*

Finlay, R. D. (1992). Uptake and mycelial translocation of nutrients by ectomycorrhizal fungi. In *Mycorrhiza in Ecosystems*, ed. D. J. Read, D. H. Lewis, A. H. Fitter & I. J. Alexander. Wallingford, UK: CAB International, pp. 91–97.

Finlay, R. D. (1995). Interactions betweeen soil acidification, plant growth and nutrient uptake in ectomycorrhizal associations of forest trees: problems and progress. In *Ecological Bulletin*, Vol. 44, *Effects of Acid Deposition and Ozone on the Terrestrial Environment in Sweden*, ed. H. Staaf & G. Tyler. Copenhagen: Blackwell, pp. 197–214.

Finlay, R. D. & Söderström, B. (1992). Mycorrhiza and carbon flow to the soil. In *Mycorrhizal Functioning*, ed. M. F. Allen. New York: Chapman and Hall, pp. 134–60.

Finlay, R. D., Chalot, M., Brun, A. & Söderström, B. (1996). Interactions in the carbon and nitrogen metabolism of ectomycorrhizal associations. In *Mycorrhizas in Integrated Systems – From Genes to Plant Development*, ed. J. M. Barea, C. Azcón-Aguilar, R. Azcón & J. A. Ocampo. Brussels: European Commission Report EUR 16728 EN, pp. 279–84.

Frey-Klett, P., Chavatte, M., Clausse, M.-L. *et al.* (2005). Ectomycorrhizal symbiosis affects functional diversity of rhizosphere fluorescent pseudomonads. *New Phytologist*, 165, 317–28.

Gadd, G. M. (1993). Interactions of fungi with toxic metals. *New Phytologist*, 124, 25–60.

Gadd, G. M. (1999). Fungal production of citric and oxalic acid: importance in metal speciation, physiology and biogeochemical processes. *Advances in Microbial Physiology*, 41, 48–91.

Genet, P., Prevost, A. & Pargney, J. C. (2000). Seasonal variations of symbiotic ultrastructure and relationships of two natural ectomycorrhizae of beech (*Fagus sylvatica/ Lactarius blennius* var. *viridis* and *Fagus sylvatica/ Lactarius subdulcis*). *Trees*, 14, 465–74.

Gharieb, M. M., Sayer, J. A. & Gadd, G. M. (1998). Solubilization of natural gypsum ($CaSO_4*2H_2O$) and the formation of calcium oxalate by *Aspergillus niger* and *Serpula himantioides*. *Mycological Research*, 102, 825–30.

Graustein, W. C., Cromack, K. Jr. & Sollins, P. (1977). Calcium oxalate: occurrence in soils and effect on nutrient and geochemical cycles. *Science*, 198, 1252–4.

Griffiths, R. P., Baham, J. E. & Caldwell, B. A. (1994). Soil solution chemistry of ectomycorrhizal mats in forest soil. *Soil Biology and Biochemistry*, 26, 331–7.

Hagerberg, D., Thelin, G. & Wallander, H. (2003). The production of ectomycorrhizal mycelium in forests: relation between forest nutrient status and local mineral sources. *Plant and Soil*, 252, 279–90.

Hamel, R., Levasseur, R. & Appanna, V. D. (1999). Oxalic acid production and aluminium tolerance in *Pseudomonas fluorescens*. *Journal of Inorganic Biochemistry*, 76, 99–104.

Haselwandter, K. & Winkelmann, G. (2002). Ferricrocin – an ectomycorrhizal siderophore of *Cenococcum geophilum*. *BioMetals*, 15, 73–7.

Heinonsalo, J., Jorgensen, K. S. & Sen, R. (2001). Microcosm-based analyses of Scots pine seedling growth, ectomycorrhizal fungal community growth and bacterial carbon utilisation profiles in boreal forest humus and underlying illuvial mineral horizons. *FEMS Microbiology Ecology*, 36, 73–84.

Henderson, M. E. K. & Duff, R. B. (1963). The release of metallic and silicate ions from minerals, rock, and soils by fungal activity. *Journal of Soil Science*, 14, 236–46.

Hibbett, D. S., Gilbert, L.-B. & Donaghue, M. J. (2000). Evolutionary instability of ectomycorrhizal symbioses in basidiomycetes. *Nature*, 407, 506–8.

Hirsch, P., Eckhardt, F. E. W. & Palmer, J. R. J. (1995). Fungi active in weathering of rock and stone monuments. *Canadian Journal of Botany*, 73 (Suppl. 1), S1384–90.

Hodge, A., Campbell, C. D. & Fitter, A. H. (2001). An arbuscular mycorrhizal fungus accelerates decomposition and acquires nitrogen directly from organic material. *Nature*, 413, 297–9.

Högberg, P., Nordgren, A., Buchmann, N. *et al.* (2001). Large-scale forest girdling shows that current photosynthesis drives soil respiration. *Nature*, **411**, 789–92.

Holmström, S. J. M., Lundström, U. S., Finlay, R. D. & van Hees, P. A. W. (2004). Siderophores in forest soil solution. *Biogeochemistry*, **71**, 247–58.

Horton, T. R. & Bruns, T. D. (2001). The molecular revolution in ectomycorrhizal ecology: Peeking into the black-box. *Molecular Ecology*, **10**, 1855–71.

Jackson, R. B., Canadell, J., Ehleringer, J. R. *et al.* (1996). A global analysis of root distribution for terrestrial biomes. *Oecologia*, **108**, 389–411.

Jacobs, H., Boswell, G. P., Ritz, K., Davidson, F. A. & Gadd, G. M. (2002). Solubilization of calcium phosphate as a consequence of carbon translocation by *Rhizoctonia solani. FEMS Microbiology Ecology*, **40**, 65–71.

Jentschke, G., Brandes, B., Kuhn, A. J. *et al.* (2000). The mycorrhizal fungus *Paxillus involutus* transports magnesium to Norway spruce seedlings. Evidence from stable isotope labeling. *Plant and Soil*, **220**, 243–6.

Jentschke, G., Brandes, B., Kuhn, A. J., Schröder, W. H. & Godbold, D. L. (2001). Interdependence of phosphorus, nitrogen, potassium and magnesium translocation by the ectomycorrhizal fungus *Paxillus involutus*. *New Phytologist*, **149**, 327–37.

Johnson, D., Leake, J. R., Ostle, N., Ineson, P. & Read, D. J. (2002). *In situ* (CO_2)-^{13}C pulse-labeling of upland grassland demonstrates a rapid pathway of carbon flux from arbuscular mycorrhizal mycelia to the soil. *New Phytologist*, **153**, 327–34.

Jones, D. L., Dennis, P. G., Owen, A. G. & van Hees, P. A. W. (2003). Organic acid behaviour in soils – misconceptions and knowledge gaps. *Plant and Soil*, **248**, 31–41.

Jones, E. B. G. (1994). Fungal adhesion. *Mycological Research*, **98**, 961–81.

Jongmans, A. G., van Breemen, N., Lundström, U. *et al.* (1997). Rock eating fungi. *Nature*, **389**, 682–3.

Klironomos, J. N. & Hart, M. M. (2001). Food-web dynamics. Animal nitrogen swap for plant carbon. *Nature*, **410**, 651–2.

Landeweert, R., Hoffland, E., Finlay, R. D. & van Breemen, N. (2000). Linking plants to rocks: ectomycorrhizal fungi mobilize nutrients from minerals. *Trends in Ecology and Evolution*, **16**, 248–54.

Landeweert, R., Leeflang, P., Kuyper, T. W. *et al.* (2003). Molecular identification of ectomycorrhizal mycelium in soil horizons. *Applied and Environmental Microbiology*, **69**, 327–53.

Leake, J. R., Donnelly, D. P., Saunders, E. M., Boddy, L. & Read, D. J. (2001). Rates and quantities of carbon flux to ectomycorrhizal mycelium following ^{14}C pulse labeling of *Pinus sylvestris* seedlings: effects of litter patches and interaction with a wood-decomposer fungus. *Tree Physiology*, **21**, 71–82.

Leake, J. R., McKendrick, S. L., Bidartondo, M. & Read, D. J. (2004). Symbiotic germination and development of the myco-heterotroph *Monotropa hypopitys* in nature and its requirement for locally distributed *Tricholoma* spp. *New Phytologist*, **163**, 405–23.

Leyval, C. & Berthelin, J. (1989). Interactions between *Laccaria laccata-Agrobacterium radiobacter* and beech roots influence on phosphorus, potassium, magnesium and iron mobilisation from minerals and plant growth. *Plant and Soil*, **117**, 103–10.

Lindahl, B., Stenlid, J., Olsson, S. & Finlay, R. D. (1999). Translocation of ^{32}P between interacting mycelia of a wood decomposing fungus and ectomycorrhizal fungi in microcosm systems. *New Phytologist*, **144**, 183–93.

Lindahl, B., Stenlid, J. & Finlay, R. D. (2001). Effects of resource availability on mycelial interactions and ^{32}P transfer between a saprotrophic and an ectomycorrhizal fungus in soil microcosms. *FEMS Microbiology Ecology*, **38**, 43–52.

Lindahl, B., Taylor, A. F. S. & Finlay, R. D. (2002). Defining nutritional constraints on carbon cycling – towards a less 'phytocentric' perspective. *Plant and Soil*, **242**, 123–35.

Lindahl, B. D., Finlay, R. D. & Cairney, J. W. G. (2005). Enzymatic activities of mycelia in mycorrhizal fungal communities. In *The Fungal Community: its Organization and Role in the Ecosystem*. ed. J. Dighton, P. Oudemans & J. White. New York: Marcel Dekker, pp. 331–48.

Little, B. J. & Wagner, P. A. (1997). Spatial relationships between bacteria and mineral surfaces. *Reviews in Mineralogy*, **35**, 123–60.

Lussenhop, J. & Fogel, R. (1999). Seasonal change in phosphorus content of *Pinus strobus-Cenococcum geophilum* mycorrhizae. *Mycology*, **91**, 742–6.

Ma, J. F., Zheng, S. J., Matsumoto, H. & Hiradate, S. (1997). Detoxifying aluminium with buckwheat. *Nature*, **390**, 569–70.

Mahmood, S., Finlay, R. D., Erland, S. & Wallander, H. (2001). Solubilization and colonisation of wood ash by ectomycorrhizal fungi isolated from a wood ash fertilised spruce forest. *FEMS Microbiology Ecology*, **35**, 151–61.

Marschner, H. (1998). Soil-root interface: biological and biochemical processes. In *Soil Chemistry and Ecosystem Health*, Special Publication no. 52. Madison: Soil Science Society of America, pp. 191–231.

Näsholm, T., Ekblad, A., Nordin, A. *et al.* (1998). Boreal forest plants take up organic nitrogen. *Nature*, **392**, 914–16.

Nurmiaho-Lassila, E. L., Timonen, S., Haahtela, K. & Sen, R. (1997). Bacterial colonisation patterns of intact *Pinus sylvestris* mycorrhizospheres in dry pine forest soil: an electron microscopy study. *Canadian Journal of Microbiology*, **43**, 1017–35.

Olsson, P. A., Chalot, M., Bååth, E., Finlay, R. D. & Söderström, B. (1996). Reduced bacterial activity in a sandy soil with ectomycorrhizal mycelia growing with *Pinus contorta* seedlings. *FEMS Microbiology Ecology*, **21**, 77–86.

Palfreyman, J. W., Phillips, E. M. & Staines, H. J. (1996). The effect of calcium ion concentration on the growth and decay capacity of *Serpula lacrymans* (Schumacher ex Fr.) Gray and *Coniophora puteana* (Schumacher ex Fr.) Karst. *Holzdorschung*, **50**, 3–8.

Paris, F., Bonnaud, P., Ranger, J. & Lapeyrie, F. (1995). In vitro weathering of phlogopite by ectomycorrhizal fungi. I. Effect of K^+ and Mg^{2+} deficiency on phyllosilicate evolution. *Plant and Soil*, **177**, 191–205.

Paris, F., Botton, B. & Lapeyrie, F. (1996). *In vitro* weathering of phlogopite by ectomycorrhizal fungi II. Effect of K^+ and Mg^{2+} deficiency and N source on accumulation of oxalate and H^+. *Plant and Soil*, **179**, 141–50.

Perez-Moreno, J. & Read, D. J. (2000). Mobilisation and transfer of nutrients from litter to tree seedlings via vegetative mycelium of ectomycorrhizal plants. *New Phytologist*, **145**, 301–9.

Perez-Moreno, J. & Read, D. J. (2001a). Nutrient transfer from soil nematodes to plants: a direct pathway provided by the mycorrhizal mycelial network. *Plant Cell and Environment*, **24**, 1219–26.

Perez-Moreno, J. & Read, D. J. (2001b). Exploitation of pollen by mycorrhizal mycelial systems with special reference to nutrient cycling in boreal forests. *Proceedings of the Royal Society of London, Series B*, **268**, 1329–35.

Piotrowski, J. S., Denich, T., Klironomos, J. N., Graham, J. M. & Rillig, M. C. (2004). The effects of arbuscular mycorrhizas on soil aggregation depend on the interaction between plant and fungal species. *New Phytologist*, **164**, 365–73.

Querejeta, J. I., Egerton-Warburton, L. M. & Allen, M. F. (2003). Direct nocturnal water transfer from oaks to their mycorrhizal symbionts during severe soil drying. *Oecologia*, **134**, 55–64.

Rambelli, A. (1973). The rhizosphere of mycorrhizae. In *Ectomycorrhizae*, ed. G. L. Marks & T. T. Koslowski, New York: Academic Press, pp. 299–343.

Read, D. J. (1991). Mycorrhizas in ecosystems. *Experientia*, **47**, 376–91.

Read, D. J. & Perez-Moreno, J. (2003). Mycorrhizas and nutrient cycling in ecosystems – a journey towards relevance? *New Phytologist*, **157**, 475–92.

Richards, B. N. (1987). *The Microbiology of Terrestrial Ecosystems*. New York: Longman, Harlow.

Robinson, D. & Fitter, A. H. (1999). The magnitude and control of carbon transfer between plants linked by a common mycorrhizal network. *Journal of Experimental Botany*, **50**, 9–13.

Rosling, A., Landeweert, R., Lindahl, B. D. *et al.* (2003). Vertical distribution of ectomycorrhizal fungal taxa in a podzol profile determined by morphotyping and genetic verification. *New Phytologist*, **159**, 775–83.

Rosling, A., Lindahl, B. D. & Finlay, R. D. (2004a). Carbon allocation to ectomycorrhizal roots and mycelium colonizing different mineral substrates. *New Phytologist*, **162**, 795–802.

Rosling, A., Lindahl, B. D., Taylor, A. F. S. & Finlay, R. D. (2004b). Mycelial growth and substrate acidification of ectomycorrhizal fungi in response to different minerals. *FEMS Microbiology Ecology*, **47**, 31–7.

Ruijter, G. J. G., van de Vondervoort, P. J. I. & Visser, J. (1999). Oxalic acid production by *Aspergillus niger*: an oxalate-non-producing mutant produces citric acid at pH 5 and the presence of manganese. *Microbiology*, **145**, 2563–76.

Ryan, P. R., Delhaize, E. & Jones, D. L. (2001). Function and mechanism of organic anion exudation from plant roots. *Annual Review of Plant Physiology and Plant Molecular Biology*, **52**, 527–60.

Sandnes, A., Eldhuset, T. D. & Wollebæk, G. (2005). Organic acids in root exudates and soil solution of Norway spruce and silver birch. *Soil Biology and Biochemistry*, **37**, 259–69.

Schimel, J. P. & Bennett, J. (2004). Nitrogen mineralisation: challenges of a changing paradigm. *Ecology*, **85**, 591–602.

Sen, R. (2000). Budgeting for the wood-wide web. *New Phytologist*, **145**, 161–5.

Shenker, M., Ghirlando, R., Oliver, I. *et al.* (1995). Chemical structure and biological activity of a siderophore produced by *Rhizopus arrhizus*. *Soil Science Society of America Journal*, **59**, 837–43.

Smith, S. E. & Read, D. J. (1997). *Mycorrhizal Symbiosis*. San Diego: Academic Press.

Smits, M. M., Hoffland, E., Jongmans, A. G. & van Breemen, N. (2005). Contribution of feldspar tunneling by fungi in weathering. *Geoderma*, **125**, 59–69.

Sterflinger, K. (2000). Fungi as geologic agents. *Geomicrobiology Journal*, **17**, 97–124.

Sun, Y.-P., Unestam, T., Lucas, S. D. *et al.* (1999). Exudation-reabsorption in mycorrhizal fungi, the dynamic interface for interaction with soil and other microorganisms. *Mycorrhiza*, **9**, 137–44.

Sverdrup, H. & Warfvinge, P. (1995). Estimated field weathering rates using laboratory kinetics. *Reviews in Mineralogy*, **31**, 485–541.

Taylor, A. F. S. (2002). Fungal diversity in ectomycorrhizal communities: sampling effort and species detection. *Plant and Soil*, **244**, 19–28.

Timonen, S. & Sen, R. (1998). Heterogeneity of fungal and plant enzyme expression in intact Scots pine-*Suillus bovinus* and -*Paxillus involutus* mycorrhizospheres developed in natural forest humus. *New Phytologist*, **138**, 355–66.

Timonen, S., Jørgensen, K., Haahtela, K. & Sen, R. (1998). Bacterial community structure at defined locations of the *Pinus sylvestris-Suillus bovinus* and -*Paxillus involutus* mycorrhizospheres in dry forest humus and nursery peat. *Canadian Journal of Microbiology*, **44**, 499–513.

Tisdall, J. M. & Oades, J. M. (1979). Stabilisation of soil aggregates by the root systems of ryegrass. *Australian Journal of Soil Research*, **17**, 429–41.

Treseder, K. K. & Allen, M. F. (2000). Mycorrhizal fungi have a potential role in soil carbon storage under elevated CO_2 and nitrogen deposition. *New Phytologist*, **147**, 189–200.

Ullman, W. J., Kirchman, D. L., Welch, S. A. & Vandervivere, P. (1996). Laboratory evidence for microbially mediated silicate mineral dissolution in nature. *Chemical Geology*, **132**, 11–17.

Unestam, T. & Sun, Y.-P. (1995). Extramatrical structures of hydrophobic and hydrophilic ectomycorrhizal fungi. *Mycorrhiza*, **5**, 301–11.

van Breemen, N., Finlay, R. D., Lundström, U. *et al.* (2000a). Mycorrhizal weathering: a true case of mineral nutrition? *Biogeochemistry*, **49**, 53–67.

van Breemen, N., Lundström, U. S. & Jongmans, A. G. (2000b). Do plants drive podzolization via rock-eating mycorrhizal fungi? *Geoderma*, **94**, 163–71.

van Hees, P. A. W., Lundström, U. & Giesler, R. (2000). Low molecular weight organic acids and their Al-complexes in soil solution – composition, distribution and seasonal variation in three podzolized soils. *Geoderma*, **94**, 173–200.

van Hees, P. A. W., Jones, D. L., Jentschke, G. & Godbold, D. L. (2003). Mobilization of aluminium, iron and silicon by *Picea abies* and ectomycorrhizas in a forest soil. *European Journal of Soil Science*, **55**, 101–11.

van Hees, P. A. W., Godbold, D. L., Jentschke, G., & Jones, D. L. (2005a). Impact of ectomycorrhizas on the concentration and biodegradation of simple organic acids in a forest soil. *European Journal of Soil Science*, **54**, 697–706.

van Hees, P. A. W., Jones, D. L., Finlay, R. D., Godbold, D. L. & Lundström, U. S. (2005b). The carbon we do not see: do low molecular weight compounds have a significant impact on carbon dynamics and respiration in forest soils? *Soil Biology and Biochemistry*, **37**, 1–13.

Villarréal-Ruiz, L., Anderson, I. C. & Alexander, I. J. (2004). Interaction between an isolate from the *Hymenoscyphus ericae* aggregate and roots of *Pinus* and *Vaccinium*. *New Phytologist*, **164**, 183–92.

Watmough, S. A. & Dillon, P. J. (2003). Mycorrhizal weathering in base-poor forests. *Nature*, **423**, 823–4.

3

Fungal roles in transport processes in soils

KARL RITZ

Introduction

Fundamentally, biogeochemical cycling involves the transformation of compounds between various forms, and a movement of such compounds within and between compartments of the biosphere and geosphere.
These processes operate across a wide range of spatial and temporal scales,
from micrometres to kilometres, from seconds to centuries. In terrestrial
systems, transformations and movement of materials below-ground are
governed by the spatial organization of the soil system, and particularly
the architecture of the pore network. This 'inner space' provides the
physical framework in and through which the majority of soil-based
processes occur. The labyrinthine nature of the pore network, and the
exchange properties of associated surfaces, strongly modulates the transport of materials through the soil matrix. From a physicochemical perspective soil structure generally retards transport processes for two main
reasons: the complex geometry of the pore network increases path lengths,
both for diffusive and bulk-flow movement; and charged mineral and
organic constituents in the soil act as exchange surfaces which bind transportable compounds to varying degrees. Transport processes may also be
accelerated by structural properties, for example if solutes or particulates
are carried via preferential and bypass-flow channels of water through
macropores.

Soil organisms play a key role in driving terrestrial nutrient cycling, and
play both direct and indirect roles in effecting and affecting transport
processes. Fungi contribute a particularly wide range of functions relating
to nutrient cycling. Eucarpic fungi are well adapted for living in spatially
structured environments such as soils, since the filamentous growth form
allows an effective exploration of space for resources, and an exploitation

Fungi in Biogeochemical Cycles, ed. G. M. Gadd. Published by Cambridge
University Press. © British Mycological Society 2006.

of such substrate when located. In this chapter I review how the spatial organization of physical and nutritional factors affect mycelial morphology of soil fungi, the transport of materials via such networks and the implications for nutrient cycling in soil systems.

Indirect effects on nutrient transport

Fungi indirectly affect nutrient transport in soils via the myriad of anabolic and catabolic reactions they carry out. Decomposition of substrate by fungi is largely extracellular, and the simpler molecular forms that arise from the breakdown of complex polymers tend to be inherently more mobile and prone to dissolution and transport by diffusion or mass flow as solutes. By contrast, elements incorporated into structural components of fungal biomass will generally be less prone to further transport until faunal grazing or their decomposition occurs. Some metal elements bind strongly to fungal cell walls, and others are immobilized inside hyphae. Extracellular precipitation of the wide variety of metals rendered insoluble via reaction with protons or organic acids released from fungi strongly retards transport of such elements (see Fomina *et al.*, Chapter 10, this volume). Soil structure influences how fungi grow and are organized in soils, but fungi in turn also impact upon soil structural dynamics (Ritz & Young, 2004). Thus, fungi also indirectly affect transport processes in soils by modulating the soil architecture, which in turn affects the transport mechanisms discussed above.

Direct effects on nutrient transport

Eucarpic fungi play direct roles in the transport of elements through soils by virtue of the mycelium. This structure, created by the fundamentally simple process of apical extension and periodic branching of hyphae has two key properties of relevance to transport. Firstly, it is very efficient at filling three-dimensional space and ramifying through porous media; a small amount of resource goes a long way when formed into cylinders a few micrometres in diameter. Indeed, the fractal dimension of the mass distribution of hyphae in many mycelia (Ritz & Crawford, 1990; Bolton & Boddy, 1993; Mihail *et al.*, 1995) reveals that space-filling efficiency is optimal given the quantity of material used and area (volume) occupied. This growth form thus enables the fungal organism to locate spatially distributed substrate via explorative or foraging growth, and to colonize such food resources when encountered by a more prolific branching and thus exploitative growth phase. Secondly, since hyphae are hollow cylinders, the mycelium is an interconnected pipe network through which materials can be transported. When hyphae encounter each other, if they are derived from the same parent mycelium or from somatically compatible individuals, they may fuse (anastomose) and increase the degree of interconnectivity of the network. Thus, materials required for growth can be delivered to the apices of extending hyphae, but anastomosis also provides the possibility for larger-scale redistribution of materials within mycelia in many directions. This enables the fungus to spatially reallocate resources according to local demands, for example if growing through nutrient-impoverished zones where local substrate is unavailable, or where particular nutrients are limiting in one region but available in other distal zones. Substances transported within hyphae are essentially separated from the bulk soil phase and are thus unconstrained by factors such as pore tortuosity, exchange surfaces or assimilation by other organisms. This has profound consequences for transport processes in soil systems.

Spatial distribution of fungi in soils

Fungally mediated translocation paths through soils will be defined by the spatial location and organization of mycelia. The spatial organization of fungi in soil is governed by many factors. Fundamentally, there are intrinsic properties of the fungus, and these largely have a genetic basis. These include the inherent extension rates of hyphae, the propensity for branching, the range in branch angles and the ability to form higher-order structures such as cords or rhizomorphs by hyphal aggregation. These properties define the basic form of the mycelium and vary widely within and between species, as is clearly apparent by the great variety of colony forms that are manifest when different fungi are grown on culture media. Whatever the intrinsic ground rules, the extrinsic environment plays a governing role upon mycelial distribution. Firstly, the basic architecture of the soil defines the physical framework through which hyphae must grow. In organic soils and litter horizons, this is unlikely to be particularly constraining, but in mineral soils it is of significance. In such structured soils, the labyrinthine pore network is a complex three-dimensional maze with varying degrees of connectivity and tortuosity, and ultimately defines the available paths. In principle, hyphae will be unable to penetrate pores narrower than their diameters, although they may be able to physically separate soil particles to create paths available for growth. Very little is known about whether this actually occurs in soils. Individual hyphae can exert pressures of $50 \, \mathrm{mN \, m^{-2}}$ (Money, 2004), and where hyphal aggregation occurs pressures up to $1.3 \, \mathrm{kN \, m^{-2}}$ can be

generated and may be spectacularly manifest as the emergence of basidiocarps through pavements (Niksic *et al.*, 2004). Soil bulk density has an
effect upon the extent to which soils are colonized by mycelia. It might be
intuitively expected that the higher the bulk density, and hence concomitantly smaller relative volume of pore space, the more fungal growth would
be constrained. Harris *et al.* (2003) mapped the hyphal distribution of
Rhizoctonia solani in a mineral soil packed at different bulk densities and
found the converse was the case, with fungal extent significantly greater at
higher bulk densities. There is also a significant interaction between the
soil water content, the pore network and fungal growth. Filamentous fungi
are predominantly aerobes and water-filled pores are generally oxygen-
starved, which curtails hyphal extension. Otten *et al.* (1999) demonstrated
how sensitive growth of *R. solani* is to air-filled porosity and determined
that there are threshold values below which growth is rapidly curtailed.
The distribution of nutrient resources within soils has a large effect upon
the spatial organization of fungal mycelia. Generally, hyphae proliferate
and mycelia are denser in the presence of nutrients, and sparse in nutrient-
impoverished zones (Rayner, 1994; Ritz *et al.*, 1996). This relates to the
exploitative and explorative growth phases outlined above, but foraging
paths will be constrained by the physical restrictions soil structure imparts.
Mycelial connections between such patchy resources set up source:sink
relationships that affect transport within mycelia and are discussed in
more detail below (see p. 60). Fungal distribution in soils is also affected
by the presence of other organisms in the soil, particularly other fungi. The
spatial organization of fungi in soils, and therefore potential pathways of
transport, is thus not random but governed by a range of intrinsic and
extrinsic factors.

Transport mechanisms

Transport of materials by fungi has been recognized since the
earliest days of mycology and has been reviewed in many contexts
(Schütte, 1956; Wilcoxson & Sudia, 1968; Jennings *et al.*, 1974; Howard,
1978; Jennings, 1987; Cairney, 1992; Lindahl & Olsson, 2004). It is still a
poorly understood process, and tends to be dealt with superficially in most
fungal textbooks and mycological fora. As this review shows, it is a
complex process but of great significance both for the fungal organism
and ecosystem functioning.

Transport can potentially occur via fungal hyphae by both passive and
active mechanisms, and there is experimental evidence that both modes
are manifest under different circumstances. Simple diffusion may occur

along the outer hyphal wall, where the filaments essentially act as wicks (Read & Stribley, 1975). Such movement is outwith the direct control of the organism and will be modulated by the extent to which hydrophobic compounds are deposited within the hyphal wall. Simple diffusion may also occur within hyphae, as an inevitable consequence of concentration gradients prevailing within the mycelium. However, where such transport occurs in the context of living protoplasm it will be prone to organismal control, and active mechanisms may operate to accelerate or retard translocation within mycelia. Empty, highly vacuolated or moribund hyphae may act as diffusion wicks, but poorly insulated forms are unlikely to prevail in soils since they will be prone to rapid decomposition in the absence of recalcitrant polymeric (hydrophobic) coatings, and are therefore unlikely to contribute significantly to transport within soils under most circumstances.

That active transport occurs within mycelia is well established, and invoked when measured rates of translocation are too rapid to be accounted for by diffusion alone. In many cases, diffusion-driven translocation alone can be adequate to account for observed translocation rates, but several models of fungal growth confirm the necessity for active transport in order to account for observed translocation rates, hyphal extension rates or mycelial growth rates (Davidson & Olsson, 2000; Boswell *et al.*, 2002, 2003).

Active transport is postulated to occur via a number of mechanisms and is poorly understood, most likely because no single mechanism is dominant, and different modes operate under different circumstances, notwithstanding variation between species. Cytoplasmic streaming within hyphae is a phenomenon that has been noted since the earliest days of microscopic observation of fungi, and intuitively suggests the primary mode of transport. However, this term is vague since it confounds the movement of liquid cytosol and associated water, organelles and other particulate inclusions within hyphae which may be occurring simultaneously via a number of different mechanisms. Simultaneous bidirectional movement of cytoplasm and inclusions, at different velocities, is often manifest, which further suggests a variety of controlling mechanisms. These include pressure-driven mass flow (Jennings, 1987; Amir *et al.*, 1995), peristalsis via tubular vacuolar systems (Shepherd *et al.*, 1993; Cole *et al.*, 1998) and vesicle translocation on cytoskeletal elements (Steinberg, 1998; Bago *et al.*, 2002). The degree of metabolic control that the fungal organism can exert over these mechanisms is also likely to vary. Mass flow may be driven by osmotic gradients, metabolic sinks for water (for example at the growing front), or transpiration from uninsulated regions of mycelia. Basidiocarps set up strong transpiration streams for water, and here flow rates will be modulated by the geometry and structure of the fruit body and extrinsic factors such as the relative humidity of the atmosphere (Schütte, 1956). Thrower and Thrower (1968) reported translocation-driven streaming occurring at smaller scales via water loss from sporangiophores of *Phycomyces blakesleeanus*. Mechanisms involving the transport of membrane-bound vesicles along the cytoskeleton will afford a high degree of control, albeit at some energetic cost. Clearly, if substances are transported within mycelia against concentration gradients, or counter to mass-flow phenomena, this can only be achieved by active processes that involve energy-consuming metabolic processes. Prevailing translocation modes may relate to the overall energetic status of the mycelium in the context of existing source:sink relationships and the nutritional status of the environment. Hence active transport may occur during the expansive phase of mycelial growth when substrate is available, but within established mycelia, passive transport is adequate to serve the basal metabolic needs of the organism (Gray *et al.*, 1995; Jacobs *et al.*, 2004). Experimental evidence supported by modelling for this was found by Boswell *et al.* (2003) who proposed that active translocation occurred in response to hyphal tip densities within the mycelium. An important point is that the mycelia of many fungi are coherent structures and that integrated behaviour that balances spatial heterogeneity in sources, sinks and energetic status is feasible. Mycelial-level transduction mechanisms may include translocation-based routes. Using real-time imaging techniques, Tlalka *et al.* (2002) have demonstrated pulsatile fluxes of translocation of ^{14}C-labelled amino-acid analogues in mycelia of *Phanerochaete velutina*. High-speed electrochemical mechanisms akin to action potentials in nerve fibres have also been reported (Olsson & Hansson, 1995). This concept of high-level mycelial integration is increasingly acknowledged (Rayner, 1991, 1994; Cairney & Burke, 1996; Boddy, 1999) but is complex and remains poorly understood. It is certainly of great consequence for governing translocation processes within eucarpic fungi, and a more detailed understanding is arguably one of the most significant challenges in contemporary mycology.

Transport distances and rates

Some authors who have screened various fungi for transport phenomena report that fungal species tend to be 'translocating' or 'non-translocating types' (Schütte, 1956; Olsson, 1995). However, all eucarpic fungi must be capable of transporting nutrients along their hyphae to some extent if they are able to traverse any non-nutritive substratum or air gaps, such as is common in soil, since there is no immediately local source to fuel tip growth. It is rather an issue of the distance over which materials can apparently be transported, and here a considerable range is manifest from micrometres to decametres. Undifferentiated mycelia tend to transport shorter distances, due to their inherently smaller size. Arbuscular mycorrhizas are commonly demonstrated to transport elements over tenths of metres. Rhizomorphs and cords are differentiated structures specifically adapted to long-distance transport of water and substrate. There is also great variation in measured transport rates, from micrometres to metres per hour, between and within species, and depending upon context. As would be anticipated, there appears to be a relationship between rates and attainable distances, in that the greatest rates are found to occur in cord-forming species. Olsson (1995) demonstrated that bidirectional transport over a few centimetres of either carbon or inorganic nutrients, or both, could occur within some fungal species to satisfy local demand, but that some species could not apparently translocate either the carbon, the inorganic nutrients, or both such forms.

Transportable elements
   There is a large and varied literature reporting elemental transport within fungi, which collectively suggests that some elements are inherently prone to translocation within mycelia, and some are not (Table 3.1). Unsurprisingly, the major nutrients carbon, nitrogen, phosphorus and sulphur always appear to be highly mobile. Early demonstrations of large-scale translocation were made by observing movement of dissolved organic stains through mycelia and basidiocarps (Schütte, 1956). Demonstration of transport involves a robust physical separation of a traceable elemental source (e.g. radioactive or stable isotopic form) and its subsequent appearance in distal regions of the mycelium. This is only really unequivocal if net transport is confirmed by double-labelling to demonstrate that simple isotopic exchange is not occurring. Such checks are rarely included in studies, but are particularly important where there is likely to be a significant presence of other isotopic forms of the element under consideration within the fungus, such as major nutrients. A clear example of such bidirectional exchange was provided by Lindahl *et al.* (2001) who grew *Hypholoma fasciculare* from a point inoculum toward a discrete bait in a linear system. The inoculum was labelled with ^{33}P, and the bait with ^{32}P. The location of the isotopes was visualized

Table 3.1. *Some examples of translocatable and apparently non-translocatable elements within fungal mycelia, based on detection of radioactive- or stable-isotope forms of the elements within fungal structures physically separated from the sources of such elements*

Translocation Detected	Reference
Ag	Anderson *et al.* (1997)
C	Wells *et al.* (1995); Olsson & Gray (1998); Tlalka *et al.* (2002)
Ca	Harley & Smith (1983)
Cs	Gray *et al.* (1995); Anderson *et al.* (1997); Declerck *et al.* (2003)
Fe	Caris *et al.* (1998)
K	Brownlee & Jennings (1982)
N	George *et al.* (1992); Johansen *et al.* (1993)
Na	Suzuki *et al.* (2001)
P	Read & Stribley (1975); Wells & Boddy (1995a); Lindahl *et al.* (1999); Jansa *et al.* (2003)
Pb	Kirchner & Daillant (1998)
Ra	Kirchner & Daillant (1998)
Rb	Suzuki *et al.* (2001)
S	Harley & Smith (1983)
Se	Suzuki *et al.* (2001)
Sr	Suzuki *et al.* (2001)
U	Rufyikiri *et al.* (2003)
Y	Suzuki *et al.* (2001)
Zn	Suzuki *et al.* (2001); Jansa *et al.* (2003)

Translocation Not Detected	Reference
Be	Suzuki *et al.* (2001)
Co	Suzuki *et al.* (2001)
Cr	Suzuki *et al.* (2001)
Fe	Caris *et al.* (1998)
Mn	Suzuki *et al.* (2001)
Sc	Suzuki *et al.* (2001)
Zr	Suzuki *et al.* (2001)

non-destructively using an electronic autoradiographic scanning system. Following contact with the bait, ^{33}P derived from the inoculum was visualized therein, which would be expected given the growth direction of the mycelium. However, ^{32}P derived from the bait was also rapidly detected in the inoculum zone as well as the extending mycelial front growing in the opposite direction (Fig. 3.1). This phenomenon is of less significance where biologically rare elements such as heavy-metal radio-nuclides are apparently transported, since they are unlikely to be pre-existent in the detection region. The lack of reported translocation of an element does not mean that such transport does not occur, rather that it has not been observed to date. Some elements, such as iron, have been reported as translocatable by some authors, but not recorded as such by others (Table 3.1). This demonstrates that such transport may also be context-dependent, both in relation to fungal species and environmental circumstances, for example where local soil conditions may strongly affect bioavailability. There is an apparent paradox as to why elements of no nutritive value, or indeed of potential toxicity, such as silver, or uranium may occur, and why they are apparently mobile within hyphae (e.g. Anderson *et al.*, 1997; Rufyikiri *et al.*, 2003).

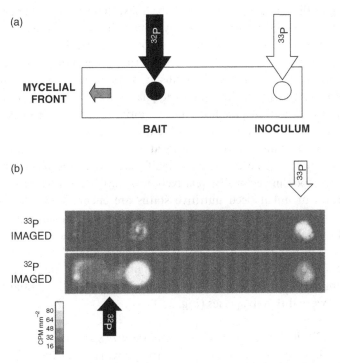

Fig. 3.1. Simultaneous bidirectional transport of phosphorus within mycelium of *Hypholoma fasciculare* demonstrated using double-labelling and digital autoradiography. (a) Experimental design. (b) Imaging of isotope distribution 29 days after addition of labels. Microcosms are 20 cm long. Derived from Lindahl *et al.* (2001).

Transport directions

The direction of transport of materials and elements within fungi is basically governed by source:sink relationships, and for nutritive elements generally follows biologically intuitive patterns. This implies that the fungal organism is by and large in control of translocation within its thallus, and that translocation is an intrinsic part of fungal growth and development. As discussed above (p. 54), such control is achieved through active transport processes augmented by passive mechanisms that involve less use of energy. The ranges of transport routes are detailed in the following sections.

To the mycelial front

Most mycelia show polarity in their growth. This is a natural consequence of the apical extension of hyphae, and there is thus a very strong sink for substrate at the hyphal tip. If particular nutrients are lacking in the immediate vicinity of the apex, these can be augmented by import from distal regions of the mycelium; this has been demonstrated many times at all spatial scales and across the eucarpic fungal kingdom from undifferentiated moulds to cord-forming basidiomycetes (Olsson, 1995; Boddy, 1999). Obligate mutualistic forms such as arbuscular mycorrhizas must translocate carbon to hyphal apices since the sole carbon source is their autotrophic host. The majority of arbuscular mycorrhizal fungi are coenocytic, and it is possible that the lack of septa may lead to more efficient translocation. Translocation of a range of nutrients to the growing front will be of greatest significance when the mycelium is in foraging mode in regions of relatively low nutritive status. However, if substrates of unbalanced nutritive status are encountered, the missing components may be imported from elsewhere to augment exploitative growth. Wells and Boddy (1995b) demonstrated this where ^{32}P was shown to be rapidly translocated to wood blocks of high C/P ratio. They further demonstrated how the relative magnitude of the source (inoculum) and sink (bait block) affected such translocation and differential behaviour between different species (Fig. 3.2).

To the future – reproductive and survival structures

For many organisms, reproductive growth is a substrate-demanding process and fungal reproductive structures such as perithecia, conidia and basidiocarps act as strong sinks for nutrients. Survival structures such as sclerotia also act as nutrient sinks. Ritz (1995) showed this using a tessellated agar tile system where sclerotia of *Rhizoctonia solani* formed exclusively in tiles of tap-water agar by importing substrate from adjacent nutrient agar; studies using radiotracers have also demonstrated sclerotial sinks for phosphorus (Littlefield, 1967). Basidiocarps are apparently particularly strong sinks, and not just for the nutritive elements their development requires, which can be substantial for larger forms. Accumulation and hence concentration of radionuclides such as ^{134}Cs, ^{137}Cs, ^{210}Pb and ^{226}Ra in fruit bodies emanating from contaminated soils is well established (Haselwandter *et al.*, 1988; Bakken & Olsen, 1990; Kirchner & Daillant, 1998) and represents a significant radiological and toxicological consequence of fungal transport in the soil. Translocation to basidiocarps may be particularly efficient due to mass flow within translocation streams.

(a)

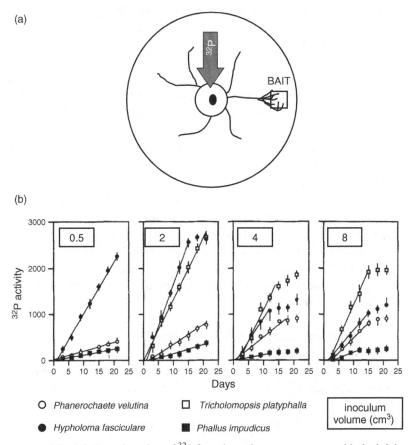

(b)

Fig. 3.2. Translocation of ^{32}P from inoculum source to wood bait sink by basidiomycetes. (a) Experimental design. (b) Effect of inoculum volume upon translocation rates. Derived from Wells and Boddy (1995b).

Such mass flow may be the primary mechanism by which non-nutritive elements such as these radionuclides are apparently so efficiently concentrated in fruit bodies. However, Gray *et al.* (1995) demonstrated that even the primordia of basidiocarps forming within mycelia in vitro are strong sinks for caesium.

To the host – mycorrhizal transport

While there is clearly an efficient translocation of carbon from the autotrophic host to the fungal partner in mycorrhizal systems as discussed above, there is a concomitant transport of elements taken up from the soil by the fungus and translocated through the extra radical mycelium to the host plant (Smith & Read, 1997). This has been extensively studied and demonstrated for a wide range of both nutritive and non-nutritive elements (Table 3.1) for both arbuscular and ectomycorrhizal types. Such transport is typically demonstrated by growing plants in compartmentalized sand or soil media and constraining roots behind meshes of a gauge that prevents roots from passing but are open to mycorrhizal hyphae. Soil in the mycorrhizal compartment is then dosed with isotopic forms of the element under investigation and its presence assayed in host plant tissue. More robust experimental designs include an airgap between the root-excluding mesh and the hyphal compartment to prevent any diffusive movement of the isotopes through the soil. There can be complex interactions between host, fungus, element type and prevailing nutrient status of the environment. For example, Caris *et al.* (1998) showed that arbuscular mycorrhizas significantly improved uptake of ^{32}P from such hyphal compartments by both peanut (*Arachis hypogea*) and sorghum (*Sorghum bicolor*) plants. However, in the same experimental system mycorrhizas had no significant effect upon ^{59}Fe uptake by the peanut plants, but increased the concentration of iron in the sorghum, contingent upon the iron status of the soils (Fig. 3.3).

Between the hosts – inter-plant transfer

Mycorrhizal fungi show varying degrees of specificity in relation to the host plants they can infect (Smith & Read, 1997). Where there is low host specificity, it is possible that individual plants may be colonized by the same mycorrhizal network and thus become connected. This has been confirmed by direct observation both for ectomycorrhizal (Finlay & Read, 1986a) and arbuscular mycorrhizal (Heap & Newman, 1980) types. Such infection establishes potential pathways for direct transport of material between roots via mycelial bridges, even of different plant

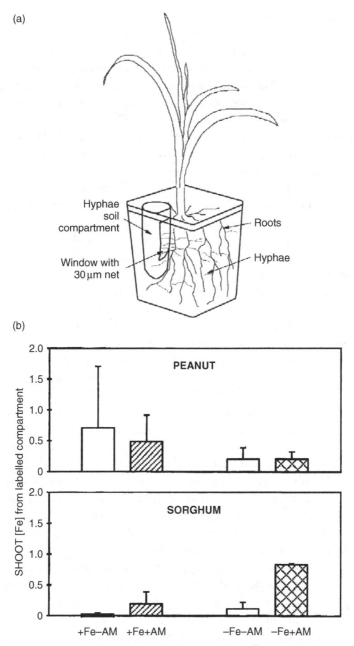

Fig. 3.3. Uptake and translocation patterns of ^{59}Fe in the presence or absence of arbuscular mycorrhizas (AM), and with or without iron fertilization of soil, to peanut or sorghum plants. (a) Experimental design – ^{59}Fe or ^{32}P (data not shown) added to hyphal compartment. (b) Amount of isotope detected in shoot of plant. Derived from Caris *et al.* (1998).

species. It also allows for the concept of higher-order levels of organization within plant communities where nutrients are transported between individuals, short-circuiting retardant bulk soil phases and attendant competition from soil organisms. This has proved a particularly attractive hypothesis to ecologists and there have been many experimental demonstrations of the existence of apparent inter-plant transfer using labelled forms of carbon (Hirrel & Gerdemann, 1979; Finlay & Read, 1986a; Watkins et al., 1996), nitrogen (Bethlenfalvay et al., 1991; Arnebrant et al., 1993) and phosphorus (Finlay & Read, 1986b; Newman & Eason, 1993). The basic approach is to label a donor plant independently of the soil and to measure the concentration of the element in neighbouring receiver plants, in the presence and absence of mycorrhizas. The general observation is that of enhanced transfer of elements between plants in the presence of mycorrhizas. However, very few such studies have included a crucial control treatment to check for and take account of isotope exchange as opposed to net transfer, despite the early warning of its necessity by Ritz and Newman (1984), reinforcement by Newman (1988) and reiteration by Smith and Read (1997). A rare exception is a field study by Simard et al. (1997) who used a double-labelling approach with ^{13}C and ^{14}C and detected net transfer of carbon between birch (Betula papyrifera) and Douglas fir (Pseudotsuga menziesii) colonized by shared ectomycorrhizas. Graves et al. (1997) used an innovative method based on exposing mature swards of sheeps fescue (Festuca ovina) to CO_2 depleted in ^{13}C, and measuring the isotopic ratio of carbon in neighbouring swards. They determined that in the absence of mycorrhiza, no carbon was present in the receiver sward, but some 41% of the ^{13}C-depleted carbon exported from leaves to roots of donor swards was transported to the receiver sward. However, they did not include a bidirectional control and admit that this does not demonstrate unequivocal one-way transfer. However, an important observation was that there was no evidence that such transported carbon had left fungal structures within the receiver roots, i.e. it was simply present in the roots as fungal tissue. This was evidence for a high degree of mycorrhizal interconnectivity between the swards, and despite this there was no detectable import of such carbon into the receiver symplast. A lack of carbon transfer from fungus to host has recently been confirmed in further dual-labelling studies (Pfeffer et al., 2004). While such direct inter-plant transfer pathways undoubtedly exist, no studies to date have demonstrated an unequivocal effect on the growth or nutrient status of putative receiver plants, even where source:sink relationships are apparently heavily in favour of a 'recipient'

(Ritz & Newman, 1986). However, if a plant dies in the presence of mycorrhizal networks the source:sink relationship is very different, and mycorrhizal systems that have colonized the dead plant and are well established in proximal plants significantly enhance transport of nutrients to such living neighbours. This was demonstrated by Johansen and Jensen (1996) who grew pea (*Pisum sativum*) and barley (*Hordeum vulgare*) in split-chamber systems with and without arbuscular mycorrhiza (Fig. 3.4a).

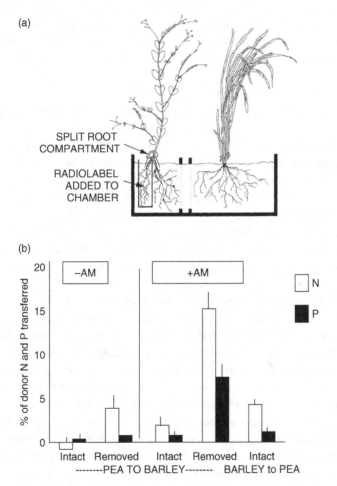

Fig. 3.4. Transfer of nitrogen and phosphorus between pea and barley plants in the presence or absence of arbuscular mycorrhizas (AM), and where peas shoots were removed. (a) Experimental design (b) Amount of transfer under the different circumstances. Derived from Johansen and Jensen (1996).

Very little inter-plant transfer of ^{32}P or ^{15}N occurred from pea to barley in non-mycorrhizal systems with either intact plants or where the pea plants had their shoots removed. In mycorrhizal systems, significantly greater transfer occurred from decapitated pea plants (Fig. 3.4b). Whilst an elegant demonstration of such source:sink relationships, this experimental design also suffered from the lack of reciprocal controls. Furthermore, an alternative mechanism for all such inter-plant transfer may simply be a more efficient uptake of elements liberated from donor plant roots by exudation or sloughing, by mycorrhizas associated with the receivers that are simply in the vicinity of the donor rhizosphere. None of the large number of inter-plant experiments reported actually discriminate between these alternative mechanisms.

To each other – inter-mycelial transfer

Since fungi are such spatially exploitative organisms, there are inevitably many instances where mycelia encounter each other in soils and litters. The outcomes of such interactions have been extensively studied (e.g. Rayner & Boddy, 1988; Stahl & Christensen, 1992), and a wide variety of outcomes from mycelial interactions are manifest, ranging from the mycelia intermingling and essentially ignoring each other, through 'deadlock' where mycelia deposit resistant barriers at the interaction zones (leading to a stasis), to total annihilation of one or other fungus. Such interactions inevitably also involve transfer of nutrients between mycelia. The nature and extent of such transfer appears to follow source:sink patterns that map to the combative relationship between the fungi. For example, transfer of phosphorus from *R. solani* to the *Arthrobotrys oligosporathe* occurs when the latter parasitizes the former (Olsson & Persson, 1994). Lindahl *et al.* (1999) established linear microcosms planted with pine (*Pinus sylvestris*) seedlings infected by the ectomycorrhizal fungus *Suillus variegatus* at one end, and the saprophyte *Hypholoma fasciculare* inoculated onto wood blocks at the opposite end. By reciprocal labelling of either end with ^{32}P, and autoradiographic imaging of the subsequent location of the isotope, they demonstrated that in this case there was no transfer of isotope from the mycorrhiza to the saprophyte, but clear transfer from the latter to the former. In many instances, the outcome of combative interactions between mycelia relate to the relative sizes of the mycelia, with larger mycelia often outcompeting smaller competitors. More internal resources will be available to larger mycelia at the interaction front, particularly if they can be efficiently translocated there.

Through the ecosystem – bulk transfer of elements

The great majority of experimental studies into fungal transloca-tion phenomena involve the study of individual mycelia of single species of fungi, usually in simple contexts of restricted numbers of other organisms (fungi, host plants) or controlled nutrient depots (wood blocks, plant material). Field studies also tend to focus on translocation in individual fungi. However, whilst such simple systems are necessary to elucidate basic processes, in most natural circumstances whole communities of fungi will also operate and the wide range of mechanisms discussed above will be operative. Very few studies have formally measured the extent to which gross fungal activity transports materials at an ecosystem level. Hart *et al.* (1993) estimated nitrogen flow from mineral soil to decomposing surface residues as 0.29 and $0.02 \, \mathrm{g \, N \, m^{-2} \, y^{-1}}$ for a grassland and forest respectively, and postulated this occurred predominantly via fungal trans-location. Frey *et al.* (2000) made such measurements in a no-till agricul-tural system by inhibiting fungi by application of fungicide, and measuring transport of inorganic ^{15}N from the soil to surface residues. Fungicide application reduced fungal biomass and reduced total nitrogen flux to surface residues by 52%–86% compared to controls. They estimated fungal translocation of $2.4 \, \mathrm{g \, N \, m^{-2}}$ over the growing season. In a further study, Frey *et al.* (2003) demonstrated that decomposer fungal commu-nities translocate litter-derived carbon into the underlying soil whilst simultaneously transporting soil-derived inorganic nitrogen into the litter layer. In an experiment involving ^{14}C-labelled straw applied to the surface of a soil pre-labelled with inorganic ^{15}N, they measured reciprocal transfer of significant quantities of carbon into the soil and nitrogen into the straw. Fungal inhibition reduced fungal biomass and the bidirectional carbon and nitrogen flux by some 50%. Such community-level transport of elements has also been demonstrated in forest systems. For example, Rafferty *et al.* (1997) detected an increase in radiocaesium concentration of litter bags in coniferous forests in Ireland and the Ukraine, which they attributed to the import of the ^{137}Cs by decomposer fungi colonizing the bags.

Conclusions

Fungi clearly play significant direct and indirect roles in effecting and modulating transport of a wide range of elements in soil systems. The principal transport modes that they are implicated in can be summarized as dispersion, concentration, inter-organism and bulk transfer (Fig. 3.5).

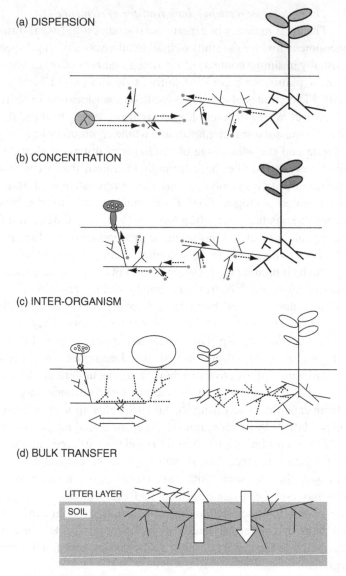

Fig. 3.5. The four primary modes of nutrient transport processes in soils governed by fungi. (a) Dispersion from point sources (left) or host plants (right – thicker line represents roots) to zones within mycelia and soil phases associated with hyphae. (b) Concentration from diffuse sources to fungal structures (left) or host plants (right – thicker line represents roots). (c) Inter-organism between fungi (left) or between plants (right – broken line represents fungal mycelium, solid lines represent roots). (d) Bulk transfer of elements from entire soil zones by various members of fungal community.

In any given ecosystem at any particular time, fungi will be effecting these modes to varying degrees resulting in very complex patterns of transport. It is arguable that the most significant role they play is in the short-circuiting of diffusion pathways that prevail in the soil, and thus establish closed-circuit movement through the matrix. Where active transport is involved, such movement is largely under the control of the fungal organism. Fungally mediated transport is largely ignored by the majority of nutrient transport models in soils, which is a significant omission. A greater awareness and appreciation of the fact that fungi act as spatial integrators, or 'the great connectors' of biogeochemical cycling in soils is needed.

References

Amir, R., Steudle, E., Levanon, D., Hadar, Y. & Chet, I. (1995). Turgor changes in *Morchella esculenta* during translocation and sclerotial formation. *Experimental Mycology*, **19**, 129–36.

Anderson, P., Davidson, C. M., Littlejohn, D. *et al.* (1997). The translocation of caesium and silver by fungi in some Scottish soils. *Communications in Soil Science and Plant Analysis*, **28**, 635–50.

Arnebrant, K., Ek, H., Finlay, R. D. & Söderström, B. (1993). Nitrogen translocation between *Alnus glutinosa* (L.) Gaertn. seedlings inoculated with *Frankia* sp. and *Pinus contorta* Doug. ex Loud seedlings connected by a common ectomycorrhizal mycelium. *New Phytologist*, **124**, 213–42.

Bago, B., Pfeffer, P. E., Zipfel, W., Lammers, P. & Shachar-Hill, Y. (2002). Tracking metabolism and imaging transport in arbuscular mycorrhizal fungi. Metabolism and transport in AM fungi. *Plant and Soil*, **244**, 189–97.

Bakken, L. R. & Olsen, R. A. (1990). Accumulation of radiocaesium in fungi. *Canadian Journal of Microbiology*, **36**, 704–10.

Bethlenfalvay, G. J., Reyes-Solis, M. G., Camel, S. B. & Ferrera-Cerrato, R. (1991). Nutrient transfer between the root zones of soybean and maize plants connected by a common mycorrhizal mycelium. *Physiologia Plantarum*, **82**, 423–32.

Boddy, L. (1999). Saprotrophic cord-forming fungi: meeting the challenge of heterogeneous environments. *Mycologia*, **91**, 13–32.

Bolton, R. G. & Boddy, L. (1993). Characterisation of the spatial aspects of foraging mycelial cord systems using fractal geometry. *Mycological Research*, **97**, 762–8.

Boswell, G. P., Jacobs, H., Davidson, F. A., Gadd, G. M. & Ritz, K. (2002). Functional consequences of nutrient translocation in mycelial fungi. *Journal of Theoretical Biology*, **217**, 459–77.

Boswell, G. P., Jacobs, H., Davidson, F. A., Gadd, G. M. & Ritz, K. (2003). Growth and function of fungal mycelia in heterogeneous environments. *Bulletin of Mathematical Biology*, **65**, 447–77.

Brownlee, C. & Jennings, D. H. (1982). Long-distance translocation in *Serpula lacrymans*: velocity estimates and the continuous monitoring of induced perturbations. *Transactions of the British Mycological Society*, **79**, 143–8.

Cairney, J. W. G. (1992). Translocation of solutes in ectomycorrhizal and saprotrophic rhizomorphs. *Mycological Research*, **96**, 135–41.

Cairney, J. W. G. & Burke, R. M. (1996). Physiological heterogeneity within fungal mycelia: an important concept for a functional understanding of the ectomycorrhizal symbiosis. *New Phytologist*, **134**, 685–95.

Caris, C., Hordt, W., Hawkins, H. J., Romheld, V. & George, E. (1998). Studies of iron transport by arbuscular mycorrhizal hyphae from soil to peanut and sorghum plants. *Mycorrhiza*, **8**, 35–9.

Cole, L., Orlovich, D. A. & Ashford, A. E. (1998). Structure, function and motility of vacuoles in filamentous fungi. *Fungal Genetics and Biology*, **24**, 86–100.

Davidson, F. A. & Olsson, S. (2000). Translocation induced outgrowth of fungi in nutrient-free environments. *Journal of Theoretical Biology*, **205**, 73–84.

Declerck, S., Dupre de Boulois, H., Bivort, C. & Delvaux, B. (2003). Extraradical mycelium of the arbuscular mycorrhizal fungus *Glomus lamellosum* can take up, accumulate and translocate radiocaesium under root-organ culture conditions. *Environmental Microbiology*, **5**, 510–16.

Finlay, R. & Read, D. J. (1986a). The structure and function of the vegetative mycelium of ectomycorrhizal plants.1. Translocation of ^{14}C-labelled carbon between plants interconnected by a common mycelium. *New Phytologist*, **103**, 143–56.

Finlay, R. & Read, D. J.(1986b). The structure and function of the vegetative mycelium of ectomycorrhizal plants. 2. The uptake and distribution of phosphorus by mycelium interconnecting host plants. *New Phytologist*, **103**, 157–65.

Frey, S. D., Elliott, E. T., Paustian, K. & Peterson, G. A. (2000). Fungal translocation as a mechanism for soil nitrogen inputs to surface residue decomposition in a no-tillage agroecosystem. *Soil Biology and Biochemistry*, **32**, 689–98.

Frey, S. D., Six, J. & Elliott, E. T. (2003). Reciprocal transfer of carbon and nitrogen by decomposer fungi at the soil-litter interface. *Soil Biology and Biochemistry*, **35**, 1001–4.

George, E., Haussler, K. U., Vetterlein, D., Gorgus, E. & Marschner, H. (1992). Water and nutrient translocation by hyphae of *Glomus mosseae*. *Canadian Journal of Botany*, **70**, 2130–7.

Graves, J. D., Watkins, N. K., Fitter, A. H., Robinson, D. & Scrimgeour, C. (1997). Intraspecific transfer of carbon between plants linked by a common mycorrhizal network. *Plant and Soil*, **192**, 153–9.

Gray, S. N., Dighton, J., Olsson, S. & Jennings, D. H. (1995). Real-time measurement of uptake and translocation of 137Cs within mycelium of *Schizophyllum commune* Fr by autoradiography followed by quantitative image-analysis. *New Phytologist*, **129**, 449–65.

Harley, J. L. & Smith, S. E. (1983). *Mycorrhizal Symbiosis*. London: Academic Press.

Harris, K., Young, I. M., Gilligan, C. A., Otten, W. & Ritz, K. (2003). Effect of bulk density on the spatial organization of the fungus *Rhizoctonia solani* in soil. *FEMS Microbiology Ecology*, **44**, 45–56.

Hart, S. C., Firestone, M. K., Paul, E. A. & Smith, J. L. (1993). Flow and fate of soil nitrogen in an annual grassland and a young mixed-conifer forest. *Soil Biology and Biochemistry*, **25**, 431–42.

Haselwandter, K., Berrek, M. & Brunner, P. (1988). Fungi as bioindicators of radiocaesium contamination: pre- and post-Chernobyl activities. *Transactions of the British Mycological Society*, **90**, 171–4.

Heap, A. J. & Newman, E. I. (1980). Links between roots by hyphae of vesicular-arbuscular mycorrhizas. *New Phytologist*, **85**, 169–71.

Hirrel, M. C. & Gerdemann, J. W. (1979). Enhanced carbon transfer between onions infected with a vesicular-arbuscular mycorrhizal fungus. *New Phytologist*, **83**, 731–8.

Howard, A. J. (1978). Translocation in fungi. *Transactions of the British Mycological Society*, **70**, 265–9.

Jacobs, H., Boswell, G. P., Scrimgeour, C. M. *et al.* (2004). Translocation of carbon by *Rhizoctonia solani* in nutritionally-heterogeneous microcosms. *Mycological Research*, **108**, 453–62.

Jansa, J., Mozafar, A. & Frossard, E. (2003). Long-distance transport of P and Zn through the hyphae of an arbuscular mycorrhizal fungus in symbiosis with maize. *Agronomie*, **23**, 481–8.

Jennings, D. H. (1987). Translocation of solutes in fungi. *Biological Reviews*, **62**, 215–243.

Jennings, D. H., Thornton, J. D., Galpin, M. F. J. & Coggins, C. R. (1974). Translocation in fungi. In *Transport at the Cellular Level*, ed. M. A. Sleigh & D. H. Jennings. Cambridge: Cambridge University Press, pp. 139–56.

Johansen, A. & Jensen, E. S. (1996). Transfer of N and P from intact or decomposing roots of pea to barley interconnected by an arbuscular mycorrhizal fungus. *Soil Biology and Biochemistry*, **28**, 73–81.

Johansen, A., Jakobsen, I. & Jensen, E. S. (1993). Hyphal transport by a vesicular-arbuscular mycorrhizal fungus of N applied to the soil as ammonium or nitrate. *Biology and Fertility of Soils*, **16**, 66–70.

Kirchner, G. & Daillant, O. (1998). Accumulation of ^{210}Pb, ^{226}Ra and radioactive cesium by fungi. *Science of the Total Environment*, **222**, 63–70.

Lindahl, B. D. & Olsson, S. (2004). Fungal translocation – creating and responding to environmental heterogeneity. *Mycologist*, **18**, 79–88.

Lindahl, B., Stenlid, J., Olsson, S. & Finlay, R. (1999). Translocation of ^{32}P between interacting mycelia of a wood-decomposing fungus and ectomycorrhizal fungi in microcosm systems. *New Phytologist*, **144**, 183–93.

Lindahl, B., Finlay, R. & Olsson, S. (2001). Simultaneous, bidirectional translocation of ^{32}P and ^{33}P between wood blocks connected by mycelial cords of *Hypholoma fasciculare*. *New Phytologist*, **150**, 189–94.

Littlefield, L. J. (1967). Phosphorus-32 accumulation in *Rhizoctonia solani* sclerotia. *Phytopathology*, **57**, 1053–5.

Mihail, J. D., Obert, M., Bruhn, J. N. & Taylor, S. J. (1995). Fractal geometry of diffuse mycelia and rhizomorphs of *Armillaria* species. *Mycological Research*, **99**, 81–8.

Money, N. P. (2004). The fungal dining habit – a biomechanical perspective. *Mycologist*, **18**, 71–6.

Newman, E. I. (1988). Mycorrhizal links between plants: their functioning and ecological significance. *Advances in Ecological Research*, **18**, 243–70.

Newman, E. I. & Eason, W. R. (1993). Rates of phosphorus transfer within and between ryegrass (*Lolium perenne*) plants. *Functional Ecology*, **7**, 242–8.

Niksic, M., Hadzic, I. & Glisic, M. (2004). Is *Phallus impudicus* a mycological giant? *Mycologist*, **18**, 21–2.

Olsson, S. (1995). Mycelial density profiles of fungi on heterogeneous media and their interpretation in terms of nutrient reallocation patterns. *Mycological Research*, **99**, 143–53.

Olsson, S. & Gray, S. N. (1998). Patterns and dynamics of ^{32}P-phosphate and labelled 2-aminoisobutyric acid (C^{14}-AIB) translocation in intact basidiomycete mycelia. *FEMS Microbiology Ecology*, **26**, 109–20.

Olsson, S. & Hansson, B. S. (1995). Action potential-like activity found in fungal mycelia is sensitive to stimulation. *Naturwissenschaften*, **82**, 30–1.

Olsson, S. & Persson, Y. (1994). Transfer of phosphorus from *Rhizoctonia solani* to the mycoparasite *Arthrobotrys oligospora*. *Mycological Research*, **98**, 1065–8.

Otten, W., Gilligan, C. A., Watts, C. W., Dexter, A. R. & Hall, D. (1999). Continuity of air-filled pores and invasion thresholds for a soil-borne fungal plant pathogen, *Rhizoctonia solani*. *Soil Biology and Biochemistry*, **31**, 1803–10.

Pfeffer, P. E., Douds, D. D., Bucking, H., Schwartz, D. P. & Shachar-Hill, Y. (2004). The fungus does not transfer carbon to or between roots in an arbuscular mycorrhizal symbiosis. *New Phytologist*, **163**, 617–27.

Rafferty, B., Brennan, M. & Kliashtorin, A. (1997). Decomposition in two pine forests: the mobilisation of ^{137}Cs and K from forest litter. *Soil Biology and Biochemistry*, **29**, 1673–81.

Rayner, A. D. M. (1991). The challenge of the individualistic mycelium. *Mycologia*, **83**, 48–71.

Rayner, A. D. M. (1994). Pattern-generating processes in fungal communities. In *Beyond the Biomass: Compositional and Functional Analysis of Soil Microbial Communities*, ed. K. Ritz, J. Dighton & K. E. Giller. Chichester, UK: John Wiley, pp. 247–58.

Rayner, A. D. M. & Boddy, L. (1988). Fungal communities in the decay of wood. *Advances in Microbial Ecology*, **10**, 155–63.

Read, D. J. & Stribley, D. P. (1975). Diffusion and translocation in some fungal culture systems. *Transactions of the British Mycological Society*, **64**, 381–8.

Ritz, K. (1995). Growth responses of some soil fungi to spatially heterogeneous nutrients. *FEMS Microbiology Ecology*, **16**, 269–80.

Ritz, K. & Crawford, J. W. (1990). Quantification of the fractal nature of colonies of *Trichoderma viride*. *Mycological Research*, **94**, 1138–42.

Ritz, K. & Newman, E. I. (1984). Movement of ^{32}P between intact grassland plants of the same age. *Oikos*, **43**, 138–42.

Ritz, K. & Newman, E. I. (1986). Nutrient transport between ryegrass plants differing in nutrient status. *Oecologia*, **70**, 128–31.

Ritz, K. & Young, I. M. (2004). Interactions between soil structure and fungi. *Mycologist*, **18**, 52–9.

Ritz, K., Millar, S. M. & Crawford, J. W. (1996). Detailed visualisation of hyphal distribution in fungal mycelia growing in heterogeneous nutritional environments. *Journal of Microbiological Methods*, **25**, 23–8.

Rufyikiri, G., Thiry, Y. & Declerck, S. (2003). Contribution of hyphae and roots to uranium uptake and translocation by arbuscular mycorrhizal carrot roots under root-organ culture conditions. *New Phytologist*, **158**, 391–9.

Schütte, K. H. (1956). Translocation in the fungi. *New Phytologist*, **55**, 164–82.

Shepherd, V. A., Orlovich, D. A. & Ashford, A. E. (1993). A dynamic continuum of pleiomorphic tubules and vacuoles in growing hyphae of a fungus. *Journal of Cell Science*, **104**, 495–507.

Simard, S., Perry, D. A., Jones, M. D. *et al.* (1997). Net transfer of carbon between ectomycorrhizal tree species in the field. *Nature*, **388**, 579–82.

Smith, S. E. & Read, D. J. (1997). *Mycorrhizal Symbiosis*. London: Academic Press.

Stahl, P. D. & Christensen, M. (1992). In vitro mycelial interactions among members of a soil micro-fungal community. *Soil Biology and Biochemistry*, **24**, 309–16.

Steinberg, G. (1998). Organelle transport and molecular motors in fungi. *Fungal Genetics and Biology*, **24**, 161–77.

Suzuki, H., Kumagai, H., Oohashi, K. *et al.* (2001). Transport of trace elements through the hyphae of an arbuscular mycorrhizal fungus into marigold determined by the multitracer technique. *Soil Science and Plant Nutrition*, **47**, 131–7.

Thrower, L. B. & Thrower, S. L. (1968). Movement of nutrients in fungi. II. The effect of reproductive structures. *Australian Journal of Botany*, **16**, 81–7.

Tlalka, M., Watkinson, S. C., Darrah, P. R. & Fricker, M. D. (2002). Continuous imaging of amino acid translocation in intact mycelia of *Phanerochaete velutina* reveals rapid, pulsatile fluxes. *New Phytologist*, **153**, 173–84.

Watkins, N. K., Fitter, A. H., Graves, J. D. & Robinson, D. (1996). Quantification using stable carbon isotopes of carbon transfer between C3 and C4 plants linked by a common mycorrhizal network. *Soil Biology and Biochemistry*, **28**, 471–7.

Wells, J. M. & Boddy, L. (1995a). Phosphorus translocation by saprotrophic basidiomycete mycelial cord systems on the floor of a mixed deciduous woodland. *Mycological Research*, **99**, 977–80.

Wells, J. M. & Boddy, L. (1995b). Translocation of soil-derived phosphorus in mycelial cord systems in relation to inoculum resource size. *FEMS Microbiology Ecology*, **17**, 67–75.

Wells, J. M., Boddy, L. & Evans, R. (1995). Carbon translocation in mycelial cord systems of *Phanerochaete velutina* (DC.: Pers.) Parmasto. *New Phytologist*, **129**, 467–76.

Wilcoxson, R. D. & Sudia, T. W. (1968). Translocation in fungi. *Botanical Reviews*, **34**, 32–50.

4

Water dynamics of mycorrhizas in arid soils

MICHAEL F. ALLEN

Introduction

The interaction of mycorrhizas and water in understanding plant water dynamics has been relevant since Frank (1885) first coined the term mykorhiza, a plant–fungus mutualism. He described an ectomycorrhiza as a 'wet-nurse' to the host in that water and nutrients must flow through the hyphae to the plant root tip. Stahl (1900) proposed that mycorrhizas increased water throughput, depositing greater amounts of nutrients in the roots resulting in the improved growth. We now know that carbon and nutrient exchange is an active process, regulated by both plant and fungal genes, and requiring substantial inputs of energy from the host and concentrating mechanisms in the fungus. However, water movement is a passive process. That is, it flows in response to energy gradients, without regard to active processes. Because it is a passive process, in mesic regions a large amount of water flows through a relatively saturated soil around the fungal hypha into the rather high area of root surface. This occurs at rates which would not be affected by the comparatively small surface area of the hypha–root interface. The focus of studies on mycorrhizas and water relations has been on whether mycorrhizas enhance plant water uptake with drought. This becomes rather critical in that past studies have often misinterpreted data of mycorrhizas and water flux, or designed studies measuring water fluxes in materials such as sand, or in limited potting volumes (relative to root length) that place unreasonable constraints on mycorrhizal response. Others have postulated that ectomycorrhizas (EM) and arbuscular mycorrhizas (AM) have fundamentally different impacts because EM enclose a root and AM only intermittently penetrate a root. Although many types of mycorrhizas exist, I will focus only on EM and AM, for which data on water relations exist. The focus of mycorrhizal

Fungi in Biogeochemical Cycles, ed. G. M. Gadd. Published by Cambridge University Press. © British Mycological Society 2006.

research has been on interfaces, but to understand mycorrhizal functioning we need to understand the role and nature of the external hyphae (Hatch, 1937; Leake *et al.*, 2004).

In EM, comparative studies of water relations of mycorrhizal and nonmycorrhizal roots were inconsistent. In part, this was because root length (and thus surface area) was stunted in an EM plant. Ectomycorrhizas thereby comparatively lowered root water uptake. However, Duddridge *et al.* (1980) showed that some of the rhizomorphic hyphae are highly specialized to transport water rapidly and for long distances. These data also demonstrated the limits under which most studies were undertaken (greenhouse pots) compared with the field conditions under which mycorrhizas exist.

In AM, Safir *et al.* (1971) demonstrated that mycorrhizas increase water flow, but then suggested that the increasing water uptake by mycorrhizas was a consequence of increased phosphorus status, an indirect response (Safir *et al.* 1972). That hypothesis was based on benomyl studies and the hypothesis that hyphae were too small. The study provided an interesting insight into how water uptake might be affected by AM. In their 1972 study, the application of benomyl blocked the enhanced phosphorus, but not water uptake. They concluded that because phosphorus uptake and not water uptake was influenced, and the hyphae were presumably killed, that water uptake was not a function of the hyphae. However, an alternative explanation is that phosphorus uptake, an *active* process, was stopped because the hyphae were killed. Since water transport is a *passive* process, and dead hyphae themselves could still be conducting by providing a direct pathway for water along and through the mycelium for a time period before degradation. Phosphorus or nitrogen would *not* be expected to continue to be transported. Hardie and Layton (1981) then suggested that the hyphae were involved by measuring the changes in whole plant conduction (L_p) associated with AM and Allen (1982) calculated hyphal transport based on difference equations. Others continued to argue that the hyphae were too small (Tinker & Nye, 2000) (the same argument has been made for root hairs). Fluxes are related to the radius to the 4th power, where: flux $= (\pi r^4/8\nu)\ dP/dx$, where r is radius, ν is viscosity and dP/dx is the pressure gradient (Tinker & Nye, 2000). The terminal hypha of the network is large when it enters the root (10 µm in diameter) compared with the tips (2 µm in diameter) that have radiated from the 6-order network (Friese & Allen, 1991). Further, that hyphal network may extend over 100 cm penetrating several centimetres into the soil from the point of entry into the root. This means a hypha entering a root would have an order of magnitude greater water flux rates compared with the hyphal tips, which might account for a relatively large exchange between plant and fungus (a flux rate of 5^4x (or 625x) would occur at the root entry, compared with 64 hyphal tips, each with a flux of 1^4x (or 64x)). Subsequently, Hardie (1985) undertook an experiment and clipped the entry hyphae. In doing so, she reduced the rates of water uptake of a mycorrhizal plant to that of the non-mycorrhizal control.

These problems illuminate the difficulty in studying hyphal fluxes. The passive nature of transport means that both dead and live hyphae may play a role, and the 'standing-dead' crop of fungi may exceed 90% of the total length. Furthermore, transport can be in either direction, depending on the driving gradient, and is continually shifting through both day and night. Despite the importance of water fluxes, there are few direct data on water fluxes within mycorrhizal hyphae that help us assess the overall dynamics. Because hyphae are small in comparison to the footprint of current measurement technology, theoretical calculations of potential transports are necessary to tease apart likely flow pathways. For the remainder of this chapter, the role of mycorrhizas in water fluxes will be examined in both directions as well as the importance of hyphal transport of water to nutrient dynamics in arid ecosystems.

Water in arid and semi-arid soils and fungi

Most studies of mycorrhizas and water fluxes centre around difference estimates under relatively saturated conditions. Under these conditions, the importance of mycorrhizal fungal hyphal transport often may be relatively small simply because the total root surface area is very large compared to the area of entry points provided by hyphae, despite the added surface area provided by the hyphae. For example, in seedlings of *Bouteloua gracilis*, the root surface area (including root hairs) was $1070\,mm^2$ but, with 1600 entry points and a hyphal diameter of $10\,\mu m$, the total entry point area would only be $0.13\,mm^2$ (data from Allen, 1982). In a field study when soils were either saturated or when extremely dry, no effect of mycorrhizas on transpiration was measured (Allen & Allen, 1986). It was during the transition from wet to dry soils when mycorrhizas made a difference in water loss and, correspondingly, carbon gain. It is this period, the 'ecological crunch', in which mycorrhizas make their greatest contribution.

Much of the world's agriculture, particularly small grain farming and grazing, and a large fraction of the world's wildlands exist in arid to semi-arid regions. Most arid to semi-arid perennial plants form mycorrhizas.

Under drying conditions, mycorrhizas can make an important contribution to the plants' water balance (Auge', 2001). Further, much of the plant growth occurs from the time following a precipitation event, until water is no longer available. Thus, short-term dynamics also regulate overall photosynthetic patterns in these regions. However, definitions of available water are often inadequate. 'Permanent wilting point' is defined by many texts to be at a soil water potential (ψ_s) of -1.5 MPa. Down to this level, most soil pores still contain water. Below this level, soil pores begin increasingly to contain air instead of water. As these pores continue to dry, water is reduced to a meniscus covering soil particles or organic material. To move water from one point to another requires that it follow a tortuous pathway along the individual soil particles. Between saturation and wilting point (<20% soil moisture in loamy soils), hydraulic conductivity (K) drops an order of magnitude. In loamy soils, K drops another order of magnitude between 20% and 10% soil water (Hanks & Ashcroft, 1980). Much of the time, semi-arid soils are less than 10% water content. As this happens, soil resistance (R_s) dramatically increases relative to root resistance (R_p). For example, under saturated conditions, R_p/R_s is 250, but as the soil dries to -1 MPa, the R_p/R_s ratio is only 5.2 (Tinker & Nye, 2000). By the time soils reach most drought conditions, R_p/R_s goes to less than 1. Alternatively, a root, root hair, or hypha that crosses a gap forms a shortened pathway across that gap, as long as the hypha is saturated, and the meniscus remains.

In many situations, ψ_s is at or below -1.5 MPa much of the growing season, and the high levels of evaporative demand means that ψ_p of un-irrigated plants remains below the 'permanent wilting point' much of the growing season. Thus, new growth in the form of both hyphal tips and root intercepts becomes water-limited. Often, plants in these areas are dependent upon groundwater reserves accessible to deep-rooted plants. In many arid regions of the world, it rarely rains more than a single event within a two-week period. Precipitation in the semi-arid areas such as southwestern USA is both spatially and temporally patchy (Fig. 4.1). What happens at a patch and deeper in the soil becomes absolutely critical to the health of the mycorrhizal plant. Hydraulic lift is the process whereby deep water is tapped by plants and transported to surface soil layers during the night when stomata are closed (Richards & Caldwell, 1987). These patterns suggest an important complex role of mycorrhizas, including bidirectional transport, water uptake by mycorrhizal hyphae and hydraulic-lift water transported from plant to fungus, as mechanisms for sustaining the turgor of mycorrhizal fungal hyphae (Querejeta *et al.*, 2003).

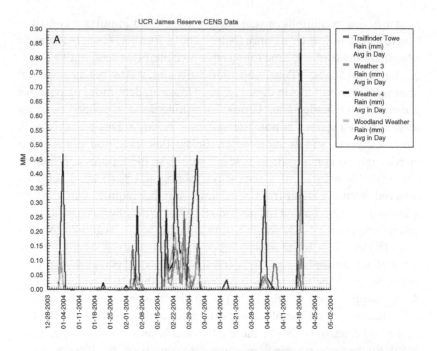

Fig. 4.1. (cont.)

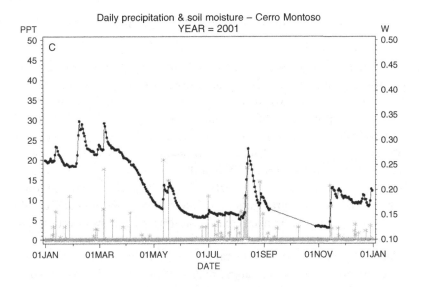

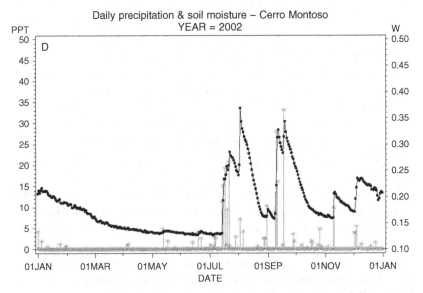

Fig. 4.1. Spatial and temporal variation in precipitation and soil moisture. Shown in panel A is the variation in precipitation between four locations located within 20 m of each other at the James Reserve <http://www.jamesreserve.edu>, above Riverside CA, USA. Panel B shows the spatial variation in moisture input simply based on the snow patterns at the Sevilleta National Wildlife Refuge, in New Mexico, USA (http://www.lternet.edu/sites/sev/). Panels C and D show the variation in precipitation and soil moisture through two successive growing seasons (2001, 2002) at the pinyon-juniper site of the Sevilleta National Wildlife Refuge, NM, USA. (Graphs prepared by D. Moore, Sevilleta LTER).

Structure and water relations of the mycorrhizal mycelium

Individual hyphae

One of the most important issues in studying mycorrhizas is the need for reverse flows of carbon and nutrients. Carbon must move from plant to fungus, and nutrients from fungus to plant. How can these flows be reversed especially as both resources move in an aqueous medium? Cairney (1992) postulated that there could be a spatial separation of water and elements in that materials transported by bulk flow in water might be transported through the walls of the hyphae and nutrients through the cytoplasm. A modification of this hypothesis could be found in the diurnal shifts in gradients of ψ. These hypotheses are *not* mutually exclusive. But both mechanisms require shifting water-flow directions and high rates of flow.

An individual mycorrhizal fungal hypha is structured and behaves in a similar manner to saprobic fungal hyphae. The major difference is that one end of the mycorrhizal fungal hypha has a direct connection to a simple and large carbon source, the plant, while the other end extends into the soil matrix. In a saprobic fungus, the hypha is within both the carbon and nutrient patch. Although cords stretch across soil and organic matter patches, they are primarily structured to get from a depleted patch to an unexploited patch, but rarely to transport materials back to the same point over long time periods. The infecting end of the mycorrhizal fungal mycelium itself comprises a major plant carbon sink. The other end is immersed in a source of nutrients both for the fungus and for its host. Thus, it is critical that the mycorrhizal fungus be thought of, and studied, as a single unit.

In studying mycorrhizal fungi, the repeated evolution of a mycorrhizal condition necessarily means that virtually all iterations of the basic fungal structure need to be considered. Arbuscular mycorrhizal hyphae are coenocytic and have only infrequent, adventitious septa. The symplast is continuous from the arbuscules or coils in the root cortex through to the tips of the extramatrical fine hyphae and the large runner (or arterial) hyphae. When an adventitious septum forms, it constricts the cytoplasm between cells within the hypha restricting symplastic, but not necessarily apoplastic flow (this will become important later). Ectomycorrhizas are formed by Endogonaceae, ascomycetes and basidiomycetes. This means that hyphae with few or no septa, simple pore- and doliopore septa are present depending on the taxon of the fungus. Membranes do not necessarily separate individual cells, only the protoplasmic material from the external wall (Fig. 4.2). Nuclei and other organelles migrate through hyphae so that the only place where there is a certain plasma membrane partitioning the flow is in the Hartig net within the root cortex and at the tip of the extramatrical hypha – just as in the AM. However, individual hyphae can be plugged by structures such as Woronin bodies (in ascomycetes) that reduce or prevent flow of material through the hypha. Also important may be the cellular structure of a hypha. Microtubules and actin fibres are oriented longitudinally through the hypha. These facilitate the flow of materials and water.

Flows within a hypha can be rather high. Cooke and Whipps (1993) and Eamus and Jennings (1986) reported cytoplasmic flows of 25 cm h^{-1}. For a 10 μm diameter hypha, this means a flux rate of 19.5 nl h^{-1}. These flow rates occur across longitudinal gradients of 0.5 MPa (in *Neurospora crassa*) and 1.0 MPa (in *Armillaria*) in non-mycorrhizal fungi measured in culture. In some cases, hyphae can survive and even grow at extremely low ψ_s (Griffin, 1972; Jennings, 1995).

The plasma membrane is a crucial feature in the regulation of water and nutrient fluxes. The presence of the membrane itself creates the hydrostatic pressure of fungal hyphal tip growth. Calcium, potassium and sodium concentrate at the hyphal tip reducing the $\psi_{cytoplasm}$ and creating a water potential gradient from the older part of the hypha. Also at the tip is the Spitzenkorper, which produces new vesicles and organelles regulating growth patterns. By reducing rapid loss of water from the tip, the turgor pressures aiding growth are built up. Growth rates can be rather rapid. Prosser (1983) noted that the leading hypha of *Mucor*, 13 μm in diameter, grew 38 μm per minute and the primary branch, 8.9 μm in diameter grew at 27.4 μm per minute. On the other hand, the limiting factor to transport between plant and fungus becomes the plasma membrane, which separates

Fig. 4.2. Water transport within a hypha. Water moving between points A and B (plant:fungus) or E and F (soil:fungus) must pass through a fungal membrane, a major limit to transport (see text). However in fungi, water may not necessarily have to transport through a membrane between cells as no membrane separates individual 'cells'. Thus, rapid equilibration in ψ can occur both within cells (e.g. B–C or D–E) and between C and D. Structures as large nuclei can be seen moving between cells squeezing between the walls (see Alexopolis *et al.*, 1996).

the cytoplasm and vacuoles from the wall. This represents the greatest obstacle to water transport in that the hydraulic conductivity is quite low compared to other materials. The L_w (hydraulic conductivity) across the plasma membrane in plants is 1×10^{-2} cm (s MPa)$^{-1}$ (Nobel, 1974). As we shall see shortly, this limitation could dramatically reduce plant–hyphal exchange.

One additional feature to consider is the wall structure itself. The wall is comprised of a complex layering oriented along the hypha. The tip is generally about 50 nm thick and comprised of chitin. As one moves away from the tip, wall layers are added (Fig. 4.3). This layering contains both hydrophobic and hydrophilic regions. The wall layers provide structure to the hypha and protection from the surrounding soil or plant tissue. Wall layers can be up to several microns thick in coarse AM fungal hyphae and covering chlamydospores. The hyphal wall thickness is variable

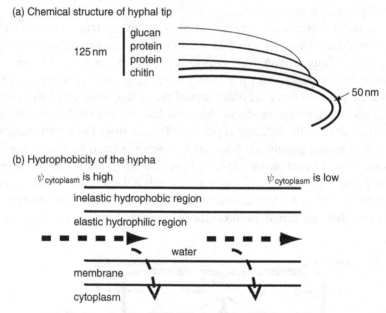

Fig. 4.3. Chemical and physical structure of the individual hypha and the hyphal tip. The layering chemistry and structure of the hypha (a) creates an inelastic outer hydrophobic layer and elastic hydrophilic regions (b). At the tip, the layering has not yet developed, creating hydrophilic tip regions allowing water uptake or loss depending on the gradient in ψ. However, the membrane limits water loss from within the hypha due to low L_w (see text). The entire structure allows water flow through the walls along whatever ψ gradient is formed. (model derived from Bartnicki-Garcia, 2002).

among species and conditions, but is generally multilayered with an outer wall thickness of 25 to 40 nm, and an inner wall of 0.1 to 5 μm thickness. Along the hypha, the wall tends to be rigid and often impermeable to water (Unestam, 1991). At the tip, the hyphal wall is elastic and generally hydrophilic. As individual hyphae grow, the chitin microfibrils expand and new wall is produced as the osmotic pressure pulls water from the interior of the hypha and, in appropriate cases, through the membrane of the tip itself. Because these fibrils are elastic, individual hyphae can force through small openings in many different types of substrates. The pressure exerted by the hyphal tip can be rather extraordinary. Hyphae break through cracks in soil, through hard organic materials and even through cracks in concrete and rock (see Gorbushina (Chapter 11) & Fomina *et al.* (Chapter 10), this volume).

Cytoplasmic flow is constricted at septa, but still continuous through long portions of the hyphal length. Water and nutrients must attempt to equilibrate along that length. However, these demands form gradients between the ends. During daylight, within the plant, there is a demand within the roots caused by evaporation from open stomata. Here, the $\psi_p < \psi_s$, or the water potential at location A < that of location F (Fig. 4.2). At the other end of the hypha, the tip is within the soil matrix. In wetter regions, this means that these tips are located in continuous moisture. This demand creates the drivers for water uptake (F to A). However, in arid and semi-arid regions, hyphal tips can be located in soils with $\psi_s < -3$ MPa. Where surface $\psi_s < -3$ MPa, the ψ_s can often be lower than the rehydrating ψ_l (leaf water potential) during the night, creating a reverse gradient, where the location F has a greater demand than location A (Fig. 4.2). Thus, at night, these demands can reverse. As the plant leaves equilibrate with the soil moisture throughout the rooting system, the ψ_l can exceed that of the surface ψ_s. Water moves passively in response to ψ gradients. Thus, hydraulic lift (A to F) can occur driving flow in the reverse direction (Querejeta *et al.*, 2003). These demands vary at both daily and hourly time scales. Using dyes, all water transport from deep roots went to the leaves during daylight when stomata were open. However, at night, those dyes were redistributed through the root tips, through the mycorrhizal hyphae and even exuded out of hyphal tips (Fig. 4.4).

It is under unsaturated conditions that the structure of the AM fungal mycelium becomes of particular interest. The hydraulic conductivity of the entire hypha (L_p) of fungal hyphae of *Phycomyces blakesleeanus* was measured at 4.6×10^{-5} cm (s MPa)$^{-1}$ (Cosgrove *et al.*, 1987) but that value includes the membrane resistance. Water conduction (L_w) is drastically limited by membranes while being transported across the plasma membrane into the cytoplasm (Fig. 4.2). Thus, water could move and equilibrate rapidly within the cytoplasm but is restricted from being exchanged with the plant or the soil. Water fluxes from soil to hyphae to plant (J_w) across a plasma membrane can be estimated using the formulation of water conduction as:

$$J_w = L_w \, \Delta \, \psi,$$

where $L_w = 1 \times 10^{-2}$ cm (s MPa)$^{-1}$, $\Delta\psi = -0.5$ MPa, and the hyphal diameter of 10 µm, the individual hyphal transport was 0.14 nl h^{-1}. Thus, the limiting step in water transport is the membrane resistance to water flux. The flux through the membrane would dramatically affect fluxes observed by Eamus and Jennings (1986) (25 cm h^{-1}) or Cowan *et al.* (1972) (131 nl h^{-1}), or my estimates of AM fungal hyphal fluxes of up to 100 nl h^{-1} (Allen, 1982). In the cases of flux through the saprobes, the water is moving from the source to the hyphal tips without encountering a membrane, until the sporophore. However, in an AM fungus, water flowing through the cytoplasm must be exchanged through the membrane at

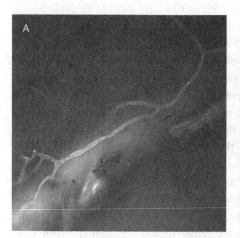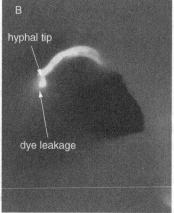

Fig. 4.4. Lucifer yellow carbohydrazide dye indicating water flow applied to a deep-root chamber, and detected in mycorrhizal hyphae in a hyphal chamber after crossing airgaps that restrict diffusion. Panel A is an AM hypha that transported the dye during the night (hydraulic lift). Panel B shows a mycorrhizal fungus hydrophilic tip, with hydraulically lifted water exuding out the tip onto a piece of organic detritus. Photographs by Louise Egerton-Warburton and details of the experiment can be found in Querejeta *et al.* (2003).

some point. However, L_p is 10^4 to 10^5 times greater than L_w. The majority of the hypha is hydrophobic, except for the tips, both inside the root, and in the soil (Fig. 4.2). If water were to move through the elastic, hydrophilic portion of the walls, then L_p of the hyphal wall network would increase dramatically. I have not found measurements of L_p in these regions, although L_p values of plant vessel elements range from 0.2 to $10\,cm\,s^{-1}\,MPa$ (Jury *et al.*, 1981). As shown in Fig. 4.3, the hydrophilic part of the wall can be up to 125 nm thick. Again using:

$$J_w = L_p \, \Delta \, \psi$$

where $L_p = 1\,cm\,(s\,MPa)^{-1}$, and a wall layer of 0.1 μm (100 nm). With a hyphal diameter of 10 mm, then water transport under a 0.5 MPa gradient would be $5.4\,nl\,h^{-1}$, and with an L_p of $10\,cm\,(s\,MPa)^{-1}$, then transport could reach as high as $54\,nl\,h^{-1}$.

If we add together viscosity along the hyphal surfaces (Allen, 1996), flow through the walls and symplastic flow, this would account for a significant rate of water flux through AM hyphae. It certainly can account for any reverse flows such as hydraulic lift. Our most recent estimates suggest that the reverse water flows into soil are much smaller than the flow rates associated with transpiration (Querejeta, Egerton-Warburton & Allen, in preparation). This pattern of water fluxes through the apoplastic system of the fungal hyphae can be visualized. During dye studies of hydraulic lift, the apoplastic flow of dyes can be seen proceeding but subsequently equilibrating with the cytoplasm of individual hyphae (Fig. 4.5).

Structure of the mycelium

Water fluxes through an entire mycelium differ in important ways from flows through individual hyphae. Individual hyphae branch at rather regular intervals, the pattern of which depends on the group. In the case of AM hyphae, they bifurcate at angles somewhat less than 90°, and in the case of some ascomycetes and basidiomycetes, can branch at right angles. Arbuscular mycorrhizal fungi (AMF) form an absorptive network branching at regular intervals, with each branch diminishing in diameter. In the material studied by Friese and Allen (1991), the branches were regular to the point that a simple model of hyphal expansion (h) was created (Allen *et al.* 2003) where:

$$h = \Sigma_1^n [a(r^n - 1)]/(r - 1)$$

Where a = segment length = 0.5 cm, r = branch ratio = 2, and n = number of branches, which ranged from 1 to 8 (maximum). This length is maximal at 122 cm hypha per entry point (Friese & Allen, 1991). This also expands the absorbing area beyond the root surface by 2 to 3 cm.

Bago *et al.* (1998) expanded this basic model in that they found that the absorbing hyphae could also branch off from the runner hyphae. Thus, as a runner hypha spreads into the soil, multiple absorbing networks form. This would have the effect of expanding the absorbing area of AM hyphae several-fold. We do not know the extent of this expansion. The depletion zones of

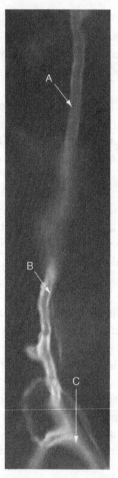

Fig. 4.5. Flow of dyes from hydraulically lifted water into an AM fungal hypha. Shown is the dye moving through the hyphal wall, and equilibrating with the cytoplasm. Photograph by Louise Egerton-Warburton and details of the staining process can be found in Querejeta *et al.* (2003).

phosphorus by AMF extend 2–4 cm beyond the root surface (Owusu-Bennoah & Wild, 1979) and direct uptake and transport of phosphorus and calcium has been measured up to 6 cm (Rhodes & Gerdemann, 1975).

At the point where the fungus emerges from the root into the soil, single hyphae may emerge, or multiple hyphae can wrap around each other (Fig. 4.6). In AM, this may not be very common but we have observed this phenomenon, as have others (e.g. David Stribley, personal communications). In some cases, if a patch of material or growing root is nearby, numerous hyphae may bridge the gap wrapping around each other forming a structure resembling a very primitive 'cord'. Both primitive and advanced cords consist of multiple hyphae, often wrapped around each other (Fig. 4.7). When these hyphae touch, or nearly touch, water forms a layer between them, the size of which and flow rates depend upon the local ψ, and the gradient in ψ. This process, viscosity, can play an especially critical role in water transport. In a relatively wet system along a glass plate, a viscous layer of water could be observed moving from soil to root during the day (Allen, 1996). During the night, in hydraulic-lift studies, water moved between these rhizomorph hyphae (Querejeta *et al.*, 2003) at both faster rates and in greater amounts than through single hyphae (Querejeta, Egerton-Warburton & Allen, in preparation).

With EM fungi (EMF) a net mycelium can form a more complex arrangement in soil because branches can occur at right angles and

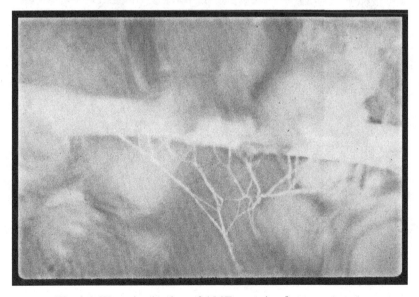

Fig. 4.6. Wrapping hyphae of AMF emerging from roots.

anastomosis is relatively common. In some fungi, complex 'vessel rhizo-morphs' are formed that act like vessel elements in plants, transporting water up to 27 cm h^{-1} (Duddridge *et al.* 1980). Up to 20 to 80 m of hyphae can extend from a single mycorrhizal tip (Duddridge *et al.*, 1980). In some cases, the individualistic hyphal structure is also common in ascomycetous mycorrhizas such as *Cenococcum*. Particularly in basidiomycetes, the emerging hyphae can become extremely complex forming rhizomorphs with vessel-like elements embedded within the radiating hyphae. These rhizomorphs can extend up to several metres across the soil radiating out from a single plant into the rhizosphere of surrounding vegetation. *Pisolithus* and *Boletus* rhizomorphs can be detected up to 10 m from the host roots (M. F. Allen, unpublished observations).

When viewed as a mycelium, there are also distinct structural/functional differences along the hyphal network, particularly for a mycorrhizal fungus. The development of a hyphal network is integral to understanding how transport actually works. There are fundamental differences between AMF and EMF, but within that structure there are analogous features unique to a mycorrhiza. The fungus starts external to the plant root, whether from a germinating spore or from an existing hypha. Infection

Fig. 4.7. Intact rhizomorph of an EM fungus from a minirhizotron image.

develops at an initiation point (the infection process is well described, see Smith and Read, 1997 for a comprehensive review). The external mycelium expands outward into the soil, looking for new roots to infect, and absorbing both water and nutrients (Allen, 1991). It is important to note that in extremely wet soils, particularly in clay soils, O_2 can limit fungal growth and thereby reduce mycorrhizal effectiveness. At the other textural extreme, in wet sandy soils, water fills most pore spaces and nutrients are readily available, again reducing mycorrhizal effectiveness as bulk flow and diffusion supplies virtually everything needed by the plant. Some experiments in which roots already completely occupy a tube or pot, especially with sandy soils, often fail to show a mycorrhizal response because the critical feature of a mycorrhiza, the external hyphal area, is no longer contributing by accessing new soil patches. These conditions describe many experiments studying mycorrhizas and water relations. However, in open systems, as soils begin to dry, airgaps replace the water-filled pores, particularly at $\psi_s < -1.5\,\text{MPa}$.

Hyphal transport in unsaturated conditions

As described above, flow of water slows down dramatically as soils dry, particularly below $-1.5\,\text{MPa}$, as airgaps begin dominating soil pore space. This is the time when fungal hyphae could become especially important in water transport, and hydraulic lift could be important in supporting a hyphal network. The direction and magnitude depends on the spatial structure of dry zones. Two processes occur of importance. Firstly, a hypha may grow into a saturated soil pore, extracting the water during the day and transferring the water to the plant. Secondly, at night as the deep roots become more hydrated than surface soil and stomata close, water is taken up from deep in the profile by roots, and (because the stomata are not open, and the driving gradient is now into dry surface soils) transferred to the fungus allowing it to continue growing into the next pore. The hyphae bridge airgaps and water moves along the hyphae from the saturated pore, and along the soil pore surface. The distance is far less along the relatively straight surface of the hypha compared with the tortuous pathway along the pore surface.

Putting the system together

Sustaining the plants and hyphae through the dry season

Flux rates are complex in that they represent feedback between the atmosphere (the ultimate driver) and internal leaf water. It is important to remember that even 'dry' leaves, those with a $\psi_l < -2\,\text{MPa}$ have water contents greater than 95%. This means that the dynamics of water flow are continually shifting, within a day, across seasonal transitions and across variable years. For example, even during the wet season in the tropics, stomata can close as radiant energy loading on leaf surfaces is high causing a lower ψ_l, even in saturated soils. This causes a midday wilting as a means to rehydrate, reducing water loss but simultaneously stopping carbon gain. The stomata also close during the night, again allowing rehydration. Thus, there is a continually shifting gradient, and even a shifting direction to the gradient as water-flow patterns change to try to equilibrate to the changing demands.

The transition from the wet to dry season is a critical period for all plants, and mycorrhizas may play key roles during this period. An example is shown in Allen and Allen (1986). In this case, as soil water declines during the drought season, the daily stomatal conductance shows a mid-day decline in non-mycorrhizal plants corresponding to a midday rise in ψ_l. The mycorrhizal plants did not show such a decline maintaining higher water inflows as shown in the higher ψ_l and higher leaf conductance. This response only occurred for approximately two weeks, but was critical to the plants' life-cycle in that this allowed the plants to continue to photo-synthesize for an additional two weeks, increasing seed set and, likely, tillering, in preparation for the next growing season.

A plant mycorrhizal system integrates a large volume of soil. In small plants, mycorrhizal roots extend laterally for nutrients and water. In *Atriplex gardneri*, a small chenopodiaceous shrub, mycorrhizal formation with functioning arbuscules was limited to only a few weeks (Allen, 1983), but was nevertheless effective in nutrient and carbon exchange (Allen & Allen, 1990). In shrubs and trees in extreme arid lands, the volume accessed by mycorrhizal fungi can extend for tens of metres horizontally, and greater than 2 to 5 m deep. Virginia *et al.* (1986) reported that mycorrhizas in *Prosopis* in the low deserts were found just above the water table, several metres deep in sandy washes. They postulated that these mycorrhizas were crucial for phosphorus uptake, because the soils above this level were too dry for nutrient uptake. In alluvial soils of oak-savanna riparian zones, mycorrhizas could be found 4 or 5 m deep to the water table (Querejeta & Allen, unpublished data).

In southern California hillsides, the soil is not deep enough to hold enough water to support evergreen trees such as live oaks and chamise through the entire dry season. As much as 86% of the available soil water in a profile is located in the granite matrix below the soil (Bornyasz *et al.*

2005). Roots of these trees and shrubs may go several metres into rock fractures to find water (Fig. 4.8). Mycorrhizal roots extend into fractures in the granite caused by earthquakes. Extending out from those tips, mycorrhizal hyphae penetrate into the granite matrix (Egerton-Warburton *et al.*, 2003; Bornyasz *et al.*, 2005). There are limited or no nutrient or carbon resources in this rock matrix, but there is considerable water. Evidence for rock penetration by mycorrhizal fungal hyphae in other systems is accumulating, including feldspar (Hoffland *et al.*, 2003) and limestone (Querejeta, Estrada, Snyder, Jimenez, and Allen, unpublished observations; see also Fomina *et al.*, (Chapter 10) this volume). Stable-isotope data demonstrated that during summer drought, oak trees are obtaining water from this granite matrix (Fig. 4.9). As this matrix is

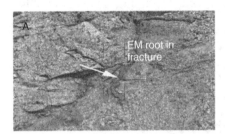

Fig. 4.8. Ectomycorrhizas in the granite bedrock. Shown in panel A is the EM root extending into a fracture in the rock, and in panel B is the EM hyphae extending into the granite matrix, following chipping with a rock hammer (photo by Margaret Bornyasz, details in Bornyasz *et al.*, 2005).

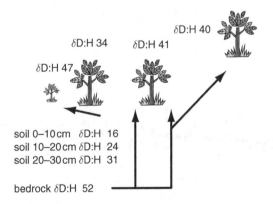

Fig. 4.9. Water from the soil layers and the granite matrix, and the values of mature and oak seedlings in May, late in the growing season. Isotopic signatures of $\delta D_2 O$ are expressed as $\delta D{:}H$ ratio in % (from Bornyasz, Querejeta & Allen, unpublished data).

unsaturated, flow through the matrix itself is very slow because of the complex pore structure of the material (Hubbert *et al.*, 2001). However, flow along or through the fungal hypha can be very rapid. Again, the Ψ of the granite would be relatively high compared with the surface soils during the drought period, and with the leaves. Since there is a relatively large amount of hyphae (2–5 m cm^{-3}, Bornyasz *et al.*, 2005), this source is likely very important in these ecosystems.

These responses are also of critical importance for the fungi themselves. While the minimum plant ψ_s may be only −2 or −3 MPa, the surface soil containing many of the fungi may drop to −5 or −10 MPa or even lower.

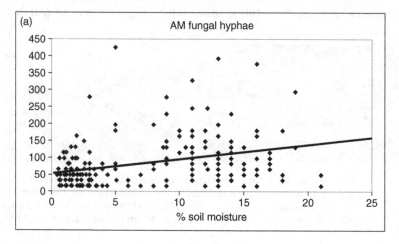

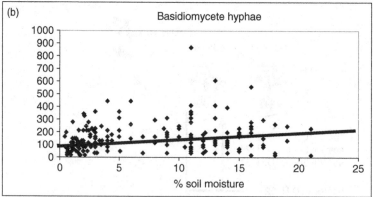

Fig. 4.10. Persistence of fungal hyphae of both AMF (panel a) and basidiomycetes (panel b). Many of the basidiomycetes in this system appear to be EM (Allen *et al.*, 1999). ψ_s reaches permanent wilting point ($\psi_s < -1.5$ MPa at 10%–12%).

Yet, individual mycorrhizal root tips persist through extreme dry seasons (Allen *et al.*, 2003) with some tips surviving up to two or three years including years of severe drought (Treseder *et al.*, 2004). *Pisolithus* sp. fruits under oaks in southern California near the end of extreme drought periods (soil moisture levels of 2%–5%, < −5 MPa), usually in October or November. The sporocarps themselves are saturated and obtaining water directly from hydraulic lift (likely), perched water pockets (possible) or from a very large soil volume (unlikely). In developing an absorbing network, individual mycelia spread out in a fan, pulling nutrients back into cords or rhizomorphs that connect to the mycorrhizal root tip (Donnelly *et al.*, 2004). This entire system, energetically, is not recreated following each drought event. In pinyon pinelands (where hydraulic lift has been measured), mycorrhizal rhizomorphs persist for several months, beginning following spring snowmelt, extending through the summer dry season, and then into the primary monsoonal fall growing season (Treseder *et al.*, 2005). During the dry period, these rhizomorphs survived even in soil moisture levels around 2%. In chaparral vegetation, there was little change in hyphal lengths for either AMF or basidiomycete fungi (Fig. 4.10). Arbuscular mycorrhizal fungal hyphae were found even in desert phreatophytes in Baja California where rainfall has not occurred for several years (Allen, unpublished observations). In both cases, hydraulic lift is a likely mechanism supporting the survival of the fungal hyphae.

Conclusions

Mycorrhizas are critical to plant survival and production in arid soils. Hyphae transport water in both directions. Water from patches of moist soils to the plant is provided by fungal hyphae and rhizomorphs that explore large volumes of soil ranging outward from the canopy edge. Mycorrhizal hyphae even extend from deep roots into the bedrock to access sources of water that cannot be reached or transported in the time scales necessary to be important to plant water balance. Mycorrhizal hyphae also benefit from hydraulically lifted water from plants that have deep root systems that reach groundwater or perched water tables.

Mycorrhizas are best understood in the context of nutrient uptake, and as a carbon sink. However, mycorrhizal formation is coupled to water content (e.g. Mexal & Reid, 1973; Reid & Bowen, 1979a) and nutrient uptake is dependent firstly upon adequate water for uptake and transport (Reid & Bowen, 1979b; Wells *et al.*, 2001). The mycelial network must be maintained through drought periods because reconstructing that network following each drought event would be energetically prohibitive. Once water is adequate, the dynamic interplay between nutrient and carbon dynamics can be studied using stoichiometric approaches (Treseder & Allen, 2002; Allen *et al.*, 2003). Because of the dynamic nature of hyphal water relations, any measurements reflect only a fraction of the conditions in which a hypha exists through its life. Thus, this analysis is largely based on only a few measurements, and must begin to depend upon extensive modelling of the dynamic aspects of water fluxes. In fact, ecosystems are rarely in equilibrium (Seastedt & Knapp, 1993). Mycorrhizas (Allen, 2001; Allen *et al.*, 2003) are highly dynamic, changing on a daily basis, and carbon allocation (Wang *et al.*, 1989) and water fluxes (Querejeta *et al.*, 2003) shift on an hourly basis. Clearly, there is a need to develop instrumentation and experimental approaches that can visualize these processes at the scale at which they occur, and visualize them in the field. Only then will we be able to completely understand the dynamic relationships between soil water, mycorrhizal fungi and plants.

Acknowledgements

I thank William Swenson, Jose' Ignacio Querejeta, Louise Egerton-Warburton and Margaret Bornyasz for their inputs to this programme. This research was funded by grants from the US National Science Foundation grants EF0410408, DEB 9981548, and DEB 0217774 from the Biocomplexity and LTER programs.

References

Alexopolis, C. J. Mims, C. W. & Blackwell, M. (1996). *Introductory Mycology*, 4th edn. New York: John Wiley & Sons, Inc.

Allen, M. F. (1982). Influence of vesicular-arbuscular mycorrhizas on water movement through *Bouteloua gracilis*. *New Phytologist*, **91**, 191–6.

Allen, M. F. (1983). Formation of vesicular-arbuscular mycorrhizas in *Atriplex gardneri* (Chenopodiaceae): seasonal response in a cold desert. *Mycologia*, **75**, 773–6.

Allen, M. F. (1991). *The Ecology of Mycorrhizas*. New York: Cambridge University Press.

Allen, M. F. (1996). The ecology of arbuscular mycorrhizas: a look back into the 20th century and a peek into the 21st. *Mycological Research*, **100**, 769–82.

Allen, M. F. (2001). Modelling arbuscular mycorrhizal infection: is % infection an appropriate variable? *Mycorrhiza*, **10**, 255–8.

Allen, E. B. & Allen, M. F. (1986). Water relations of xeric grasses in the field: Interactions of mycorrhizas and competition. *New Phytologist*, **104**, 559–71.

Allen, E. B. & Allen, M. F. (1990). Carbon source of VA mycorrhizal fungi associated with Chenopodiaceae from a semi-arid steppe. *Ecology*, **71**, 2019–21.

Allen, M. F., Egerton-Warburton, L. M., Allen, E. B. & Karen, O. (1999). Mycorrhizas in *Adenostoma fasciclatum* Hook. & Arn: a combination of unusual ecto- and endo-forms. *Mycorrhiza*, **8**, 225–8.

Allen, M. F., Swenson, W. Querejeta, J. L., Egerton-Warburton, L. M. & Treseder, K. K. (2003). Ecology of mycorrhizas: a conceptual framework for complex interactions among plants and fungi. *Annual Review of Phytopathology*, **41**, 271–303.

Auge', R. M. (2001). Water relations, drought and vesicular-arbuscular mycorrhizal symbiosis. *Mycorrhiza*, **11**, 3–42.

Bago, B., Azcon-Aguilar, C. & Piche, Y. (1998). Architecture and developmental dynamics of the external mycelium of the arbuscular mycorrhizal fungus *Glomus intraradices* grown under monoxenic conditions. *Mycologia*, **90**, 52–62.

Bartnicki-Garcia, S. (2002). Hyphal tip growth: outstanding questions. In *Molecular Biology of Fungal Development*, ed. H. D. Osiewacz. New York: Marcel Dekker, pp. 29–58.

Bornyasz, M. A., Graham, R. & Allen, M. F. (2005). Ectomycorrhizas in a soil-weathered granitic bedrock regolith: linking matrix resources to plants. *Geoderma*, **126**, 141–60.

Cairney, J. W. G. (1992). Translocation of solutes in ectomycorrhizal and saprotrophic rhizomorphs. *Mycological Research*, **96**, 135–41.

Cooke, R. C. & Whipps, J. M. (1993). *Ecophysiology of Fungi*. New York: Blackwell.

Cosgrove, D. J., Ortega, J. K. E. & Shropshire, W., Jr. (1987). Pressure probe study of the water relations of *Phycomyces blakesleeanus* sporangiophores. *Biophysical Journal*, **51**, 413–24.

Cowan, M. C., Lewis, B. G. & Thain, J. F. (1972). Uptake of potassium by the developing sporangiophore of *Phycomyces blakesleeanus*. *Transactions of the British Mycological Society*, **58**, 113–26.

Donnelly, D. D., Boddy, L. & Leake, J. R. (2004). Development, persistence and regeneration of foraging ectomycorrhizal mycelial systems in soil microcosms. *Mycorrhiza*, **14**, 37–45.

Duddridge, J. A., Malibari, A. & Read, D. J. (1980). Structure and function of mycorrhizal rhizomorphs with special reference to their role in water transport. *Nature (London)*, **287**, 834–6.

Eamus, D. & Jennings, D. H. (1986) Water, turgor and osmotic potentials of fungi. In *Water, Fungi and Plants*, ed. P. G. Ayres & L. Boddy. Cambridge: Cambridge University Press, pp. 27–48.

Egerton-Warburton, L. M., Graham, R. C. & Hubbert, K. R. (2003). Spatial variability in mycorrhizal hyphae and nutrient and water availability in a soil-weathered bedrock profile. *Plant and Soil*, **249**, 331–42.

Frank, A. B. (1885). Ueber die auf Wurzelsymbiose beruhende Erhaehrung gewisser Baume durch unterirdische Pilze. *Berichte der Deutsche Botanische Gesellschaft*, **3**, 128–45.

Friese, C. F. & Allen, M. F. (1991). The spread of VA mycorrhizal fungal hyphae in the soil: inoculum types and external hyphal architecture. *Mycologia*, **83**, 409–18.

Griffin, D. M. (1972). *The Ecology of Soil Fungi*. Syracuse, NY: Syracuse University Press.

Hanks, R. J. & Ashcroft, G. L. (1980). *Applied Soil Physics*. New York: Springer-Verlag.

Hardie, K. (1985). The effect of removal of extraradical hyphae on water uptake by vesicular-arbuscular mycorrhizal plants. *New Phytologist*, **101**, 677–84.

Hardie, K. & Layton, L. (1981). The influence of vesicular-arbuscular mycorrhiza on growth and water relations of red clover. I. In phosphate deficient soil. *New Phytologist*, **89**, 599–608.

Hatch, A. B. (1937). The physical basis of mycotrophy in *Pinus*. *The Black Rock Forest Bulletin*, **6**, 1–168.

Hoffland, E., Giesler, R., Jongmans, A. G. & van Breemen, N. (2003). Feldspar tunneling by fungi along natural productivity gradients. *Ecosystems*, **6**, 739–46.

Hubbert, K. R., Beyers, J. L. & Graham, R. C. (2001). Roles of weathered bedrock and soil in seasonal water relations of *Pinus jeffreyi* and *Arctostaphylos patula*. *Canadian Journal of Forest Research*, **31**, 1947–57.

Jennings, D. H. (1995). *The Physiology of Fungal Nutrition*. Cambridge: Cambridge University Press.

Jury, W. A., Letey, J. Jr, & Stolzy, L. H. (1981). Flow of water and energy under desert conditions. In *Water in Desert Ecosystems*, ed. D. D. Evans & J. L. Thames. New York: Dowden, Hutchinson & Ross, pp. 92–113.

Leake, J., Johnson, D., Donnelly, D. *et al.* (2004). Networks of power and influence: the role of mycorrhizal mycelium in controlling plant communities and agroecosystem functioning. *Canadian Journal of Botany*, **82**, 1016–45.

Mexal, J. & Reid, C. P. P. (1973). The growth of selected mycorrhizal fungi in response to induced water stress. *Canadian Journal of Botany*, **51**, 1579–88.

Nobel, P. S. (1974). *Biophysical Plant Physiology*. New York: Freeman Press.

Owusu-Bennoah, E. & Wild, A. (1979). Autoradiography of the depletion zone of phosphate around onion roots in the presence of vesicular arbuscular mycorrhiza. *New Phytologist*, **82**, 133–40.

Prosser, J. I. (1983). Hyphal growth patterns. In *Fungal Differentiation: A Contemporary Synthesis*, ed. J. E. Smith. New York: Marcel Dekker, Inc., pp. 357–418.

Querejeta, J. I., Egerton-Warburton, L. & Allen, M. F. (2003). Direct nocturnal water transfer from oaks to their mycorrhizal symbionts during severe soil drying. *Oecologia*, **134**, 55–64.

Reid, C. P. P. & Bowen, G. D. (1979a). Effects of soil moisture on V/A mycorrhiza formation and root development in Medicago. In *The Soil-Root Interface*, ed. J. L. Harley & R. S. Russell. London: Academic Press, pp. 211–19.

Reid, C. P. P. & Bowen, G. D. (1979b). Effect of water stress on phosphorus uptake by mycorrhizas of *Pinus radiata*. *New Phytologist*, **83**, 103–7.

Rhodes, L. H. & Gerdemann, J. W. (1975). Phosphate uptake zones of mycorrhizal and non-mycorrhizal onions. *New Phytologist*, **75**, 555–61.

Richards, J. J. & Caldwell, M. M. (1987). Hydraulic lift: substantial nocturnal water transport between soil layers by *Artemisia tridentata* roots. *Oecologia*, **73**, 486–9.

Safir, G. R., Boyer, J. S. & Gerdemann, J. W. (1971). Mycorrhizal enhancement of water transport in soybeans. *Science*, **172**, 581–3.

Safir, G. R., Boyer, J. S. & Gerdemann, J. W. (1972). Nutrient status and mycorrhizal enhancement of water transport in soybeans. *Plant Physiology*, **49**, 700–3.

Seastedt, T. R. & Knapp, A. K. (1993). Consequences of nonequilibrium resource availability across multiple time scales: the transient maxima hypothesis. *American Naturalist*, **141**, 621–33.

Smith, S. E. & Read, D. J. (1997). *Mycorrhizal Symbiosis*, 2nd edn. New York: Academic Press.

Stahl, E. (1900). Der sinn der mycorrhizenbildung. *Jahrbucher fuer wissenschaftliche Botanik*, **34**, 539–668.

Tinker, P. B. & Nye, P. H. (2000). *Solute Movement in the Rhizosphere*. New York: Oxford University Press.

Treseder, K. K. & Allen, M. F. (2002). Evidence for direct N and P limitation of arbuscular mycorrhizal fungi. *New Phytologist*, **155**, 507–15.

Treseder, K. K., Masiello, C. A., Lansing, J. L. & Allen, M. F. (2004). Species-specific measurements of ectomycorrhizal turnover under N-fertilization: combining isotopic and genetic approaches. *Oecologia*, **138**, 419–25.

Treseder, K. K., Allen, M. F., Ruess, R. W., Pregitzer, K. S. & Hendrick, R. L. (2005). Lifespans of fungal rhizomorphs under nitrogen fertilization in a pinyon-juniper woodland. *Plant and Soil*, **270**, 249–55.

Unestam, T. (1991). Water repellency, mat formation, and leaf-stimulated growth of some ectomycorrhizal fungi. *Mycorrhiza*, **1**, 13–20.

Virginia, R. A., Jenkins, M. B. & Jarrell, W. M. (1986). Depth of root symbionts occurrence in soil. *Biology and Fertility of Soils*, **2**, 127–30.

Wang, G. M., Coleman, D. C., Freckman, D. W. *et al.* (1989). Carbon partitioning patterns of mycorrhizal versus non-mycorrhizal plants: real-time dynamic measurements using $^{11}CO_2$. *New Phytologist*, **112**, 489–93.

Wells, J. M., Thomas, J. & Boddy, L. (2001). Soil water potential shifts: developmental responses and dependence on phosphorus translocation by the saprotrophic, cord-forming basidiomycete *Phanerochaete velutina*. *Mycological Research*, **105**, 859–67.

5

Integrating ectomycorrhizal fungi into quantitative frameworks of forest carbon and nitrogen cycling

ERIK A. HOBBIE AND HÅKAN WALLANDER

Introduction

Ecosystem ecologists have calculated carbon and nitrogen budgets for a variety of forest ecosystems. Despite a growing awareness of the importance of mycorrhizal fungi in nitrogen uptake, as carbon sinks for photosynthate and as conduits for carbon from plants to the below-ground community, few ecosystem ecologists have incorporated mycorrhizal fungi in their conceptual models of how forests function. Longstanding difficulties in assessing the presence and quantity of mycorrhizal fungi in soil, in identifying mycorrhizal fungi to species, and in assessing the mycorrhizal role in carbon and nitrogen cycling, have probably limited the willingness and ability of ecosystem ecologists to incorporate mycorrhizal fungi into their research. In particular, ecosystem models have not yet included mycorrhizal fungi, despite the key role of mycorrhizal fungi at the interface of plants, the soil and microbial communities below-ground.

In this review we will focus on ectomycorrhizal fungi that form symbioses with many of the dominant trees of temperate and boreal forests, particularly in trees of the Pinaceae, Fagaceae, Betulaceae and Salicaceae. Ectomycorrhizal fungi also form symbioses with many tropical trees, including the Dipterocarpaceae of southeast Asia and *Eucalyptus* of Australia. We will lay out the current state of knowledge of the functioning of ectomycorrhizal fungi in carbon and nitrogen cycling of forest ecosystems as inferred from field and laboratory studies. Finally, we will discuss progress in integrating mycorrhizal fungi into quantitative frameworks of forest ecosystem function. Additional viewpoints on placing mycorrhizal fungi in an ecosystem context are found in recent reviews by Lindahl *et al.* (2002), Simard *et al.* (2002), and Read & Perez-Moreno (2003).

Fungi in Biogeochemical Cycles, ed. G. M. Gadd. Published by Cambridge University Press. © British Mycological Society 2006.

In addition, books by Waring and Running (1998) on forest ecosystems, Coleman *et al.* (2004) on soil ecology and Smith and Read (1997) on mycorrhizal symbioses address many of the issues surrounding the interactions among mycorrhizal fungi, plants and the soil microbial community.

Role of mycorrhizal fungi in nitrogen acquisition

The role of ectomycorrhizal fungi in plant nutrition has been discussed for well over a century. Because most forests dominated by ectomycorrhizal trees are nitrogen-limited, much work has focused on the ability of ectomycorrhizal fungi to acquire nitrogen in different forms and transfer a portion to the host plant. Such studies have used various techniques, including ^{15}N labelling, tests for gene expression and enzyme assays. Ectomycorrhizal fungi form a protective sheath over most fine roots in nitrogen-limited forests, ensuring that most nitrogen must pass through ectomycorrhizal fungal tissue prior to plant acquisition. This joint structure, termed the ectomycorrhiza, is augmented by fungal hyphae extending beyond the root tissue (extraradical hyphae) into the surrounding soil. The low diffusivity of ammonium and amino acids in the soil solution and the much greater surface area for absorption of extraradical hyphae than of mycorrhizas suggest that nitrogen may be primarily taken up by extraradical hyphae rather than by mycorrhizas (Wallenda *et al.*, 2000). Although a few studies have also shown that ectomycorrhizal fungi can provide host plants with access to unusual forms of nitrogen, including nematodes (Perez-Moreno & Read, 2001), powdered animal hide, chitin (Dighton *et al.*, 1987) and ammonium in rocks (Paris *et al.*, 1995; Landeweert *et al.*, 2001), most studies have focused on the use of ammonium, nitrate or amino acids by ectomycorrhizal fungi.

Inorganic nitrogen use

Ectomycorrhizal fungi can take up nitrate (Ho & Trappe, 1980), but appear to prefer ammonium as a nitrogen source. For example, colonization by *Hebeloma crustuliniforme* of *Picea sitchensis* and *Tsuga heterophylla* did not increase nitrate uptake relative to uncolonized plants (Rygiewicz *et al.*, 1984), and incorporation of ^{15}N-labelled ammonium in extraradical mycelia and mycorrhizal root tips was about twice that of incorporation of ^{15}N-labelled nitrate after applying labelled nitrogen to the extraradical hyphae of *Paxillus involutus* in symbiosis with *Fagus sylvatica, Betula pendula* and *Picea abies* (Finlay *et al.*, 1989; Ek *et al.*, 1994). These authors suggested that *P. involutus* assimilated ammonium into glutamine prior to any plant transfer whereas nitrate was passed along as nitrate to the host plant. Studies of respiration rates of *Pisolithus tinctorius–Picea abies* associations supplied with ammonium or nitrate support this suggestion, as respiration rates per unit fungal biomass were 11 times higher than plant roots when supplied with ammonium, versus only 3.4 times higher when supplied with nitrate (Eltrop & Marschner, 1996). However, respiration rates did not differ in *P. involutus* supplied with ammonium or nitrate (Ek, 1997). Sangtiean and Schmidt (2002) suggested that nitrate use by ectomycorrhizal fungi should be more prevalent in ecosystems such as rainforests with high levels of soil nitrate, and presented data that one of the two rainforest taxa they sampled (*Gymnoboletus*) grew as well on nitrate as on other nitrogen sources. Given the generally low levels of free nitrate in most ectomycorrhizally dominated forests, a preference of ectomycorrhizal fungi for ammonium is hardly unexpected.

In contrast to the great number of laboratory studies on ammonium versus nitrate preferences, few studies have examined such preferences in the field. In a study in which ^{15}N was applied as either ammonium or nitrate in Norway spruce stands, mycorrhizal fungi were the only pool sampled in which ^{15}N retention was higher from ammonium than from nitrate (Buchmann et al., 1996). This may possibly indicate that fungi preferentially transfer nitrate over ammonium to plants, or could also indicate that roots preferentially assimilated nitrate whereas mycorrhizal fungi preferentially assimilated ammonium. However, laboratory studies indicating that conifers preferentially assimilate ammonium over nitrate (Ingestad, 1979) would argue against the latter interpretation.

Organic nitrogen use

The prevailing paradigm in much research on forest ecosystems is that trees only take up nitrogen after release of inorganic nitrogen by microbes. Attacking this paradigm has become quite popular (Read, 1991; Schimel & Bennett, 2004), in part because it offers such an easy target! For example, rates of ammonium and nitrate production estimated from enclosed cores probably greatly overestimate the true production of inorganic nitrogen in forest soils because such cores restrict labile carbon inputs that serve to maintain nitrogen in microbial biomass (Lindahl et al., 2002). Several studies have shown that inorganic nitrogen fluxes are insufficient to account for nitrogen losses from litter or account for plant nitrogen demands (Kielland, 1994; Bending & Read, 1995; Lipson & Näsholm, 2001). Chapin (1980) pointed out that fluxes of organic nitrogen should always be greater than fluxes of inorganic nitrogen because decomposition of complex substances first releases small organic nitrogen molecules such as amino acids or oligopeptides. These organic nitrogen molecules are readily taken up by soil microbes, and net release of inorganic nitrogen only occurs once the C/N ratio of soil organic matter drops below a critical level. Thus, organic nitrogen can cycle several times through soil organic matter and microbes before inorganic nitrogen is produced.

Researchers are increasingly examining how mycorrhizal fungi may provide plants with access to organic forms of nitrogen, thereby short-circuiting the conventional cycle of inorganic nitrogen release and uptake (Read, 1991). In the laboratory, numerous studies have examined the ability of mycorrhizal fungi alone or in symbiotic culture to use organic nitrogen sources such as amino acids or proteins (Abuzinadah & Read, 1986a; Baar *et al.*, 1997). In the field, natural abundance measurements of the ^{15}N content in ectomycorrhizal fungi also appear potentially useful in indicating organic nitrogen use, as high ^{15}N content in ectomycorrhizal fruiting bodies correlates with proteolytic capabilities (Lilleskov *et al.*, 2002b). Several short-term field studies using [15]N- and ^{13}C-labelled amino acids have demonstrated that ectomycorrhizal plants can take up amino acids (Kielland, 1994; Näsholm *et al.*, 1998; McKane *et al.*, 2002), presumably with much of this uptake facilitated by mycorrhizal fungi. However, the direct application of labelled amino acids may create artificially high amino acid concentrations that allow direct uptake by roots, as suggested by similar levels of amino acid incorporation into arbuscular mycorrhizal, ectomycorrhizal and ericoid mycorrhizal plants in some studies (Persson *et al.*, 2003). Continued emphasis on field studies with labelled material will advance our understanding, particularly if natural substances such as leaf or root litter can be simultaneously labelled with ^{13}C and ^{15}N and the isotopic label tracked into different ecosystem materials over several growing seasons.

A nice example of the potential of such isotopic labelling to provide new insights into mycorrhizal function under natural conditions was provided by Zeller *et al.* (2000, 2001), in which ^{15}N-labelled beech litter was applied to a beech forest and the ^{15}N label followed into different ecosystem pools for 3 years. Mycorrhizas extracted from the top 2 cm of soil appeared to be in isotopic equilibrium with the soil, suggesting that mycorrhizas took up litter-derived nitrogen in a similar form to which it moved from litter to the topsoil, presumably as organic nitrogen. In contrast, leaves accumulated ^{15}N about one-third as quickly as mycorrhizas in the top 2 cm of soil, indicating that much plant nitrogen is derived from below the top 2 cm of soil. These results suggest that any contribution of nitrogen to mycorrhizas from extraradical hyphae must be primarily restricted to hyphae living in the same soil layer as the mycorrhiza. Similar conclusions on the nitrogen sources for fine roots and ectomycorrhizal fungi being restricted to the horizon in which they occur have been reached from natural abundance ^{15}N patterns (Högberg et al., 1996; Wallander et al., 2004).

Although most studies of organic nitrogen use in ectomycorrhizal fungi have focused on amino acids, there is some evidence that small peptides are also good nitrogen sources for some fungi. From the limited studies available, it appears that many fungi (including ectomycorrhizal fungi) can readily take up short-chain oligopeptides up to five amino acids in length (Abuzinadah & Read, 1986b; Jennings, 1995; Krznaric, 2004; Fig. 5.1). Recently, specific transporters for oligopeptide uptake have been identified in the ectomycorrhizal fungus *Hebeloma cylindrosporum* (Marmeisse

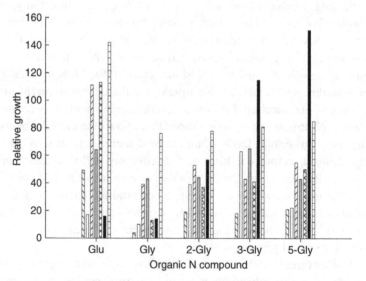

Fig. 5.1. Growth of seven ectomycorrhizal taxa on ammonium, glutamate, glycine, diglycine, triglycine and pentaglycine, with growth on ammonium set to 100. Fungi were supplied with carbon (glucose plus amino acid) at a concentration of 170 mM and nitrogen at a concentration of 4.4 mM and harvested after 12–28 days. Glu = glutamate, Gly = glycine, 2-Gly = diglycine, 3-Gly = triglycine and 5-Gly = pentaglycine. Symbols for different taxa are: *Paxillus involutus*, reverse slash; *Thelephora terrestris*, white fill; *Suillus luteus* DK, forward slash; *Suillus luteus* LK, gray fill; *Suillus bovinus*, crosshatched; *Pisolithus tinctorius*, black fill; *Rhizopogon luteolus*, horizontal bars. Data from Krznaric (2004).

et al., 2004). Such uptake capabilities would be particularly important to ectomycorrhizal fungi possessing proteolytic enzymes. These uptake capabilities would allow them to compete with saprotrophic fungi for the initial products of proteolysis without requiring additional peptide hydrolysis.

Interactions of ectomycorrhizal and saprotrophic fungi influencing nitrogen cycling

Interactions among mycorrhizal fungi and other taxonomic groups in soils may influence nitrogen cycling and availability. Many saprotrophic fungi decompose plant litter more effectively than mycorrhizal fungi (Colpaert & van Laere, 1996; Lindahl *et al.*, 2002), suggesting that interactions between saprotrophic and mycorrhizal fungi influence nitrogen supply to plants. Lindahl *et al.* (2002) argued persuasively that nitrogen transfer from plant litter is primarily mediated by saprotrophic fungi because of their extensive enzymatic capabilities, and that little ammonium will be released during primary decomposition by macrofungi because decomposition is usually a nitrogen-requiring process, as shown by the immobilization of nitrogen during plant litter decomposition. Because concentrations of inorganic nitrogen will be very low in nitrogen-limited forests, uptake of amino acids and oligopeptides may dominate, with proteolytic activities tightly coupled to uptake of organic nitrogen forms. Grazing on fungi and bacteria by protozoa, nematodes and other soil taxa are important modes of release of nitrogen from these primary decomposers to other functional groups. Such grazing may therefore facilitate the transfer of nitrogen from primary decomposers to mycorrhizal fungi (Coleman *et al.*, 2004).

Early studies of interactions between saprotrophs and ectomycorrhizal fungi indicated that ectomycorrhizal fungi retarded decomposition rates. This was explained by nutrient uptake by the mycorrhizal fungi resulting in a nutrient limitation of decomposition (Gadgil & Gadgil, 1975). Similar studies in both pine roots (Parmelee *et al.*, 1993) and mycorrhizas (Koide & Wu, 2003) found that decomposition varied with water content, suggesting that the extraction of water from the soil by ectomycorrhizal fungi reduced decomposition rates by co-occurring saprotrophic microbes.

Role of mycorrhizal fungi in carbon cycling

Field studies

Ectomycorrhizal fungi depend on their hosts for carbon and energy supply. This access to host carbon makes them unique within the soil microbial community since availability of carbon and energy usually limits growth for soil microorganisms (Aldén *et al.*, 2001). In most boreal and temperate forests dominated by trees associated with ectomycorrhizal fungi, nitrogen limits tree growth (Tamm, 1991). Carbon supplied by host trees to ectomycorrhizal fungi presumably allows these fungi to supply their host trees with soil-derived nitrogen that would otherwise be unavailable for plant uptake (Read, 1991).

Although large amounts of carbon are assumed to be allocated below-ground to roots and associated ectomycorrhizal fungi, only a few studies have actually quantified such allocation in the field (Table 5.1). All studies to date have been on common species used for timber production in the *Pinaceae*, with four of the six studies listed in Table 5.1 on *Pinus* species. None of the estimates in Table 5.1 are for ectomycorrhizal forests dominated by deciduous trees. Clearly, future work should focus on obtaining a greater variety of field-based estimates.

In field studies, allocation to roots as a percentage of net primary production (NPP) ranged from 10% to 65% and varied inversely with nitrogen availability (Beets & Whitehead, 1996; Waring & Running, 1998). In a nitrogen-limited Scots pine forest studied by the Swedish Coniferous Forest Project, below-ground carbon allocation was calculated at 63% of NPP (Ågren *et al.*, 1980). A follow-up study estimated that root production

Table 5.1. *Field-based estimates of carbon allocation to ectomycorrhizal fungi as a percentage of net primary production*

Forest type	Percentage of NPP	Notes	Reference
Abies amabilis	15%[1]		Vogt *et al.* (1982)
Pseudotsuga menziesii	28%[1]		Fogel & Hunt (1983)
Pinus sylvestris	14%–25%	830 kg ha^{-1} yr^{-1}	Finlay & Söderström (1992)
P. sylvestris	5%–25%	includes exudates	Janssens *et al.* (2002)
Pinus radiata (control)	18%[2]	allocation/GPP	Ryan *et al.* (1996)
P. radiata (irrigated)	13%[2]	allocation/GPP	Ryan *et al.* (1996)

[1] No allocation to extraradical hyphae.
[2] 'Unknown' allocation in Ryan *et al.* (1996) invoked to close mass balance was interpreted by Waring and Running (1998) as allocation to mycorrhizal fungi and exudates. NPP is net primary production, GPP is gross primary production.

accounted for only 25% of this below-ground allocation, with the remaining 75% presumably allocated to respiration and allocation to ectomycorrhizal fungi (Högberg et al., 2002). In general, proportionally less carbon is allocated below-ground to roots and their associated fungal symbionts at higher growth rates (Ingestad et al., 1986). For example, relative carbon allocation to roots in a 10-year-old Scots pine stand declined from 46% to 30% after irrigation and fertilization increased production from $5.7\,Mg\,ha^{-1}\,yr^{-1}$ to $12.4\,Mg\,ha^{-1}\,yr^{-1}$ (Axelsson & Axelsson, 1986). None of these field studies directly quantified allocation of carbon to mycorrhizal fungi.

Early attempts to quantify mycorrhizal biomass in soil include work by Vogt et al. (1982) and Fogel and Hunt (1979). Vogt et al. (1982) estimated that mycorrhizal fruit bodies, ectomycorrhizal sheaths around fine roots and ectomycorrhizal sclerotia made up $4150\,kg\,ha^{-1}$ in a mature *Abies amabilis* stand in Washington, USA. Although this contributed only 1% of total biomass in the stand the mycorrhizal contribution to NPP was 15%. Using similar techniques, Fogel and Hunt (1979) estimated a mycorrhizal biomass of $12\,860\,kg$ (4.3% of total biomass) in a Douglas fir forest in Oregon, or about twice the biomass of soil hyphae in this forest. This estimate included dead mycorrhizas. In both of these studies extraradical mycorrhizal mycelia were not quantified since methods were not available at that time to separate mycorrhizal from saprotrophic mycelia. A follow-up paper by Fogel and Hunt (1983) estimated the allocation to ectomycorrhizal fungi at 28% of NPP.

More recent studies have confirmed that large amounts of carbon are allocated to roots and their associated ectomycorrhizal fungi. A large-scale girdling experiment in northern Sweden stopped carbon flux below-ground and dramatically reduced total soil respiration by over 50% within 2 weeks (Högberg et al., 2001; Bhupinderpal-Singh et al., 2003), indicating that recent allocation to roots is a key driver of soil respiration. The amount of microbial biomass in soil also declined by one-third after tree girdling, indicating that ectomycorrhizal fungi constitute a large part of the soil microbial biomass (Högberg & Högberg, 2002). Earlier studies have also shown that respiration and fruiting by ectomycorrhizal fungi were highly dependent on supplies of recent photosynthate (Last et al., 1979; Söderström & Read, 1987; Lamhamedi et al., 1994). A recent modification of the girdling technique restricts below-ground carbon flux by chilling tree stems using flexible piping within which antifreeze is circulated (R. Waring, personal communication). Although the procedure does not completely eliminate below-ground flow, the reversible nature of chilling in contrast to girdling promises new insights into how environmental conditions alter below-ground fluxes, including allocation to mycorrhizal fungi.

Bååth et al. (2004) recently developed a new approach to estimate biomass of ectomycorrhizal fungi in soil, in which the extent of degradation of fungal biomass in soil samples during laboratory incubations is used to estimate ectomycorrhizal biomass. The method assumes that ectomycorrhizal mycelia in soil samples degrade when deprived of their energy source (tree-derived carbohydrates) while the biomass of saprotrophic fungi is maintained during the incubation. One caveat is that free-living fungi dependent on recent photosynthate will also be classified as mycorrhizal by this method. Using this approach, Wallander et al. (2004) estimated ectomycorrhizal biomass in spruce forests and mixed spruce–oak forests (down to 70 cm soil depth) in southern Sweden to be around 5000 kg ha^{-1}. These high values suggest that much of the ectomycorrhizal biomass must persist in the soil for several years, as has also been suggested from ^{14}C measurements on ectomycorrhizas in pines (Treseder et al., 2004).

Dramatic shifts in forest productivity over short gradients of nutrient availability allow the effects of nutrient availability in the field on plant–mycorrhizal interactions to be studied without experimental manipulation and without the confounding effects of climatic differences. Such gradients have been studied in northern Sweden where vegetation changes over a distance of 100 m from low productivity, nitrogen-limited pine forests, to high productivity, phosphorus-limited spruce forests (Giesler et al., 1998). Tree productivity along one such gradient increased from 2.8 m^3 ha^{-1} yr^{-1} to 6.0 m^3 ha^{-1} yr^{-1} (1.1–2.4 Mg ha^{-1} yr^{-1}, assuming a specific gravity for wood of 0.4). The soil fungal biomass along this gradient declined considerably, presumably due to reduced ectomycorrhizal biomass (Högberg et al., 2003). Nilsson et al. (2005) used the soil incubation technique (Bååth et al., 2004) to quantify ectomycorrhizal biomass in this and three additional gradients. With increasing forest productivity, total fungal biomass in soil samples declined and the ratio of mycorrhizal fungi (including both ectomycorrhizas and ericoid mycorrhizas) to total soil fungal biomass declined from 54% to 34%.

Nilsson and Wallander (2003) incubated soil cores in the field to estimate the influence of nitrogen fertilization on ectomycorrhizal biomass in a Norway spruce forest in southern Sweden. Ectomycorrhizal biomass in the humus layer declined from 800 kg ha^{-1} to 300 kg ha^{-1} after adding 100 kg N ha^{-1} yr^{-1} for 10 years. These estimates of ectomycorrhizal biomass in soils of Swedish forests are much larger than the amounts present in similar forests as ectomycorrhizal mantles (150 kg ha^{-1}, Kårén & Nylund, 1997) or fruit bodies (1.5–2.5 kg ha^{-1}, Wiklund *et al.*, 1995), but lower than or similar to estimates of mycorrhizal sclerotia and sheaths from forests in the western USA discussed above (Fogel & Hunt, 1979; Vogt *et al.*, 1982).

To estimate production of new ectomycorrhizal mycelia in soil, Wallander *et al.* (2001) developed another technique, in which sand-filled ingrowth mesh bags are buried in the soil to allow ingrowth of fungal hyphae but not of roots. Ectomycorrhizal mycelia can be partitioned from saprotrophic mycelia by using trenched plots and by using the different carbon isotopic composition of ectomycorrhizal and saprotrophic fungi (Hobbie *et al.*, 1999; Högberg *et al.*, 1999; Kohzu *et al.*, 1999). Studies using this technique indicate that ectomycorrhizal fungi produce between 100 and 600 kg mycelia ha^{-1} yr^{-1} in the soil (Wallander *et al.*, 2001, 2004; Hagerberg & Wallander, 2002; Hagerberg *et al.*, 2003; Nilsson & Wallander, 2003; Nilsson *et al.*, 2005). These estimates are generally low compared to biomass estimates using the soil incubation technique described above (Bååth *et al.*, 2004). However, growth in sand-filled mesh bags is unlikely to represent growth in forest soil. The ingrowth mesh-bag method is thus more suitable for comparing the effects of different treatments on the relative growth of ectomycorrhizal fungi than for estimating the absolute growth of ectomycorrhizal mycelia. For example, in agreement with results from laboratory incubations, fertilization with 1000 kg nitrogen in the field over a 10-year period reduced ectomycorrhizal growth by about 50% in a spruce forest (Nilsson & Wallander, 2003). Similarly, adding apatite as a phosphorus source stimulated ectomycorrhizal growth in mesh bags incubated in a phosphorus-poor forest but not in a forest with adequate phosphorus (Hagerberg *et al.*, 2003). Relative ectomycorrhizal growth was also reduced along the short productivity gradients discussed above (Nilsson *et al.*, 2005) and in oak forests along a nitrogen deposition gradient in southern Sweden (Nilsson, 2004).

Culture studies

The partitioning of carbon between host plants and associated fungal symbionts is much easier to study in the laboratory than in the field (Table 5.2), and allocation to mycorrhizal fungi in culture studies is therefore known with greater accuracy than in field studies. Carbon allocation to ectomycorrhizal fungi has generally been measured in three different pools: fungal carbon in fine roots or mycorrhizas, extraradical mycelia and respiration. An additional important parameter is carbon use efficiency, or the ratio of growth to supplied carbon.

Table 5.2. *Estimates of carbon or biomass allocation to ectomycorrhizal fungi (EMF) as a proportion of total net primary production in laboratory studies. The proportion of total NPP directed below-ground is also given. NM = non-mycorrhizal*

System	%NPP to EMF	%Below-ground allocation	Reference
Pinus sylvestris (NM)	–	50%[1], 61%[2]	Hobbie & Colpaert (2003)
Thelephora terrestris	15%[1], 17%[2]	56%[1], 65%[2]	Hobbie & Colpaert (2003)
Suillus luteus	8%[1], 18%[2]	54%[1], 68%[2]	Hobbie & Colpaert (2003)
NM	–	44%[1] 49%[2]	Colpaert et al. (1996)
Suillus bovinus	9%[1], 8%[2]	46%[1], 49%[2]	Colpaert et al. (1996)
T. terrestris	16%[1], 9%[2]	50%[1], 50%[2]	Colpaert et al. (1996)
Scleroderma citrinum	13%[2]	51%[2]	Colpaert et al. (1996)
T. terrestris	5%	64%	Colpaert et al. (1992)
Laccaria laccata	6%	61%	Colpaert et al. (1992)
Scleroderma citrinum	18%	65%	Colpaert et al. (1992)
Paxillus involutus	12%	58%	Colpaert et al. (1992)
S. luteus (2 strains)	18%	59%	Colpaert et al. (1992)
S. bovinus (3 strains)	16%	60%	Colpaert et al. (1992)
Hebeloma crustuliniforme	4 ± 0.8%	41%	Wallander & Nylund (1992)
Laccaria bicolor	9 ± 0.6%	47%	Wallander & Nylund (1992)
S. bovinus	11 ± 1.7%	57%	Wallander & Nylund (1992)
Pinus muricata			
Rhizopogon (3 strains)	3.7%[3]	44.3%	Bidartondo et al. (2001)
P. involutus	9.9%[3]	53.5%	Bidartondo et al. (2001)
Suillus pungens	1.9%[3]	43.1%	Bidartondo et al. (2001)
Pinus taeda–Pisolithus tinctorius	16 ± 5%[4]	40%[4]	Reid et al. (1983)
Pinus ponderosa (NM)	–	33.4%	Rygiewicz & Andersen (1994)
Hebeloma crustuliniforme	7%	41.2%	Rygiewicz & Andersen (1994)
Pinus contorta (pH 3.8)	10%[3]	59%	Erland et al. (1991)
P. contorta (pH 5.2)	10%[3]	53%	Erland et al. (1991)

Picea abies-Pisolithus tinctorius			
ammonium supply	1.7%	42%	Eltrop & Marschner (1996)
nitrate supply	1.0%	39%	Eltrop & Marschner (1996)
Betula pendula			
P. involutus	20%, 23%, 29%[5]	no data	Ek (1997)
Betula papyrifera (NM)	–	34%	Jones & Hutchinson (1988)
Lactarius rufus	3.5%[6]	35%	Jones & Hutchinson (1988)
Scleroderma flavidum	10.6%[6]	50%	Jones & Hutchinson (1988)
Salix viminalis (NM)	–	40.6%[7]	Durall et al. (1994)
T. terrestris	7.3%	47.9%[7]	Durall et al. (1994)
Salix viminalis (NM)	–	38.5%	Tinker et al. (1990)
Laccaria proxima	7%[8]	45.2%	Tinker et al. (1990)
T. terrestris	12%[8]	50.9%	Tinker et al. (1990)

[1] High nitrogen supply.
[2] Low nitrogen supply.
[3] Extraradical allocation only.
[4] Based on 53 hour ^{14}C pulse-chase study.
[5] Includes fungal respiration.
[6] Mycorrhizas only.
[7] Mean of harvests at 50, 60, 85 and 98 days.
[8] Based on increase in below-ground allocation after colonization.

The proportion of fungal biomass or carbon in fine roots is an important parameter for estimating the carbon allocation to mycorrhizal fungi, and will vary with the intensity of mycorrhizal development of the root system, the proportion of mycorrhizas consisting of fungal biomass and the operational definition of fine roots used. Although some estimates of ecosystem-scale allocation to mycorrhizal fungi have assumed the proportion of fungal matter in fine roots to be as high as 40% (Finlay & Söderström, 1992), researchers should be careful to avoid equating ecosystem measurements of fine-root biomass with biomass of mycorrhizas. For example, fine roots are often operationally defined as roots 2 mm or less in diameter, whereas most ectomycorrhizas are 1.0 mm in diameter or less.

In cultures of *Picea sitchensis–Lactarius rufus*, Alexander (1981) suggested that 20% of root-system weight might be from fungal matter. Mycorrhizas in this system were 250 µm in radius, with a sheath thickness of 15–20 µm, so treating the mycorrhizas as perfect cylinders means that 12%–23% of mycorrhiza volume consists of fungal material. Harley and Smith (1983) indicated that mycorrhizas of *Fagus* and *Nothofagus* had sheath thicknesses of 39 µm and 30 µm, respectively, whereas sheath tissue was 32% of total weight of *Fagus* mycorrhizas, according to Lewis and Harley (1965). In *Pinus sylvestris* cultures, *Thelephora terrestris* mycorrhizas were 22% and 23% fungal matter at high and low nitrogen availabilities, respectively, whereas *Suillus luteus* mycorrhizas were 14% and 20% fungal matter (Hobbie & Colpaert, 2003). In *Picea abies* colonized by *Pisolithus tinctorius*, the proportion of fungal matter in roots was only 1.55% for ammonium-supplied plants (71% colonization) and 0.96% for nitrate-supplied plants (55% colonization) (Eltrop & Marschner, 1996). In contrast, nitrate-supplied *Pinus pinaster* roots colonized by *H. cylindrosporum* consisted of 20% fungal matter (Plassard *et al.*, 1994).

Ectomycorrhizal fungi also use host carbon for respiration and for assimilating inorganic nitrogen into amino acids, of which a portion are subsequently translocated to the host. How the carbon is partitioned between these different sinks will vary with species and nutrient conditions. For example, the proportion of carbon used to translocate amino acids to the host increases at elevated nitrogen levels, resulting in less carbon available for producing external mycelia in the soil (Wallander, 1995). The proportion of carbon used in respiration can also vary considerably between species. A strain of *Rhizopogon* respired 7.1% of total assimilated carbon in microcosm systems while the same figure for *Suillus pungens* was 4.9% (Bidartondo *et al.*, 2001). In the same experiment *S. pungens* used 3.5% of assimilated carbon for producing biomass whereas *Rhizopogon* used 3.2%. Addition of nitrogen increased respiration of *Rhizopogon* and *S. pungens* by 205% and 328%, respectively.

In culture studies, allocation to fungal biomass as a percentage of NPP has ranged from 1% to 20%, depending on fungal–plant association and treatment (Table 5.2). In one of the most careful studies, Rygiewicz and Andersen (1994) constructed a complete carbon budget for a mycorrhizal system consisting of *Pinus ponderosa* seedlings colonized by *Hebeloma crustuliniforme* and found that 7% of the assimilated carbon was allocated to the fungal symbiont. Colpaert *et al.* (1992) grew 6 different ectomycorrhizal species (9 isolates) with pine seedlings and found that allocation to different species varied from 5% to 18% (Table 5.2). If the allocation to ectomycorrhizal fungi from Table 5.2 is plotted versus the below-ground allocation (Fig. 5.2), the resulting regression equation has a slope of 0.43, meaning that 43% of increases in below-ground allocation are allocated to the fungal symbiont. If this relationship between below-ground allocation and allocation to ectomycorrhizal fungi can be validated in field studies, then ecosystem ecologists will be able to scale from measurements of total below-ground allocation to allocation to ectomycorrhizal fungi.

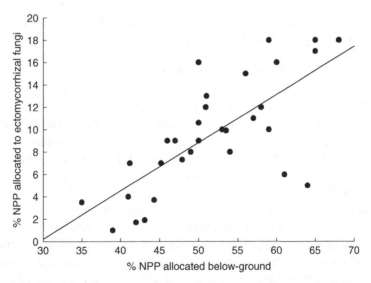

Fig. 5.2. Allocation to ectomycorrhizal fungi correlates with below-ground allocation in culture studies. Values are given as a percentage of net primary production (NPP), and are listed in Table 5.2. A linear regression gives an equation of $\%NPP_{fungi} = 0.43 \times \%NPP_{below-ground} - 12.7\%$, $r^2 = 0.52$, $p < 0.001$. One study (Reid *et al.*, 1983) was excluded since the methodology (pulse-chase) was very different than in any other study.

Ectomycorrhizal colonization can affect the growth of host plants in various ways. In many studies in which the rate of nutrient supply is balanced with plant growth rates (also termed exponential growth), ecto-mycorrhizal colonization cannot improve the nutrient status of host plants. Under these conditions, ectomycorrhizal colonization reduced the growth of the host seedlings (Ingestad *et al.*, 1986; Hobbie & Colpaert, 2003) and fungal development and host growth were negatively correlated (Colpaert *et al.*, 1992), reflecting the large carbon drain by the fungus. This large carbon sink by ectomycorrhizal fungi has also stimulated host photosynthesis in many studies independent of any effects of mycorrhizal colonization on host nutrition (Nylund & Wallander, 1989; Dosskey *et al.*, 1990; Colpaert *et al.*, 1996). In several of the ectomycorrhizal species studied by Colpaert *et al.* (1992) (especially *Scleroderma citrinum*), nitrogen retention in fungal tissue severely limited nitrogen supply and reduced growth in the host pine seedlings. In a later study, Colpaert *et al.* (1996) estimated that *S. citrinum* retained 32% of added nitrogen, *Suillus bovinus* 19% and *T. terrestris* 18%. Under conditions of restricted nitrogen supply such as used in these studies, increased nitrogen retention in fungal biomass coupled to the lower C/N ratio of fungal biomass compared to plant biomass must decrease overall biomass accumulation of the mycorrhizal symbiosis.

Translating measurements of carbon accumulation of fungal biomass into the true allocation cost to plant hosts requires knowledge of the efficiency of conversion of photosynthate into fungal matter. In Table 5.3, we have calculated microbial efficiencies from various studies and also report values of carbon use efficiency of intact plant–mycorrhizal systems. At the ecosystem scale, carbon use efficiency is equivalent to the ratio of NPP to gross primary production (GPP). Modelling studies have generally used values of carbon use efficiency of 45% to 50% for plants (Landsberg & Waring, 1997, R. McKane, personal communication), whereas field measurements in four temperate forests gave values of 41% ± 13% (standard deviation) (Waring & Schlesinger, 1985). Later work in various temperate forests suggested that the carbon use efficiency (NPP/GPP) is relatively invariant at about 47% (summarized in Waring & Running, 1998), although carbon use efficiency in boreal forests may be somewhat lower (Ryan *et al.*, 1997). Reported values of efficiency in Table 5.3 range from 10% to 84%, with only one value above 65%, a value of 84% for *Paxillus involutus* in Bidartondo *et al.* (2001). Since the theoretical maximum efficiency is only 75% and the efficiency of leaf production itself is only 65% (Komor, 2000), we believe that the

Table 5.3. *Estimates of carbon use efficiency in biomass production*

System	Efficiency	Reference
Pure culture		
3 fungal species[1]	54%–22%	Henn & Chapela (2000)
soil microbial community[2]	50%[5], 61%[6]	Blagodatskiy et al. (1993)
ectomycorrhizal species[2]	35%–40%	Lindeberg & Lindeberg (1977)
Trametes versicolor[3]	15%–21%	Kohzu et al. (1999)
Poria oleracea[3]	19%	Lekkerkerk et al. (1990)
Phanerochaete chrysosporium[3]	12%	Lekkerkerk et al. (1990)
microbes (0–50 days)[4]	50%–60%	Hart et al. (1994)
microbes (100–450 days)[4]	15%–25%	Hart et al. (1994)
Symbiotic culture		
Betula pendula–Paxillus involutus	34%, 36%, 57%	Ek (1997)
Eucalyptus coccifera (NM)	43% ± 2.2%[7]	Jones et al. (1998)
Thelephora terrestris	55.7% ± 1.7%[7]	Jones et al. (1998)
Pinus sylvestris (3 fungi, 8 treatments)	40%–48%	Colpaert et al. (1996)
Pinus muricata		
Rhizopogon (3 strains)	31%[8], 54%[8], 64%[8]	Bidartondo et al. (2001)
Paxillus involutus	84%[8]	Bidartondo et al. (2001)
Suillus pungens	41%[8]	Bidartondo et al. (2001)
P. ponderosa–Hebeloma crustuliniforme	39% (48%roots)	Rygiewicz & Andersen (1994)
Salix viminalis–Thelephora terrestris	43%[7]	Durall et al. (1994)
Pinus contorta (unknown isolate)	41% (pH 3.8)	Erland et al. (1991)
P. contorta (unknown isolate)	51% (pH 5.2)	Erland et al. (1991)
Field measurements		
Pinus radiata	43%–50%[9]	Ryan et al. (1996)
Populus tremuloides	34%	Ryan et al. (1997)
Picea mariana	34%	Ryan et al. (1997)
Pinus banksiana	39%	Ryan et al. (1997)
Modelling (NPP/GPP)		
Pinus sylvestris stand	26.4%	Janssens et al. (2002)
Modelling parameter	45%	Landsberg & Waring (1997)

[1] Grown on 9:1 sucrose:malt extract.
[2] Grown on glucose.
[3] Grown on wood.
[4] Grown on forest soil.
[5] N-deficient.
[6] N-sufficient.
[7] 9 d pulse ^{14}C labelling.
[8] Efficiency for mycorrhizal fungi only.
[9] Above-ground only.

anomalously high value reported for *P. involutus* probably arose from using an ergosterol to biomass conversion factor that was too low.

The lowest values of carbon use efficiency (10%–26%) are reported from saprotrophic fungi growing on complex substrates under nutrient-limited conditions (Table 5.3; Lekkerkerk *et al.*, 1990; Kohzu *et al.*, 1999). Under these conditions, both ectomycorrhizal and saprotrophic fungi appear to shunt much of their metabolism through alternate oxidase pathways with the concomitant production of large quantities of organic acids (Coleman & Harley, 1976; Jennings, 1995). These organic acids may complex with aluminium and iron in the soil to increase the solubility and bioavailability of phosphorus, iron and base cations (Dutton & Evans, 1996). Thus, low efficiencies may be a common result of nutrient deficiencies of some elements.

Researchers have employed a variety of measurement approaches to estimate mycorrhizal biomass in the laboratory or in the field. Estimates are easiest in inert media such as sand or perlite when a fungal compartment can be physically isolated from roots by using fine mesh or other physical barriers (Rygiewicz & Andersen, 1994; Wallander *et al.*, 2001). If physical barriers are not used to separate fungal from plant material, or if complex media such as soil are the growth media, then various markers for fungal biomass are used. The ratio of fungal biomarker to fungal biomass is often derived from pure culture studies. Fungal biomass has been estimated by using such biomarkers as chitin concentration (Eltrop & Marschner, 1996), ergosterol (Wallander & Nylund, 1992; Hobbie & Colpaert, 2003), phospholipid fatty acids (Wallander *et al.*, 2001), the nitrogen concentration of mycorrhizas (Hobbie & Colpaert, 2003), or even the loss on ignition of inert media (Wallander *et al.*, 2004). Interestingly, laboratory studies also indicate a tight correlation between carbon allocation to ectomycorrhizal fungi and ^{15}N content of host seedlings (Fig. 5.3; Hobbie & Colpaert, 2003, J. Colpaert, unpublished), presumably because carbon and nitrogen cycling are tightly coupled in the mycorrhizal symbiosis and ectomycorrhizal fungi preferentially transfer ^{15}N-depleted amino acids to plants. This suggests that ^{15}N measurements in ectomycorrhizal plants in the field could potentially indicate carbon allocation to ectomycorrhizal fungi, although more study is needed on what factors control ^{15}N patterns in ectomycorrhizal plants in natural ecosystems. To expand the usefulness of such measurements to realistic field conditions, more work should be done on comparing different methods. For example, in a study using ingrowth cores, Wallander *et al.* (2001) reported that fungal biomass as calculated from ergosterol measurements was only 60% to 78% of that calculated from phospholipid fatty acids.

Nutrients other than nitrogen may sometimes limit growth and influence carbon allocation patterns in trees. In laboratory experiments by Ekblad *et al.* (1995), all six macronutrients were varied in a reduced factorial design that included both *Alnus* and *Pinus* seedlings colonized by *P. involutus*. Phosphorus limitation increased carbon allocation to the mycorrhizal symbiont while potassium limitation reduced carbon allocation to the fungal symbiont. The effect of nitrogen on carbon allocation to the mycorrhizal symbiont was difficult to interpret because of the strong growth response to added nitrogen. The other tested nutrients (calcium, magnesium and sulphur) did not significantly affect carbon allocation.

Ericsson (1995) also studied the influence of nutrient limitation on carbon allocation in tree seedlings. In this study, the influence of growth-limiting availabilities of nitrogen, phosphorus, potassium, magnesium, iron, sulphur and manganese on carbon allocation to roots of non-mycorrhizal birch seedlings was tested. Limitation by nitrogen, phosphorus, sulphur and iron increased allocation to roots whereas limitation by potassium, magnesium and manganese reduced the allocation to roots. These different carbon allocation patterns at different nutrient limitations have implications for predicting carbon cycling in future forest ecosystems in response to anthropogenic activities that may shift forests from nitrogen

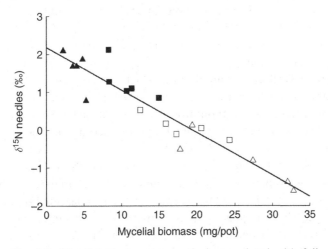

Fig. 5.3. Mycelial biomass is negatively correlated with foliar ^{15}N in mycorrhizal *Pinus sylvestris*. Fungal biomass in perlite calculated from ergosterol measurements and appropriate conversion factors for *Thelephora* or *Suillus*. $r^2 = 0.90$, $p < 0.001$. High nitrogen supply, filled symbols; low nitrogen supply, empty symbols; triangles, *Suillus*; squares, *Thelephora*. From Hobbie and Colpaert (2003).

limitation to limitation by other nutrients such as potassium or phosphorus. Whether potassium or phosphorus becomes limiting in a forest receiving large amounts of anthropogenic nitrogen depends on the composition of the parent material of the soil. As we have seen, potassium and phosphorus limitations are likely to have very different influences on carbon cycling of forest ecosystems (see Wallander, Chapter 14, this volume). It is therefore important to refine forest productivity models to accurately predict how forest management and other anthropogenic activities will influence the nutrient status of forest ecosystems in the future.

Hyphal turnover and incorporation into soil organic matter

The large amount of ectomycorrhizal biomass in the soil and the potentially high production rate of ectomycorrhizal mycelia suggest that ectomycorrhizal fungi contribute considerably to soil organic matter. Their contribution will depend on the highly uncertain lifespan and decomposition rate of soil hyphae. Longevity of mycorrhizal roots based on minirhizotron studies is rather high, 200–250 days (Majdi et al., 2001), and the estimated age of carbon in mycorrhizal roots using radiocarbon indicates a mean age of 4 years, also suggesting a long lifespan for mycorrhizal roots (Treseder et al., 2004). The rather different estimates of root lifespan derived from radiocarbon studies and minirhizotron studies partly arise from assuming in radiocarbon studies that the probability of root death is constant throughout root lifespan, whereas the probability of root death probably is much higher in younger fine roots than in older fine roots (Tierney & Fahey, 2002).

Most fungal cell walls contain chitin, which is a nitrogen-rich compound that degrades slowly in the soil. Mycorrhizal roots usually decompose more slowly than roots without mycorrhizas (Langley & Hungate, 2003), perhaps because the chitin in fungal mantles surrounding the fine roots retards decomposition. Langley and Hungate (2003) suggested that diversion of labile root exudates to produce ectomycorrhizal hyphae would reduce soil microbial activity and thereby increase inputs of carbon to soil organic matter. Indeed Olsson et al. (1996) demonstrated that growth of ectomycorrhizal hyphae reduced activity of soil bacteria. The strong nitrogen retention capacity of ectomycorrhizal mycelia discussed above and the slow decomposition rate of chitin may indicate that ectomycorrhizal mycelia could contribute significantly to both the nitrogen and carbon pools of forest soil.

Based on what is currently known about the production, lifespan and decomposition rate of mycorrhizal roots, we can safely conclude that these roots may significantly influence soil carbon processing rates (Langley & Hungate, 2003). In contrast, there are no good estimates of hyphal lifespan for ectomycorrhizal mycelia, although values as short as one week have been used in some initial estimates of carbon allocation to ectomycorrhizal fungi (Finlay & Söderström, 1992). The current scarcity of information about lifetime and turnover rates of mycorrhizal hyphae indicates that this research field should be a priority in future studies.

Responses to nitrogen deposition and elevated CO_2

The response of ectomycorrhizal fungi to environmental gradients has recently been reviewed by Erland and Taylor (2002) and the influence of nitrogen specifically on ectomycorrhizal fungi was reviewed by Wallenda and Kottke (1998). Soil acidification and nitrogen fertilization resulted in lower fine-root densities and lower ectomycorrhizal species diversity (Jonsson *et al.*, 2000; Peter *et al.*, 2001), whereas liming and elevated CO_2 increased fine-root densities and increased species diversity of ectomycorrhizal fungi (Andersson & Söderström, 1995; Fransson *et al.*, 2001). However, Erland and Taylor (2002) stressed that changes in ecto-mycorrhizal species diversity must be treated with caution since reliable sampling strategies for studying non-randomly distributed ectomycor-rhizal fungi in soil are lacking.

After nitrogen addition to forest ecosystems, almost all root tips remain colonized by mycorrhizal fungi but allocation to extraradical hyphae and sporocarps appears to decline (Wallenda & Kottke, 1998). The species diversity of ectomycorrhizal communities also usually declines (Taylor *et al.*, 2000). The consequences of such changes in species composition for nutrient cycling and tree growth are difficult to predict since little is known about nutrient uptake capabilities in different species. Ectomycorrhizal species with a high capacity to use complex organic nitrogen sources decreased in abundance in ecosystems exposed to nitro-gen additions along a nitrogen gradient in Europe (Taylor *et al.*, 2000). Similarly, along a nitrogen deposition gradient in Alaska, 70% of root tips were colonized by ectomycorrhizas with proteolytic capabilities at low nitrogen deposition versus only 7% at the highest nitrogen deposition (Lilleskov *et al.*, 2002a). This type of response probably occurs because the production of nitrogen-mobilizing enzymes uses assimilate and would be of reduced benefit with increased concentrations in the soil of inorganic nitrogen. Experiments of Bidartondo *et al.* (2001) with ectomycorrhizal pine suggest an alternate explanation for different tolerances to increased nitrogen levels among ectomycorrhizal fungi. In their experiments, addition of nitrogen increased respiration of *Rhizopogon* by 205% but increased respiration by *P. involutus* by only 68%. Therefore, some species may respire proportionally more carbon at elevated nitrogen levels than others, thereby reducing carbon available for producing new biomass.

A decline in the ability to use organic nitrogen sources with elevated nitrogen deposition may not greatly influence nitrogen dynamics and tree growth because the increasing availability of inorganic nitrogen will reduce the benefits of using organic nitrogen sources by the trees. However, uptake of other nutrients and water may be adversely affected by an overall decline in uptake capability often accompanying increased nitrogen deposition.

Because many important soil processes, such as soil carbon storage and nitrogen leaching, are mediated by interactions among plants, mycorrhizal fungi and the soil, the effects of elevated CO_2 and nitrogen deposition on mycorrhizal fungi deserve continued study. Elevated CO_2 generally results in larger assimilation rates of plants and a greater allocation of carbon to roots and mycorrhizal fungi (Treseder & Allen, 2000; Kubiske & Godbold, 2001). Elevated CO_2 has increased hyphal growth in laboratory experiments (Rouhier & Read, 1998; Fransson, 2002). Whether this increased input of carbon to the soil will increase soil carbon storage depends on the fate of this carbon over time, such as whether turnover rates of mycorrhizal mycelia are influenced by elevated CO_2. Treseder and Allen (2000) stress the importance of learning more about survivorship and decomposition of mycorrhizal biomass in soil. Preliminary data suggest that hyphal turnover increases after nitrogen deposition, which may result in less carbon sequestration by ectomycorrhizal fungi. Elevated CO_2 and nitrogen may thereby have contrasting influences on soil carbon sequestration (Treseder & Allen, 2000).

Linking carbon and nitrogen cycling in ectomycorrhizal fungi through ecosystem models

Over the last ten years, researchers have continued to improve techniques to estimate carbon allocation to mycorrhizal fungi, to assess the importance of labile organic nitrogen for mycorrhizal and plant nitrogen budgets, and to evaluate interactions between mycorrhizal fungi and the saprotrophic community. This progress suggests that ecosystem models could soon explicitly include mycorrhizal fungi as a competing sink with plant roots for plant sugars, an important conduit for carbon from plants to soil organic matter, a competitor with free-living bacteria and fungi for organic nitrogen in the soil, an important source for plant nitrogen supply and a potential mechanism for plants to directly access labile organic nitrogen. A possible modelling framework is depicted in Fig. 5.4 that explicitly incorporates new knowledge about access of mycorrhizal fungi to organic nitrogen, carbon allocation to mycorrhizal fungi from plants, and the competition for resources between mycorrhizal fungi and the saprotrophic community. Important changes from existing ecosystem models include adding ectomycorrhizal fungi and including amino acids and peptides as potential nitrogen sources available for direct uptake. In addition, ectomycorrhizal fungi are now a significant sink for plant carbon. Saprotrophic microbes are explicitly included and compete with ectomycorrhizal fungi for uptake of inorganic nitrogen and labile organic nitrogen. For clarity, CO_2 fluxes and macromolecules other than protein are not included in the figure.

Some outstanding issues to be addressed in order to facilitate modelling include the following: Accuracy of estimates of biomass of mycorrhizal fungi that rely on girdling (field studies, Högberg *et al.*, 2001) or soil samples (Bååth *et al.*, 2004). Both of these techniques actually measure the importance of recent plant-derived carbon, and cannot distinguish between mycorrhizal fungi and free-living microbes in the rhizosphere that depend on supplies of labile carbon from plants or mycorrhizal fungi to fuel metabolism. Combined with other techniques such as litter

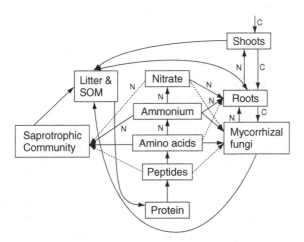

Fig. 5.4. Proposed model structure for integrating mycorrhizal fungi into an ecosystem framework of carbon and nitrogen cycling. Unless otherwise noted, both carbon and nitrogen fluxes are implied. Dashed lines indicate fluxes of lesser or uncertain importance. SOM = soil organic matter.

fall and soil CO_2 flux measurements (Davidson *et al.*, 2002), the use of ingrowth cores may finally allow ecologists to realistically assess carbon allocation to ectomycorrhizal fungi in the field. Comparisons with ecosystem-scale assessments of carbon flux patterns are necessary for validation. Such assessments would be particularly valuable in deciduous forests, for which there are currently no good estimates of ectomycorrhizal allocation. These studies may enable researchers to test the simple relationship between total below-ground allocation and ectomycorrhizal allocation shown in Fig. 5.2.

Extent of retranslocation of nitrogen in fungal mycelia. A simple economic model of mycorrhizal symbioses depicts carbon allocated to mycorrhizal fungi from plants as accumulating any surplus soil-derived nitrogen in excess of what is required for fungal growth. The ability of mycorrhizal fungi to translocate nitrogen from older tissues to regions of rapid growth should increase the overall C/N ratio of fungal biomass and thereby allow a larger fraction of nitrogen taken up by mycorrhizal hyphae to be transferred to plant hosts. The extent of translocation in natural ecosystems is not known. Recent estimates of C/N ratios of ectomycorrhizal mycelia by direct harvesting mycelia from ingrowth mesh bags suggest a rather high C/N ratio (around 20) (Wallander *et al.*, 2003) compared to estimates of 7–11 for microbial biomass from incubating coniferous forest soils (Hart *et al.*, 1994). Data from the girdling experiment in northern Sweden by Högberg *et al.* (2003) also suggest that ectomycorrhizal fungi have a higher carbon:nitrogen ratio than the total microbial biomass.

What constitutes 'available' nitrogen in the soil and litter? Realistic estimates are necessary of the mix of nitrogen sources available to roots or ectomycorrhizal fungi, including nitrate, ammonium, amino acids and small peptides. Given the tremendous surface area for adsorption in soils, the extent to which the available nitrogen pool includes compounds complexed with organic colloids or clay particles is potentially important. Measuring available nitrogen pools may indicate that insufficient nitrogen is available in the soil solution to supply plant and mycorrhizal demand, thereby allowing researchers to indirectly calculate the importance of access via hydrolytic enzymes to complex organic nitrogen forms such as protein or chitin. Such analyses may shed light on the relative importance of amino acids versus oligopeptides in ectomycorrhizal nutrition.

Including food-web processes in quantitative models of carbon and nitrogen cycling of forest ecosystems. The competition between saprotrophic and mycorrhizal fungi for resources is a key aspect of below-ground functioning (Lindahl *et al.*, 2002). Quite detailed computer models of soil food-web functioning have been implemented in grasslands and agroecosystems that include fungi, bacteria, protozoa, nematodes and higher trophic levels (Moore *et al.*, 1996). However, the required intensity of data collection for such modelling will restrict the ability to carry out such modelling in forests for the foreseeable future. Nonetheless, given that one of the main reasons that trees associate with ectomycorrhizal fungi is to allow them to compete by proxy against free-living microbes in the soil, explicitly modelling the competition between saprotrophic and ectomycorrhizal communities for various resources appears important. Using ^{13}C- and ^{15}N-labelled litter and tracking the resulting label into ectomycorrhizal and saprotrophic communities may be a particularly powerful approach to examine interactions between these two functional groups.

The lifespan and decomposition rate of ectomycorrhizas and extraradical ectomycorrhizal hyphae are poorly known. Such values are crucial to estimating carbon and nitrogen fluxes of ectomycorrhizal fungi in ecosystems. Uncertain turnover times for fine roots and hyphae mean that estimates of biomass cannot currently be used to estimate total flux. Estimates derived from various biomarkers may be useful in this regard, as different biomarkers turn over at different rates within living hyphae and will also degrade at different rates after hyphal death.

Conclusions

Mycorrhizal fungi remain difficult to study in forest ecosystems but the last ten years has seen an explosion of promising new techniques to probe their functioning, including isotopic analyses at natural abundance and tracer levels, genetic techniques, compound-specific measurements and a variety of field-based manipulative approaches. As outlined in this chapter, the tools appear to be in place to assess carbon and nitrogen budgets of ectomycorrhizal fungi at the ecosystem level. Such assessments should include efforts to explicitly incorporate ectomycorrhizal fungi into ecosystem-scale models. At this point, the barriers to further progress are as much institutional and informational as scientific. As emphasized by Read and Perez-Moreno (2003), mycorrhizal research always involves a trade-off between the precision with which systems can be measured and the relevance of that system to the actual ecosystems. Both mycorrhizal researchers and ecosystem ecologists have struggled with incorporating each other's worldview and the potential loss of precision such attempts could bring, but the increased relevance of such a potential synthesis to our understanding of forest functioning is clear now, and will undoubtedly become only clearer in the years ahead.

Acknowledgements
This article was supported by the US National Science Foundation, award DEB-0235727. We thank John Hobbie for comments on an earlier version of this chapter.

References

Abuzinadah, R. A. & Read, D. J. (1986a). The role of proteins in the nitrogen nutrition of ectomycorrhizal plants. III. Protein utilization by Betula, Picea, and Pinus in mycorrhizal association with *Hebeloma crustuliniforme*. *New Phytologist*, **103**, 507–14.

Abuzinadah, R. A. & Read, D. J. (1986b). The role of proteins in the nitrogen nutrition of ectomycorrhizal plants. I. Utilization of peptides and proteins by ectomycorrhizal fungi. *New Phytologist*, **103**, 481–93.

Ågren, G. I., Axelsson, B., Flower-Ellis, J. G. K. *et al.* (1980). Annual carbon budget for a young Scots pine. In *Ecological Bulletin Vol. 32. Structure and Function of Northern Coniferous Forests – An Ecosystem Study*, ed. T. Persson. Stockholm: Swedish Natural Science Research, pp. 307–13.

Aldén, L., Demoling, F. & Bååth, E. (2001). Rapid method of determining factors limiting bacterial growth in soil. *Applied and Environmental Microbiology*, **67**, 1830–8.

Alexander, I. J. (1981). *Picea sitchensis* and *Lactarius rufus* mycorrhizal association and its effects on seedling growth and development. *Transactions of the British Mycological Society*, **76**, 417–23.

Andersson, S. & Söderström, B. (1995). Effects of lime (CaCO₃) on ectomycorrhizal colonization of *Picea abies* (L.) Karst. seedlings planted in a spruce forest. *Scandinavian Journal of Forest Research*, **10**, 149–54.

Axelsson, E. & Axelsson, B. (1986). Changes in carbon allocation patterns in spruce and pine trees following irrigation and fertilization. *Tree Physiology*, **2**, 189–204.

Baar, J., Comini, B., Elferink, M. O. & Kuyper, T. W. (1997). Performance of four ectomycorrhizal fungi on organic and inorganic nitrogen sources. *Mycological Research*, **101**, 523–9.

Bååth, E., Nilsson, L. O., Göransson, H. & Wallander, H. (2004). Can the extent of degradation of soil fungal mycelium during soil incubation be used to estimate ectomycorrhizal biomass in soil? *Soil Biology and Biochemistry*, **36**, 2105–9.

Beets, P. N. & Whitehead, D. (1996). Carbon partitioning in *Pinus radiata* in relation to foliage nitrogen status. *Tree Physiology*, **16**, 131–8.

Bending, G. D. & Read, D. J. (1995). The structure and function of the vegetative mycelium of ectomycorrhizal plants. V. Foraging behaviour and translocation of nutrients from exploited litter. *New Phytologist*, **130**, 401–9.

Bhupinderpal-Singh, Nordgren, A., Ottoson-Löfvenius, M. *et al.* (2003). Tree root and soil heterotrophic respiration as revealed by girdling of boreal Scots pine forest: extending observations beyond the first year. *Plant, Cell and Environment*, **26**, 1287–96.

Bidartondo, M. I., Ek, H., Wallander, H. & Söderström, B. (2001). Do nutrient additions alter carbon sink strength of ectomycorrhizal fungi? *New Phytologist*, **151**, 543–50.

Blagodatskiy, S. A., Larionova, A. A. & Yevdokimov, I. V. (1993). Effect of mineral nitrogen on the respiration rate and growth efficiency of soil microorganisms. *Eurasian Soil Science*, **25**, 85–95.

Buchmann, N., Gebauer, G. & Schulze, E.-D. (1996). Partioning of ^{15}N-labeled ammonium and nitrate among soil, litter, below- and above-ground biomass of trees and understory in a 15-year-old *Picea abies* plantation. *Biogeochemistry*, **33**, 1–23.

Chapin, F. S. III. (1980). The mineral nutrition of wild plants. *Annual Review of Ecology and Systematics*, **11**, 233–60.

Coleman, D. C., Crossley, D. A. Jr & Hendrix, P. F. (2004). *Fundamentals of Soil Ecology*, 2nd edn. Boston: Elsevier.

Coleman, J. O. D. & Harley, J. L. (1976). Mitochondria of mycorrhizal roots of *Fagus sylvatica*. *New Phytologist*, **76**, 317–30.

Colpaert, J. V. & van Laere, A. (1996). A comparison of the extracellular enzyme activities of two ectomycorrhizal and a leaf-saprotrophic basidiomycete colonizing beech leaf litter. *New Phytologist*, **134**, 133–41.

Colpaert, J., van Assche, J. A. & Luijtens, K. (1992). The growth of the extramatrical mycelium of ectomycorrhizal fungi and the growth response of *Pinus sylvestris* L. *New Phytologist*, **120**, 127–35.

Colpaert, J., van Laere, A. & van Assche, J. A. (1996). Carbon and nitrogen allocation in ectomycorrhizal and non-mycorrhizal *Pinus sylvestris* L. seedlings. *Tree Physiology*, **16**, 787–93.

Davidson, E. A., Savage, K., Bolstad, P. *et al.* (2002). Belowground carbon allocation in forests estimated from litterfall and IRGA-based soil respiration measurements. *Agricultural and Forest Meteorology*, **113**, 39–51.

Dighton, J., Thomas, E. D. & Latter, P. M. (1987). Interactions between tree roots, mycorrhizas, a saprotrophic fungus and the decomposition of organic substrates in a microcosm. *Biology and Fertility of Soils*, **4**, 145–50.

Dosskey, M. G., Linderman, R. G. & Boersma, L. (1990). Carbon-sink stimulation of photosynthesis in Douglas fir seedlings by some ectomycorrhizas. *New Phytologist*, **115**, 269–74.

Durall, D. M., Jones, M. D. & Tinker, P. B. (1994). Allocation of ^{14}C-carbon in ectomycorrhizal willow. *New Phytologist*, **128**, 109–14.

Dutton, M. V. & Evans, C. S. (1996). Oxalate production by fungi: its role in pathogenicity and ecology in the soil environment. *Canadian Journal of Microbiology*, **42**, 881–95.

Ek, H. (1997). The influence of nitrogen fertilization on the carbon economy of *Paxillus involutus* in ectomycorrhizal association with *Betula pendula*. *New Phytologist*, **135**, 133–42.

Ek, H., Sjögren, M., Arnebrant, K. & Söderström, B. (1994). Extramatrical mycelial growth, biomass allocation and nitrogen uptake in ectomycorrhizal systems in response to collembolan grazing. *Applied Soil Ecology*, **1**, 155–69.

Ekblad, A., Wallander, H., Carlsson, R. & Huss-Danell, K. (1995). Fungal biomass in roots and extramatrical mycelium in relation to macronutrients and plant biomass of ectomycorrhizal *Pinus sylvestris* and *Alnus incana*. *New Phytologist*, **131**, 443–51.

Eltrop, L. & Marschner, H. (1996). Growth and mineral nutrition of non-mycorrhizal and mycorrhizal Norway spruce (*Picea abies*) seedlings grown in semi-hydroponic sand culture. II. Carbon partitioning in plants supplied with ammonium or nitrate. *New Phytologist*, **133**, 479–86.

Ericsson, T. (1995). Growth and shoot:root ratio of seedlings in relation to nutrient availability. *Plant and Soil*, **169**, 205–14.

Erland, S. & Taylor, A. F. S. (2002). Diversity of ecto-mycorrhizal fungal communities in relation to the abiotic environment. In *Mycorrhizal Ecology, Ecological Studies, 157*, ed. M. G. A. Van der Heijden & I. Sanders. Berlin: Springer-Verlag, pp. 163–224.

Erland, S., Finlay, R. & Söderström, B. (1991). The influence of substrate pH on carbon translocation in ectomycorrhizal and non-mycorrhizal pine seedlings. *New Phytologist*, **119**, 235–42.

Finlay, R. D. & Söderström, B. (1992). Mycorrhiza and carbon flow to the soil. In *Mycorrhizal Functioning*, ed. M. F. Allen. New York: Chapman and Hall, pp. 134–60.

Finlay, R. D., Ek, H., Odham, G. & Söderström, B. (1989). Uptake, translocation and assimilation of nitrogen from ^{15}N-labelled ammonium and nitrate sources by intact ectomycorrhizal systems of *Fagus sylvatica* infected with *Paxillus involutus*. *New Phytologist*, 113, 47–55.

Fogel, R. & Hunt, G. (1979). Fungal and arboreal biomass in a western Oregon Douglas-fir ecosystem: distribution patterns and turnover. *Canadian Journal of Forest Research*, 9, 245–56.

Fogel, R. & Hunt, G. (1983). Contribution of mycorrhizae and soil fungi to nutrient cycling in a Douglas-fir ecosystem. *Canadian Journal of Forest Research*, 13, 219–32.

Fransson, P. M. A. (2002). Responses of ectomycorrhizal fungi to changes in carbon and nutrient availability. Unpublished Ph.D. thesis, Swedish University of Agricultural Sciences, Uppsala, Sweden.

Fransson, P. M. A., Taylor, A. F. S. & Finlay, R. D. (2001). Elevated atmospheric CO_2 alters root symbiotic community structure in forest trees. *New Phytologist*, 152, 431–42.

Gadgil, R. L. & Gadgil, P. D. (1975). Suppression of litter decomposition by mycorrhizal roots of *Pinus radiata*. *New Zealand Journal of Forest Science*, 5, 33–41.

Giesler, R., Högberg, M. & Högberg, P. (1998). Soil chemistry and plants in Fennoscandian boreal forest as exemplified by a local gradient. *Ecology*, 79, 119–37.

Hagerberg, D. & Wallander, H. (2002). The impact of forest residue removal and wood ash amendment on the growth of the ectomycorrhizal external mycelium. *FEMS Microbiology Ecology*, 39, 139–46.

Hagerberg, D., Thelin, G. & Wallander, H. (2003). The production of ectomycorrhizal mycelium in forests: relation between forest nutrient status and local mineral sources. *Plant and Soil*, 252, 279–90.

Harley, J. L. & Smith, S. E. (1983). *Mycorrhizal Symbiosis*. Oxford: Academic Press.

Hart, S. C., Nason, G. E., Maryold, D. D. & Perry, D. A. (1994). Dynamics of gross nitrogen transformations in an old-growth forest: the carbon connection. *Ecology*, 75, 880–91.

Henn, M. R. & Chapela, I. H. (2000). Differential C isotope discrimination by fungi during decomposition of C_3- and C_4-derived sucrose. *Applied and Environmental Microbiology*, 66, 4180–6.

Ho, I. & Trappe, J. M. (1980). Nitrate reductase activity of nonmycorrhizal Douglas-fir rootlets and some associated mycorrhizal fungi. *Plant and Soil*, 54, 395–8.

Hobbie, E. A. & Colpaert, J. V. (2003). Nitrogen availability and colonization by mycorrhizal fungi correlate with nitrogen isotope patterns in plants. *New Phytologist*, 157, 115–26.

Hobbie, E. A., Macko, S. A. & Shugart, H. H. (1999). Insights into nitrogen and carbon dynamics of ectomycorrhizal and saprotrophic fungi from isotopic evidence. *Oecologia*, 118, 353–60.

Högberg, M. N. & Högberg, P. (2002). Extramatrical ectomycorrhizal mycelium contributes one-third of microbial biomass and produces, together with associated roots, half the dissolved organic carbon in a forest soil. *New Phytologist*, 154, 191–5.

Högberg, M. N., Bååth, E., Nordgren, A., Arnebrant, K. & Högberg, P. (2003). Field evidence of opposing effects of nitrogen availability on carbon supply to ectomycorrhizal fungi and decomposers in boreal forest. *New Phytologist*, 160, 225–38.

Högberg, P., Högbom, L., Schinkel, H. *et al.* (1996). [15]N abundance of surface soils, roots and mycorrhizas in profiles of European forest soils. *Oecologia*, 108, 207–14.

Högberg, P., Plamboeck, A. H., Taylor, A. F. S. & Fransson, P. M. A. (1999). Natural ^{13}C abundance reveals trophic status of fungi and host-origin of carbon in mycorrhizal fungi in mixed forests. *Proceedings of the National Academy of Sciences of the United States of America*, **96**, 8534–9.

Högberg, P., Nordgren, A., Buchmann, N. *et al.* (2001). Large-scale forest girdling shows that current photosynthesis drives soil respiration. *Nature*, **411**, 789–92.

Högberg, P., Nordgren, A. & Ågren, G. I. (2002). Carbon allocation between tree root growth and root respiration in a boreal pine forest. *Oecologia*, **132**, 579–81.

Ingestad, T. (1979). Mineral nutrient requirements of *Pinus sylvestris* and *Picea abies* seedlings. *Physiologia Plantarum*, **45**, 373–80.

Ingestad T., Arveby, A. S. & Kähr, M. (1986). The influence of ectomycorrhiza on nitrogen nutrition and growth of *Pinus sylvestris* seedlings. *Physiologia Plantarum*, **68**, 575–82.

Janssens, I. A., Sampson, D. A., Curiel-Yuste, J., Carrara, A. & Ceulemans, R. (2002). The carbon cost of fine root turnover in a Scots pine forest. *Forest Ecology and Management*, **168**, 231–40.

Jennings, D. H. (1995). *The Physiology of Fungal Nutrition*. Cambridge: Cambridge University Press.

Jones, M. D. & Hutchinson, T. C. (1988). Nickel toxicity in mycorrhizal birch seedlings infected with *Lactarius rufus* or *Scleroderma flavidum*. I. Effects on growth, photosynthesis, respiration and transpiration. *New Phytologist*, **108**, 451–9.

Jones, M. D., Durall, D. M. & Tinker, P. B. (1998). A comparison of arbuscular and ecto-mycorrhizal *Eucalyptus coccifera*: growth response, phosphorus uptake efficiency and external hyphal production. *New Phytologist*, **140**, 125–34.

Jonsson, L., Dahlberg, A. & Brandrud, T.-E. (2000). Spatiotemporal distribution of an ectomycorrhizal community in an oligotrophic Swedish *Picea abies* forest subjected to experimental nitrogen addition: above and below-ground views. *Forest Ecology and Management*, **132**, 143–56.

Kårén, O. & Nylund, J.-E. (1997). Effects of ammonium sulphate on the community structure and biomass of ectomycorrhizal fungi in a Norway spruce stand in southwestern Sweden. *Canadian Journal of Botany*, **75**, 1628–43.

Kielland, K. (1994). Amino acid absorption by arctic plants: implications for plant nutrition and nitrogen cycling. *Ecology*, **75**, 2373–83.

Kohzu, A., Yoshioka, T., Ando, T. *et al.* (1999). Natural ^{13}C and ^{15}N abundance of field-collected fungi and their ecological implications. *New Phytologist*, **144**, 323–30.

Koide, R. T. & Wu, T. (2003). Ectomycorrhizas and retarded decomposition in a *Pinus resinosa* plantation. *New Phytologist*, **158**, 401–7.

Komor, E. (2000). Source physiology and assimilate transport: the interaction of sucrose metabolism, starch storage and phloem export in source leaves and the effects on sugar status in phloem. *Australian Journal of Plant Physiology*, **27**, 497–505.

Krznaric, E. (2004). Uptake of amino acids and oligopeptides in mycorrhizal *Pinus sylvestris*. Unpublished MS thesis, Katholieke Hogeschool Limburg.

Kubiske, M. E. & Godbold, D. L. (2001). Influence of CO_2 on the growth and function of roots and root systems. In *The Impact of Carbon Dioxide and Other Greenhouse Gases on Forest Ecosystems*, ed. D. F. Karnovsky, R. Ceulemans, G. Scarascia-Mugnozza & J. L. Innes. Wallingford, UK: CAB International, pp. 147–91.

Lamhamedi, M. S., Godbout, C. & Fortin, J. A. (1994). Dependence of *Laccaria bicolor* basidiome development on current photosynthesis of *Pinus strobus* seedlings. *Canadian Journal of Forest Research*, **24**, 1797–804.

Landeweert, R., Hoffland, E., Finlay, R. D., Kuyper, T. W. & van Breemen, N. (2001). Linking plants to rocks: ectomycorrhizal fungi mobilize nutrients from minerals. *Trends in Ecology and Evolution*, **16**, 248–54.

Landsberg, J. J. & Waring, R. H. (1997). A generalised model of forest productivity using simplified concepts of radiation-use efficiency, carbon balance and partitioning. *Forest Ecology and Management*, **95**, 209–28.

Langley, J. A. & Hungate, B. A. (2003). Mycorrhizal controls on belowground litter quality. *Ecology*, **84**, 2302–12.

Last, F. T., Pelham, J., Mason, P. A. & Ingleby, K. (1979). Influence of leaves on sporophore production by fungi forming sheathing mycorrhizas with *Betula* spp. *Nature*, **280**, 168–9.

Lekkerkerk, L., Lundkvist, H., Ågren, G. I., Ekbohm, G. & Bosatta, E. (1990). Decomposition of heterogeneous substrates: an experimental investigation of a hypothesis on substrate and microbial properties. *Soil Biology and Biochemistry*, **22**, 161–7.

Lewis, D. H. & Harley, J. L. (1965). Carbohydrate physiology of mycorrhizal roots of beech. II. Utilization of exogenous sugars by uninfected and mycorrhizal roots. *New Phytologist*, **64**, 238–55.

Lilleskov, E. A., Fahey, T. J., Horton, T. R. & Lovett, G. M. (2002a). Belowground ectomycorrhizal fungal community change over a nitrogen deposition gradient in Alaska. *Ecology*, **83**, 104–15.

Lilleskov, E. A., Hobbie, E. A. & Fahey, T. J. (2002b). Ectomycorrizal fungal taxa differing in response to nitrogen deposition also differ in pure culture organic nitrogen use and natural abundance of nitrogen isotopes. *New Phytologist*, **154**, 219–31.

Lindahl, B. O., Taylor, A. F. S. & Finlay, R. D. (2002). Defining nutritional constraints on carbon cycling in boreal forests-towards a less 'phytocentric' perspective. *Plant and Soil*, **242**, 123–35.

Lindeberg, G. & Lindeberg, M. (1977). Pectinolytic ability of some mycorrhizal and saprophytic hymenomycetes. *Archives of Microbiology*, **115**, 9–12.

Lipson, D. & Näsholm, T. (2001). The unexpected versatility of plants: organic nitrogen use and availability in terrestrial ecosystems. *Oecologia*, **128**, 305–16.

McKane, R. B., Johnson, L. C., Shaver, G. R. *et al.* (2002). Resource-based niches provide a basis for plant species diversity and dominance in arctic tundra. *Nature*, **415**, 68–71.

Majdi, H., Damm, E. & Nylund, J.-E. (2001). Longevity of mycorrhizal roots depends on branching order and nutrient availability. *New Phytologist*, **150**, 195–202.

Marmeisse, R., Guidot, A., Gay, G. *et al.* (2004). *Hebeloma cylindrosporum* – a model species to study ectomycorrhizal symbiosis from gene to ecosystem. *New Phytologist*, **163**, 481–98.

Moore, J. C., de Ruiter, P. C., Hunt, H. W., Coleman, D. C. & Freckman, D. W. (1996). Microcosms and soil ecology: critical linkages between field studies and modelling food webs. *Ecology*, **77**, 694–705.

Näsholm, T., Ekblad, A., Nordin, A. *et al.* (1998). Boreal forest plants take up organic nitrogen. *Nature*, **392**, 914–16.

Nilsson, L. O. (2004). External mycelia of mycorrhizal fungi – responses to elevated N in forest ecosystems. Unpublished Ph.D. thesis, Lund University, Sweden.

Nilsson, L. O. & Wallander, H. (2003). Production of external mycelium by ectomycorrhizal fungi in a Norway spruce forest was reduced in response to nitrogen fertilization. *New Phytologist*, **158**, 409–16.

Nilsson, L. O., Giesler, R., Bååth, E. & Wallander, H. (2005). Growth and biomass of mycorrhizal mycelia in coniferous forests along short natural nutrient gradients. *New Phytologist*, **165**, 613–22.

Nylund, J.-E. & Wallander, H. (1989). Effects of ectomycorrhiza on host growth and carbon balance in a semi-hydroponic cultivation system. *New Phytologist*, **112**, 389–98.

Olsson, P. A., Chalot, M., Bååth, E., Finlay, R. D. & Söderström, B. (1996). Ectomycorrhizal mycelia reduce bacterial activity in a sandy soil. *FEMS Microbiology Ecology*, **21**, 77–86.

Paris, F., Bonnaud, P., Ranger, J., Robert, M. & Lapeyrie, F. (1995). Weathering of ammonium- or calcium-saturated 2:1 phyllosilicates by ectomycorrhizal fungi *in vitro*. *Soil Biology and Biochemistry*, **27**, 1237–44.

Parmelee, R. W., Ehrenfeld, J. G. & Tate, R. L. III. (1993). Effects of pine roots on microorganisms, fauna, and nitrogen availability in two soil horizons of a coniferous forest spodosol. *Biology and Fertility of Soils*, **15**, 113–19.

Perez-Moreno, J. & Read, D. J. (2001). Nutrient transfer from soil nematodes to plants: a direct pathway provided by the mycorrhizal mycelial network. *Plant, Cell and Environment*, **24**, 1219–26.

Persson, J., Högberg, P., Ekblad, A. *et al.* (2003). Nitrogen acquisition from inorganic and organic sources by boreal forest plants in the field. *Oecologia*, **137**, 252–7.

Peter, M., Ayer, F. & Egli, S. (2001). Nitrogen addition in a Norway spruce stand altered macromycete sporocarp production and below-ground ectomycorrhizal species composition. *New Phytologist*, **149**, 311–25.

Plassard, C., Barry, D., Eltrop, L. & Moussin, D. (1994). Nitrate uptake in maritime pine (*Pinus sylvestris* Soland in Ait.) and the ectomycorrhizal fungus *Hebeloma cylindrosporum*: effect of ectomycorrhizal symbiosis. *Canadian Journal of Botany*, **72**, 189–97.

Read, D. J. (1991). Mycorrhizas in ecosystems. *Experientia*, **47**, 376–91.

Read, D. J. & Perez-Moreno, J. (2003). Mycorrhizas and nutrient cycling in ecosystems – a journey towards relevance? *New Phytologist*, **157**, 475–92.

Reid, C. P. P., Kidd, F. A. & Ekwebelam, S. A. (1983). Nitrogen nutrition, photosynthesis and carbon allocation in ectomycorrhizal pine. *Plant and Soil*, **71**, 415–32.

Rouhier, H. & Read, D. J. (1998). Plant and fungal responses to elevated atmospheric carbon dioxide in mycorrhizal seedlings of *Pinus sylvestris*. *Environmental and Experimental Botany*, **40**, 237–46.

Ryan, M. G., Hubbard, R. M., Pongracic, S., Raison, R. J. & McMurtrie, R. E. (1996). Foliage, fine-root, woody-tissue and stand respiration in *Pinus radiata* in relation to nitrogen status. *Tree Physiology*, **16**, 333–44.

Ryan, M. G., Lavigne, M. B. & Gower, S. T. (1997). Annual carbon cost of autotrophic respiration in boreal forest ecosystems in relation to species and climate. *Journal of Geophysical Research*, **102**, 28 871–83.

Rygiewicz, P. T. & Andersen, C. P. (1994). Mycorrhizae alter quality and quantity of carbon allocated below ground. *Nature*, **369**, 58–60.

Rygiewicz, P. T., Bledsoe, C. S. & Zasoski, R. J. (1984). Effects of ectomycorrhizae and solution pH on ^{15}N-nitrate uptake by coniferous seedlings. *Canadian Journal of Forest Research*, **14**, 893–9.

Sangtiean, T. & Schmidt, S. (2002). Growth of subtropical ECM fungi with different nitrogen sources using a new floating culture technique. *Mycological Research*, **16**, 74–85.

Schimel, J. & Bennett, J. (2004). Nitrogen mineralization: challenges of a changing paradigm. *Ecology*, **85**, 591–602.

Simard, S. W., Durall, D. & Jones, M. (2002). Carbon and nutrient fluxes within and between mycorrhizal plants. In *Mycorrhizal Ecology*, ed. M. G. A. Van der Heijden & I. R. Sanders. Berlin: Springer-Verlag, pp. 33–74.

Smith, S. E. & Read, D. J. (1997). *Mycorrhizal Symbiosis*, 2nd edn. New York: Academic Press.

Söderström, B. & Read, D. J. (1987). Respiratory activity of intact and excised ectomycorrhizal mycelial systems growing in unsterilised soil. *Soil Biology and Biochemistry*, **19**, 231–236.

Tamm, C.-O. (1991). *Nitrogen in Terrestrial Ecosystems. Questions of Productivity.* Berlin: Springer-Verlag.

Taylor, A. F. S., Martin, F. & Read, D. J. (2000). Fungal diversity in ectomycorrhizal communities of Norway spruce (*Picea abies* (L.) Karst.) and beech (*Fagus sylvatica* L.) along North-South transects in Europe. In *Carbon and Nitrogen Cycling in European Forest Ecosystems, Ecological Studies 142*, ed. E. D. Schulze. Berlin: Springer-Verlag, pp. 343–65.

Tierney, G. L. & Fahey, T. J. (2002). Fine root turnover in a northern hardwood forest: a direct comparison of the radiocarbon and minirhizotron methods. *Canadian Journal of Forest Research*, **32**, 1692–7.

Tinker, P. B., Jones, M. D. & Durall, D. M. (1990). Phosphorus and carbon relationships in willow ectomycorrhizae. *Symbiosis*, **9**, 43–9.

Treseder. K. K. & Allen, M. F. (2000). Mycorrhizal fungi have a potential role in soil carbon storage under elevated CO_2 and nitrogen deposition. *New Phytologist*, **147**, 189–200.

Treseder, K. K., Masiello, C. A., Lansing, J. L. & Allen, M. F. (2004). Species-specific measurements of ectomycorrhizal turnover under N-fertilization: combining isotopic and genetic approaches. *Oecologia*, **138**, 419–25.

Vogt, K. A., Grier, C., Meier, C. E. & Edmonds, R. L. (1982). Mycorrhizal role in net primary production and nutrient cycling in *Abies amabilis* ecosystems in western Washington. *Ecology*, **63**, 370–80.

Wallander, H. (1995). A new hypothesis to explain allocation of dry matter between ecto-mycorrhizal fungi and pine seedlings. *Plant and Soil*, **168–169**, 243–8.

Wallander, H. & Nylund, J-E. (1992). Effects of excess nitrogen and phosphorus starvation on the extramatrical mycelium of ectomycorrhizas of *Pinus sylvestris* L. *New Phytologist*, **120**, 495–503.

Wallander, H., Nilsson, L. O., Hagerberg, D. & Bååth, E. (2001). Estimation of the biomass and seasonal growth of external mycelium of ectomycorrhizal fungi in the field. *New Phytologist*, **151**, 753–60.

Wallander, H., Nilsson, L.-O., Hagerberg, D. & Rosengren, U. (2003). Direct estimates of C:N ratios of ectomycorrhizal mycelia collected from Norway spruce forest soils. *Soil Biology and Biochemistry*, **35**, 997–9.

Wallander, H., Göransson, H. & Rosengren, U. (2004). Production, standing biomass and natural abundance of ^{15}N and ^{13}C in ectomycorrhizal mycelia collected at different soil depths in two forest types. *Oecologia*, **139**, 89–97.

Wallenda, T. & Kottke, I. (1998). N deposition and ectomycorrhizas. *New Phytologist*, **139**, 169–87.

Wallenda, T., Stober, C., Högbom, L. *et al.* (2000). Nitrogen uptake processes in roots and mycorrhizas. In *Carbon and Nitrogen Cycling in European Forest Ecosystems, Ecological Studies 142*, ed. E. D. Schulze. Berlin: Springer-Verlag, pp. 122–43.

Waring, R. H. & Running, S. W. (1998). *Forest Ecosystems: Analysis at Multiple Scales*. New York: Academic Press.

Waring, R. H. & Schlesinger, W. H. (1985). *Forest Ecosystems: Concepts and Management*. New York: Academic Press.

Wiklund, K., Nilsson, L.-O. & Jacobsson, S. (1995). Effect of irrigation, fertilization, and drought on basidioma production in a Norway spruce stand. *Canadian Journal of Botany*, **73**, 200–8.

Zeller, B., Colin-Belgrand, M., Dambrine, E., Martin, F. & Bottner, P. (2000). Decomposition of ^{15}N-labelled beech litter and fate of nitrogen derived from litter in a beech forest. *Oecologia*, **123**, 550–9.

Zeller, B., Colin-Belgrand, M., Dambrine, E. & Martin, F. (2001). Fate of nitrogen released from ^{15}N-labeled litter in European beech forests. *Tree Physiology*, **21**, 153–62.

6

Role of arbuscular mycorrhizal fungi in carbon and nutrient cycling in grassland

DAVID JOHNSON, JONATHAN R. LEAKE
AND DAVID J. READ

Introduction

Arbuscular mycorrhizal fungi (AMF) are the most ancient, widespread and ubiquitous of all the groups of mycorrhiza: they have a global distribution in widely contrasting plant communities including the Tropics, the Boreal forest, arctic tundra and all types of grassland. Considerable effort has been made in recent years in order to set AMF within a robust phylogeny. Recent advances in molecular biological techniques have enabled scientists to place AMF in a new division, the Glomeromycota. At present, this division contains only about 150 species, which is remarkable given the enormous number of plant species the fungi readily colonize. The mutualistic symbioses that AMF form with their host plants give rise to a number of important benefits to both the plant and fungus. A brief glance at a standard mycorrhizal text will list many ecologically important attributes, such as improved disease resistance, water uptake, nutrient transfer and the ability of the fungus to be a major sink for photosynthate. Indeed, the importance of AMF for nutrient uptake and carbon allocation has been recognized for decades. The ability of AMF (and other mycorrhizal types) to utilize recent plant photosynthate and thus have access to a near continuous supply of energy immediately gives them a potential advantage over saprotrophic microorganisms, which are forced to obtain their energy in the highly carbon-limited heterogeneous soil environment.

However, despite the small number of species and broad geographical and ecological distribution, AMF continue to represent something of an enigma for the ecologist because the real importance of these organisms in

Fungi in Biogeochemical Cycles, ed. G. M. Gadd. Published by Cambridge University Press. © British Mycological Society 2006.

carbon and nutrient cycling in their *natural* environment remains poorly understood. This chapter attempts to draw together some of the recent advances made in understanding how AMF operate in terms of carbon and nutrient cycling but, where possible, will place emphasis on their function in nature. Particular emphasis will also be placed on the external mycelium of AMF, since this is one of the least studied but functionally most important components of the symbiosis. The ecology of these organisms in their full complement of habitats is beyond the scope of one chapter so we will focus on AMF in temperate grassland systems.

The nature of AMF in grassland soils

The roots of virtually all plants in temperate grassland are heavily colonized by AMF (Read *et al.*, 1976; Sparling & Tinker, 1978). The morphology of AMF within their host plants' roots is well known and has been described in detail since the turn of the twentieth century (Gallaud, 1905). Whilst Gallaud's beautifully executed line drawings highlighted important differences in the nature of the fungal mycelium in the roots themselves, he was unable to accurately describe the external mycelium. Even today, about 100 years later, the external mycelium is still regarded as the 'hidden half of the symbiosis' (Leake *et al.*, 2004). Despite this the external mycelium is of fundamental importance, being the key interface between the fungus and the bulk soil through which carbon and nutrients must pass, and knowledge of its distribution, biomass and physiology is vital in order to understand the role of AMF in biogeochemical cycling.

The fact that AMF are obligately mycotrophic means that progress in understanding the key functional attributes of the external mycelium has lagged behind that of ectomycorrhizal and saprotrophic fungi. Many of the basic physiological experiments that would normally be undertaken in culture conditions simply cannot be done. In addition, the external mycelium of AMF is particularly fine, which makes them prone to damage from disturbance. The diameter of typical arbuscular mycorrhizal (AM) fungal hyphae is about 2–3 μm, although some studies have reported that their hyphae can penetrate 0.45 μm mesh (Fitter *et al.*, 1998). It is of course the fine scale and highly branched nature of the external AM fungal mycelium that permits it to so effectively exploit the heterogeneous soil environment.

The extent of extraradical mycelial networks

The distribution of external AM fungal hyphae throughout natural soils is undoubtedly considerable. This realization has led (or hopefully should lead), at least for most plant communities, to the commonly used term 'rhizosphere' becoming largely redundant and being replaced by the more correct term 'mycorrhizosphere'. The spread of AM fungal mycelium through soil is similar or more rapid than non-mycorrhizal fungi. Jakobsen *et al.* (1992) and Powell (1979) measured growth rates between 0.8 and 8.8 mm d^{-1} using *Trifolium* plants colonized mainly by *Glomus* spp. and *Acaulospora* spp., whilst rates of 0.5–1 mm d^{-1} for non-mycorrhizal fungi have been observed (Wagner, 1974). However, the growth rates of AM fungal mycelium in the field situation where root densities can be considerable, particularly in the surface fermentation (F) and humic (H) horizons, is largely unknown, and it seems likely that growth will be considerably more restricted. Similar field information is also required for the extension of hyphae belonging to particular species of AMF beyond the root surface. Laboratory evidence has clearly shown that some species of AMF can extend further away from the root surface than others, with corresponding implications for resource capture (Jakobsen *et al.*, 1992). Whether such interspecific differences are realized in the field, especially in soil horizons where root densities reach 5 m cm^{-3} soil, remains to be seen.

Reasonably reliable field data exist for the total length of AM fungal mycelium in soil. Considerable progress has been made in the development of visual methods to determine the total length of extraradical mycelium and the methods of its extraction from the soil (Boddington *et al.*, 1999). Several studies have quantified hyphal lengths in a range of temperate grassland plant communities and have highlighted that AM fungal mycelial systems are extensive (up to 100 m g soil^{-1}) but can also be highly variable (Table 6.1). One of the major difficulties with such observations however is to try and calculate the proportion of the total hyphal length that is active. Vital stains, for example tetrazolium red, have typically been used to do this (Sylvia, 1988), although criticism is often levelled because false-positives can be detected as a result of bacterial contamination of the hyphal surface. Isotope labelling may provide a more rigorous approach to resolving this issue (see later section, p. 134). Rather less information is available concerning the distribution of external mycelium of AMF with depth. In deeper horizons, where root densities are smaller, AM fungal mycelial growth is likely to be more akin to that in laboratory experiments. While the length and biomass (on a volume basis) of AMF have been shown to be less in subsurface than surface soil horizons (Nehl *et al.*, 1999), the total biomass, and hence total pool of carbon and nutrients, may be considerable in relatively deep soils.

Table 6.1. *Length of extraradical AM fungal mycelium in field soils, expressed per unit weight of soil. Note values are rounded to the nearest whole number. Where original values were expressed per cm³ of soil these were expressed per gram of soil assuming a typical bulk density of 1.32 g cm⁻³ for cultivated mineral soils (*) and a typical bulk density of 1.10 g cm⁻³ for permanent grassland (‡). Modified from* Leake et al. *(2004)*

Culture conditions and plant species	Hyphal length g⁻¹ dry wt soil	Reference
Zea mays, preceded by fallow or cover crop	10–35 m g⁻¹	Kabir & Koide (2002)
Zea mays, conventional, reduced and zero tillage	2–4 m g⁻¹*	Kabir *et al.* (1998)
Native grassland, Serengeti	0–7 m g⁻¹	McNaughton & Oestenheld (1990)
Prairie – 11 years old	68–101 m g⁻¹‡	Miller *et al.* (1995)
Ungrazed pasture – 17 years old	45–74 m g⁻¹‡	Miller *et al.* (1995)
Avena barbata 2-year	6 m g⁻¹	Rillig *et al.* (2002)
Aegilops triuncialis 2-year	4 m g⁻¹	Rillig *et al.* (2002)
Trifolium microcephalum 2-year	4 m g⁻¹	Rillig *et al.* (2002)
Amsinckia douglasiana 2-year	3 m g⁻¹	Rillig *et al.* (2002)
Taeniatherum caput-medusae 2-year	5 m g⁻¹	Rillig *et al.* (2002)
Upland acid pasture	6–7.5 m g⁻¹	Staddon *et al.* (2003a)
Calcareous grassland	3–10 m g⁻¹	Staddon *et al.* (2003c)
Sown *Lolium perenne*	14 m g⁻¹	Tisdall & Oades (1979)
Trifolium repens	3 m g⁻¹	Tisdall & Oades (1979)

What is the biomass of AMF in soil?

This apparently straightforward and fundamental question seems to have been largely ignored by many researchers studying AMF. In contrast, soil microbiologists have for years spent considerable effort in quantifying the biomass of microorganisms in a variety of agricultural and semi-natural systems. This effort was given a considerable boost by the development of chloroform fumigation incubation (Jenkinson & Powlson, 1976), chloroform fumigation extraction (Brookes *et al.*, 1982, 1985), and substrate-induced respiration (Anderson & Domsch, 1978). These methods have enabled soil scientists to estimate the amount of soil carbon (and nitrogen and phosphorus in the case of fumigation extraction) stored in the microbial community and its response to a range of perturbations. A brief search of the literature database reveals over 3700 papers supplying this information since 1981, and 500 specifically related to grassland. There can be no doubt that the microbial biomass as a whole is a hugely important component of total soil carbon, nitrogen and phosphorus.

It is somewhat surprising therefore that none of these studies or methods have been able to partition these data to give information on the amount of carbon held in AM fungal biomass. Such data would seem of fundamental importance to understanding the role of the key components of the soil microbial community (which of course includes AMF) in carbon cycling. The nature of the methods used to measure soil microbial biomass in their present forms, however, fails to target AM (and other groups of mycorrhiza) fungal biomass. The commonly used substrate-induced respiration method (Anderson & Domsch, 1978) relies entirely on the ability of microbes to immediately utilize glucose. There is no evidence to show that AMF have the ability to do this, as would be deduced from their obligately mycotrophic habit. Indeed, Pfeffer *et al.* (1999) demonstrated that ^{13}C-labelled glucose and fructose were not taken up by the external mycelium of *Glomus intraradices*.

Quantification of ergosterol in soil has often been used to determine fungal biomass, and has on occasions been applied specifically to AMF (Hart & Reader, 2002). However, the production of ergosterol by AMF differs considerably between species, and experimental systems need to be aseptic in order to prevent contamination from saprotrophs that contain large amounts of the sterol compared to AMF (Olsson *et al.*, 2003). For this reason it seems unlikely that ergosterol can be considered to be a reliable indicator of AM fungal biomass.

One of the main problems that prevent the fumigation-based methods from accounting for AM fungal biomass is the preparation of the soil samples prior to measurement. Most protocols state the soil should be sieved and allowed to equilibrate for several hours or days afterwards. This procedure will clearly destroy the extensive mycelial networks formed by AMF (and other fungi) and stimulate their decomposition by other organisms. Soil scientists have rarely acknowledged such limitations in these methods, probably in part because the methods were originally developed for agricultural soils where many species of AMF are likely to be minor components of the microbial biomass. However, the increasing adoption of low-intensive and organic agricultural practices may result in AMF becoming a more important constituent of microbial biomass in agricultural soils.

Olsson *et al.* (1995) have pioneered the use of phospholipid fatty acids (PLFAs) to estimate the biomass of AMF in grassland systems.

The PLFAs can be readily extracted from soils and quantified by gas chromatography. Arbuscular mycorrhizal fungal biomass is determined by relying on the observation that the PLFA 16:1ω5 is mycorrhiza specific. This was first reported by Olsson *et al.* (1995) who grew mycorrhizal plants in compartmentalized pots containing autoclaved calcareous sand from a dune grassland and found that the PLFA 16:1ω5, and to a lesser extent 18:1ω5, were significantly more enriched in the hyphal compartment compared to the non-mycorrhizal compartment, thus demonstrating that these PLFAs can be good indicators of AM fungal biomass. Furthermore, the amount of 16:1ω5 from the neutral lipid fraction decreased markedly during storage of soil from the hyphal compartment, indicating a decrease in the amount of fungal storage lipids. The authors subsequently applied this methodology to the field situation and estimated AM fungal biomass to be in the order 34 μg dry AM fungal biomass g^{-1} soil (Olsson & Wilhelmsson, 2000) and constituted approximately 30% of the total microbial biomass.

While the PLFA approach shows considerable promise and should be complimented for being one of the first attempts to quantify AM fungal biomass, it should be remembered that the extrapolation of the assumptions from the original pot experiment (Olsson *et al.*, 1995) into the field situation should always be undertaken with caution, especially in soils with large organic matter contents. For example, the PLFA 16:1ω5 has been found in a range of environments, such as marine sediments, where AMF are not likely to predominate, although in grassland soil, its production may still be dominated by AMF. Some evidence that the pathway of carbon into 16:1ω5 is unlike other PLFAs, and which therefore could support the notion that this PLFA is either specific to AMF or produced in relatively large quantities by AMF, is gained from determination of its natural abundance $\delta^{13}C$ signature. Johnson *et al.* (unpublished data) showed 16:1ω5 to be significantly more enriched in ^{13}C compared to 18 other PLFAs extracted from an upland grassland. This leads to the intriguing possibility of using $^{13}CO_2$ pulse-labelling (see later section) to trace labelled carbon into PLFAs in mycorrhizal and non-mycorrhizal systems, perhaps using experimental approaches similar to those in Olsson and Wilhelmsson (2000).

How much carbon is allocated to AMF?

Isotope pulse-labelling experiments have been the main technique by which carbon can be traced through the plant:mycorrhizal:soil system in order to quantify its allocation to AMF. By providing plants with

$^{14}CO_2$, Ho and Trappe (1973) reported one of the first experiments that demonstrated that recent plant photosynthate is allocated to roots colonized by AMF and into chlamydospores of *Endogone mosseae* (now *Glomus mosseae*). Later more sophisticated experiments were able to partition carbon allocation into the external mycelium. For example, Paul and Kucey (1981) showed that *Vicia faba* plants allocated about 3% of fixed carbon to the external AM fungal mycelium. Similar methods have subsequently been used to trace recent assimilate into the roots of mycorrhizal plants connected via a common mycelial network (Francis & Read, 1984), into the roots of mycorrhizal plants grown in intact grassland microcosms (Grime *et al.*, 1987) and into hyphal compartments containing natural non-sterile soil (Johnson *et al.*, 2001). These and other similar studies have vastly improved our understanding of the rates and quantities of carbon allocated into AMF during short time periods in a range of conditions and plant communities. The range of values of the proportion of carbon allocated to AMF reported thus far range from 1%–26%. It should be noted that the often-quoted headline-hitting upper values (20%–26%) are derived from pot experiments using species such as cucumber (*Cucumis sativus*) as the host plant. Extrapolation of such data to natural grassland plant communities must therefore be undertaken with great caution.

A number of studies have specifically investigated carbon fluxes from grassland plants to soil microbiota using $^{14}CO_2$ (e.g. Martin & Merckx, 1992; Saggar *et al.*, 1997; Kuzyakov *et al.*, 1999), but in circumstances where AM associations were either absent or ignored. As a result, most recent models of carbon fluxes between plants and the soil have been based upon the assumption that root exudation, sloughed cells and dead roots provide the only significant pathways for the supply of plant-fixed carbon to the free-living microbial populations in soils (e.g. Toal *et al.*, 2000; Kuzyakov *et al.*, 2001). Increasing awareness of the extent of the AM fungal mycelial network in grassland ecosystems exposes the limitations of these models and highlights the need for quantification of its involvement in the plant-soil carbon flux pathway.

More recent pulse-labelling studies have utilized the stable isotope ^{13}C as a tracer. This has considerable advantages over ^{14}C in that it has no safety restrictions and can thus be used in the field. The difficulties in providing plants with $^{13}CO_2$ at ambient CO_2 concentrations in the field have been overcome by Ostle *et al.* (2000), although bottled 99 air containing atm% $^{13}CO_2$ at 350 ppm is now becoming more widely available. Recent field-based pulse-labelling studies, which have used mesh cores filled with non-sterile natural soil, have provided good evidence that carbon allocation to external AM fungal mycelium is rapid (Johnson *et al.*, 2001). In an acid upland grassland, about 6%–8% of fixed carbon was respired by AM fungal mycelium 24 hours after the end of pulse-labelling, and the greatest values were measured only 14 hours after the end of labelling. These results provide the first estimates of respiratory carbon release by external mycelium under field conditions and highlight the requirement to sample during the first few hours post-labelling. The rapidity of carbon transfer to mycorrhizal hyphae has also been shown for ectomycorrhizal fungi colonizing pine seedlings in which the maximum rates of transfer to the external mycelium occurred after 14 hours (Leake *et al.*, 2001).

It is also surprising that few, if any studies, have attempted to consider the temporal aspects of carbon allocation to AMF in the field. One suspects that the incentive to pulse-label plants in 'gloomy' conditions is considerably less than in ideal conditions where rates of photosynthesis and the likelihood of obtaining the largest values of below-ground allocation are maximized. Certainly evidence from pulse-labelling studies in subarctic tundra systems that are dominated by ericaceous vegetation have shown that the amount of ^{14}C allocated to fine ericoid mycorrhizal hair roots can vary tenfold between July and September (Olsrud & Christensen, 2004). Do some species of AMF have greater sink strength earlier or later in the growing season, perhaps reflecting the growth strategies of the host plants? What is the average annual mass of carbon allocated to AMF in grassland? These are vital questions that need to be addressed in order to get to grips with the role of AMF in biogeochemical cycling in nature.

The main strength in using $^{14}CO_2$ is the sensitivity of its detection but, being radioactive, it is rarely used in field situations. In addition, most labelling experiments have supplied the plants with a pulse of $^{14}CO_2$ in a sealed static environment and allowed the plants to deplete the CO_2 concentration of the chamber well-below ambient ($\sim$350 ppm). The consequences of growing the plants, even for short periods of time, in such unnatural carbon-stressed conditions are rarely discussed. There is growing evidence (e.g. Holtum & Winter, 2003) that the dynamic responses of photosynthesis to fluctuations in CO_2 concentration (from high to very low) are different from steady state rates of CO_2 fixation. Because labelling experiments are concerned with the latter, future experiments should, where possible, be undertaken using ambient concentrations of $^{14}CO_2$.

Since the Second World War (WW2), air has become contaminated with trace quantities of ^{14}C derived from the detonation of atomic weapons and can thus be used as a natural tracer. Staddon *et al.* (2003b) elegantly exploited this property by growing *Plantago lanceolata* seedlings in the mycorrhizal condition in an atmosphere of ^{14}C-depleted air, which had been derived from fossil fuel combustion. The plants were subsequently transferred to an atmosphere of present-day air and micro-hyphal samples ($\sim$50 µg) were taken during a 30-day period. The hyphae were analysed for ^{14}C content by accelerator mass spectrometry (AMS), which showed their ^{14}C concentration increased rapidly to post-WW2 levels during the first 6 days.

The rapid allocation of trace quantities of ^{14}C to external AM fungal hyphae provides independent support for the observations by Johnson *et al.* (2002a, b) that carbon transfer to mycorrhizas occurs very quickly. Staddon *et al.* (2003b) concluded that the turnover rate of AMF hyphae was likely to be just 5–6 days. This corresponds well with the turnover rates by absorptive hyphal structures of 5–7 days observed in *Artemisia tridentata* and *Oryzopsis hymenoides* communities (Friese & Allen, 1991). However, it is not known from the Staddon *et al.* (2003b) study if the carbon allocated to the AMF was used for production of mycelial biomass or allocated into compounds that are rapidly respired, in accordance with the observations of Johnson *et al.* (2002a, b).

The increasing application of pulse-labelling experiments to mycorrhizal systems has resulted in a greater understanding of short-term carbon fluxes to AMF. While this information is crucially important, the role of AMF in long-term carbon storage must also be remembered. This is particularly crucial in the context of global climate change and elevated concentrations of atmospheric CO_2. Given the size of long-term soil carbon pools, it has been postulated that AMF may have an important role in soil carbon sequestration under elevated CO_2 (Treseder & Allen, 2000). Subsequent experiments by Treseder *et al.* (2003) revealed that the amount of carbon and the amount of hyphae within water-stable aggregates both increased in chaparral ecosystems subjected to elevated CO_2.

Very few experiments have been undertaken to provide reliable data on the rate of carbon allocation to recalcitrant compounds, which contribute to long-term soil carbon pools, specifically related to AMF. The combination of stable-isotope labelling and its analysis by compound specific mass spectrometry may shed some light on the quantities of carbon allocated to specific compounds. Recent research has led to the discovery of the glycoprotein glomalin, which is thought to be specific to AMF. Glomalin may represent between 30% and 60% of soil carbon in undisturbed soils and thus is likely to be an important source of recalcitrant soil carbon (Treseder & Allen, 2000). Given the potential importance and ubiquity of glomalin, it is surprising that its concentration has not been determined in a wider range of temperate grassland soils. In fact, its analysis seems to have been entirely restricted to tropical forest and steppe AM systems. This may be due in part to the complex extraction procedures necessary for its determination and the fact that glomalin is only operationally defined, but this is a clear line of research that needs to be more fully exploited.

Transport of carbon to external mycelium

An understanding of the biochemistry of carbon transport from the internal arbuscular mycelium to the external mycelium is essential, particularly in the light of current debates about the significance of interplant transfer of photosynthate. Isotope-labelling techniques have enabled researchers to determine that the storage lipid triacylglycerol is the key carbon-rich molecule exported to the external mycelium (Pfeffer *et al.*, 1999), although, more recently, export of carbohydrates has been shown to occur (Bago *et al.*, 2003). Arbuscular mycorrhizal fungi contain very large amounts of lipid, possibly as much as 25% of their dry weight. Lipids form compact molecules that can be easily transported throughout the extensive mycelial systems characteristic of AMF. While the net direction of lipid transfer is towards the growing tips, significant bidirectional transfer has also been observed (Bago *et al.*, 2002). The key step for lipid synthesis seems to be the conversion of plant-derived hexose by up-regulation of acetyl-coenzyme A dehydrogenase. Within the internal mycelium, hexose is also converted to glycogen, trehalose and incorporated into the oxidative pentose phosphate pathway. The synthesis of lipid in the internal mycelium and its subsequent conversion to carbohydrate via the glyoxylate cycle in the external mycelium seems energetically inefficient, but the fact that lipids can be translocated very readily throughout the mycelium is likely to compensate for this.

Soil-mediated controls on carbon turnover in AMF

In most temperate grassland, AMF thrive in soils that are chemically, physically and biologically diverse. The nature and properties of soils must therefore have a huge influence on AM fungal morphology and physiology. The fact that many studies of the functioning of AMF have used growth media that is far removed from natural soil, and may thus give a poor indication of mycorrhizal functioning in nature, has been highlighted (Read, 2002), and here we can do little more than to re-emphasize this important message. In the Staddon *et al.* (2003b) study, the plants were grown in a sand-based medium with low organic matter content. Zhu and

Miller (2003) have pointed out that physical protection of hyphae plays an important role in its susceptibility to decomposition by other micro- and macro-organisms (Fig. 6.1), so the value of hyphal turnover given by Staddon *et al.* (2003b) may be an underestimate. This may well be the case, although the fact that the results tie in with measurements made in natural soil and mycorrhizal communities suggests otherwise. Nevertheless, there is considerable scope in the future to take forward the Staddon study by increasing the ecological relevance of the experimental design to determine how more complex soil physical environments affect hyphal turnover. Collaboration between soil physicists and mycorrhizal ecologists is essential if we are to more fully understand the interactions between AM fungal hyphae and soil particles.

A second important factor that may affect carbon turnover from AMF is the composition of the mycorrhizal communities themselves. Arbuscular mycorrhizal fungal species diversity has important consequences for eco-system functioning (Van der Heijden *et al.*, 1998) and some studies have highlighted that local diversity and host specificity may be considerably greater than previously thought (Vandenkoornhuyse *et al.*, 2002). The chemical composition and growth morphologies of different species of AMF have been little studied in natural populations, primarily because of the inability to culture AMF. Some species of AMF that are more amenable to manipulation in trap cultures have been shown to produce different quantities of thicker-walled foraging hyphae. Thicker-walled hyphae, which contain greater quantities of chitin, are likely to be more resistant to decomposition (Fig. 6.1). Glomalin has also been shown to

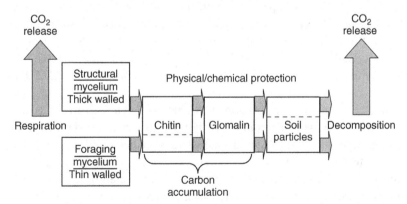

Fig. 6.1. Schematic diagram showing the fate of carbon allocated to external AM fungal mycelium in relation to the soil physical environment and the morphology and chemical composition of the hyphae (modified from Zhu & Miller, 2003).

have a strong influence on the soil physical environment where it increases the stability of soil particles. One of its key roles may therefore be to provide additional physical protection to AM fungal hyphae (Fig. 6.1). It remains to be seen whether different species of AMF can produce contrasting quantities of glomalin as an active defence mechanism.

Chitin is a major component of fungal hyphae that can account for up to 60% of their cell walls, or $0.3 \, \mu g \, m^{-1}$ of external hyphae (Frey *et al.*, 1994), and because it is highly recalcitrant, it may be an important contributor to long-term soil carbon pools (Fig. 6.1). The fact that roots containing intraradical AM hyphae can contain even greater quantities of chitin (Frey *et al.*, 1994) suggests that AMF have the potential to alter fine-root decomposition rates (Langley & Hungate, 2003). Several studies have found chitinolytic microorganisms to be associated with AMF. For example, Ames *et al.* (1989) found that about 60% of AM fungal spores extracted from a calcareous grassland had chitin decomposers associated with them, such as *Streptomyces, Nocardia* and *Streptosporangium*, and Linkins and Neal (1982) observed that soil chitinase activity was greater in the presence of mycorrhizal shrubs in arctic tundra. The observation in upland acid grassland that both chitin decomposition and soil chitinase activity is seasonal, being greater in summer and autumn than in winter (Metcalfe *et al.*, 2002), indicates a coupling with plant and mycorrhizal growth. However, the crucial missing information is the rates and quantities of carbon allocation into AM fungal chitin and its subsequent rate of turnover in soil.

Because it is now established that the amounts of carbon allocated from plants to AMF vary depending upon both the plant and fungal species (Lerat *et al.*, 2003), future pulse-labelling studies also need to quantify carbon fluxes to the external mycelium of specific fungi. Stable-isotope probing (Radajewski *et al.*, 2000) is a promising new technology that has the potential to be developed for the determination of species of AMF that are actively receiving plant photosynthate to lesser or greater degrees. The method relies on the incorporation of ^{13}C, provided as a pulse to plants, into microbial DNA, which is subsequently separated by ultra-centrifugation from unlabelled ^{12}C-DNA and characterized using standard molecular biology techniques. In order to provide useful information on relatively slow-growing fungal populations (in contrast to bacteria), it is necessary to extract and separate ^{13}C-RNA, which is produced independent of cell replication. This has been achieved in rice fields where recently fixed carbon was detected in *Fusarium* and *Aspergillus* species (Lueders *et al.*, 2004), but has yet to be successfully applied to ecosystems dominated by AMF.

The fate of AM fungal carbon in soils

Because the quantities of carbon held within AMF are very large, their biomass must represent an attractive energy source for a range of organisms. This is borne out by the estimates of rapid hyphal turnover discussed already, which suggest at least some of the carbon held in AM fungal mycelium is decomposed rapidly. In the case of ectomycorrhizas, large numbers of bacteria have been observed (using scanning electron microscopy) on hyphal surfaces, and distinct bacterial communities have been characterized in the mycorrhizosphere. The obvious hypothesis to explain this phenomenon is that the carbon-rich mycorrhizal hyphae stimulate proliferation of bacteria. The situation with AMF is somewhat different with conflicting evidence as to the influence of AM fungal hyphae on bacterial populations. In some experiments distinct microbial communities have been found whilst others have shown no effect, although virtually all of these studies were undertaken before molecular microbial ecology techniques were in routine use. Stable-isotope probing may again be able to provide some of the answers by targeting mycorrhizosphere organisms directly involved in carbon uptake.

The nature of the extensive mycelial networks typical of AMF is such that they are prone to grazing by mycophagous soil invertebrates. The numbers of mesofauna in grassland can be enormous. In the UK, upland grassland typically contains between 20 000 and 160 000 collembola, mites, enchytraeids and nematodes per m^2. Collembola have been implicated in consumption of AM fungal hyphae for some time. McGonigle and Fitter (1988) and Warnock *et al.* (1982) found fragments of *Glomus*-like hyphae in the guts of the 'model' collembolan *Folsomia candida*. More recently, Johnson *et al.* (unpublished) found that the collembolan *Protaphorura armata*, which is commonly found in upland grassland, reduced respiration of recently fixed carbon from AMF by over 30%. There is no doubt that one of the major challenges for ecologists during the coming years is to untangle the biological pathways of carbon flux in complex soil food webs.

Breakdown of organic material and nutrient uptake

Possibly one of the most significant findings in plant and fungal ecology during the last decade or so was the unequivocal demonstration that ectomycorrhizal and ericoid mycorrhizal fungi could access organic forms of nutrients directly, and thus bypass the traditional view of the nitrogen cycle. The organic compounds are either taken up directly or chemically altered into simpler forms for uptake. Therefore, this process not only has a significant bearing on nutrient cycling pathways but also on the carbon cycle. The situation for AMF is less clear, primarily because much research has focused on their role in uptake of inorganic nutrients, especially phosphorus. The apparent inability of AMF to have any saprotrophic capability also suggested they could not utilize organic compounds. Furthermore, unlike ecto- and ericoid mycorrhizas, AMF tend to occur in more mineral soils, which lends support to this view. However, while on a global level this is clearly the case, much upland temperate grassland soil is highly enriched in organic matter and parallels can be drawn with ecto- and ericoid mycorrhizal habitats. In these often acidic grasslands, the plant communities frequently comprise long-lived species with relatively high C/N ratios (e.g. *Nardus*, *Molinia*), which in combination with low temperatures results in slow rates of litter decomposition. These conditions result in a large proportion of the total pool of nutrients to being in the organic form. The selection pressures placed on AMF that inhabit these systems are therefore likely to result in species that are capable of accessing organic forms of nitrogen and phosphorus. A growing body of literature is providing good evidence that plants can access organic forms of nitrogen. Glycine, which is one of the more labile amino acids, is often used as a tracer in such studies. Weigelt *et al.* (2003) injected a solution containing double-labelled (i.e. both ^{13}C- and ^{15}N-labelled) glycine into pots containing the grasses *Holcus lanatus*, *Agrostis capillaris* and *Lolium perenne* grown in soil taken from low- and high-nutrient-status upland grassland. They demonstrated that all species were able to take up the glycine intact, but that the greatest uptake was seen in *A. capillaris*, which typically inhabits low-fertility soils.

The pathways of this organic nitrogen assimilation from soil have yet to be established. Certainly, the role of saprotrophic microorganisms is of importance. Similar tracer studies have shown that significant quantities of labelled nitrogen are immobilized by the microbial biomass (at least, the potassium sulphate extractable biomass susceptible to chloroform fumigation), and, it is hypothesized, that the subsequent turnover of this labile pool is likely to be a major source of plant-available nitrogen (Bardgett *et al.*, 2003). The involvement of AMF in this process has rarely been studied. Hodge *et al.* (2001) were able to provide some evidence that AMF were able to stimulate release of organic nitrogen, in this case ^{13}C/^{15}N-labelled grass leaves, from a nutrient-rich patch in an inorganic soil. The presence of AM fungal mycelium was found to accelerate decomposition of the grass leaves in proportion to the biomass of AM mycelium, which resulted in greater uptake of labelled nitrogen. It should be stressed, however, that nitrogen was not shown to be taken up in the organic form directly.

A similar experiment, in which plants were grown in the mycorrhizal condition with three species of fungi or in the non-mycorrhizal condition, demonstrated that AMF did not respond to the presence of labelled glycine and did not result in greater uptake of nitrogen. One explanation for the inability of AMF to have major impacts on plant uptake of intact forms of organic nitrogen in these experiments may be the choice of fungal species or isolates used (*Glomus hoi, G. mosseae* and *Scutellospora dipurpurescens*). While these are common species, acid upland grassland may support more specialist AMF species or individuals that have adapted to the unique (by AMF standards) climatic and edaphic conditions. Given the mounting evidence that plants from acid upland grassland can take up organic nitrogen directly (Weigelt *et al.*, 2003), it seems highly improbable that AMF are not intimately involved in this process.

There can be no doubt that improved acquisition of inorganic phosphorus is a key function of AMF as demonstrated by the 300 or so papers that have dealt with this topic in the last 20 years. But how much closer are we in understanding the importance of inorganic phosphorus acquisition by AMF in nature, particularly in semi-natural grassland? Uptake of ^{33}P-labelled orthophosphate by the external mycelium of natural communities of AMF within intact upland grassland turfs has been shown to be extremely rapid, and a factor of ten greater than for plants that did not have mycelial access to the isotope (Johnson *et al.*, 2001). This and similar studies confirm the importance of the external mycelium in inorganic phosphorus uptake but unfortunately tells us little about the role of individual species. Early work has shown that some species of AMF, such as *Acaulospora laevis*, are able to forage several centimetres away from host roots, whilst other species, such as *Glomus* spp., concentrated their external mycelium in the rhizoplane (Jakobsen *et al.*, 1992). The recent move towards the view that AMF have greater host specificity than was hitherto thought, brought about by more thorough molecular analysis of communities of AMF in the field (Vandenkoornhuyse *et al.*, 2001), highlights the need to understand their functional diversity. But how can we test the hypothesis that one species of unculturable mycorrhizal fungus has a greater propensity of nutrient uptake than another? One approach may be to develop ^{33}P-isotope probing. Plant and mycorrhizal communities could be supplied with a large pulse of ^{33}P and the mycorrhizal DNA that is enriched, and therefore heavier, in ^{33}P separated from DNA that is not enriched. A potential constraint may be the amount of isotope needed versus the damage caused to DNA by its ionizing radiation. However, enrichment of DNA by ^{32}P, a far more energetic radionuclide than ^{33}P, has for many years been used to produce isotopic-ally distinct nucleic acid in bacteria (Hershey & Chase, 1952).

There is an increasing recognition that intraspecific functional variation may be considerable in AMF. This was elegantly demonstrated recently by Munkvold *et al.* (2004). These authors grew cucumber in association with 24 isolates of 4 species of *Glomus* and determined the degree of hyphal proli-feration and the amount of radioactive phosphorus taken up by the plants from root-free soil compartments. They found that there was considerable intraspecific variation in hyphal uptake of ^{33}P, which was explained by similar differences in hyphal proliferation. When expressed as ^{33}P uptake per length of hyphae, there was little intraspecific variation. In other words, the key physiological activities involved in phosphorus uptake and transport in the different isolates are similar, but the genetic or environmental cues that regulate hyphal growth differ. Such marked intraspecific differences may compensate for the comparatively small diversity of AMF seen in grasslands.

For many years, AM research centred on capture and translocation of inorganic phosphorus, largely, one suspects, because of the obvious impli-cations for agriculture. Research on the utilization of organic forms of phosphorus by AMF has lagged behind considerably, despite the well-known recognition that organic phosphorus constitutes a major compo-nent of total soil phosphorus in many semi-natural grassland soils. The ability of AMF to hydrolyse different organic phosphorus compounds by production of extracellular phosphatase activities has been the focus of much of this research to date. The outcome of these experiments still leaves many questions unanswered. Tarafdar and Marschner (1994) demon-strated that the mycorrhizal contribution to plant phosphorus uptake was 22%–33% when the nutrient was supplied in the inorganic form compared to 48%–59% when it was supplied as sodium phytate. In con-trast, Joner *et al.* (1995) found that AM fungal hyphae had no effect on soil phosphatase activity. Joner *et al.* (2000) suggest that the production of phosphatases by other soil organisms far exceeds that by AM fungal myce-lium and that the release of phosphatases into soil by AMF is likely to be incidental. It is therefore surprising that the extraradical mycelium of *Glomus intraradices* was shown to hydrolyse 5-bromo-4-chloro-indolyl phosphate and phenolphthalein diphosphate in axenic conditions (Koide & Kabir, 2000). It was also found that uptake of inorganic phosphorus was eight times greater than from organic sources; a not inconsiderable difference considering the potential reserves of organic phosphorus in many soils. To what extent the physiology of the single *G. intraradices* isolate used reflects this species, or indeed other species of AMF, as a whole remains to be seen.

Like many soil organisms, AMF have been shown to have some ability to alter the chemical environment in their immediate vicinity. A major control of mineral availability in soil is pH; aluminium and iron phosphates generally increase in availability with increasing pH, while the opposite is the case for calcium phosphates. Several studies have shown that hyphae of AMF can create localized zones of acidification. One of the most important mechanisms by which pH is lowered is through the release of H^+ ions after uptake of ammonium. This is however only likely to occur in the immediate vicinity of hyphae, as was the case when *G. mosseae* was grown with white clover (Li *et al.*, 1991). As previously discussed, AM fungal hyphae receive significant quantities of plant photosynthate, some of which is rapidly respired as CO_2. This CO_2 can dissolve in soil water to form a weak acid thus reducing soil pH and accelerating mineral weathering. Exudation of organic acids, such as oxalic acid, which has been demonstrated to occur from ectomycorrhizal mycelium (van Hees *et al.*, 2003), has yet to be seen in AMF. Calcium oxalate deposits have been observed in soils containing wheat plants and a native mix of *Glomus* species (Knight *et al.*, 1989), but later studies were unable to establish the plants or the fungus as the source of the oxalate (Knight *et al.*, 1992). Clearly, this is potentially an important process by which sparingly soluble elements could be made available by AMF and more work must be done in this area. Although there is an extensive literature on the role of AMF in mineral uptake from soil solution (see Clark & Zeto, 2000), to our knowledge, there are no studies that have investigated the impact of AMF on mineral weathering. This is not the case for ectomycorrhizal fungi, which have been implicated in dissolution of apatite-P and biotite-K. They are also thought to be responsible for the formation of micropores in feldspars and hornblendes in subsurface podzolic soil horizons and bedrock under coniferous and ericaceous plant communities (Jongmans *et al.*, 1997). In deeper soils, it seems unlikely that AM fungal mycelium associated with grassland plants would penetrate particularly far down the profile. However, some soils, for example rendzinas, are extremely shallow and AM fungal mycelium could easily come into contact with bedrock or large mineral fragments. In the case of rendzinas, phosphorus is almost certainly the limiting nutrient and dissolution of phosphates from the parent material may be an important pathway of phosphorus uptake.

Conclusions

Significant progress in recent years, particularly as a result of the widespread use and availability of isotopes and molecular techniques, has

146   *D. Johnson* et al.

been made in understanding the fate of carbon and nutrients in the natural environment and the specific roles played by AMF. However, the construction of realistic carbon budgets that include the contribution of AMF must remain a key priority. There is a danger that some of the values of carbon transfer to mycorrhizal fungi obtained from the pioneering but simplified experimental systems of the past become regarded as the norm for all AMF. There is an underlying feeling from the literature that scientists involved in ectomycorrhizal research are continually one step ahead of their colleagues working on AMF in the quest for understanding complex systems: for example, several studies have addressed how competition between ectomycorrhizal and saprotrophic species affects carbon and nutrient transfer. These interactions will certainly occur in grassland, which supports a healthy collection of saprotrophic fungi. This is not a criticism but merely highlights the relative difficulties of working on AMF. This should not be seen as an excuse but as a challenge in order to disentangle the undoubted importance of these organisms in biogeochemical cycling in nature.

Acknowledgements

The authors would like to thank Dr C. Osborne for comments on this chapter. DJ is supported by a Natural Environment Research Council (NERC) Advanced Fellowship.

References

Ames, R. N., Mihara, K. L. & Baynes, H. G. (1989). Chitin-decomposing actinomycetes associated with a vesicular arbuscular mycorrhizal fungus from a calcareous grassland. *New Phytologist*, **111**, 67–71.
Anderson, J. P. E. & Domsch, K. H. (1978). A physiological method for the quantitative measurement of microbial biomass in soils. *Soil Biology and Biochemistry*, **10**, 215–21.
Bago, B., Zipfel, W., Williams, R. M. *et al.* (2002). Translocation and utilization of fungal storage lipid in the arbuscular mycorrhizal symbiosis. *Plant Physiology*, **128**, 108–24.
Bago, B., Pfeffer, P. E., Abubaker, J. *et al.* (2003). Carbon export from arbuscular mycorrhizal roots involves the translocation of carbohydrate as well as lipid. *Plant Physiology*, **131**, 1496–507.
Bardgett, R. D., Streeter, T. C. & Bol, R. (2003). Soil microbes compete effectively with plants for organic-nitrogen inputs to temperate grasslands. *Ecology*, **84**, 1277–87.
Boddington, C. L., Bassett, E. E., Jakobsen, I. & Dodd, J. C. (1999). Comparison of techniques for the extraction and quantification of extraradical mycelium of arbuscular mycorrhizal fungi in soils. *Soil Biology and Biochemistry*, **31**, 479–82.
Brookes, P. C., Powlson, D. S. & Jenkinson, D. S. (1982). Measurement of microbial biomass phosphorus in soil. *Soil Biology and Biochemistry*, **14**, 319–29.

Brookes, P. C., Landman, A., Pruden, G. & Jenkinson, D. S. (1985). Chloroform fumigation and the release of soil-nitrogen – a rapid direct extraction method to measure microbial biomass nitrogen in soil. *Soil Biology and Biochemistry*, 17, 837–42.

Clark, R. B. & Zeto, S. K. (2000). Mineral acquisition by arbuscular mycorrhizal plants. *Journal of Plant Nutrition*, 23, 867–902.

Fitter, A. H., Graves, J. D., Watkins, N. K., Robinson, D. & Scrimgeour, C. (1998). Carbon transfer between plants and its control in networks of arbuscular mycorrhizas. *Functional Ecology*, 12, 406–12.

Francis, R. & Read, D. J. (1984). Direct transfer of carbon between plants connected by vesicular-arbuscular mycorrhizal mycelium. *Nature*, 307, 53–6.

Frey, B., Vilarino, A., Schuepp, H. & Arines, J. (1994). Chitin and ergosterol content of extraradical and intraradical mycelium of the vesicular-arbuscular mycorrhizal fungus *Glomus intraradices*. *Soil Biology and Biochemistry*, 26, 711–17.

Friese, C. F. & Allen, M. F. (1991). The spread of VA mycorrhizal fungal hyphae in soil: inoculum type and external hyphal architecture. *Mycologia*, 83, 409–18.

Gallaud, I. (1905). Etudes sur les mycorrhizes endotrophs. *Revue Générale de Botanique*, 17, 5–48, 66–83, 123–35, 223–39, 313–25, 425–33, 479–500.

Grime, J. P., Mackey, J. M. L., Hillier, S. H. & Read, D. J. (1987). Floristic diversity in a model system using experimental microcosms. *Nature*, 328, 420–2.

Hart, M. M. & Reader, R. J. (2002). Taxonomic basis for variation in the colonization strategy of arbuscular mycorrhizal fungi. *New Phytologist*, 153, 335–44.

Hershey, A. D. & Chase, M. (1952). Independent functions of viral protein and nucleic acid in growth of bacteriophage. *Journal of General Physiology*, 36, 39–56.

Ho, I. & Trappe, J. M. (1973). Translocation of ^{14}C from *Festuca* plants to their endomycorrhizal fungi. *Nature New Biology*, 244, 30–1.

Hodge, A., Campbell, C. D. & Fitter, A. H. (2001). An arbuscular mycorrhizal fungus accelerates decomposition and acquires nitrogen directly from organic material. *Nature*, 413, 297–9.

Holtum, J. A. M. & Winter, K. (2003). Photosynthetic CO_2 uptake in seedlings of two tropical tree species exposed to oscillating elevated concentrations of CO_2. *Planta*, 218, 152–8.

Jakobsen, I., Abbott, L. K. & Robson, A. D. (1992). External hyphae of vesicular-arbuscular mycorrhizal fungi associated with *Trifolium subterraneum* L. 1. Spread of hyphae and phosphorus inflow into roots. *New Phytologist*, 120, 371–80.

Jenkinson, D. S. & Powlson, D. S. (1976). The effects of biocidal treatments on metabolism in soil V. A method for measuring soil biomass. *Soil Biology and Biochemistry*, 8, 209–13.

Johnson, D., Leake, J. R. & Read, D. J. (2001). Novel in-growth core system enables functional studies of grassland mycorrhizal mycelial networks. *New Phytologist*, 152, 555–62.

Johnson, D., Leake, J. R., Ostle, N., Ineson, P. & Read, D. J. (2002a). In situ $^{13}CO_2$ pulse-labelling of upland grassland demonstrates a rapid pathway of carbon flux from arbuscular mycorrhizal mycelium to the soil. *New Phytologist*, 153, 327–34.

Johnson, D., Leake, J. R. & Read, D. J. (2002b). Transfer of recent photosynthate into mycorrhizal mycelium of an upland grassland: short-term respiratory losses and accumulation of ^{14}C. *Soil Biology and Biochemistry*, 34, 1521–4.

Joner, E. J., Magid, J., Gahoonia, T. S. & Jakobsen, I. (1995). P depletion and activity of phosphatases in the rhizosphere of mycorrhizal and non-mycorrhizal cucumber (*Cucumis sativus* L). *Soil Biology and Biochemistry*, 27, 1145–51.

Joner, E. J., van Aarle, I. M. & Vosatka, M. (2000). Phosphatase activity of extraradical arbuscular mycorrhizal hyphae: a review. *Plant and Soil*, 226, 199–210.

Jongmans, A. G., van Breemen, N., Lundstrom, U. *et al.* (1997). Rock-eating fungi. *Nature*, 389, 682–3.

148   *D. Johnson* et al.

Kabir, Z. & Koide, R. T. (2002). Effect of autumn and winter mycorrhizal cover crops on soil properties, nutrient uptake and yield of sweet corn in Pennsylvania, USA. *Plant and Soil*, **238**, 205–15.

Kabir, Z., O'Halloran, I. P., Widden, P. & Hamel, C. (1998). Vertical distribution of arbuscular mycorrhizal fungi under corn (*Zea mays* L.) in no-till and conventional tillage systems. *Mycorrhiza*, **8**, 53–5.

Knight, W. G., Allen, M. F., Jurinak, J. J. & Dudley, L. M. (1989). Elevated carbon-dioxide and solution phosphorus in soil with vesicular-arbuscular mycorrhizal western wheatgrass. *Soil Science Society of America Journal*, **53**, 1075–82.

Knight, W. G., Dudley, L. M. & Jurinak, J. J. (1992). Oxalate effects on solution phosphorus in a calcareous soil. *Arid Soil Research Rehabilitation*, **6**, 11–20.

Koide, R. T. & Kabir, Z. (2000). Extraradical hyphae of the mycorrhizal fungus *Glomus intraradices* can hydrolyse organic phosphate. *New Phytologist*, **148**, 511–17.

Kuzyakov, Y., Kretzschmar, A. & Stahr, K. (1999). Contribution of *Lolium perenne* rhizodeposition to carbon turnover of pasture soil. *Plant and Soil*, **213**, 127–36.

Kuzyakov, Y., Ehrensberger, H. & Stahr, K. (2001). Carbon partitioning and below-ground translocation by *Lolium perenne*. *Soil Biology and Biochemistry*, **33**, 61–74.

Langley, J. A. & Hungate, B. A. (2003). Mycorrhizal controls on below-ground litter quality. *Ecology*, **84**, 2302–12.

Leake, J. R., Donnelly, D. P., Saunders, E. M., Boddy, L. & Read, D. J. (2001). Rates and quantities of carbon flux to ectomycorrhizal mycelium following ^{14}C pulse labelling of *Pinus sylvestris* seedlings: effects of litter patches and interaction with a wood-decomposer fungus. *Tree Physiology*, **21**, 71–82.

Leake, J. R., Johnson, D., Donnelly, D. P. et al. (2004). Networks of power and influence: the role of mycorrhizal mycelium in controlling plant communities and agroecosystem function. *Canadian Journal of Botany*, **82**, 1016–45.

Lerat, S., Lapointe, L., Piche, Y. & Vierheilig, H. (2003). Variable carbon-sink strength of different *Glomus mosseae* strains colonizing barley roots. *Canadian Journal of Botany*, **81**, 886–9.

Li, X. L., George, E. & Marschner, H. (1991). Phosphorus depletion and pH decrease at the root soil and hyphae soil interfaces of VA mycorrhizal white clover fertilized with ammonium. *New Phytologist*, **119**, 397–404.

Linkins, A. E. & Neal, J. L. (1982). Soil cellulase, chitinase, and protease activity in *Eriophorum vaginatum* tussock tundra in Eashle Summit, Alaska. *Holarctic Ecology*, **5**, 135–8.

Lueders, T., Wagner, B., Claus, P. & Friedrich, M. W. (2004). Stable isotope probing of rRNA and DNA reveals a dynamic methylotroph community and trophic interactions with fungi and protozoa in oxic rice field soil. *Environmental Microbiology*, **6**, 60–72.

McGonigle, T. P. & Fitter, A. H. (1988). Ecological consequences of arthropod grazing on VA mycorrhizal fungi. *Proceedings of the Royal Society of Edinburgh B*, **94**, 25–32.

McNaughton, S. J. & Oestenheld, M. (1990). Extramatrical mycorrhizal abundance and grass nutrition in a tropical grazing ecosystem, the Serengeti National Park, Tanzania. *Oikos*, **59**, 92–6.

Martin, J. K. & Merckx, R. (1992). The partitioning of photosynthetically fixed carbon within the rhizosphere of mature wheat. *Soil Biology and Biochemistry*, **24**, 1147–56.

Metcalfe, A. C., Krsek, M., Gooday, G. W., Prosser, J. I. & Wellington, E. M. H. (2002). Molecular analysis of a bacterial chitinolytic community in an upland pasture. *Applied and Environmental Microbiology*, **68**, 5042–50.

Miller, R. M., Reinhardt, D. R. & Jastrow, J. D. (1995). External hyphal production of vesicular-arbuscular mycorrhizal fungi in pasture and tallgrass prairie communities. *Oecologia*, **103**, 17–23.

Munkvold, L., Kjøller, R., Vestberg, M., Rosendahl, S. & Jakobsen, I. (2004). High functional diversity within species of arbuscular mycorrhizal fungi. *New Phytologist*, **164**, 357–64.

Nehl, D. B., McGee, P. A., Torrisi, V., Pattinson, G. S. & Allen, S. J. (1999). Patterns of arbuscular mycorrhiza down the profile of a heavy textured soil do not reflect associated colonization potential. *New Phytologist*, **142**, 495–503.

Olsrud, M. & Christensen, T. R. (2004). Carbon cycling in subarctic tundra; seasonal variation in ecosystem partitioning based on *in situ* ^{14}C pulse-labelling. *Soil Biology and Biochemistry*, **36**, 245–53.

Olsson, P. A. & Wilhelmsson, P. (2000). The growth of external AM fungal mycelium in sand dunes and in experimental systems. *Plant and Soil*, **226**, 161–9.

Olsson, P. A., Bååth, E., Jakobsen, I. & Söderström, B. (1995). The use of phospholipid and neutral lipid fatty acids to estimate biomass of arbuscular mycorrhizal fungi in soil. *Mycological Research*, **99**, 623–9.

Olsson, P. A., Larsson, L., Bago, B., Wallander, H. & van Aarle, I. M. (2003). Ergosterol and fatty acids for biomass estimation of mycorrhizal fungi. *New Phytologist*, **159**, 7–10.

Ostle, N., Ineson, P., Benham, D. & Sleep, D. (2000). Carbon assimilation and turnover in grassland vegetation using an *in situ* ^{13}CO$_2$ pulse labelling system. *Rapid Communications Mass Spectrometry*, **14**, 1345–50.

Paul, E. A. & Kucey, R. M. N. (1981). Carbon flow in plant microbial associations. *Science*, **213**, 473–4.

Pfeffer, P. E., Douds, D. D., Becard, G. & Shachar-Hill, Y. (1999). Carbon uptake and the metabolism and transport of lipids in an arbuscular mycorrhiza. *Plant Physiology*, **120**, 587–98.

Powell, C. L. (1979). Spread of mycorrhizal fungi through soil. *New Zealand Journal of Agricultural Research*, **22**, 335–9.

Radajewski, S., Ineson, P., Parekh, N. R. & Murrell, J. C. (2000). Stable-isotope probing as a tool in microbial ecology. *Nature*, **403**, 646–9.

Read, D. J. (2002). Towards ecological relevance – progress and pitfalls in the path towards an understanding of mycorrhizal functions in nature. In *Mycorrhizal Ecology*, ed. M. G. A. Van der Heijden & I. R. Sanders. Berlin: Springer-verlag, pp. 3–24.

Read, D. J., Koucheki, H. K. & Hodgson, J. (1976). Vesicular-arbuscular mycorrhiza in natural vegetation systems. I. The occurrence of infection. *New Phytologist*, **77**, 641–53.

Rillig, M. C., Wright, S. F. & Eviner, V. T. (2002). The role of arbuscular mycorrhizal fungi and glomalin in soil aggregation: comparing effects of five plant species. *Plant and Soil*, **238**, 325–33.

Saggar, S., Hedley, C. & Mackay, A. D. (1997). Partitioning and translocation of photosynthetically fixed ^{14}C in grazed hill pastures. *Biology and Fertility of Soils*, **25**, 152–8.

Sparling, G. P. & Tinker, P. B. (1978). Mycorrhizal infection in Pennine grassland. I. Levels of infection in the field. *Journal of Applied Ecology*, **15**, 943–950.

Staddon, P. L., Ostle, N., Dawson, L. A. & Fitter, A. H. (2003a). The speed of soil carbon throughput in an upland grassland is increased by liming. *Journal of Experimental Botany*, **54**, 1461–1469.

Staddon, P. L., Ramsey, C. B., Ostle, N., Ineson, P. & Fitter, A. H. (2003b). Rapid turnover of hyphae of mycorrhizal fungi determined by AMS microanalysis of ^{14}C. *Science*, **300**, 1138–1140.

Staddon, P. L., Thompson, K., Jakobsen, I. *et al.* (2003c). Mycorrhizal fungal abundance is affected by long-term climatic manipulations in the field. *Global Change Biology*, **9**, 186–94.

Sylvia, D. M. (1988). Activity of external hyphae of vesicular-arbuscular mycorrhizal fungi. *Soil Biology and Biochemistry*, **20**, 39–43.

Tarafdar, J. C. & Marschner, H. (1994). Phosphatase-activity in the rhizosphere and hyphosphere of VA mycorrhizal wheat supplied with inorganic and organic phosphorus. *Soil Biology and Biochemistry*, **26**, 387–95.

Tisdall, J. M. & Oades, J. M. (1979). Stabilization of soil aggregates by the root systems of ryegrass. *Australian Journal of Soil Research*, **17**, 429–41.

Toal, M. E., Yeomans, C., Killham, K. & Meharg, A. A. (2000). A review of rhizosphere carbon flow modelling. *Plant and Soil*, **222**, 263–81.

Treseder, K. K. & Allen, M. F. (2000). Mycorrhizal fungi have a potential role in soil carbon storage under elevated CO_2 and nitrogen deposition. *New Phytologist*, **147**, 189–200.

Treseder, K. K., Egerton-Warburton, L. M., Allen, M. F., Cheng, Y. F. & Oechel, W. C. (2003). Alteration of soil carbon pools and communities of mycorrhizal fungi in chaparral exposed to elevated carbon dioxide. *Ecosystems*, **6**, 786–96.

Van der Heijden, M. G. A., Klironomos, J. N., Ursic, M. *et al.* (1998). Mycorrhizal fungal diversity determines plant biodiversity, ecosystem variability and productivity. *Nature*, **396**, 69–72.

van Hees, P. A. W., Godbold, D. L., Jentschke, G. & Jones, D. L. (2003). Impact of ectomycorrhizas on the concentration and biodegradation of simple organic acids in a forest soil. *European Journal of Soil Science*, **54**, 697–706.

Vandenkoornhuyse, P., Leyval, C. & Bonnin, I. (2001). High genetic diversity in arbuscular mycorrhizal fungi: evidence for recombination events. *Heredity*, **87**, 243–53.

Vandenkoornhuyse, P., Baldauf, S. L., Leyval, C., Straczek, J. & Young, J. P. W. (2002). Extensive fungal diversity in plant roots. *Science*, **295**, 2051.

Wagner, G. (1974). Observation of fungal growth in soil using a capillary pedoscope. *Soil Biology and Biochemistry*, **6**, 327–33.

Warnock, A. J., Fitter, A. H. & Usher, M. B. (1982). The influence of a springtail *Folsomia candida* (Insecta, Collembola) on the mycorrhizal association of leek *Allium porrum* and the vesicular-mycorrhizal endophyte *Glomus fasciculatus*. *New Phytologist*, **90**, 285–92.

Weigelt, A., King, R., Bol, R. & Bardgett, R. D. (2003). Inter-specific variability in organic nitrogen uptake of three temperate grassland species. *Journal of Plant Nutrition and Soil Science*, **166**, 606–11.

Zhu, Y. G. & Miller, R. M. (2003). Carbon cycling by arbuscular mycorrhizal fungi in soil-plant systems. *Trends in Plant Science*, **8**, 407–9.

7

The role of wood decay fungi in the carbon and nitrogen dynamics of the forest floor

SARAH WATKINSON, DAN BEBBER,
PETER DARRAH, MARK FRICKER,
MONIKA TLALKA AND LYNNE BODDY

Introduction

The mycelium of woodland fungi can act both as a reservoir and as a distribution system for nutrients, owing to its physiological and developmental adaptations to life at the interface between organic and mineral soil horizons. The mobility of accumulated nitrogen and phosphorus within the mycelial networks of cord-forming wood decay fungi and ectomycorrhiza enables fungi to play key roles as wood decomposers and root symbionts. The dynamics of nitrogen movement have been less investigated than phosphorus owing to lack of a suitable tracer. We have developed a new technique for tracing nitrogen translocation in real time, using ^{14}C as a marker for nitrogen by incorporating it into a non-decomposed amino acid that tracks the mycelial free amino acid pool. Its movement can be imaged by counting photon emissions from a scintillant screen in contact with the mycelial system. This method allows real-time imaging at high temporal and spatial resolution, for periods of weeks and areas up to $1\,m^2$, in microcosms that mimic the mineral/organic soil interface of the forest floor. The results reveal a hitherto unsuspected dynamism and responsiveness in amino acid flows through mycelial networks of cord-forming, wood-decomposing basidiomycetes. We interpret these in the light of current understanding of the pivotal role of fungi in boreal and temperate forest floor nutrient cycling, and attempt to formulate key questions to investigate the effects of mycelial nitrogen translocation on forest floor decomposition and nitrogen absorption.

Fungi in Biogeochemical Cycles, ed. G. M. Gadd. Published by Cambridge University Press. © British Mycological Society 2006.

Adaptive ecophysiology of wood decay fungi

Physiological adaptation for nitrogen scavenging

Wood, the main carbon and energy source of wood decay fungi, is carbon rich but nitrogen poor, with a C/N ratio of up to 1250/1. The physiological adaptations of wood decay fungi were investigated in a classic series of papers (see Levi & Cowling, 1969). In defined media with carbon and nitrogen sources (glucose and asparagine) in solution, a carbon/energy supply was shown to be essential for nitrogen uptake, and a proportion of available nitrogen remained unassimilated by the mycelium when the C/N ratio was below 40/1. In wood, both carbon and nitrogen are in insoluble form, protected from decomposition by substantial barriers. Cellulose, which is the main utilizable carbon component, is masked by a lignin coating in the S1 and S3 cell wall layers, while the sparse nitrogen content is embedded as protein (Rayner & Boddy, 1988). Sequential activities of extracellular enzyme complexes, and the cell membrane transport systems for the uptake of resulting monomers, are induced and repressed during fungal degradation of plant cell walls.

Fungi specialized to utilize wood show a number of nitrogen-conserving strategies, including nitrogen recycling from old to new mycelium (Lilly *et al.*, 1991), selective depletion of cell materials under nitrogen starvation and the ability to utilize and select between many different chemical compounds of nitrogen (Caddick, 2002, 2004). From gross measurements, mycelial nitrogen appears to be retrieved from old mycelium and preferentially allocated to sites of growth, for example nitrogen is moved from decayed tree trunks into sporophores (Merrill & Cowling, 1966). Nitrogen can be concentrated from wood, e.g. the mycelium of *Serpula lacrymans*, growing out over a non-nutrient surface from wood blocks with an initial nitrogen concentration of 0.38 and 0.67 mg g^{-1} dry wt contained 16.4 and 36.8 mg N g^{-1} dry wt (Watkinson *et al.*, 1981). Levi and Cowling (1969) measured a 40-fold difference, 0.2% to 8%, in total nitrogen content between mycelium of a white-rot wood decay fungus growing on nitrogen-poor and nitrogen-rich defined media.

Nitrogen scavenging involves metabolic regulation both for opportunistic accumulation and storage, and for adjustment to starvation. Surplus nitrogen is stored in an expandable free amino acid pool (Venables & Watkinson, 1989; Griffin, 1994; Hanks *et al.*, 2003), partly in the cytoplasm and partly as basic amino acids in vacuoles (Klionsky *et al.*, 1990), which may have a role in translocation (Bago *et al.*, 2001; Govindarajulu *et al.*, 2005). Protein accumulated by fungi when nitrogen exceeds requirements may include specific storage proteins (e.g. Rosen *et al.*, 1997). Nitrogen incorporated into chitin in cell walls during mycelial growth may also serve as a reserve. Other nitrogen compounds stored in mycelium have been less well investigated, but include nucleic acids, polyamines (Davis, 1996) and allantoin (Cooper, 1996). Under conditions of nitrogen starvation, metabolic and transport pathways for nitrogen acquisition, and for catabolism of reserves, are activated by global regulators of gene expression (Marzluf, 1996; Caddick, 2002, 2004), which control cellular sensing and response to nitrogen starvation, sensed by uncharged tRNA when intracellular levels of amino acid fall to very low levels (Marzluf, 1996; Caddick, 2002, 2004; Winderickx & Taylor, 2004).

Development and function of translocating networks

The ability to develop and operate complex, extensive and persistent translocation networks (Leake *et al.*, 2004; Fricker *et al.*, 2005) underlies the unique ecosystem function of woodland basidiomycetes (Fig. 7.1). Sugars and sugar alcohols (Jennings, 1987, 1995), amino acids (Tlalka *et al.*, 2002, 2003) phosphate (Wells *et al.*, 1998; 1999) and oxygen (Pareek, Allaway & Ashford, unpublished) can all be translocated. Phosphate translocation over areas in the order of square metres has been measured in *Hypholoma fasciculare*, *Phanerochaete velutina* and *Phallus impudicus* cord systems growing in woodland (Wells & Boddy, 1995). Translocation pathways link sites of supply and demand to facilitate processes such as advance over non-nutrient surfaces (Paustian & Schnurer, 1987; Davidson & Olsson, 2000), the secretion and activities of lignocellulolytic enzymes in massive woody remains (Sinsabaugh & Liptak, 1997), solubilization of phosphate compounds in soil (Jacobs *et al.*, 2002), and the supply of energy to hyphal tips to maintain the hydrostatic pressure for invasive hyphal growth, since 25% of hyphal ATP is used to drive active uptake (Wessels, 1999). The mechanisms that allow both a circulatory translocation system (Olsson & Gray, 1998; Wells *et al.*, 1998; Lindahl *et al.*, 2001), and mass flow (Jennings, 1987; Tlalka *et al.*, 2002) remain obscure (Cairney, 1992).

Continuous remodelling of the mycelium, with development of cords and rhizomorphs (Cairney *et al.*, 1989; Cairney, 1992; Fig. 7.2), opens new nutrient flow pathways in response to the geometry of supply and demand. The responsive growth of fungi searching for and colonizing woody resource units is termed 'foraging' (Boddy, 1993, 1999; Olsson, 2001), and involves coordinated context-dependent modulation of the network (Fig. 7.3). It is not known how local capture of a fresh carbon resource cues a global reorganization of the colony but it is evident, from the simultaneous coordinated responses of the whole system, that local sensing generates a propagated signal that elicits site-specific developmental responses. There is hyphal aggregation and differentiation along those hyphae that have made contact with the fresh resource, forming mycelial cords connecting old and new resources. At the same time, distant mycelium that has not colonized a new resource dies back. Changes in

Fig. 7.1. Diagram illustrating the translocation demands on a mycelial network. (a) Hypothetical microcosm, with mycelium growing from a central resource, encountering carbon or nitrogen resources, or autolysing in regions of the network where no further resources have been acquired; (b) Spatial relationships of the mycelial network and its resources in the forest floor.

intracellular hydrostatic pressure, intracellular amino acid concentration, and electrical potentials, have been suggested as translocatable stimuli that cue developmental changes (Rayner, 1991; 1994; Olsson & Hansson, 1995; Rayner *et al.*, 1995; Watkinson, 1999). Measuring the intracellular protein-ase expression cued by remote resource contact could provide a marker to report reception of a propagated stimulus. Starvation-induced proteinases have been investigated with a view to developing a suitable reporter for identifying sites in the mycelium where autolysis and regression are induced. A serine proteinase activity was identified in *Serpula lacrymans* with characteristics consistent with a role in mobilizing intracellular protein stores from cytoplasm, as well as a lysosomal-type activity optimal at low pH (Watkinson *et al.*, 2001; Wadekar *et al.*, 1995). Genes for serine

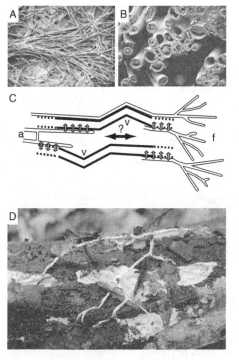

Fig. 7.2. The structure of the translocation pathway in mycelial cords. (A) Hyphae fanning out at the distal end of a cord of *Phanerochaete velutina* (scanning electron microscopy by A. Yarwood); (B) Internal structure of a cord of *Serpula lacrymans*, showing vessels and cytoplasm-filled hyphae and extracellular matrix material. (C) Diagram of the components of the translocation pathway (adapted from Cairney, 1992): v, vessel hypha; f, foraging front; a, anastomosis; (D) A cord system in beech woodland showing both corded mycelium and diffuse growth in contact with the wood substrate.

proteinases of several cord-forming wood decay fungi have been cloned (Eastwood & Tlalka, unpublished; Eastwood & Higgins, unpublished) and show homology to other fungal serine proteinases with putative roles in internal protein mobilization (Kingsnorth *et al.*, 2001). Autolysis associated with development in fungi has features associated with apoptosis, and caspase-like activity has been demonstrated in *Aspergillus nidulans* (Thrane *et al.*, 2004), but it is not known if the autolysis that occurs during the development of cord formers is apoptotic.

Cord development is triggered not only by connection between resources, but also by the nutrient status of the mycelium. Developmental responses in saprotrophic cord formers may be caused by the changes in C/N balance in individual hyphae resulting from a local rise in nutrient levels due to uptake and translocation of nutrients within some hyphae but not others. Development and biomass production are differentially affected by carbon and nitrogen content and the C/N ratio in defined media (Watkinson, 1999; Fig. 7.4). Biomass increased with both sucrose and aspartate, while

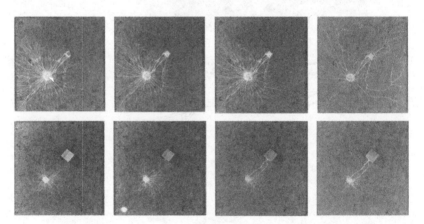

Fig. 7.3. The effect of resource size on foraging patterns. The images show the development of *Phanerochaete velutina* extending from 0.5 cm³ beech (*Fagus sylvatica*; resource on left) wood resources to 0.5 cm³ (top row) and 8 cm³ (bottom row), in 24 × 24 cm trays of non-sterile soil after 11, 15, 20 and 27 days (from left to right). Note that when the new resource is the same size as the inoculum (top row), the former is colonized and there is then continued outgrowth from this resource. There are extensive tangential anastomoses, joining mycelial cords to the new resource. When the new resource is considerably larger than the inoculum, there is considerable thickening of cords joining the two resources, and regression of much of the non-connective mycelium. Colonization of and subsequent outgrowth from the large new resource (not shown in this series) takes considerably longer than from the small new resource. Digital images were obtained from photographs taken by Rory Bolton.

cord development was greatest at the highest sucrose and lowest aspartate content, suggesting that nitrogen limitation in the presence of surplus carbon could be a stimulus to development. Nitrate, in contrast to glutamate, suppressed secondary metabolism and accelerated the extension rate, cord formation with the production of extracellular matrix material, and the onset of autolysis. In heterogeneous microcosms, with cellulose as sole carbon source and nitrate as nitrogen source (Watkinson *et al.*, 1981), cellulolysis was nitrogen-limited at the lowest nitrate levels but hyphal extension was not. The rate of colony radial extension from a defined resource over an adjacent non-nutrient surface was affected by nitrogen in the resource, being maximal at intermediate levels. High nitrogen levels in the resource induced abundant localized growth, and low levels supported only a very sparse outgrowth.

The genetics of the morphogenetic effect of intracellular nutrient levels has been investigated mainly in *Saccharomyces cerevisiae* and in mammals (Cooper, 2004; Roosen *et al.*, 2004; Winderickx & Taylor, 2004). Yeast and mammalian models have identified the proteins GCN2, GCN4, elF2 and TOR (Target of Rapamycin) as key components of intracellular nutrient sensing and signalling to elicit metabolic and developmental cell responses, such as pseudomycelium formation in yeast cells. The *GCN4* encodes a transcription factor with homology to mammalian GATA factors, which increases transcription of amino acid biosynthetic genes.

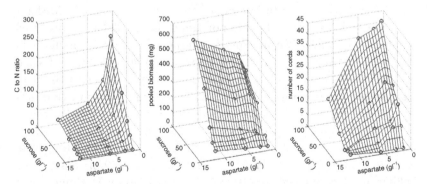

Fig. 7.4. Differential effect of nutrient supply on biomass and cord development in *Serpula lacrymans*. The fungus was inoculated centrally on uniform defined agar media in a factorial experiment with 5, 10, 20, 40 and 80 g/l sucrose and 1.5, 3, 6 and 12.5 g/l sodium aspartate. Biomass was measured as oven dry weight at 8 weeks, cords were counted crossing a circular transect 3 cm from the inoculum disc. Left to right: C/N ratio of each of the twenty combinations of carbon and nitrogen concentration; biomass; cord development. (Adapted from Watkinson, 1975).

The transcription regulator elF2 is implicated in sensing intracellular amino acid via levels of uncharged tRNA. Target of Rapamycin has a role in inducing autophagy via sensing intracellular amino acid starvation. Specific amino acids may give developmental cues in fungi (e.g. Watkinson, 1977), and these can be sensed inside or outside the cell, although the receptor–transduction–effector pathways are obscure. In fungi on uniformly nutrient-rich medium, the intracellular amino acid concentration and content rises and falls with that of the medium (Griffin, 1994), but in the absence of a supply in the immediate environment, hyphae that form a translocation pathway will become relatively nutrient-rich, and may support further growth and development in their immediate vicinity.

Mycelial nitrogen accumulation and distribution in the forest floor

The forest floor

The forest floor litter layers, also termed the 'O' horizon (Currie, 1999; Currie et al., 1999), are persistent and stratified under conditions where slow decomposition favours the accumulation of organic matter overlying the mineral soil beneath. Litter depth varies, reflecting the decay rate. Three layers can typically be distinguished: the intact surface litter (L) with coarse woody particulate residues, lying above the fragmented litter (F) and humus (H) sub-horizons. The main carbon input to the forest floor is from canopy litter fall. In a warm temperate pine forest, Richter et al. (1999) measured annual carbon input to the forest floor as $2.9\,\mathrm{Mg\,ha^{-2}}$ $\mathrm{yr^{-1}}$, of which 2.45 was in the form of canopy litter fall, 0.08 as dissolved organic carbon and 0.37 came from root turnover. Carbon accumulated in the forest floor was estimated as 13 years' accretion, consisting of accumulated woody litter and recalcitrant organic compounds. There is abundant carbon in the form of dead leaves, seeds, branches and tree trunks (Harmon et al., 1986), which give the surface layers of the forest floor their physical structure, and plant nutrients flow into it from rain washing through the canopy and surface litter (Table 7.1; Heal & Dighton, 1986; Richter & Markewitz, 2001). The forest floor organic horizons are the habitat of a highly specialized flora and fauna. In lower F, H and upper A layers, cord systems of saprotrophs (Thompson, 1984) intermingle with ectomycorrhizal mycelium (Agerer, 2001) and rootlets. Ectomycorrhizal basidiomycetes are closely related to saprotrophic species from which they have probably evolved relatively recently (Hibbett et al., 2000), and can have the capacity to utilize plant litter, as well as living roots, as both a nitrogen and a carbon source (Read, 1991; Bending & Read, 1995;

Table 7.1. *Typical data for boreal and temperate forest floor nitrogen pools and fluxes (g m^{-2}; g m^{-2} yr^{-1} respectively), to illustrate potential capacity of forest floor mycelium to absorb atmospheric reactive nitrogen*

Source or pool for nitrogen	Quantity	Reference
Rain	NH_4^+ 6.4 g m^{-2} yr^{-1} NO_3^- 6.6 ''	Bringmark (1980), cited in
Canopy through-fall	NH_4^+ 4.2 g m^{-2} yr^{-1} NO_3^- 6.3 ''	Heal & Dighton (1986)
Litter leachate at 6 cm depth	NH_4^+ 1.1 g m^{-2} yr^{-1} NO_3^- 0.55 ''	
Total solution input to forest floor	0.8–1.4 g m^{-2} yr^{-1}	Richter & Markewitz, (2001)
Typical atmospheric pollution	0.5–3.0 g m^{-2} yr^{-1}	Sievering (1999)
Fungal biomass, living and dead, in boreal forest soil O and upper A horizons	102 g dry wt m^{-2}	Bååth & Söderström (1979)
Total N in living and dead hyphae, assuming ambient levels typical of N-limited forest soil	2.7 g m^{-2}	Bååth & Söderström (1979)
Total N in living hyphae (with fungal biomass estimated from ergosterol assay)	2.2–3.2 g m^{-2}	Markkola *et al.*, (1995)
Total retention in forest floor litter	15 g m^{-2}	Currie (1999)
Typical input as agricultural fertilizer	12.4 g m^{-2}	Dafodu (2003)

Perez-Moreno & Read, 2000). Decomposer fauna abound, including mites and collembola, many of which are mycophagous (Maraun *et al.*, 2003).

Most boreal and temperate forests are nitrogen-limited, becoming more so with prolonged lack of disturbance (Vitousek & Howarth, 1991; Aber, 1992; Hart *et al.*, 1994; Richter & Markewitz, 2001; Wardle, 2002; Wardle *et al.*, 2004b). In mature boreal forest the nitrogen present in the soil is held in the biota, plant roots and microbial biomass, or is in the form of chemical compounds such as amino-phenolic polymers that are highly resistant to enzymic degradation with half-lives of many decades (Northup *et al.*, 1995). Nitrogen compounds in canopy through-fall are immediately absorbed by plants and microbes (Bringmark, 1980; Heal & Dighton, 1986; Downs *et al.*, 1996). Amino acids (utilizable by boreal forest plants, Nasholm *et al.*, 1998) or inorganic ions, are available to soil microbes and plants only sporadically, as a result of disturbances such as fire, wet–dry cycles, damage to individual trees, or clear-cutting. Available nitrogen is competed for by plants and microbes (Tamm, 1982; Kaye & Hart, 1997; Hodge *et al.*, 2000; Guo *et al.* 2004). Nitrate is rapidly removed (Cain *et al.*, 1999) and may remain in the soil for periods of hours only (Stark & Hart, 1997; Berntson & Aber, 2000). A pulsed addition of $^{15}NO_3^-$ gave 16% enrichment of microbial biomass in 2 hours (Zogg *et al.*, 2000), and such rapid uptake may be critical in competition between microbes and plants. Hirobe *et al.* (2003) found a very low net rate of nitrification in upland moder soils under *Cryptomeria* forest, which they ascribed to dynamic internal nitrogen cycling with immediate microbial consumption of the low levels of nitrate produced. Freshly collected soil samples had no measurable nitrate content, although after incubation (without further carbon input) between 50 and 100 mg nitrate N per kg of soil was detected.

Nitrogen limitation affects the rate of litter decomposition. As a general rule applicable to temperate forests, the higher the C/N ratio and lignin content of plant litter, the more slowly it decomposes (Aber & Melillo, 1982; Dighton, 1997, 2003). The reasons for this are not straightforward, and at the whole-organism level the C/N ratio has complex interactive effects, e.g. raised ambient nitrogen may increase cellulase activity (Watkinson *et al.*, 1981) and decrease lignin degradation (Keyser *et al.*, 1978; Kirk & Fenn, 1982; Carreiro *et al.*, 2000). In the field, application of nitrogen has variable effects on litter decomposition (Neff *et al.*, 2002; Micks *et al.*, 2004), and on enzymic activities of decomposers, including polyphenol oxidation and polysaccharide hydrolysis (Sinsabaugh *et al.*, 2002; Waldrop *et al.*, 2004).

Decomposition of woody litter in the slow-decomposing, nitrogen-limited environment typical of most temperate forest soils does not release inorganic nutrients directly into the soil (Wardle, 2002; Lal, 2004; Wardle *et al.*, 2004a, b). Lindahl *et al.* (2002) have proposed a model of boreal forest floor nutrient cycling in which nitrogen from decomposing plant litter is not released as nitrate, but retained within mycelium of decomposer fungi and translocated to nitrogen-poor, carbon-rich litter. This contrasts with the conventional model for the role of decomposer microbes in nitrogen cycling, based on agricultural soils, in which mineral nitrogen is thought to be released once the C/N ratio of the substrate falls below a critical level (Swift *et al.*, 1979). In the Lindahl model, nitrogen remains in the decomposer mycelium until it is captured by other organisms such as antagonistic ectomycorrhizal or saprotrophic fungi or grazing inverte-brates. When nitrogen is released to the soil solution remains unclear

(Dighton & Boddy, 1989), but loss can occur during interspecific interactions (J. M.Wells & L. Boddy, unpublished), and as a result of invertebrate grazing (Anderson *et al.*, 1983).

The fungal nitrogen reservoir

Nitrogen is scavenged by wood decay fungi from many sources (Cromack & Caldwell, 1992), some discrete such as plant litter, colonies of nitrogen-fixing bacteria, nematodes (Barron, 1992), collembola (Klironomos & Hart, 2001), animal remains and excreta, and others diffuse such as dilute solutions perfusing through the forest floor mycelial network from canopy through-fall and leachates from surface litter layers (Heal & Dighton, 1986; Richter & Markewitz, 2001). Phosphorus transfer occurs between ectomycorrhizal and saprotrophic networks in microcosms (Lindahl *et al.*, 1999), but it is not known whether nitrogen is also exchanged in these interactions. The different ^{15}N depletion signatures (Hobbie *et al.*, 1999) of saprotrophic and ectomycorrhizal mycelium suggest that exchange may not be significant. In microcosm studies using ^{15}N there was no evidence of capture of nitrogen by the saprotroph *Phanerochaete velutina* or the ectomycorrhizal *Paxillus involutus* during interactions when either fungus contained the isotope (Donnelly *et al.*, unpublished observations). However, the nitrogen isotope was taken up from sites of point application in soil and allocated to mycelium away from interaction zones by both fungi.

The amounts of nitrogen that can be held in mycelium in the forest floor are ecologically significant, in the sense that they potentially regulate productivity via effects on plant growth and microbial decomposition (Table 7.1). Most of the nitrogen in mature, unpolluted boreal forest soils is biologically inert, being held in the biota or in the form of recalcitrant soil organic compounds with long turnover times (Davidson *et al.*, 1992; Richter & Markewitz, 2001), so that nitrogen within fungal networks probably represents a large fraction of the more rapidly turning-over soil nitrogen. Fungal biomass can be up to $100\,\mathrm{g\,dry\,wt\,m^{-2}}$ of mycelium in soil of mature boreal forest, in which basidiomycete hyphae predominate and can reach hyphal lengths of up to $18\,000\,\mathrm{m\,(g\,dry\,wt)^{-1}}$ soil (Bååth & Söderström, 1979; Markkola *et al.*, 1995). Bååth and Söderström (1979) calculated that this mycelium can contain up to 20% of total soil nitrogen, based on volume estimates from forest soil hyphal lengths, mean diameter and nitrogen contents from mycelium cultured on media with nitrogen concentration equivalent to that measured in litter. The nitrogen contents of rhizomorphs and cords were 1%–2% in conifer litter (Stark, 1972) and 3.5% in broadleaf litter (Frankland, 1982), and sporophores up to 6.6.% nitrogen (Lindahl *et al.*, 2002). Variations in nitrogen input into the forest floor occur seasonally with litter fall and, in some areas from atmospheric deposition of nitrogen compounds. Because the nitrogen content of mycelium of saprotrophic woodland fungi can vary up to 40-fold in response to variations in external supply, as described above (see p. 152), fungal mycelium could act as a fluctuating storage reservoir that holds varying levels of available nitrogen according to ambient supply, and thus have effects on rates of nitrogen-limited ecosystem processes.

A third ecological effect of nitrogen limitation and fungal immobilization has recently come to the fore. Some forests are now receiving damaging nitrogen inputs in the form of atmospheric deposition of pollutant reactive nitrogen (Fenn *et al.*, 1997; Galloway *et al.*, 2003). Initially this can be accumulated and immobilized in the soil, but once saturation point is reached, at a critical soil C/N ratio, nitrification produces nitrate, which is mobile in the soil (Currie, 1999) and is leached out (Tietema *et al.*, 1998) as the forest ceases to absorb input. Damaging effects include eutrophication of watercourses and loss of ectomycorrhizal diversity (Arnolds, 1997; Wallenda & Kottke, 1998), probably resulting from inhibition of mycorrhiza development associated with raised nitrogen content of the host tree (Wallander & Nylund, 1991). The mechanisms by which a low C/N ratio in soil leads to the symptoms of nitrogen saturation are not understood.

The pre-saturated forest floor has a remarkable ability to respond to nitrogen pollution by increasing its nitrogen-absorptive capacity. Currie (1999) describes an experiment carried out in a Swedish forest, where inorganic nitrogen input was artificially increased 6-fold over the ambient level of $0.9 \, \mathrm{g \, N \, m^{-2} \, yr^{-1}}$, resulting in a 6-fold increase in nitrogen retention in the forest floor. Currie *et al.* (1999) hypothesized that ectomycorrhizal fungi might be able to acquire the necessary carbon from their hosts to support enhanced nitrogen uptake. Saprotrophic fungi were excluded from consideration, because the amount of carbon in litter was not considered adequate to support the observed nitrogen assimilation. This conclusion was based on the model of Swift *et al.* (1979), which states that litter decomposers release nitrogen as soon as the C/N ratio of their substrate falls. However, in the alternative model of tightly coupled internal nitrogen cycling in the boreal forest floor proposed by Lindahl *et al.* (2002), inorganic nitrogen is not released by mycelium but instead is accumulated, translocated and imported into woody litter, thus enabling it to be utilized to support further fungal growth. There seems no reason, therefore, to reject the possibility that mycelium of saprotrophic wood decay fungi could also contribute to responsive forest floor nitrogen absorption, since their ability to translocate nitrogen into the abundant forest floor woody litter gives them access to a virtually unlimited supply of carbon and energy. The expansion of nitrogen storage capacity in response to added nitrogen described by Currie (1999) could arise because up to a point (perhaps related to the nitrogen saturation point) the added nitrogen could release the nitrogen-limitation of woody litter decomposition, allowing the fungal network to grow and thus increase its volume for nitrogen uptake. Micks *et al.* (2004) analysed the fate of ^{15}N tracer added to various types of litter during a long-term experiment, the Harvard Forest Nitrogen Saturation Experiment, which compared the effects of ambient and elevated nitrogen inputs on litter decomposition in hardwood and pine forest. Elevated nitrogen led to enhanced retention of tracer in the litter, as well as greater decomposition of the hardwood litter, although the effects on leaf litter decomposition were variable between ecosystems. They suggested that differences might be due to variability in the microbial decomposer community, citing Blagodatskaya and Anderson, (1998) who demonstrated that fungi predominate over bacteria among beech forest litter decomposers at lower soil pH.

The fungal nitrogen distribution system

Mycelium is a significant absorptive nutrient reservoir at the O/A interface of undisturbed soil. Its geometry, lying in the plane of the soil horizons, and its connectivity via cords within the organic horizons, mean that it is well positioned to forage for wood resources and to capture nutrients from solution (Fig. 7.1). Nitrogen passes vertically into this system, from above in canopy and litter through-fall and atmospheric deposition (Richter & Markewitz, 2001), and from below by upward translocation from the soil H and A horizons (Hart & Firestone, 1991). Because they can take up and translocate nitrogen from soil, hyphae and cords provide a route for soil nutrients from mineral soil into the microbial decomposer community in the litter layers above. Spatial redistribution of nitrogen through hyphae could influence the outcome of competition for nitrogen between decomposer fungi and plant roots. In microcosms with spatially heterogeneous distribution of nitrogen-poor and nitrogen-rich carbon sources, containing competing decomposer microbes and plants, the balance of nitrogen acquisition between the competitors was affected by the distance between them (Wang & Bakken, 1997), with a distance of 3–6 mm being critical (Korsaeth *et al.*, 2001). Fungi that can rapidly sequester available nitrogen and control its spatial redistribution probably have a competitive advantage in nitrogen capture from soils with patchy nitrogen distribution. The ability to translocate nitrogen into resources where high C/N ratio limits decomposition is an important distinction between non-resource-restricted fungi and resource-restricted fungi or unicellular microorganisms as decomposers.

The wood wide web of lateral connections between ectomycorrhizal host plants and nutrient sources in soil carries carbon (Simard *et al.*, 1997), phosphorus (Read, 1991) and nitrogen (Perez-Moreno & Read 2000, 2001a, b; Leake *et al.*, 2004) through translocation pathways. The connectivity of networks of wood decay saprotrophs is less well described, though there is extensive work on the translocation of phosphorus in soil microcosms (Wells *et al.*, 1998, 1999; Boddy, 1999), some information on mycelial connectedness in the field (Thompson, 1984) and evidence of translocation of phosphorus through saprotrophic cord systems in woodland (Wells & Boddy, 1995). It has been shown that translocation through mycelial connections can relieve nitrogen limitation of cellulolysis in microcosms with heterogeneous resource distribution (Watkinson *et al.*, 1981). Further, in the field nitrogen translocation from soil into surface litter can directly affect the rate of litter decomposition (Frey *et al.*, 2000, 2003). In the words of Lindahl *et al.* (2002) 'the ability to transport nutrients and energy-rich carbohydrates implies that the acquisition of energy and nutrients is spatially uncoupled'. In agricultural ecosystems without tillage where low levels of soil disturbance allow hyphae to persist across the O and A horizons, coarse crop residues lying on the soil surface can be invaded by hyphae of mycelial networks that scavenge and import nitrogen from the underlying mineral soil, and are the primary decomposers under this management regime (Beare *et al.*, 1992; Frey *et al.*, 2000). Relatively stable spatial relationships develop in less disturbed ecosystems and functional interactions can thus develop between above- and belowground biota (Ettema & Wardle, 2002). Nitrogen imported into leaf litter in forests increases the decay rate (Lodge, 1993) and binds litter physically (Lodge & Asbury, 1988). In forests where levels of disturbance are low, and available soil nitrogen is scarce, hyphal systems persist (Miller & Lodge, 1997), and the rate of decomposition of coarse woody litter on the surface may be affected by fungal import of nitrogen from the underlying soil.

Translocation has important ecological effects over a range of scales relating to the different adaptive niches of soil fungi. Different foraging strategies endow some litter decomposer species with a propensity to exploit large woody fragments separated by distances of metres (for example, see Fig. 7.3) and others, finely comminuted leaf litter. The principal difference is in the size of carbon resource required to switch mycelial development between diffuse growth as separate hyphae, and the cords that act as channels for mass flow (Fig. 7.2D; Cairney *et al.*, 1989). *Phanerochaete velutina*, a basidiomycete wood decomposer, extends predominantly as corded mycelium, with hyphal aggregation starting close behind the diffuse mycelial margin, while *Agaricus bisporus* advances mainly as separate closely ranked hyphae growing mainly on leaf litter and their cords extend for relatively short distances. At smaller scales, soil microfungi such as *Rhizoctonia solani* form translocating bridges across a few micrometres between soil particles (Boswell *et al.*, 2002). Resource contact and nitrogen limitation, the cues for the foraging development of cord-forming fungi described above, appear adaptive for exploiting the upper forest floor habitat. In fungi utilizing scattered woody fragments, both colonization of fresh carbon resources, and the onset of nitrogen limitation, impose a requirement for the internal redistribution of resources.

Imaging mycelial nutrient dynamics: photon-counting scintillation imaging (PCSI)

Although there are no convenient radioactive isotopes of nitrogen for short-term dynamic studies, nitrogen translocation can be tracked using the non-metabolized amino acid, 2-aminoisobutyric acid (AIB), which is actively taken up into the mobile free amino acid pool, where it can remain unchanged for periods as long as six months, and is translocated both towards and away from the direction of growth (Watkinson, 1984). In PCSI, the fungus is placed in close proximity to a scintillation screen and the movement of the radiolabelled compound is tracked from the photons of light that result from collision of the radioactive emissions from the compound with the screen (Fig. 7.5; Olsson 2002; Tlalka *et al.*, 2002). Imaging can continue for periods of up to six weeks. The distribution of AIB can be analysed in real time, as the time and x, y coordinates of each photon are recorded. So far it has proved convenient to accumulate counts over 30–60 minutes to achieve a reasonable signal to noise ratio. The greatest resolution is constrained mainly by the spread of the radioactive emission at high magnification, or the critical pixel spacing at low magnification.

Transport of AIB has a pulsatile component (Tlalka *et al.*, 2002). We do not have a clear understanding of the origin of the pulses at present. We can exclude oscillatory behaviour through a tight coupling to an ultradian/

circadian clock as, although the oscillations persist in constant conditions, they show a marked temperature dependence (Tlalka *et al.*, 2003). Cyclical changes in colony growth might result in apparent pulsatile transport. Typically, the early growth phase in these microcosms is characterized by fairly dense, relatively infrequently branched thin hyphae. After about 3–4 days, colony growth becomes progressively more complex. Radial expansion continues at roughly the same rate, but branching becomes more frequent. Some hyphae appear to regress and the initial radial symmetry alters in favour of a more limited number of discrete point growth foci,

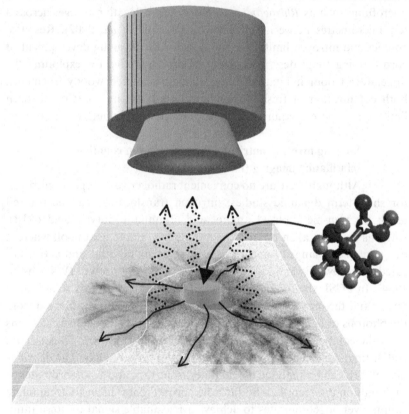

Fig. 7.5. Imaging system for photon-counting scintillation imaging (PCSI). The diagram shows mycelium pre-grown over a scintillant screen: 2-aminoisobutyric acid labelled in the carboxy–group with ^{14}C was added at the centre of the colony. Photons elicited by β-radiation from the screen are cumulatively recorded by the photon-counting camera. For use with soil systems an inverted camera is used, with the screen placed on top of the mycelium growing over soil. (see Watkinson *et al.*, 2005).

typical of the early stages in development of a corded system. There is no obvious synchronization of these events throughout the colony with a periodicity that might be associated with oscillations in ^{14}C-AIB transport. However, we cannot rule out subtle changes in growth pattern due to the technical difficulties of imaging morphological changes within the mass of thin, white hyphae superimposed on the white scintillation screen. For example, we might not detect cycles where a limited number of unsuccessful foraging hyphae regress and their contents are recycled back through the inoculum to support growth elsewhere at the margin. How the driving forces for such a cycle might operate and how such behaviour would be synchronized throughout the colony are unclear.

One current working model (Tlalka *et al.*, 2003; Fig. 7.6) suggested a dynamic system linking local uptake with outward mass flow in foraging mycelium, mediated by responses of intracellular compartmentation to amino acid level. In the model, amino acid is taken up into a (plausibly vacuolar – Klionsky *et al.*, 1990) amino acid compartment in the hyphae at the assimilation site, until a threshold level is reached. This triggers efflux into a translocation compartment in which the amino acid is exported, causing the level at the assimilation site to fall to its earlier level. Iteration of the process generates pulsatile flow. With physiological parameter values, the model generates pulses comparable with those observed in real microcosms.

^{14}C-AIB dynamics in sand microcosms

Rapid pulsatile ^{14}C-AIB transport also occurs in sand microcosms where the initial inoculum is a colonized wood block. For example, Fig. 7.7 shows the results of an experiment in which a mycelium of *Phanerochaete velutina* was grown over sand between two separate wood blocks for 2 months. After feeding at the resource block from which the mycelium has already extended, the microcosm was placed in the photon counting camera for the following 420 hours. There was extensive, rapid and non-randomly directed flow of amino acid through much of the system. Initially, a conspicuous cord forming the main route between the resource and the advancing mycelial front filled with amino acid over the space of 1 hour. A further 3 hours later, the signal trace was recordable from several more cords and appeared to have reached the edge of the advancing front where there was the appearance of accumulation, although some care is needed in this interpretation as the accumulated signal did not necessarily stoichiometrically reflect mycelial amino acid content because the mycelium was not confined at the surface. An underestimate could result by physical distance between any amino acid-loaded

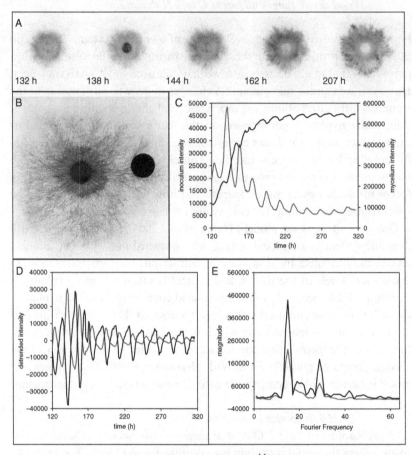

Fig. 7.6. Complementary oscillations of ^{14}C-labelled 2-aminoisobutyric acid (^{14}C-AIB) in assimilatory and foraging hyphae of *Phanerochaete velutina*. The distribution of ^{14}C-AIB was mapped using photon-counting scintillation imaging of colonies of *P. velutina* grown across a scintillation screen covered with 1.5 μm thick Mylar film in the presence of a 13 mm filter-paper 'bait'. Five scintillation images, each integrated over 60 minutes, are shown for the times indicated from a complete time series lasting over 200 hours (A). At the end of the experiment, the colony, supported on the Mylar film was photographed against a black background to visualize the growth pattern (B). The early phase of growth was dense and symmetrical followed by a switch to sparser growth. The bait was contacted around 150 hours and the beginnings of cord formation are visible by the end of the experiment. The total signal from the inoculum (C, grey trace) and foraging mycelium (C, black trace) showed pronounced oscillations with complementary profiles. This was most clearly seen after the longer-term trends were removed by subtraction of a 20 hours rolling average from each time-series (D). The maximum Fourier frequency (14) of these sequences corresponded to a period of 18.29 hours (E). Additional peaks were present in the Fourier spectrum at higher harmonics of the fundamental frequency. These reflect the underlying asymmetry in the shape of each pulse.

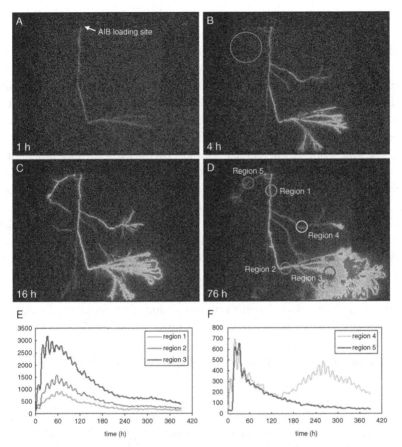

Fig. 7.7. Still images from a video showing the accumulated photon trace from a corded foraging mycelium of *Phanerochaete velutina*. The ^{14}C-AIB was added at the wood inoculum site of a mycelium that had been growing on sand for 2 months. The movement of AIB was imaged for the subsequent 420 hours. The initial 76 hours are shown in the images: (A) 1 h after loading, 4 cords filled; (B) 4 hours after loading, detectable signal in 9 cords and mycelial front; (C) 16 hours a new route opens, top left, 76 hours; (D) The mycelium has reached a wood block bait (not shown) bottom right and the top left cord signal has diminished. (E and F) pulse patterns mapped on to the mycelium in (A–D) showing, (E) synchronous pulsation in regions 1, 2 and 3 on the main cord between inoculum and bait, and progressive accumulation in the distal, advancing region; (F) The filling and subsequent emptying of the cord in region 5 where there is no bait, and the two filling phases of the cord in region 4; with initial filling, followed by loss of signal, and a refilling that probably coincided with bait contact by the mycelial margin from 180 hours.

mycelium that was below the sand surface, and the scintillant screen placed on top of the mycelium. Likewise, differential cord thicknesses will give different levels of signal attenuation. At 16 hours an additional cord at the left of the picture filled, but by 76 hours it emptied again and the signal comes mainly from the lower right hand corner of the picture, the region in which the wood bait was positioned. The real-time changes in the intensity of the signal were tracked over the experimental period in 5 sites as indicated. Within the first 10 hours of recording, a signal pulse appeared almost synchronously in regions 1, 2 and 3 on the main cord leading to the mycelial front. By 76 hours, pulsation at 1, 2 and 3 had died away, and at the same time the greater part of the signal had been conveyed to the mycelial front. A rival pathway (visible throughout as a cord leading to the left of the image) filled with amino acid and pulsed at high amplitude between 4 and 16 hours (region 5), but the pulse amplitude and total signal fell to less than half its initial value by 60 hours, and by 76 hours was almost empty. Region 4 initially showed synchronous pulsation with the other regions and then a decay in signal. However, after 130 hours the cord containing region 4 appeared to be reactivated, with an upward trend in overall signal paralleled by an increase in pulse amplitude. This route opening and route switching associated with foraging may well be relevant to redistribution of nitrogen through forest floor corded networks.

^{14}C-AIB dynamics in soil microcosms

Our results in non-sterile, soil-based microcosms are quite limited at present but are already providing some intriguing observations. For example, the total level of uptake and translocation of AIB in the well-corded system shown in Fig. 7.8, is much lower than on a comparable sand-based microcosm. Images have been integrated for considerably longer periods, although preferential transport along a subset of the available cords is still evident. The very diffuse spread of the signal also reflects a difference in the foraging behaviour of the mycelium in this system, with an abundance of fine hyphae associated with the growing mycelial front and in submarginal regions. It is possible that the low overall AIB uptake reflects competition for uptake with much higher levels of inorganic or organic nitrogen present in the soil. This may also explain the more extensive fine mycelium that is presumably scavenging for nitrogen and other nutrients, especially as in this microcosm there is a relatively large amount of carbon available from the wood blocks. As we develop more experience with such soil-based systems, it is anticipated that it will be possible to track and investigate the rate and direction of flow in response to manipulated parameters representative of field conditions. The speed and responsiveness of the foraging mycelial system and its amino acid translocation towards the advancing mycelial front, with sudden removal of amino acid from less active parts of the system, is consistent with the dynamic competitive acquisition of soil nitrogen that is considered typical of the underground biota of nitrogen-limited forests.

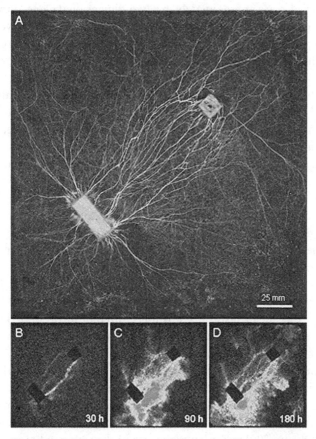

Fig. 7.8. PCSI showing the pattern of AIB translocation following addition to a wood block bait in a 25 cm square compressed-soil microcosm similar to that shown in Fig. 7.3. (A) Photograph of the mycelium at the time of adding ^{14}C-AIB. (B–D) PCSI images. Time course, with time indicated on each picture. Compared with the sand-based microcosm shown in Figure 7.7, the AIB distribution appeared weaker and more diffuse. As in sand microcosms, there was preferential filling of only a few out of several cords present, and AIB was translocated from the point of application into the bait, and beyond it to an advancing mycelial front subtended by a cord.

The potential of PCSI techniques for investigating fungal processes in carbon and nitrogen cycles

Photon-counting scintillation imaging in microcosms can contribute to modelling the forest floor responses to nitrogen and carbon inputs because, for the first time, it allows finely resolved dynamic analysis of nitrogen flows within a pre-existing mycelial system (Fig 7.8). Time-lapse two-dimensional imaging matches key microbial processes in the forest floor, which also are approximately in a two-dimensional plane at the A/O interface, and in which dynamics are critical for interactions between the biota. It is possible to combine PCSI with mass balance of total nitrogen to quantify the partitioning of nitrogen between the litter, mycelium and soil. Parameter values for mobile and immobile mycelial nitrogen can be obtained by the use of microcosms with realistic carbon and nitrogen inputs and geometry. Furthermore, it might in future be possible to relate microcosm data to field values by genomic approaches, using genes of nitrogen metabolism (Caddick 2004) and intracellular nitrogen and C/N sensing (Winderickx & Taylor, 2004) to report the nitrogen status and C/N experience of foraging mycelium.

Problems still to be overcome in using PCSI for fungal nitrogen flux analysis include the difficulty in calibrating the photon signal in terms of amino acid levels, as ^{14}C gives weak β-emission, and quenching by the tissue is therefore significant. For mass balance measurement, essential for measuring amino acid pool size and fluxes, destructive harvesting of samples at a fixed time point is still required. Aminoisobutyric acid as a marker for mycelial nitrogen may also become less representative of reality the longer experiments last, since it is not utilized or incorporated and its relative concentration within the system will rise as metabolism causes utilizable nitrogen levels to fall.

With the use of realistic soil-based microcosms, problems arise due to the intrinsic variability of real mycelial systems. The mycelium is symmetrical only in very early juvenile stages, before the onset of nutrient limitation and differentiation of separate functional states of mycelium (Fig. 7.6; Olsson, 2001). The heterogeneity of the soil/wood/mycelial system presents unique experimental difficulties. Problems of genuine variability inextricably associated with the indeterminate, context-dependent growth form of fungi have conventionally been overcome by the use of defined uniform media to ensure reproducibility. Much basic knowledge of fungal genetics and physiology has resulted from such approaches, which have been focused mainly on model fungi. However, what fungi do and their precise role in ecosystem function depends on the developmental and metabolic responsiveness of the mycelial system to the heterogeneity of its habitat. The experimental challenge is to devise rigorous quantitative experimental approaches that can be applied to real fungi from real ecosystems under realistic conditions. A key step will be to develop analysis tools that can deal with the more complex data arising from soil-based microcosms.

Conclusions

In boreal and temperate zone forests, where the greatest amount of terrestrial carbon is sequestered, mainly in the soil, (Post *et al.*, 1982; Dixon *et al.*, 1994; Schimel, 1995), fungi are the only decomposers that can degrade wood completely (Dighton & Boddy, 1989; Boddy & Watkinson, 1995). Because of their pivotal role in ecosystem productivity and nutrient cycling (Lindahl *et al.*, 2002) more certain and precise data on boreal and temperate forest floor fungi are urgently needed for dynamic predictive models of global carbon and nitrogen cycles (Falkowski *et al.*, 2000) designed to inform ecosystem management to mitigate the anthropogenic enhanced greenhouse effect. The effects on ecosystem function and diversity of increased inputs of reactive nitrogen to the nitrogen-limited northern-hemisphere forests are likely to be far-reaching (Townsend *et al.*, 1996; Torn *et al.*, 1997), and are the subject of several long-term addition studies in forests (e.g. Magill *et al.*, 2000; Aber & Magill, 2004; Vestgarden *et al.*, 2004). These are showing that the forest floor has a remarkable ability to absorb pollutant nitrogen and to increase its absorptive capacity in response to nitrogen deposition, as described above. The mechanism is not understood, but the fungal reservoir with its responsive capacity to absorb nitrogen present in excess of immediate requirements is an obvious candidate.

Translocation of nitrogen into woody litter is key to the success of saprotrophic fungi in the competition for soil nitrogen, and thus to their role as decomposers in forests. The speed of amino acid translocation during foraging in wood/soil microcosms shown by PCSI suggests that nitrogen flux into litter exceeds the quantity imported by growth alone, as a component of the invading hyphae. Directed rapid translocation induced by local acquisition of a carbon-rich, nitrogen-poor resource appears likely, but has yet to be conclusively demonstrated. Responsive import of nitrogen to carbon-rich litter would be predicted by our hypothesis that responsive nitrogen retention is a result of scavenging by foraging mycelia.

Our analysis suggests that the cause of responsive nitrogen retention should be sought in the nutrient dynamics of individual fungal networks, optimized by evolution to balance carbon and nitrogen acquisition for biomass production. Currie (1999) has recommended that the effect of soil

C/N ratio should be investigated over a range of scales to find the cause of ecosystem responses in forest floor processes. The fundamental processes in carbon and nitrogen interaction occur at the interface between the fungal mycelium, the discrete carbon resource unit, and the ambient nitrogen supply. The dynamics of nutrient-responsive foraging might be modelled in terms of these three compartments.

Key questions for experimentation based around this model are:

- How does carbon availability affect the ability of fungi to retain nitrogen? Does the capacity of a mycelium to capture and seques-ter nitrogen increase in proportion to the carbon available?
- How does the energy status of the mycelium affect its ability to concentrate nitrogen from the surrounding environment?
- At what carbon and nitrogen content and ratio do mycelia of wood-decomposing forest floor species sense nitrogen limitation, and switch from nitrogen catabolism to assimilatory metabolism?
- Is amino acid translocation preferentially directed towards a rela-tively carbon-rich, nitrogen-poor resource?
- What are the sensing and signalling mechanisms that cause global reorganization of a mycelial network following contact with resources of varying carbon content?
- How does a species' foraging strategy affect its potential for nitrogen redistribution?

These questions are amenable to various combinations of imaging, mod-elling, metabolic and gene expression studies and field observation to analyse the nutrient dynamics of the forest floor microbial decomposer subsystem, and to provide data for predictive models of forest floor responses to environmental change.

Acknowledgements
Some of the work reported in this paper was funded by the Natural Environment Research Council (A/S/2002/01005; A/S/2002/00882).

References

Aber, J. D. (1992). Nitrogen cycling and nitrogen saturation in temperate forest ecosystems. *Trends in Ecology and Evolution*, **7**, 220–3.

Aber, J. D. & Magill, A. H. (2004). Chronic nitrogen additions at the Harvard Forest (USA): the first 15 years of a nitrogen saturation experiment. *Forest Ecology and Management*, **196**, 1–5.

Aber, J. P. & Melillo, J. M. (1982). Nitrogen immobilization in decaying hardwood leaf litter as a function of initial nitrogen and lignin content. *Canadian Journal of Botany*, **60**, 2263–9.

Agerer, R. (2001). Exploration types of ectomycorrhizae. *Mycorrhiza*, **11**, 107–14.

Anderson, J. M., Ineson, P. & Huish, S. A. (1983). Nitrogen and cation mobilization by soil fauna feeding on leaf litter and soil organic matter from deciduous woodlands. *Soil Biology and Biochemistry*, **15**, 463–7.

Arnolds, E. J. M. (1997). Biogeography and conservation. In *The Mycota*, vol. IV. *Environmental and Microbial Relationships*, ed. D. T. Wicklow & B. Söderström, Berlin: Springer-Verlag, pp. 115–31.

Bååth, E. & Söderström, B. (1979). Fungal biomass and fungal immobilisation of plant nutrients in Swedish coniferous forest soils. *Revue d'Ecologie et de Biologie du Sol*, **16**, 477–89.

Bago, B., Pfeffer, P. & Shachar-Hill, Y. (2001). Could the urea cycle be translocating nitrogen in the arbuscular mycorrhizal symbiosis? *New Phytologist*, **149**, 4–8.

Barron, G. L. (1992). Ligninolytic and cellulolytic fungi as predators and parasites. In *The Fungal Community: Its Organization and Role in the Ecosystem*, ed. G. C. Carroll & D. J. Wicklow. New York: Marcel Dekker, pp. 311–54.

Beare, M. H., Parmelee, R. W., Hendrix, P. F. *et al.* (1992). Microbial and faunal interactions and effects on litter nitrogen and decomposition in ecosystems. *Ecological Monographs*, **62**, 569–91.

Bending, G. D. & Read, D. J. (1995). The structure and function of the vegetative mycelium of ectomycorrhizal plants. V. Foraging behaviour and translocation of nutrients from exploited litter. *New Phytologist*, **130**, 401–9.

Berntson, G. M. & Aber, J. D. (2000). Fast nitrate immobilisation in nitrogen saturated temperate forest soils. *Soil Biology and Biochemistry*, **32**, 151–6.

Blagodatskaya, E. & Anderson, T. H. (1998). Interactive effects of pH and substrate quality on the fungal-to-bacterial ratio and Q CO_2 of microbial communities in forest soils. *Soil Biology and Biochemistry*, **30**, 1269–74.

Boddy, L. (1993). Saprotrophic cord-forming fungi: warfare strategies and other ecological aspects. *Mycological Research*, **97**, 641–55.

Boddy, L. (1999). Saprotrophic cord-forming fungi: meeting the challenge of heterogeneous environments. *Mycologia*, **91**, 13–32.

Boddy, L. & Watkinson, S. C. (1995). Wood decomposition, higher fungi, and their role in nutrient redistribution. *Canadian Journal of Botany*, **73** (Suppl.1), S1377–83.

Boswell, G. P., Jacobs, H., Davidson, F. A., Gadd, G. M. & Ritz, K. (2002). Functional consequences of nutrient translocation in mycelial fungi. *Journal of Theoretical Biology*, **217**, 459–77.

Bringmark, L. (1980). Ion leaching through a podsol in a Scots pine stand. In *Ecological Bulletin*, Vol. 32. *Structure and Function of Northern Coniferous Forests – an Ecosystem Study*, ed. T. Persson, pp. 357–61.

Caddick, M. X. (2002). What's for dinner – what shall I choose? *Microbiology Today*, **29**, 132–4.

Caddick, M. X. (2004). Nitrogen regulation in mycelial fungi. In *The Mycota*, Vol. III. *Biochemistry and Molecular Biology*, 2nd en, ed. R. Brambl & G. A. Marzluf. Berlin: Springer-Verlag, pp. 349–68.

Cain, M. L., Subler, S., Evans, J. P. & Fortin, M.-S. (1999). Sampling spatial and temporal variation in soil nitrogen availability. *Oecologia*, **118**, 397–404.

Cairney, J. W. G. (1992). Translocation of solutes in ectomycorrhizal and saprotrophic rhizomorphs. *Mycological Research*, **96**, 135–41.

Cairney, J. W. G., Jennings, D. H. & Veltkamp, C. J. (1989). A scanning electron microscope study of the internal structure of mature linear mycelial organs of four basidiomycete species. *Canadian Journal of Botany*, **67**, 2266–71.

Carreiro, M. M., Sinsabaugh, R. L., Reperet, D. A. & Parkhurst, D. F. (2000). Microbial enzyme shifts explain litter decay responses to simulated nitrogen deposition. *Ecology*, **81**, 2359–65.

Cooper, T. G. (1996). Regulation of allantoin metabolism in *Saccharomyces cerevisiae*. In *The Mycota*, Vol. III. *Biochemistry and Molecular Biology*, ed. R. Brambl & G. A. Marzluf, Heidelberg: Springer-Verlag, pp. 139–69.

Cooper, T. G. (2004). Integrated regulation of the nitrogen-carbon interface. In: *Topics in Current Genetics*, Vol. 7. *Nutrient-induced Responses in Eukaryotic Cells*, ed. J. Winderickx & P. M. Taylor. Berlin: Springer-Verlag, pp. 225–57.

Cromack, K. & Caldwell, B. A. (1992). The role of fungi in litter decomposition and nutrient cycling. In *The Fungal Community: Its Organization and Role in the Ecosystem*, ed. G. C. Carroll & D. J. Wicklow. New York: Marcel Dekker, pp. 653–68.

Currie, W. S. (1999). The responsive C and N biogeochemistry of the temperate forest floor. *Trends in Ecology and Evolution* **14**, 316–20.

Currie, W. S., Nadelhoffer, K. J. & Aber, J. D. (1999). Soil detrital processes controlling the movement of ^{15}N tracers to forest vegetation. *Ecological Applications*, **9**, 87–102.

Dafodu, W. (2003). *The British Survey of Fertilizer Practice*. London: Crown Publications.

Davidson, E. A., Hart, S. C. & Firestone, K. (1992). Internal cycling of nitrate in soils of a mature coniferous forest. *Ecology*, **73**, 1148–56.

Davidson, F. A. & Olsson, S. (2000). Translocation induced outgrowth of fungi in nutrient-free environments. *Journal of Theoretical Biology*, **205**, 73–84.

Davis, R. H. (1996). Polyamines in fungi. In *The Mycota*, Vol. III. *Biochemistry and Molecular Biology*, ed. R. Brambl & G. A. Marzluf. Heidelberg: Springer-Verlag, pp. 347–56.

Dighton, J. (1997). Nutrient cycling by saprotrophic fungi in terrestrial habitats. In *The Mycota*, Vol. IV. *Environmental and Microbial Relationships*, ed. D. W. Wicklow & B. Söderström. Berlin: Springer-Verlag, pp. 271–93.

Dighton, J. (2003). *Fungi in Ecosystem Processes*. New York: Marcel Dekker.

Dighton, J. & Boddy, L. (1989). Role of fungi in nitrogen, phosphorus and sulphur cycling in temperate forest ecosystems. In *Nitrogen, Phosphorus and Sulphur Utilization by Fungi*, ed. L. Boddy, R. Marchant & D. J. Read. Cambridge: Cambridge University Press, pp. 269–98.

Dixon, R. K., Brown, S., Houghton, R. A. *et al.* (1994). Carbon pools and flux of global forest ecosystems. *Science*, **263**, 185–90.

Downs, M. R., Nadelhoffer, K. J., Melillo, J. M. & Aber, J. D. (1996). Immobilization of a ^{15}N-labelled nitrate addition by decomposing forest litter. *Oecologia*, **105**, 141–50.

Ettema, C. H. & Wardle, D. A. (2002). Spatial soil ecology. *Trends in Ecology and Evolution* **17**, 177–83.

Falkowski, P., Scholes, R. J., Boyle, E. *et al.* (2000). The global carbon cycle: a test of our knowledge of Earth as a system. *Science*, **290**, 291–6.

Fenn, M. E., Poth, M. A., Aber, J. D. *et al.* (1997). Nitrogen excess in North American Ecosystems: predisposing factors, ecosystem responses, and management strategies. *Ecological Applications*, **8**, 706–33.

Frankland, J. C. (1982). Biomass and nutrient cycling by decomposer basidiomycetes. In *Decomposer Basidiomycetes: Their Biology and Ecology*, ed. J. C. Frankand, J. N. Hedger & M. J. Swift. Cambridge: Cambridge University Press, pp. 241–61.

Frey, S. D., Elliott, E. T., Paustian, K. & Peterson, G. A. (2000). Fungal translocation as a mechanism for soil nitrogen inputs to surface residue decomposition in a no-tillage agroecosystem. *Soil Biology and Biochemistry*, **32**, 689–98.

Frey, S. D., Six, J. & Elliott, E. T. (2003). Reciprocal transfer of carbon and nitrogen by decomposer fungi at the soil-litter interface. *Soil Biology and Biochemistry*, **35**, 1001–4.

Fricker, M. D., Bebber, D., Tlalka, M. *et al.* (2005). Inspiration from microbes: from patterns to networks. In *Complex Systems and Inter-Disciplinary Science*, ed. B. W. Arthur, R. Axtell, S. Bornholdt *et al.* London: World Scientific Publishing Co., (in press).

Galloway, J. N., Aber, J. D., Erisman, J. W. *et al.* (2003). The nitrogen cascade. *Bioscience*, 53, 341–56.

Govindarajulu, M., Pfeffer, P. E., Jin, H. *et al.* (2005). Nitrogen transfer in the arbuscular mycorrhizal symbiosis. *Nature*, 435, 819–23.

Griffin, D. H. (1994). *Fungal Physiology*, 2nd edn. Chichester: Wiley-Liss.

Guo, D., Mou, P., Jones, R. H. & Mitchell, R. J. (2004). Spatio-temporal patterns of soil available nutrients following experimental disturbance in a pine forest. *Oecologia*, 138, 613–21.

Hanks, J. N., Hearnes, J. M., Gathman, A. C. & Lilly, W. W. (2003). Nitrogen starvation-induced changes in amino acid and free ammonium pools in *Schizophyllum commune* colonies. *Current Microbiology*, 47, 444–449.

Harmon, M. E., Franklin, J. F., Swanson, F. J. *et al.* (1986). Ecology of coarse woody debris in temperate ecosystems. *Recent Advances in Ecological Research*, 15, 133–302.

Hart, S. C. & Firestone, M. K. (1991). Forest floor-mineral soil interactions in the internal nitrogen cycle of an old-growth forest. *Biogeochemistry*, 12, 103–28.

Hart, S. C., Nason, G. E., Myrold, D. D. & Perry, D. A. (1994). Dynamics of gross nitrogen transformations in an old growth forest: the carbon connection. *Ecology*, 75, 880–91.

Heal, O. W. & Dighton, J. (1986). Nutrient cycling and decomposition in natural terrestrial ecosystems. In *Microfloral and Faunal Interactions*, ed. M. J. Mitchell & J. P. Nakas. Dordrecht: Martin Nijhoff/Dr W Junk, pp. 14–73.

Hibbett, D. S., Gilbert, L. B. & Donoghue, M. J. (2000). Evolutionary instability of ectomycorrhizal symbioses in basidiomycetes. *Nature*, 407, 506–8.

Hirobe, M., Koba, K. & Tokuchi, N. (2003). Dynamics of the internal soil nitrogen cycles under moder and mull forest floor types on a slope in a *Cryptomeria japonica* D. Don plantation. *Ecological Research*, 18, 5–64.

Hobbie, E. A., Macko, S. A. & Shugart, H. H. (1999). Insights into carbon and nitrogen dynamics of ectomycorrhizal and saprotrophic fungi from isotopic evidence. *Oecologia*, 118, 353–60.

Hodge, A., Robinson, D. & Fitter, A. (2000). Are micro-organisms more effective than plants at competing for nitrogen? *Trends in Plant Science*, 5, 304–8.

Jacobs, H., Boswell, G. P., Ritz, K., Davidson, F. A. & Gadd, G. M. (2002). Solubilisation of calcium phosphate as a consequence of carbon translocation in *Rhizoctonia solani*. *FEMS Microbiology Ecology*, 40, 65–71.

Jennings, D. H. (1987). The translocation of solutes in fungi. *Biological Reviews*, 62, 215–43.

Jennings, D. H. (1995). *The Physiology of Fungal Nutrition*. Cambridge: Cambridge University Press.

Kaye, J. P. & Hart, S. C. (1997). Competition for nitrogen between plants and soil micro-organisms. *Trends in Ecology and Evolution*, 12, 139–43.

Keyser, P., Kirk, T. K. & Zeykus, I. G. (1978). Ligninolytic system of *Phanerochaete chrysosporium*: synthesized in the absence of lignin in response to nitrogen starvation. *Journal of Bacteriology*, 135, 790–7.

Kingsnorth, C. S., Eastwood, D. C. & Burton, K. S. (2001). Cloning and post-harvest expression of serine proteinase transcripts in the cultivated mushroom *Agaricus bisporus*. *Fungal Genetics and Biology*, 32, 135–44.

Kirk, T. K. & Fenn, P. (1982). Formation and action of the ligninolytic system in basidiomycetes. In: *Decomposer Basidiomycetes*, ed. J. C. Frankland, J. N. Hedger & M. J, Swift. Cambridge: Cambridge University Press, pp. 67–90.

Klionsky, D. J., Herman, P. K. & Emr, S. D. (1990). The fungal vacuole: composition, function and biogenesis. *Microbiological Reviews*, 54, 226–92.

178   *S. Watkinson* et al.

Klironomos, J. N. & Hart, H. H. (2001). Animal nitrogen swap for plant carbon. *Nature*, **410**, 651–2.

Korsaeth, A., Molstad, L. & Bakken, L. R. (2001). Modelling the competition for nitrogen between plants and microflora as a function of soil heterogeneity. *Soil Biology and Biochemistry*, **33**, 215–26.

Lal, R. (2004). Soil carbon sequestration impacts on global climate change and food security. *Science*, **304**, 1623–7.

Leake, J., Johnson, D., Donnelly, D. & Boddy, L. (2004). Networks of power and influence: the role of mycorrhizal mycelium in controlling plant communities and agro-ecosystem functioning. *Canadian Journal of Botany*, **82**, 1016–45.

Levi, M. P. & Cowling, E. B. (1969). Role of nitrogen in wood deterioration. VII. Physiological adaptation of wood-destroying and other fungi to substrates deficient in nitrogen. *Phytopathology*, **59**, 460–8.

Lilly, W. W., Wallweber, G. J. & Higgins, S. M. (1991). Proteolysis and amino acid recycling during nitrogen deprivation in *Schizophyllum commune*. *Current Microbiology*, **23**, 27–32.

Lindahl, B., Stenlid, J., Olsson, S. & Finlay, R. (1999). Translocation of ^{32}P between interacting mycelia of a wood decomposing fungus and ectomycorrhizal fungi in microcosm systems. *New Phytologist*, **144**, 183–93.

Lindahl, B. O., Finlay, R. D. & Olsson, S. (2001). Simultaneous, bidirectional translocation of ^{32}P and ^{33}P between wood blocks connected by mycelial cords of *Hypholoma fasciculare*. *New Phytologist*, **150**, 189–94.

Lindahl, B. O., Taylor, A. F. S. & Finlay, R. D. (2002). Defining nutritional constraints on carbon cycling in boreal forests – towards a less 'phytocentric' perspective. *Plant and Soil*, **242**, 123–35.

Lodge, D. J. (1993). Nutrient cycling by fungi in wet tropical ecosystems. In *Aspects of Tropical Mycology*, ed. S. Isaac, J. C. Frankland, R. Watling & A. J. S. Whalley. Cambridge: Cambridge University Press, pp 37–57.

Lodge, D. J. & Asbury, C. E. (1988). Basidiomycetes reduce export of organic matter from forest slopes. *Mycologia*, **80**, 888–90.

Magill, A. H., Aber, J. D., Berntson, G. M. *et al.* (2000). Long-term nitrogen additions and nitrogen saturation in two temperate forests. *Ecosystems*, **3**, 238–53.

Maraun, M., Martens, H., Migge, S., Theenhaus, A. & Scheu, S. (2003). Adding to 'the enigma of soil animal diversity': fungal feeders and saprophagous soil invertebrates prefer similar food substrates. *European Journal of Soil Biology*, **39**, 85–95.

Markkola, A. M., Ohtonen, R., Tarvainen, O. & Ahonen-Jonnarth, U. (1995). Estimates of fungal biomass in Scots pine stands on an urban pollution gradient. *New Phytologist*, **131**, 139–47.

Marzluf, G. A. (1996). Regulation of nitrogen metabolism in mycelial fungi. In: *The Mycota*, Vol. III. *Biochemistry and Molecular Biology*, ed. R. Brambl & G. A. Marzluf. Berlin: Springer-Verlag, pp. 357–68.

Merrill, W. & Cowling, E. B. (1966). The role of nitrogen in wood deterioration: amount and distribution of nitrogen in tree stems. *Phytopathology*, **56**, 1085–90.

Micks, P., Downs, M. R., Magill, A. H., Nadelhoffer, K. J. & Aber, J. D. (2004). Decomposing litter as a sink for ^{15}N-enriched additions to an oak and red pine plantation. *Forest Ecology and Management*, **196**, 71–87.

Miller, R. M. & Lodge, D. J. (1997). Fungal responses to disturbance: agriculture and forestry. In *The Mycota*, Vol. IV. *Environmental and Microbial Relationships*, ed. D. T. Wicklow & B. Söderström. Berlin: Springer-Verlag, pp. 65–84.

Nasholm, T., Ekblad, A., Nordin, A. *et al.* (1998). Boreal forest plants take up organic nitrogen. *Nature*, **392**, 914–16.

Neff, J. C., Townsend, A. R., Gleixner, G., Lehmann, S. J., Turnbull, J. & Bowman, W. D. (2002). Variable effects of nitrogen addition on the stability and turnover of carbon. *Nature*, **419**, 915–17.

Northup, R. R., Yu, Z., Dahlgren, R. A. & Vogt, K. A. (1995). Polyphenol control of nitrogen release from plant litter. *Nature*, **377**, 227–9.

Olsson, S. (2001). Colonial growth of fungi. In *The Mycota*, Vol. VIII. *Biology of the Fungal Cell*, ed. R. J. Howard & N. A. R. Gow. Berlin: Springer-Verlag, pp. 125–41.

Olsson, S. (2002). Continuous imaging in fungi. *New Phytologist*, **152**, 6–7.

Olsson, S. & Gray, S. N. (1998). Patterns and dynamics of ^{32}P phosphate and ^{14}C labelled AIB translocation in intact basidiomycete mycelia. *FEMS Microbiology Ecology*, **26**, 109–20.

Olsson, S. & Hansson, B. S. (1995). The action potential-like activity found in fungal mycelium is sensitive to stimulation. *Naturwissenschaften*, **82**, 30–1.

Paustian, K. & Schnurer, J. (1987). Fungal growth response to carbon and nitrogen limitation: application of a model to field and laboratory data. *Soil Biology and Biochemistry*, **19**, 621–9.

Perez-Moreno, J. & Read, D. J. (2000). Mobilization and transfer of nutrients from litter to tree seedlings via the vegetative mycelium of ectomycorrhizal plants. *New Phytologist*, **145**, 301–9.

Perez-Moreno, J. & Read, D. J. (2001a). Nutrient transfer from soil nematodes to plants: a direct pathway provided by the mycorrhizal mycelial network. *Plant, Cell and Environment*, **24**, 1219–26.

Perez-Moreno, J. & Read, D. J. (2001b). Exploitation of pollen by mycorrhizal mycelial systems with special reference to nutrient recycling in boreal forests. *Proceedings of the Royal Society London B*, **268**, 1329–55.

Post, W. M., Emanuel, W. R., Zinke, P. J. & Stangenberger, A. G. (1982). Soil carbon pools and world life zones. *Nature*, **298**, 156–9.

Rayner, A. D. M. (1991). The challenge of the individualistic mycelium. *Mycologia*, **83**, 48–71.

Rayner, A. D. M. (1994). Pattern generating processes and fungal communities. In *Beyond the Biomass: Compositional and Functional Analysis of Microbial Communities*, ed. K. Ritz, J. Dighton & K. E. Giller. Chichester: John Wiley, pp. 247–58.

Rayner, A. D. M. & Boddy, L. (1988). *Fungal Decomposition of Wood: its Biology and Ecology*. Chichester: John Wiley International.

Rayner, A. D. M., Griffith, G. S. & Ainsworth, A. M. (1995). Mycelial interconnectedness. In *The Growing Fungus*, ed. N. A. R. Gow & G. M. Gadd. London: Chapman & Hall, pp. 21–40.

Read, D. J. (1991). Mycorrhizas in ecosystems. *Experientia*, **47**, 376–91.

Richter, D. D. & Markewitz, D. (2001). *Understanding Soil Change*. Cambridge: Cambridge University Press.

Richter, D. D., Markewitz, D., Trumbore, S. A. & Wells, G. P. (1999). Rapid accumulation and turnover of soil carbon in a re-establishing forest. *Nature*, **400**, 56–8.

Roosen, J., Oesterhelt, C., Pardons, K., Swinnen, E. & Winderickx, J. (2004). Integration of nutrient signalling pathways in the yeast *Saccharomyces cerevisiae*. In *Topics in Current Genetics*, Vol. 7. *Nutrient-induced Responses in Eukaryotic Cells*, ed. J. Winderickx & P. M. Taylor. Berlin: Springer-Verlag, pp. 277–318.

Rosen, S., Sjollema, K. S., Veenhuis, M. & Tunlid, A. (1997). A cytoplasmic lectin produced by the fungus *Arthrobotrys oligospora* functions as a storage protein during saprophytic and parasitic growth. *Microbiology*, **143**, 2593–604.

Schimel, D. S. (1995). Terrestial ecosystems and the carbon cycle. *Global Change Biology*, **1**, 77–91.

Sievering, H. (1999). Nitrogen deposition and carbon sequestration. *Nature*, **400**, 629–90.

180   S. Watkinson et al.

Simard, S. W., Perry, D. A., Jones, M. D. et al. (1997). Net transfer of carbon between ectomycorrhizal tree species in the field. Nature, 388, 579–82.

Sinsabaugh, R. L. & Liptak, M. A. (1997). Enzymatic conversion of plant biomass. In The Mycota, Vol. IV. Environmental and Microbial Relationships, ed. D. T. Wicklow & B. Söderström. Berlin: Springer-Verlag, pp. 347–57.

Sinsabaugh, R. L., Carreiro, M. M. & Repert, D. A. (2002). Allocation of extracellular enzymatic activity in relation to litter decomposition, N deposition and mass loss. Biogeochemistry, 60, 1–24.

Stark, J. M. & Hart, S. C. (1997). High rates of nitrification and nitrate turnover in undisturbed coniferous forests. Nature, 385, 61–4.

Stark, N. (1972). Nutrient cycling pathways and litter fungi. Bioscience, 22, 355–60.

Swift, M. J., Heal, O. W. & Anderson, J. M. (1979). Decomposition in Terrestrial Ecosystems. Oxford: Blackwell Scientific.

Tamm, C. O. (1982). Nitrogen cycling in undisturbed and manipulated boreal forest. Philosophical Transactions of the Royal Society, London; series B. 296, 419–25.

Thompson, W. (1984). Distribution, development and functioning of mycelial cord systems of decomposer basidiomycetes of the deciduous woodland floor. In The Ecology and Physiology of the Fungal Mycelium, ed. D. H. Jennings & A. D. M. Rayner. Cambridge: Cambridge University Press, pp. 185–214.

Thrane, C., Kaufmann, B., Stumm, B. M. & Olsson, S. (2004). Activation of caspase-like activity and poly(ADP-ribose) polymerase degradation during sporulation in Aspergillus nidulans. Fungal Genetics and Biology, 41, 361–8.

Tietema, A., Emmett, B. A., Gundersen, P., Kjonaas, O. J. & Koopmans, C. J. (1998). The fate of ^{15}N-labelled nitrogen deposition in coniferous forests. Forest Ecology and Management, 101, 19–27.

Tlalka, M., Watkinson, S. C., Darrah, P. R. & Watkinson, S. C. (2002). Continuous imaging of amino-acid translocation in intact mycelia of Phanerochaete velutina reveals rapid, pulsatile fluxes. New Phytologist, 153, 173–84.

Tlalka, M., Darrah, P. R., Hensman, D., Watkinson, S. C. & Fricker, M. D. (2003). Noncircadian oscillations in amino acid transport have complementary profiles in assimilatory and foraging hyphae of Phanerochaete velutina. New Phytologist 158, 325–35.

Torn, M. S., Trumbore, S. E., Chadwick, O. A., Vitousek, P. M. & Henricks, D. M. (1997). Mineral control of soil carbon storage and turnover. Nature, 389, 170–3.

Townsend, A. R., Braswell, B. H., Holland, E. A. & Penner, J. E. (1996). Spatial and temporal patterns in terrestrial carbon storage due to deposition of fossil fuel nitrogen. Ecological Applications, 6, 806–14.

Venables, C. E. & Watkinson, S. C. (1989). Medium-induced changes in patterns of free and combined amino acids in mycelium of Serpula lacrymans. Mycological Research, 92, 273–7.

Vestgarden, L. S., Nilsen, P. & Abramasen, G. (2004). Nitrogen cycling in Pinus sylvestris stands exposed to different nitrogen inputs. Scandinavian Journal of Forest Research, 19, 38–47.

Vitousek, P. M. & Howarth, R. W. (1991). Nitrogen limitation on land and in the sea: how can it occur? Biogeochemistry, 13, 87–119.

Wadekar, R. V., North, M. J. & Watkinson, S. C. (1995). Proteolytic enzymes in two wood decaying basidiomycete fungi, Serpula lacrymans and Coriolus versicolor. Microbiology, 141, 1575–83.

Waldrop, M. P., Zak, D. R. & Sinsabaugh, R. L. (2004). Microbial community responses to nitrogen deposition in Northern forest ecosystems. Soil Biology and Biochemistry, 36, 1443–51.

Wallander, H. & Nylund, J.-E. (1991). Effects of excess nitrogen on carbohydrate concentration and mycorrhizal development of *Pinus sylvestris* L. seedlings. *New Phytologist*, **119**, 405–11.

Wallenda, T. & Kottke, I. (1998). Nitrogen deposition and mycorrhizas. *New Phytologist*, **139**, 169–87.

Wang, J. & Bakken, L. R. (1997). Competition for nitrogen during decomposition of plant residues in soil: effect of spatial placement of N-rich and N-poor plant residues. *Soil Biology and Biochemistry*, **29**, 153–162.

Wardle, D. A. (2002). *Monographs in Population Biology*, Vol. 34. *Communities and Ecosystems: Linking the Aboveground and Belowground Components*. Princeton: Princeton University Press.

Wardle, D. A., Bardgett, R. D., Klironomos, J. N. *et al.* (2004a). Ecological linkages between aboveground and belowground biota. *Science*, **304**, 1629–33.

Wardle, D. A., Walker, L. R. & Bardgett, R. D. (2004b). Mature forest ecosystems eventually decline as soil properties deteriorate and phosphorus becomes depleted. *Science*, **305**, 509–13.

Watkinson, S. C. (1975). The relation between nitrogen nutrition and formation of mycelial strands in *Serpula lacrymans*. *Transactions of the British Mycological Society*, **64**, 195–200.

Watkinson, S. C. (1977). The effect of amino acids on coremium development in *Penicillium claviforme*. *Journal of General Microbiology*, **101**, 269–75.

Watkinson, S. C. (1984). Inhibition of growth and development of *Serpula lacrymans* by the non-metabolized amino acid analogue 2-aminoisobutyric acid. *FEMS Microbiology Letters*, **24**, 247–50.

Watkinson, S. C. (1999). Metabolism and differentiation in basidiomycete mycelium. In *The Fungal Colony*, ed. N. A. R. Gow, G. D. Robson and G. M. Gadd. Cambridge: Cambridge University Press, pp. 126–56.

Watkinson, S. C., Davison, E. M. & Bramah, J. (1981). The effect of nitrogen availability on growth and cellulolysis by *Serpula lacrymans*. *New Phytologist*, **89**, 295–305.

Watkinson, S. C., Burton, K. S. & Wood, D. A. (2001). Characteristics of intracellular peptidase and proteinase activities from the mycelium of a cord-forming wood decay fungus, *Serpula lacrymans*. *Mycological Research*, **105**, 698–704.

Watkinson, S. C., Burton, K, Darrah, P. R. *et al.* (2005). New approaches to investigating the function of mycelial networks. *Mycologist*, **19**, 11–17.

Wells, J. M. & Boddy, L. (1995). Phosphorus translocation by saprotrophic basidiomycete mycelial cord systems on the floor of a mixed deciduous woodland. *Mycological Research*, **99**, 977–80.

Wells, J. M., Boddy, L. & Donnelly, D. P. (1998). Wood decay and phosphorus translocation by the cord forming basidiomycete *Phanerochaete velutina*: the significance of local nutrient supply. *New Phytologist*, **138**, 607–17.

Wells, J. M., Harris, M. J. & Boddy, L. (1999). Dynamics of mycelial growth and phosphorus partitioning in developing mycelial cord systems of *Phanerochaete velutina*: dependence on carbon availability. *New Phytologist*, **142**, 325–34.

Wessels, J. G. H. (1999). Fungi in their own right. *Fungal Genetics and Biology*, **27**, 134–45.

Winderickx, J. G. & Taylor, P. M. (eds.) (2004). *Topics in Current Genetics*, Vol. 7, *Nutrient-induced Responses in Eukaryotic Cells*. Berlin: Springer-Verlag.

Zogg, G. P., Zak, D. R., Pregitzer, K. S. & Burton, A. J. (2000). Microbial immobilization and the retention of anthropogenic nitrate in a northern hardwood forest. *Ecology*, **81**, 1858–66.

8

Relative roles of bacteria and fungi in polycyclic aromatic hydrocarbon biodegradation and bioremediation of contaminated soils

CARL E. CERNIGLIA AND JOHN B. SUTHERLAND

Introduction

Polycyclic aromatic hydrocarbons (PAHs) are a large group of toxic compounds (Fig. 8.1) that are components of coal and petroleum and are also produced during incomplete combustion of fuels. They are introduced into the environment via many routes, including fossil-fuel combustion, automobile and diesel engine exhausts, production of manufactured gas and coal tar, wood-preservation processes and waste incineration (Harvey, 1997; Pozzoli et al., 2004). Benzenoid PAHs are thermodynamically stable, with positive bond resonance energies (Aihara, 1996), and have vapour pressures of 2.8×10^{-5} to $10.4\,Pa$ (Sonnefeld et al., 1983). The aqueous solubility of PAHs ranges from $0.2\,\mu g/l$ for indeno[1,2,3-cd]pyrene and $1.6\,\mu g/l$ for benzo[a]pyrene to $31.7\,mg/l$ for naphthalene (Lehto et al., 2003). Despite their low solubility, PAHs are widely distributed in the environment (Wilcke, 2000; Saltiene et al., 2002; Peachey, 2003; Pozzoli et al., 2004) and, as persistent organic pollutants, they are involved in biogeochemical cycling (Del Vento & Dachs, 2002; Jeon et al., 2003). The five-ring PAH, perylene, found in Jurassic sediments may even have originated from ancient fungi (Jiang et al., 2000).

Sixteen PAHs are on the lists of priority pollutants of the US Environmental Protection Agency and the European Union (Lehto et al., 2003); mixtures containing more than 50 individual PAHs have been found in sediments at hazardous waste sites (Brenner et al., 2002). Low-molecular-weight PAHs, with two or three rings, are the most volatile and usually the most abundant. High-molecular-weight PAHs, with four or more rings, are less volatile. The PAHs in contaminated soils and

Fungi in Biogeochemical Cycles, ed. G. M. Gadd. Published by Cambridge University Press. © British Mycological Society 2006.

sediments pose a significant risk to the environment since they have eco-toxic, mutagenic and in some cases also carcinogenic effects (Saltiene *et al.*, 2002; Samanta *et al.*, 2002; White, 2002).

A large number of recent reviews have been written on various aspects of the microbial degradation of PAHs (Sutherland *et al.*, 1995; Mueller *et al.*, 1996; Cerniglia, 1997; Canet *et al.*, 1999; Juhasz & Naidu, 2000; Kanaly & Harayama, 2000; Cerniglia & Sutherland, 2001, 2005; Mougin, 2002; Samanta *et al.*, 2002; Cameotra & Bollag, 2003; Habe & Omori, 2003; Makkar & Rockne, 2003; Mrozik *et al.*, 2003; Antizar-Ladislao *et al.*, 2004; Sutherland, 2004). The objective of this chapter is to review the roles of fungi and indigenous bacteria, both individually and in combina-tions, in the clean-up of PAH-contaminated soils and sediments. We also summarize the occurrence and toxicology of PAHs and describe briefly what is currently known about the fungal and bacterial degradation of these compounds.

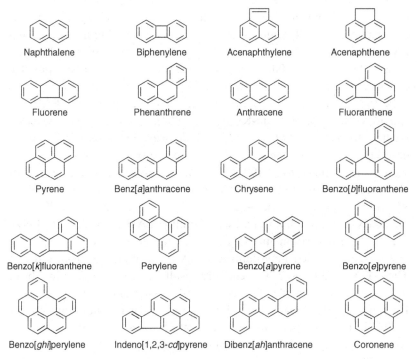

Fig. 8.1. Structures of typical PAHs.

Toxicology of PAHs

Many PAHs have ecotoxic effects and inhibit the growth of micro-organisms, plants and animals (Maliszewska-Kordybach & Smreczak, 2000; Charrois et al., 2001; Ahtiainen et al., 2002; Mendonça & Picado, 2002; Fent, 2003). The ecotoxicity of PAHs is related to water solubility; only the eight PAHs with the lowest molecular weights and highest aqueous solubilities have been shown to inhibit the growth of soil invertebrates (Sverdrup et al., 2002). The toxicity of some PAHs is increased by light and UV irradiation (Kosian et al., 1998; Yu, 2002). In humans, PAHs are more often linked to mutagenicity and carcinogenicity (Dipple et al., 1999) than to direct toxicity.

Coal tars, coal-tar pitches and tobacco smoke all contain mixtures of PAHs that are carcinogenic to humans (Culp et al., 2000; Koganti et al., 2001; Castaño-Vinyals et al., 2004; Ohura et al., 2004). There is a link between cigarette smoke and the levels of PAHs in indoor air (Ohura et al., 2004). Some high-molecular-weight PAHs, such as benz[a]anthracene, dibenz[ah]anthracene and benzo[a]pyrene, are probably carcinogenic to humans and others are possibly carcinogenic (Dipple et al., 1999; Shimada & Fujii-Kuriyama, 2004). The toxicity equivalence factor for PAHs is based on benzo[a]pyrene, which is a known carcinogen in animals (Saltiene et al., 2002; Ohura et al., 2004). Even the ubiquitous two-ring PAH, naphthalene, which is produced by at least one tropical endophytic fungus (Daisy et al., 2002), has been suggested recently to be a possible carcinogen (Preuss et al., 2003).

In the mammalian liver, cytochromes P450 may oxidize PAHs to epoxides (Shimada & Fujii-Kuriyama, 2004), which may be converted further to dihydrodiol epoxides. Biological activation mechanisms of PAHs include not only the dihydrodiol-epoxide pathway, the most significant (Dipple et al., 1999; Shimada & Fujii-Kuriyama, 2004), but also the radical-cation pathway (Majcherczyk & Johannes, 2000; Watanabe et al., 2000), the o-quinone pathway (Penning et al., 1999) and the benzylic hydroxylation pathway for alkyl-PAHs (Engst et al., 1999). All of these activation mechanisms lead to the formation of reactive intermediates, which may damage DNA by forming covalent adducts. These DNA adducts may eventually result in genetic mutations, birth defects, or cancer (Miller & Ramos, 2001; Castaño-Vinyals et al., 2004; Harrigan et al., 2004). Predictions of genotoxicity and mutagenicity of PAHs may be made using analytical chemical data, but they may correlate poorly to toxicity measured by biological effects (Ahtiainen et al., 2002).

In a two-year toxicological study (Culp *et al.*, 2000), the incidences of tumours and DNA adducts in mice fed either coal tar or pure benzo[*a*] pyrene were examined. Benzo[*a*]pyrene formed adducts readily with DNA and appeared to be responsible for forestomach tumours, like those induced by ingestion of coal tar, but not for lung tumours. Other mice that were fed coal tar (Koganti *et al.*, 2001) had DNA adducts with benzo[*a*]pyrene, benzo[*c*]fluorene and benzo[*b*]fluoranthene in the lungs, but those fed coal tar-contaminated soil had only the adducts with benzo[*c*]fluorene and benzo[*b*]fluoranthene. Although benzo[*a*]pyrene activation has been studied extensively, little is known about the activation mechanisms of benzofluorenes and benzofluoranthenes.

Human exposure to PAHs in the environment today appears to be practically unavoidable. Charcoal grilling of meat (Rivera *et al.*, 1996) and bioaccumulation of PAHs in fish (van der Oost *et al.*, 2003) are two ways by which food may become contaminated with PAHs. Workers involved in the clean-up of hazardous wastes may be at risk from PAH exposure, although smoking is also a factor (Lee *et al.*, 2002).

Several PAHs and hydroxylated or methylated PAH derivatives induce oestrogenic or dioxin-like (antioestrogenic) effects in fish and mammalian cell lines (Santodonato, 1997; Villeneuve *et al.*, 2002; Michallet-Ferrier *et al.*, 2004). For instance, benz[*a*]anthracene and dibenz[*ah*]anthracene elicit oestrogenic responses in vitro (Villeneuve *et al.*, 2002). These two PAHs and five others may also elicit dioxin-like responses, as shown by their induction of ethoxyresorufin-*O*-deethylase (EROD) activity, a biomarker for cytochrome P450 1A1 (Gravato & Santos, 2002; Villeneuve *et al.*, 2002).

Principles and challenges for bioremediation of PAH-contaminated soils

Polycyclic aromatic hydrocarbons are not completely stable in the environment. They disappear relatively slowly through physical, chemical and biological processes, some of which are mediated by bacteria and fungi. The recalcitrance of PAH residues in soils and sediments increases with molecular weight, which is toxicologically significant because many high-molecular-weight PAHs are genotoxic and carcinogenic (Lodovici *et al.*, 1998; Dipple *et al.*, 1999; Gravato & Santos, 2002). Several reasons have been proposed for the recalcitrance of PAHs in the environment (Allard & Neilson, 1997; Samanta *et al.*, 2002; Ehlers & Luthy, 2003; Huesemann *et al.*, 2003). Some of these reasons are: (1) a site may not contain microorganisms with the catabolic enzymes to degrade PAHs;

(2) the site may be deficient in nutrients; (3) the bioavailability of the PAHs may be limited; (4) there may be preferential utilization of more easily degradable substrates; (5) the site may contain toxic heavy metals; (6) a minimum threshold concentration of an inducer may be necessary for synthesis of degradative enzymes; (7) the PAHs may not be transported into the microbial cells; and (8) degradation may start but then stop due to the accumulation of toxic products. The aerobic biodegradation of PAHs by soil microorganisms uses monooxygenase, peroxidase and dioxygenase pathways; the first and third of these pathways are found in bacteria and the first and second in fungi (Fig. 8.2).

The intentional use of microbial processes to remove toxic pollutants, including PAHs, from the environment is known as bioremediation (Mueller *et al.*, 1996; Allard & Neilson, 1997; Mougin, 2002; Samanta *et al.*, 2002; Cameotra & Bollag, 2003; Mrozik *et al.*, 2003; Antizar-Ladislao *et al.*, 2004). The feasibility of this technology for clean-up of soils and sediments (Pointing, 2001) depends on its effectiveness in degrading not only PAHs but other contaminants to the levels set by regulatory authorities (Lin *et al.*, 1996; Balba *et al.*, 1998a). These target values are ideally based on toxicological and ecotoxicological risk assessments (Gaylor, 1995; White, 2002).

Recent research on the bioremediation of PAHs has been centred on four important topics: (1) the characterization of biodegradative processes useful for treatment of PAHs (Cerniglia & Sutherland, 2001; Habe & Omori, 2003); (2) the development of new methods for enhancing biodegradation (Atagana, 2004); (3) the engineering of established treatment processes to optimize biodegradation (Straube *et al.*, 1999; Samanta *et al.*, 2002; Antizar-Ladislao *et al.*, 2004); and (4) the study of ecotoxicology with risk assessment (Charrois *et al.*, 2001; Fent, 2003; van der Oost *et al.*, 2003).

When fungi are used for bioremediation of soils contaminated with PAHs, a process that has been called mycoremediation (Bhatt *et al.*, 2002), several factors must be considered. Can the fungi colonize non-sterile soil in the presence of indigenous bacteria (Novotný *et al.*, 2000)? Do the fungi produce enzymes to degrade PAHs during growth in the soil (Pointing, 2001; Rama *et al.*, 2001; Zheng & Obbard, 2001)? Do the added fungi enhance the rate of PAH degradation over that of the indigenous microflora alone (Brodkorb & Legge, 1992; Boonchan *et al.*, 2000; Andersson *et al.*, 2003)? Are the rate and extent of bioremediation sufficient to reduce PAH levels in the soil to those acceptable to regulatory agencies (Kalf *et al.*, 1997; Castaldi, 2003; Ehlers & Luthy, 2003)?

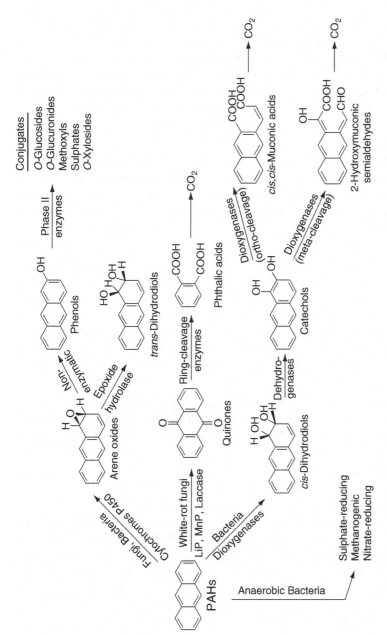

Fig. 8.2. Metabolism of typical PAHs by fungi and bacteria (after Cerniglia, 1993; Cerniglia & Sutherland, 2006).

There has long been a question of whether bioremediation can lower PAH levels enough to protect human health and the environment (Allard & Neilson, 1997). Monitoring the decrease from the levels of PAHs initially present in a polluted site does not confirm the actual extent of detoxification because some PAH metabolites themselves are toxic (Renoux *et al.*, 1999; Kazunga *et al.*, 2001). Because residual PAHs and their metabolites may still be hazardous, incomplete bioremediation of contaminated soils in the past has raised some doubts about the usefulness of this technology for meeting environmental clean-up standards (Breedveld & Karlsen, 2000; Kazunga & Aitken, 2000; Andersson *et al.*, 2003).

Polycyclic aromatic hydrocarbons in contaminated soils may be analysed chemically by standard HPLC protocols (Environmental Protection Agency, 1986). Although it is also possible to detect major PAH degradation intermediates, it is not always clear which additional tests should be done. A variety of genotoxicity assays (Gravato & Santos, 2002; White, 2002) and ecotoxicological tests (Maliszewska-Kordybach & Smreczak, 2000; Ahtiainen *et al.*, 2002; Fent, 2003) have been developed. The impact of PAH contamination on the structure and function of microbial communities may be evaluated by other methods, such as DNA microarrays (Neumann & Galvez, 2002), denaturing gradient gel electrophoresis (Leys *et al.*, 2004), or real-time PCR (Rodrigues *et al.*, 2002). A battery of tests for toxicity screening (Mendonça & Picado, 2002) may be necessary for a comprehensive evaluation of the impact of PAH contamination on the environment.

Fungi and bacteria metabolize a wide variety of PAHs (Cerniglia, 1993; Sutherland *et al.*, 1995; Juhasz & Naidu, 2000), although the principal pathways used by these groups are different. Some PAH bioactivation pathways found in mammals are also observed in microorganisms (Cerniglia & Gibson, 1980).

Factors that affect the biodegradation of PAHs in soils include soil type, pH, temperature, oxygen concentration, irradiation, carbon and nitrogen sources, inorganic nutrients and electron acceptors, as well as the solubility, volatility and sorption behaviour of the PAHs (Joshi & Lee, 1996; Breedveld & Karlsen, 2000; Atagana *et al.*, 2003; Lee *et al.*, 2003; Lehto *et al.*, 2003; Ramsay *et al.*, 2003; Rasmussen & Olsen, 2004). Bioremediation may be enhanced by bacteria that produce not only degradative enzymes but also biosurfactants or semi-colloid heteropolysaccharides (Straube *et al.*, 1999; Cameotra & Bollag, 2003). Many bacteria that can degrade PAHs adhere to solid surfaces, forming biofilms in association with PAH molecules (Bastiaens *et al.*, 2000; Johnsen & Karlson, 2004).

The addition of surfactants to soils or sediments may improve bio-remediation of PAHs by increasing aqueous solubility and bioavailability (Kotterman *et al.*, 1998; Löser *et al.*, 2000; Yang *et al.*, 2003). Although some surfactant amendments have failed in practice (Harayama, 1997; Volkering *et al.*, 1998; Pinto & Moore, 2000), other non-ionic surfactants and biosurfactants have facilitated PAH degradation by soil micro-organisms (Zheng & Obbard, 2001; Cameotra & Bollag, 2003; Makkar & Rockne, 2003; Yang *et al.*, 2003). Both the biodegradability and the toxicity of the surfactants to be used are important (Tiehm *et al.*, 1997; Volkering *et al.*, 1998; Makkar & Rockne, 2003; Yang *et al.*, 2003).

Analysis of soils before and after bioremediation usually shows the greatest reduction in two- and three-ring PAHs and the next greatest reduction in four-ring PAHs. Often, five- and six-ring PAHs will persist because of microbial limitations (Huesemann *et al.*, 2003). Removal of high-molecular-weight PAHs may be achieved by biostimulation with nutrients, especially nitrogen and phosphorus, or bioaugmentation with PAH-degrading cultures (Ruberto *et al.*, 2003; Straube *et al.*, 2003). Some other methods to achieve bioremediation include natural attenuation, landfarming, windrow composting, static bioventing piles and the use of bioreactors (Balba *et al.*, 1998b; Landmeyer *et al.*, 1998; Li *et al.*, 2002; Mendonça & Picado, 2002; Saponaro *et al.*, 2002; Straube *et al.*, 2003). Ecotoxicological and mutagenicity assays after bioremediation generally show decreases in toxicity (Mendonça & Picado, 2002). Microbial assays for toxicity of soils and sediments, however, usually require extracts to be made using water or organic solvents, which fractionate the toxic compo-nents (Kosian *et al.*, 1998; Malachová, 1999) and may underestimate or overestimate the environmental effects.

To assess the success of bioremediation, the stability of PAH residues bound to soil particles must also be determined (Eschenbach *et al.*, 1998). Changes in the solution chemistry of the humic fraction of the soil (Jones & Tiller, 1999) may release PAHs either as the free hydrocarbons or as metabolites. Little is known about the impact of bioremediation on the mutagenicity and genotoxicity of most PAHs (Malachová, 1999).

Usable methods for measuring ecotoxicity are necessary for the devel-opment of efficient strategies for bioremediation. The toxicity of soils to animals, however, cannot always be predicted by knowing only the con-taminants and their concentrations (Charrois *et al.*, 2001). There is a need for methods that will determine the risk due to contaminants that are actually bioavailable, so that the environment can be protected without unacceptably high clean-up costs. Future regulations should consider not only the amounts of PAHs in soils and sediments but also their bioavailability (Ehlers & Luthy, 2003; Huesemann *et al.*, 2003).

Fungal transformation of PAHs

Many fungi from diverse groups (Cerniglia *et al.*, 1978; Cerniglia & Sutherland, 2001, 2005; Sutherland, 2004; Verdin *et al.*, 2004) are able to metabolize PAHs. Since fungi cannot generally use PAHs as carbon and energy sources (Cerniglia & Sutherland, 2001), the culture medium must be supplied with nutrients to allow cometabolism. A few yeasts and filamentous fungi have been reported to use certain PAHs, including anthracene, phenanthrene, pyrene and even benzo[*a*]pyrene, as carbon and energy sources (Romero *et al.*, 1998, 2002; Rafin *et al.*, 2000; Lahav *et al.*, 2002; Saraswathy & Hallberg, 2002; Veignie *et al.*, 2002, 2004; Pan *et al.*, 2004). Polycyclic aromatic hydrocarbons may also be stored in intracellular lipid vesicles by fungi that do not necessarily degrade them (Verdin *et al.*, 2005).

Some fungi cometabolize PAHs to *trans*-dihydrodiols, phenols, quinones, dihydrodiol epoxides and tetraols, but they seldom degrade them completely to CO_2 (Casillas *et al.*, 1996; Cajthaml *et al.*, 2002; da Silva *et al.*, 2003). Water-soluble, detoxified conjugates of hydroxylated PAHs, including glucuronides, xylosides, glucosides and sulphates, have also been detected in culture filtrates of fungi grown with PAHs (Cerniglia, 1992, 1997; Sutherland *et al.*, 1992; Casillas *et al.*, 1996; Capotorti *et al.*, 2004). An example of a three-ring PAH that has been shown to be metabolized by several fungi is phenanthrene (Fig. 8.3) and an example of a four-ring PAH that is metabolized by fungi is pyrene (Fig. 8.4) (Cerniglia & Yang, 1984; Cerniglia *et al.*, 1986).

During the transformation of another high-molecular-weight PAH, benzo[*a*]pyrene, by various fungi (Fig. 8.5), over 13 metabolites have been identified, including *trans*-dihydrodiols, quinones, phenols, dihydrodiol epoxides, bis-*cis*-dihydrodiols, tetraols, glucuronides and sulphates (Cerniglia & Gibson, 1979; Cerniglia, 1997; Juhasz & Naidu, 2000). Little is known about the toxicity of most of these compounds, but the zygomycete *Cunninghamella elegans* produces at least one benzo[*a*]pyrene 7,8-diol-9,10-epoxide isomer that is mutagenic and carcinogenic in mammalian systems (Cerniglia & Gibson, 1980).

The high-molecular-weight PAHs perylene, indeno[1,2,3-*cd*]pyrene, dibenz[*ah*]anthracene, benzo[*ghi*]perylene and coronene are also oxidized by some fungi (Gramss *et al.*, 1999; Zheng & Obbard, 2002; Lau *et al.*, 2003; Steffen *et al.*, 2003; Verdin *et al.*, 2005). Although minor amounts of

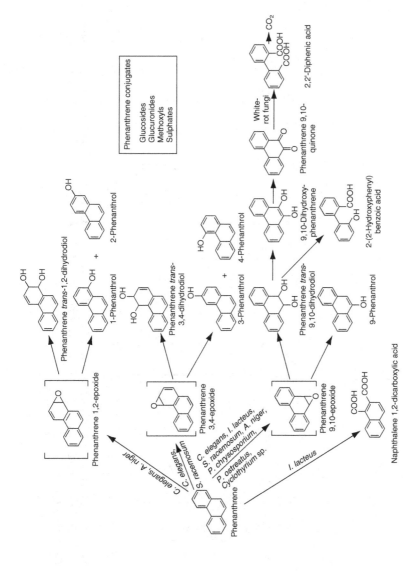

Fig. 8.3. Fungal metabolism of phenanthrene (after Cerniglia & Yang, 1984; Cerniglia & Sutherland, 2006). *A. niger*, *Aspergillus niger*; *C. elegans*, *Cunninghamella elegans*; *I. lacteus*, *Irpex lacteus*; *P. chrysosporium*, *Phanerochaete chrysosporium*; *P. ostreatus*, *Pleurotus ostreatus*; *S. racemosum*, *Syncephalastrum racemosum*.

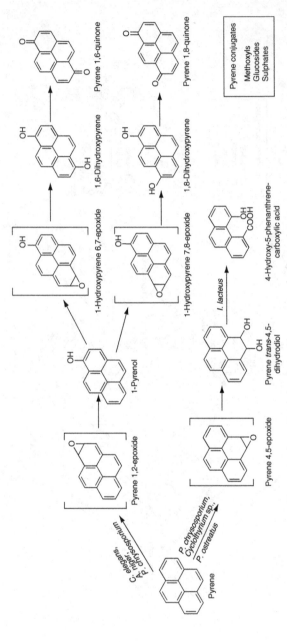

Fig. 8.4. Fungal metabolism of pyrene (after Cerniglia *et al.*, 1986; Cerniglia & Sutherland, 2005). *A. niger, Aspergillus niger; C. elegans, Cunninghamella elegans; I. lacteus, Irpex lacteus; P. chrysosporium, Phanerochaete chrysosporium; P. ostreatus, Pleurotus ostreatus.*

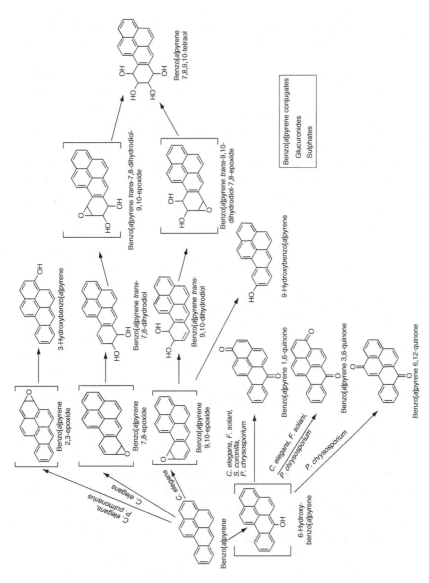

Fig. 8.5. Fungal metabolism of benzo[a]pyrene (after Cerniglia & Gibson, 1979; Cerniglia & Sutherland, 2006). *C. elegans, Cunninghamella elegans; F. solani, Fusarium solani; P. chrysosporium, Phanerochaete chrysosporium; P. pulmonarius, Pleurotus pulmonarius; S. coronilla, Stropharia coronilla.*

mutagenic or carcinogenic metabolites may be produced during the metabolism of high-molecular-weight PAHs by *C. elegans*, most of the metabolites formed by this and other non-ligninolytic fungi are less toxic than the carcinogens typically produced during mammalian metabolism (Cerniglia, 1993; Dipple *et al.*, 1999; Shimada & Fujii-Kuriyama, 2004).

The transformation of PAHs by ligninolytic, wood-decaying fungi involves several different enzymes, which depend not only on the species of fungus but also on the growth conditions (Sutherland *et al.*, 1991; Hammel *et al.*, 1992). The enzymes produced by white-rot fungi that are involved in PAH degradation (Fig. 8.2) include lignin peroxidase, manganese peroxidase, laccase, cytochrome P450 and epoxide hydrolase (Haemmerli *et al.*, 1986; Bezalel *et al.*, 1996; Cerniglia & Sutherland, 2005). Ligninolytic fungi metabolize PAHs via reactions involving reactive oxygen species to produce phenols and quinones (May *et al.*, 1997; Pickard *et al.*, 1999; Steffen *et al.*, 2003); these compounds may be degraded further by ring-fission enzymes (Hammel *et al.*, 1991; Juhasz & Naidu, 2000).

Several wood-decaying fungi, including *Bjerkandera, Coriolopsis, Irpex, Phanerochaete, Pleurotus* and *Trametes* spp., have been investigated for use in bioremediation of PAH-contaminated soils (Colombo *et al.*, 1996; Leštan & Lamar, 1996; May *et al.*, 1997; Baldrian *et al.*, 2000; Novotný *et al.*, 2000). Bench-scale trials have demonstrated their ability to degrade complex mixtures of PAHs, such as those in creosote and coal tar, but efforts at bioremediation of contaminated soils using these fungi have met with varying degrees of success (Cerniglia, 1997; Canet *et al.*, 2001; Cerniglia & Sutherland, 2001; Pointing, 2001; Hestbjerg *et al.*, 2003).

Non-ligninolytic fungi, including *Cunninghamella, Mucor, Fusarium* and *Penicillium* spp., are also being considered for use in bioremediation of PAHs even though less is known about their degradative capability in non-sterile soil (Colombo *et al.*, 1996; Cerniglia, 1997; Pinto & Moore, 2000; Rama *et al.*, 2001; Ravelet *et al.*, 2001; Saraswathy & Hallberg, 2002; Potin *et al.*, 2004). The use of fungi to treat contaminated soils remains challenging; it is still extremely difficult and time-consuming to reduce the PAH concentration in soil to meet clean-up goals.

Bacterial transformation of PAHs

Bacteria reported to metabolize PAHs include species of *Acidovorax, Acinetobacter, Actinomyces, Aeromonas, Agrobacterium, Alcaligenes, Arthrobacter, Aureobacterium, Bacillus, Brevibacterium, Burkholderia, Chryseobacterium, Comamonas, Corynebacterium, Cycloclasticus, Cyclotrophicus, Empedobacter, Flavobacterium, Flavomonas, Gordona,*

Halomonas, Lutibacterium, Marinobacter, Marinomonas, Methylococcus, Microbacterium, Micrococcus, Moraxella, Mycobacterium, Myroides, Neptunomonas, Nocardia, Nocardioides, Ochrobacter, Paenibacillus, Paracoccus, Pasteurella, Photobacterium, Polaromonas, Porphyrobacter, Pseudoalteromonas, Pseudomonas, Ralstonia, Rhodococcus, Saccharothrix, Sphingobium, Sphingomonas, Stenotrophomonas, Streptomyces, Terrabacter, Thermus, Vibrio, Weeksella and other genera (Cerniglia, 1992; Shuttleworth & Cerniglia, 1996; Šepič *et al.*, 1998; Gibson, 1999; Hedlund *et al.*, 1999; Juhasz & Naidu, 2000; Kanaly & Harayama, 2000; Chung & King, 2001; Daane *et al.*, 2002; Melcher *et al.*, 2002; Samanta *et al.*, 2002; Feitkenhauer *et al.*, 2003; Gauthier *et al.*, 2003; Habe & Omori, 2003; Hu *et al.*, 2003; Mrozik *et al.*, 2003; Jeon *et al.*, 2004; Lors *et al.*, 2004; Zhang *et al.*, 2004).

Although the *nah* genes of naphthalene-degrading pseudomonads have been studied the most extensively, other metabolic pathways are also used by bacteria for the degradation of PAHs (Habe & Omori, 2003). In Gram-negative and most Gram-positive bacteria, PAHs are oxidized first by dioxygenases that produce *cis*-dihydrodiols (Fig. 8.2), which are metabolized by dehydrogenases that produce diphenols and then by other dioxygenases that cleave the aromatic rings (Cerniglia, 1993; Sutherland *et al.*, 1995; Gibson, 1999; Brezna *et al.*, 2003; Habe & Omori, 2003). Some Gram-positive bacteria, such as *Mycobacterium vanbaalenii* and *Streptomyces flavovirens*, have cytochrome P450 monooxygenases and epoxide hydrolases (Fig. 8.2) that produce PAH *trans*-dihydrodiols (Kelley *et al.*, 1990; Sutherland *et al.*, 1990; Moody *et al.*, 2001). Even high-molecular-weight PAHs can be metabolized to both *cis*- and *trans*-dihydrodiols by some *Mycobacterium* spp. (Vila *et al.*, 2001; Moody *et al.*, 2004).

The low-molecular-weight PAHs that are oxidized by bacteria include naphthalene, acenaphthene, acenaphthylene, fluorene, phenanthrene and anthracene. Naphthalene degraders appear to be ubiquitous, but bacteria degrading the three-ring compounds are abundant only in polluted environments (Lors *et al.*, 2004) where the species diversity may be lower (Leys *et al.*, 2004). *Pseudomonas* sp. A2279, *Porphyrobacter* sp. B51 and *Sphingomonas* sp. LB126 grow on various combinations of low-molecular-weight PAHs, although no strain grows on all of them (Selifonov *et al.*, 1998; Gauthier *et al.*, 2003; van Herwijnen *et al.*, 2003). *Sphingomonas* sp. ZL5, which grows on phenanthrene but not naphthalene, has a gene for catechol 2,3-dioxygenase on a 60 kb resident plasmid (Liu *et al.*, 2004). *Mycobacterium vanbaalenii* PYR-1 cometabolizes naphthalene, phenanthrene and anthracene (Kelley et al., 1990; Moody et al., 2001). Some other Mycobacterium spp. and Saccharothrix sp. PYX-6 can grow on phenanthrene as a sole carbon source and Saccharothrix sp. can also grow on anthracene (Vila et al., 2001; Hu et al., 2003; Krivobok et al., 2003; Sho et al., 2004). Several Rhodococcus spp. have been shown to hydroxylate fluorene (Finkelstein et al., 2003).

The high-molecular-weight PAHs that are oxidized by bacteria include fluoranthene, benzo[b]fluorene, pyrene, chrysene, benz[a]anthracene, benzo [a]pyrene, benzo[e]pyrene and 3-methylcholanthrene. Some of these, such as benzo[a]pyrene, can be degraded if there are low-molecular-weight PAHs present as carbon sources (Juhasz & Naidu, 2000; Dries & Smets, 2002). Pasteurella sp. IFA, Sphingomonas paucimobilis EPA 505, Sphingomonas sp. LB126 and Pseudomonas stutzeri P16 metabolize fluoranthene by different pathways (Šepič et al., 1998; Ho et al., 2000; Kazunga et al., 2001; van Herwijnen et al., 2003). Sphingomonas yanoikuyae R1, S. paucimobilis EPA 505, Pseudomonas saccharophila P15, and P. stutzeri P16 all metabolize pyrene (Ho et al., 2000; Kazunga & Aitken, 2000). Porphyrobacter sp. B51 apparently degrades benzo[a]pyrene in the presence of chrysene (Gauthier et al., 2003). Mycobacterium vanbaalenii PYR-1 metabolizes pyrene and benzo[a]pyrene to cis- and trans-dihydrodiols with several other metabolites (Heitkamp et al., 1988a, b; Cerniglia, 1992; Kazunga & Aitken, 2000; Moody et al., 2004); it also degrades fluoranthene (Šepič et al., 1998). Several other Mycobacterium sp. strains degrade high-molecular-weight PAHs (Schneider et al., 1996; Willumsen et al., 2001; Bogan et al., 2003; Gauthier et al., 2003) and some of them can use PAHs as carbon sources (Vila et al., 2001; Krivobok et al., 2003; Sho et al., 2004). Bacillus cereus P21, Microbacterium esteraromaticum B21 and Saccharothrix sp. PYX-6 also oxidize high-molecular-weight PAHs (Kazunga & Aitken, 2000; Kazunga et al., 2001; Hu et al., 2003).

In M. vanbaalenii PYR-1, an aromatic-ring dioxygenase produces cis-dihydrodiols from PAHs; they have been identified (Heitkamp et al., 1988b; Khan et al., 2001; Moody et al., 2001, 2004) but their toxicity has not yet been determined. The cis-dihydrodiols are further metabolized to other products whose toxicity also has not been examined, such as chrysene 4,5-dicarboxylic acid, 4-formylchrysene-5-carboxylic acid and 10-oxabenzo[def]chrysen-9-one (Schneider et al., 1996; Moody et al., 2004).

Bacteria also degrade PAHs under low-oxygen (Hirano et al., 2004) and anaerobic conditions (Fig. 8.2), including sulphate-reducing, methanogenic and nitrate-reducing conditions (Coates et al., 1997; Galushko et al., 1999; Rockne et al., 2000; Chang et al., 2002; Rothermich et al.,

2002). Much less is known about the enzymes or the pathways used for anaerobic PAH degradation than about those used for aerobic degradation. Naphthalene, however, is usually metabolized anaerobically via 2-naphthoic acid (Meckenstock *et al.*, 2004).

Degradation of PAHs by mixed cultures of bacteria and fungi

Since the white-rot fungus *Phanerochaete chrysosporium* is known to degrade a variety of PAHs and even to break down ^{14}C-benzo[*a*]pyrene in liquid culture (Bumpus *et al.*, 1985), studies were done to see if this fungus could be used for bioremediation of oil tar-contaminated soil (Brodkorb & Legge, 1992). The evolution of ^{14}CO$_2$ from ^{14}C-phenanthrene in soil was enhanced almost 2-fold (from 19.5% to 37.7%) when *P. chrysosporium* was added to the indigenous soil microflora. The authors suggested that there was a synergistic relationship between the fungus and the soil microorganisms, speculating that *P. chrysosporium* metabolized phenanthrene to more water-soluble intermediates that could act as substrates for soil bacteria (Brodkorb & Legge, 1992).

The relative contribution to phenanthrene transformation of eukaryotic microorganisms, including fungi, in silty clay coastal sediments has been estimated at less than 3% of the total (MacGillivray & Shiaris, 1994), suggesting that the predominant phenanthrene degraders in the sediments must be bacteria. Recently, stable isotopic probing with ^{13}C in an aquifer contaminated by coal-tar waste has been used (Jeon *et al.*, 2003, 2004) to find a previously unknown group of bacteria responsible for the evolution of ^{13}CO$_2$ from ^{13}C-labelled naphthalene.

The utility of co-cultures of fungi with bacteria for the degradation of PAHs has been shown by Boonchan *et al.* (2000), who combined *Penicillium janthinellum* with either *Stenotrophomonas maltophilia* or a consortium of unidentified bacteria. The fungus by itself could partially degrade pyrene and benzo[*a*]pyrene but did not use either of them as a carbon source; *S. maltophilia* by itself could use pyrene as a carbon source and cometabolize benzo[*a*]pyrene. Fungal–bacterial combinations grew on pyrene, chrysene, benz[*a*]anthracene, benzo[*a*]pyrene and dibenz[*ah*]anthracene, converting 25% of the benzo[*a*]pyrene to CO$_2$ in 49 days, and they reduced the mutagenicity of contaminated soil (Boonchan *et al.*, 2000).

Combinations of white-rot fungi, including *P. chrysosporium, Pleurotus ostreatus* and *Coriolus versicolor* (=*Trametes versicolor*), were cultured on wheat straw and added to non-sterile, coal tar-contaminated soil (Canet *et al.*, 2001). There was little or no change in total PAH content over 32 weeks, even though wheat straw itself enhanced PAH degradation by the indigenous soil microflora. The fungi survived in the contaminated soil but were not active in metabolizing PAHs (Canet et al., 2001).

The relationship between fungi and indigenous soil bacteria (Meulenberg et al., 1997; Gramss et al., 1999; Andersson et al., 2000, 2003; Novotný et al., 2000; Canet et al., 2001; Saponaro et al., 2002; Hestbjerg et al., 2003) is important in PAH degradation because soil microorganisms may either inhibit the growth of the introduced mycelium or cooperate in the further metabolism of fungal PAH metabolites. Many non-ligninolytic soil fungi that have the potential to survive and compete with indigenous soil fungi and bacteria can also efficiently initiate oxidation of the aromatic rings (Cerniglia, 1997; Potin et al., 2004). These fungi may be able to enhance the degradation of PAHs by other soil microorganisms.

The white-rot fungus P. ostreatus and the brown-rot fungus Antrodia vaillantii enhanced the degradation of fluorene, phenanthrene, pyrene and benz[a]anthracene in artificially contaminated soils (Andersson et al., 2003); several PAH metabolites, including 9-fluorenone, benz[a]anthracene-7,12-dione, 4-hydroxy-9-fluorenone and 4-oxapyrene-5-one, accumulated. Unlike P. ostreatus, which inhibited the growth of indigenous soil microorganisms, A. vaillantii stimulated soil microbial activity. Phospholipid fatty acid analysis showed that P. ostreatus changed the composition of the bacterial community. This study highlights the importance of monitoring the growth of inoculated fungi and indigenous soil microorganisms during bioremediation of PAHs.

Ligninolytic fungi partially oxidize PAHs by reactions involving extracellular free radicals (Majcherczyk & Johannes, 2000), making the PAHs more water-soluble so that they are able to serve as substrates for bacterial degradation (Meulenberg et al., 1997). Partial oxidation increases the bioavailability of PAHs in most contaminated sites (Mueller et al., 1996; Meulenberg et al., 1997). Soils contaminated with substantial amounts of PAHs may contain large populations of PAH-transforming bacteria (Wrenn & Venosa, 1996; Johnsen et al., 2002) and fungi (April et al., 2000; Saraswathy & Hallberg, 2002). Combinations of several microorganisms are usually better able than pure cultures to degrade benzo[a]pyrene and other high-molecular-weight PAHs (Kanaly et al., 2000).

Less is known about the pathways of PAH degradation by co-cultures than about the pathways of degradation by individual bacteria and fungi (Juhasz & Naidu, 2000). Four bacteria (Pseudomonas aeruginosa, Pseudomonas cepacia (=Burkholderia cepacia), Pseudomonas sp. and Ralstonia pickettii) and four fungi (Alternaria tenuis, Aspergillus terreus, Trichoderma *viride* and *Penicillium* sp.) from hydrocarbon-polluted soils around oil wells and refineries in Mexico were selected for co-culture degradation experiments (Chávez-Gómez *et al.*, 2003). When the microorganisms were tested individually for the ability to degrade phenanthrene in soil containing sugarcane bagasse during 18 days of incubation, the fungi degraded 35% to 50% of the phenanthrene but the bacteria degraded only 20%. When the strains were screened in various combinations, the range of phenanthrene degradation for the co-cultures was 4% to 73%. Combinations of *Penicillium* sp. with *R. pickettii, B. cepacia*, or *P. aeruginosa* were the most successful (Chávez-Gómez *et al.*, 2003). The potential of fungal co-cultures with bacteria to remove PAHs from soil was confirmed, but the metabolites were not identified.

Conclusions

When microorganisms are being considered for bioremediation of PAH-contaminated soils and sediments, some basic and applied research issues should be addressed, especially the simultaneous or successive application of combinations of fungi and bacteria (Meulenberg *et al.*, 1997; Andersson *et al.*, 2000, 2003; Boonchan *et al.*, 2000; Chávez-Gómez *et al.*, 2003). Other factors that should be investigated include: (1) the mechanisms and pathways involved in the degradation of high-molecular-weight PAHs (Schneider *et al.*, 1996; Kazunga *et al.*, 2001; Kanaly & Watanabe, 2004); (2) the microbial metabolites that repress degradation of high-molecular-weight PAHs (Kazunga & Aitken, 2000; Kazunga *et al.*, 2001; Juhasz *et al.*, 2002); (3) the environmental fate of mixtures of PAHs and the metabolites formed from them (Selifonov *et al.*, 1998; Bouchez *et al.*, 1999; Vila *et al.*, 2001; Lotfabad & Gray, 2002); (4) the synergistic or inhibitory relationships between indigenous soil microflora and introduced mycelia (Brodkorb & Legge, 1992; Colombo *et al.*, 1996; Canet *et al.*, 2001; Andersson *et al.*, 2003); and (5) the selection of microbial strains with the ability to colonize contaminated soils and metabolize PAHs under a variety of environmental conditions (Bhatt *et al.*, 2002; Hestbjerg *et al.*, 2003; Lau *et al.*, 2003; Potin *et al.*, 2004).

References

Ahtiainen, J., Valo, R., Järvinen, M. & Joutti, A. (2002). Microbial toxicity tests and chemical analysis as monitoring parameters at composting of creosote-contaminated soil. *Ecotoxicology and Environmental Safety*, **53**, 323–9.

Aihara, J. (1996). Bond resonance energies of polycyclic benzenoid and non-benzenoid hydrocarbons. *Journal of the Chemical Society Perkin Transactions II*, **10**, 2185–95.

Allard, A.-S. & Neilson, A. H. (1997). Bioremediation of organic waste sites: a critical review of microbiological aspects. *International Biodeterioration and Biodegradation*, **39**, 253–85.

Andersson, B. E., Welinder, L., Olsson, P. A., Olsson, S. & Henrysson, T. (2000). Growth of inoculated white-rot fungi and their interactions with the bacterial community in soil contaminated with polycyclic aromatic hydrocarbons, as measured by phospholipid fatty acids. *Bioresource Technology*, **73**, 29–36.

Andersson, B. E., Lundstedt, S., Tornberg, K. *et al.* (2003). Incomplete degradation of polycyclic aromatic hydrocarbons in soil inoculated with wood-rotting fungi and their effect on the indigenous soil bacteria. *Environmental Toxicology and Chemistry*, **22**, 1238–43.

Antizar-Ladislao, B., Lopez-Real, J. M. & Beck, A. J. (2004). Bioremediation of polycyclic aromatic hydrocarbon (PAH)-contaminated waste using composting approaches. *Critical Reviews in Environmental Science and Technology*, **34**, 249–89.

April, T. M., Foght, J. M. & Currah, R. S. (2000). Hydrocarbon-degrading filamentous fungi isolated from flare pit soils in northern and western Canada. *Canadian Journal of Microbiology*, **46**, 38–49.

Atagana, H. I. (2004). Bioremediation of creosote-contaminated soil in South Africa by landfarming. *Journal of Applied Microbiology*, **96**, 510–20.

Atagana, H. I., Haynes, R. J. & Wallis, F. M. (2003). Optimization of soil physical and chemical conditions for the bioremediation of creosote-contaminated soil. *Biodegradation*, **14**, 297–307.

Balba, M. T., Al-Awadhi, N. & Al-Daher, R. (1998a). Bioremediation of oil-contaminated soil: microbiological methods for feasibility assessment and field evaluation. *Journal of Microbiological Methods*, **32**, 155–64.

Balba, M. T., Al-Daher, R., Al-Awadhi, N., Chino, H. & Tsuji, H. (1998b). Bioremediation of oil-contaminated desert soil: the Kuwaiti experience. *Environment International*, **24**, 163–73.

Baldrian, P., in der Wiesche, C., Gabriel, J., Nerud, F. & Zadražil, F. (2000). Influence of cadmium and mercury on activities of ligninolytic enzymes and degradation of polycyclic aromatic hydrocarbons by *Pleurotus ostreatus* in soil. *Applied and Environmental Microbiology*, **66**, 2471–8.

Bastiaens, L., Springael, D., Wattiau, P. *et al.* (2000). Isolation of adherent polycyclic aromatic hydrocarbon (PAH)-degrading bacteria using PAH-sorbing carriers. *Applied and Environmental Microbiology*, **66**, 1834–43.

Bezalel, L., Hadar, Y., Fu, P. P., Freeman, J. P. & Cerniglia, C. E. (1996). Metabolism of phenanthrene by the white rot fungus *Pleurotus ostreatus*. *Applied and Environmental Microbiology*, **62**, 2547–53.

Bhatt, M., Cajthaml, T. & Šašek, V. (2002). Mycoremediation of PAH-contaminated soil. *Folia Microbiologica*, **47**, 255–8.

Bogan, B. W., Lahner, L. M., Sullivan, W. R. & Paterek, J. R. (2003). Degradation of straight-chain aliphatic and high-molecular-weight polycyclic aromatic hydrocarbons by a strain of *Mycobacterium austroafricanum*. *Journal of Applied Microbiology*, **94**, 230–9.

Boonchan, S., Britz, M. L. & Stanley, G. A. (2000). Degradation and mineralization of high-molecular-weight polycyclic aromatic hydrocarbons by defined fungal-bacterial cocultures. *Applied and Environmental Microbiology*, **66**, 1007–19.

Bouchez, M., Blanchet, D., Bardin, V., Haeseler, F. & Vandecasteele, J.-P. (1999). Efficiency of defined strains and of soil consortia in the biodegradation of polycyclic aromatic hydrocarbon (PAH) mixtures. *Biodegradation*, **10**, 429–35.

Breedveld, G. D. & Karlsen, D. A. (2000). Estimating the availability of polycyclic aromatic hydrocarbons for bioremediation of creosote contaminated soils. *Applied Microbiology and Biotechnology*, **54**, 255–61.

Brenner, R. C., Magar, V. S., Ickes, J. A. *et al.* (2002). Characterization and fate of PAH-contaminated sediments at the Wyckoff/Eagle Harbor superfund site. *Environmental Science and Technology*, **36**, 2605–13.

Brezna, B., Khan, A. A. & Cerniglia, C. E. (2003). Molecular characterization of dioxygenases from polycyclic aromatic hydrocarbon-degrading *Mycobacterium* spp. *FEMS Microbiology Letters*, **223**, 177–83.

Brodkorb, T. S. & Legge, R. L. (1992). Enhanced biodegradation of phenanthrene in oil tar-contaminated soils supplemented with *Phanerochaete chrysosporium*. *Applied and Environmental Microbiology*, **58**, 3117–21.

Bumpus, J. A., Tien, M., Wright, D. & Aust, S. D. (1985). Oxidation of persistent environmental pollutants by a white rot fungus. *Science*, **228**, 1434–6.

Cajthaml, T., Möder, M., Kačer, P., Šašek, V. & Popp, P. (2002). Study of fungal degradation products of polycyclic aromatic hydrocarbons using gas chromatography with ion trap mass spectrometry detection. *Journal of Chromatography A*, **974**, 213–22.

Cameotra, S. S. & Bollag, J.-M. (2003). Biosurfactant-enhanced bioremediation of polycyclic aromatic hydrocarbons. *Critical Reviews in Environmental Science and Technology*, **30**, 111–26.

Canet, R., Lopez-Real, J. M. & Beck, A. J. (1999). Overview of polycyclic aromatic hydrocarbon biodegradation by white-rot fungi. *Land Contamination and Reclamation*, **7**, 191–7.

Canet, R., Birnstingl, J. G., Malcolm, D. G., Lopez-Real, J. M. & Beck, A. J. (2001). Biodegradation of polycyclic aromatic hydrocarbons (PAHs) by native microflora and combinations of white-rot fungi in a coal-tar contaminated soil. *Bioresource Technology*, **76**, 113–17.

Capotorti, G., Digianvincenzo, P., Cesti, P., Bernardi, A. & Guglielmetti, G. (2004). Pyrene and benzo[*a*]pyrene metabolism by an *Aspergillus terreus* strain isolated from a polycyclic aromatic hydrocarbons polluted soil. *Biodegradation*, **15**, 79–85.

Casillas, R. P., Crow, S. A., Heinze, T. M., Deck, J. & Cerniglia, C. E. (1996). Initial oxidative and subsequent conjugative metabolites produced during the metabolism of phenanthrene by fungi. *Journal of Industrial Microbiology*, **16**, 205–15.

Castaldi, F. J. (2003). Tank-based bioremediation of petroleum waste sludges. *Environmental Progress*, **22**, 25–36.

Castaño-Vinyals, G., D'Errico, A., Malats, N. & Kogevinas, M. (2004). Biomarkers of exposure to polycyclic aromatic hydrocarbons from environmental air pollution. *Occupational and Environmental Medicine*, **61**, e12 (9 pp.).

Cerniglia, C. E. (1992). Biodegradation of polycyclic aromatic hydrocarbons. *Biodegradation*, **3**, 351–68.

Cerniglia, C. E. (1993). Biodegradation of polycyclic aromatic hydrocarbons. *Current Opinion in Biotechnology*, **4**, 331–8.

Cerniglia, C. E. (1997). Fungal metabolism of polycyclic aromatic hydrocarbons: past, present and future applications in bioremediation. *Journal of Industrial Microbiology and Biotechnology*, **19**, 324–33.

Cerniglia, C. E. & Gibson, D. T. (1979). Oxidation of benzo[*a*]pyrene by the filamentous fungus *Cunninghamella elegans*. *Journal of Biological Chemistry*, **254**, 12174–80.

Cerniglia, C. E. & Gibson, D. T. (1980). Fungal oxidation of benzo[*a*]pyrene and (±)-trans-7, 8-dihydroxy-7,8-dihydro benzo[*a*]pyrene: Evidence for the formation of a benzo[*a*]pyrene 7,8-diol-9,10-epoxide. *Journal of Biological Chemistry*, **255**, 5159–63.

Cerniglia, C. E. & Sutherland, J. B. (2001). Bioremediation of polycyclic aromatic hydrocarbons by ligninolytic and non-ligninolytic fungi. In *Fungi in*

202 C. E. Cerniglia and J. B. Sutherland

Bioremediation, ed. G. M. Gadd. Cambridge: Cambridge University Press, pp. 136–87.

Cerniglia, C. E. & Sutherland, J. B. (2006). Fungal metabolism of polycyclic aromatic hydrocarbons. In *Microbial Degradation of Aromatic Compounds*, 2nd edn., ed. J. J. Kukor & G. J. Zylstra. New York: Marcel Dekker, (in press).

Cerniglia, C. E. & Yang, S. K. (1984). Stereoselective metabolism of anthracene and phenanthrene by the fungus *Cunninghamella elegans*. *Applied and Environmental Microbiology*, **47**, 119–24.

Cerniglia, C. E., Hebert, R. L., Szaniszlo, P. J. & Gibson, D. T. (1978). Fungal transformation of naphthalene. *Archives of Microbiology*, **117**, 135–43.

Cerniglia, C. E., Kelly, D. W., Freeman, J. P. & Miller, D. W. (1986). Microbial metabolism of pyrene. *Chemico-Biological Interactions*, **57**, 203–16.

Chang, B. V., Shiung, L. C. & Yuan, S. Y. (2002). Anaerobic biodegradation of polycyclic aromatic hydrocarbon in soil. *Chemosphere*, **48**, 717–24.

Charrois, J. W. A., McGill, W. B. & Froese, K. L. (2001). Acute ecotoxicity of creosote-contaminated soils to *Eisenia fetida*: a survival-based approach. *Environmental Toxicology and Chemistry*, **20**, 2594–603.

Chávez-Gómez, B., Quintero, R., Esparza-Garcia, F. *et al.* (2003). Removal of phenanthrene from soil by co-cultures of bacteria and fungi pregrown on sugarcane bagasse pith. *Bioresource Technology*, **89**, 177–83.

Chung, W. K. & King, G. M. (2001). Isolation, characterization, and polyaromatic hydrocarbon degradation potential of aerobic bacteria from marine macrofaunal burrow sediments and description of *Lutibacterium anuloederans* gen. nov., sp. nov., and *Cycloclasticus spirellensus* sp. nov. *Applied and Environmental Microbiology*, **67**, 5585–92.

Coates, J. D., Woodward, J., Allen, J., Philp, P. & Lovley, D. R. (1997). Anaerobic degradation of polycyclic aromatic hydrocarbons and alkanes in petroleum-contaminated marine harbor sediments. *Applied and Environmental Microbiology*, **63**, 3589–93.

Colombo, J. C., Cabello, M. & Arambarri, A. M. (1996). Biodegradation of aliphatic and aromatic hydrocarbons by natural soil microflora and pure cultures of imperfect and lignolitic fungi. *Environmental Pollution*, **94**, 355–62.

Culp, S. J., Warbritton, A. R., Smith, B. A., Li, E. E. & Beland, F. A. (2000). DNA adduct measurements, cell proliferation and tumor mutation induction in relation to tumor formation in B6C3F1 mice fed coal tar or benzo[*a*]pyrene. *Carcinogenesis*, **21**, 1433–40.

Daane, L. L., Harjono, I., Barns, S. M. *et al.* (2002). PAH-degradation by *Paenibacillus* spp. and description of *Paenibacillus naphthalenovorans* sp. nov., a naphthalene-degrading bacterium from the rhizosphere of salt marsh plants. *International Journal of Systematic and Evolutionary Microbiology*, **52**, 131–9.

Daisy, B. H., Strobel, G. A., Castillo, U. *et al.* (2002). Naphthalene, an insect repellent, is produced by *Muscodor vitigenus*, a novel endophytic fungus. *Microbiology*, **148**, 3737–41.

da Silva, M., Cerniglia, C. E., Pothuluri, J. V., Canhos, V. P. & Esposito, E. (2003). Screening filamentous fungi isolated from estuarine sediments for the ability to oxidize polycyclic aromatic hydrocarbons. *World Journal of Microbiology and Biotechnology*, **19**, 399–405.

Del Vento, S. & Dachs, J. (2002). Prediction of uptake dynamics of persistent organic pollutants by bacteria and phytoplankton. *Environmental Toxicology and Chemistry*, **21**, 2099–107.

Dipple, A., Khan, Q. A., Page, J. E., Pontén, I. & Szeliga, J. (1999). DNA reactions, mutagenic action and stealth properties of polycyclic aromatic hydrocarbon carcinogens (Review). *International Journal of Oncology*, **14**, 103–11.

Dries, J. & Smets, B. F. (2002). Transformation and mineralization of benzo[*a*]pyrene by microbial cultures enriched on mixtures of three- and four-ring polycyclic aromatic hydrocarbons. *Journal of Industrial Microbiology and Biotechnology*, **28**, 70–3.

Ehlers, L. J. & Luthy, R. G. (2003). Contaminant bioavailability in soil and sediment. *Environmental Science and Technology*, **37**, 295A–302A.

Engst, W., Landsiedel, R., Hermersdörfer, H., Doehmer, J. & Glatt, H. (1999). Benzylic hydroxylation of 1-methylpyrene and 1-ethylpyrene by human and rat cytochromes P450 individually expressed in V79 Chinese hamster cells. *Carcinogenesis*, **20**, 1777–85.

Environmental Protection Agency (1986). *Manual SW-846, method 8310. Polycyclic aromatic hydrocarbons*. http://www.epa.gov/epaoswer/hazwaste/test/pdfs/8310.pdf.

Eschenbach, A., Wienberg, R. & Mahro, B. (1998). Fate and stability of nonextractable residues of [^{14}C]PAH in contaminated soils under environmental stress conditions. *Environmental Science and Technology*, **32**, 2585–90.

Feitkenhauer, H., Müller, R. & Märkl, H. (2003). Degradation of polycyclic aromatic hydrocarbons and long chain alkanes at 60–70 °C by *Thermus* and *Bacillus* spp. *Biodegradation*, **14**, 367–72.

Fent, K. (2003). Ecotoxicological problems associated with contaminated sites. *Toxicology Letters*, **140–141**, 353–65.

Finkelstein, Z. I., Baskunov, B. P., Golovlev, E. L. *et al.* (2003). Fluorene transformation by bacteria of the genus *Rhodococcus*. *Microbiology (Engl. Transl.)*, **72**, 660–5.

Galushko, A., Minz, D., Schink, B. & Widdel, F. (1999). Anaerobic degradation of naphthalene by a pure culture of a novel type of marine sulphate-reducing bacterium. *Environmental Microbiology*, **1**, 415–20.

Gauthier, E., Déziel, E., Villemur, R. *et al.* (2003). Initial characterization of new bacteria degrading high-molecular weight polycyclic aromatic hydrocarbons isolated from a 2-year enrichment in a two-liquid-phase culture system. *Journal of Applied Microbiology*, **94**, 301–11.

Gaylor, D. W. (1995). Risk assessment for toxic chemicals in the environment. In *Microbial Transformation and Degradation of Toxic Organic Chemicals*, ed. L. Y. Young & C. E. Cerniglia. New York: Wiley-Liss, pp. 579–601.

Gibson, D. T. (1999). *Beijerinckia* sp. strain B1: a strain by any other name ... *Journal of Industrial Microbiology and Biotechnology*, **23**, 284–93.

Gramss, G., Voigt, K.-D. & Kirsche, B. (1999). Degradation of polycyclic aromatic hydrocarbons with three to seven aromatic rings by higher fungi in sterile and unsterile soils. *Biodegradation*, **10**, 51–62.

Gravato, C. & Santos, M. A. (2002). Juvenile sea bass liver P450, EROD induction, and erythrocytic genotoxic responses to PAH and PAH-like compounds. *Ecotoxicology and Environmental Safety*, **51**, 115–27.

Habe, H. & Omori, T. (2003). Genetics of polycyclic aromatic hydrocarbon metabolism in diverse aerobic bacteria. *Bioscience, Biotechnology and Biochemistry*, **67**, 225–43.

Haemmerli, S. D., Leisola, M. S. A., Sanglard, D. & Fiechter, A. (1986). Oxidation of benzo[*a*]pyrene by extracellular ligninases of *Phanerochaete chrysosporium*: veratryl alcohol and stability of ligninase. *Journal of Biological Chemistry*, **261**, 6900–3.

Hammel, K. E., Green, B. & Gai, W. Z. (1991). Ring fission of anthracene by a eukaryote. *Proceedings of the National Academy of Sciences of the United States of America*, **88**, 10 605–8.

Hammel, K. E., Gai, W. Z., Green, B. & Moen, M. A. (1992). Oxidative degradation of phenanthrene by the ligninolytic fungus *Phanerochaete chrysosporium*. *Applied and Environmental Microbiology*, **58**, 1832–8.

Harayama, S. (1997). Polycyclic aromatic hydrocarbon bioremediation design. *Current Opinion in Biotechnology*, **8**, 268–73.

Harrigan, J. A., Vezina, C. M., McGarrigle, B. P. *et al.* (2004). DNA adduct formation in precision-cut rat liver and lung slices exposed to benzo[a]pyrene. *Toxicological Sciences*, **77**, 307–14.

Harvey, R. G. (1997). *Polycyclic Aromatic Hydrocarbons*. Hoboken, NJ: John Wiley & Sons.

Hedlund, B. P., Geiselbrecht, A. D., Bair, T. J. & Staley, J. T. (1999). Polycyclic aromatic hydrocarbon degradation by a new marine bacterium, *Neptunomonas naphthovorans* gen. nov., sp. nov. *Applied and Environmental Microbiology*, **65**, 251–9.

Heitkamp, M. A., Franklin, W. & Cerniglia, C. E. (1988a). Microbial metabolism of polycyclic aromatic hydrocarbons: isolation and characterization of a pyrene-degrading bacterium. *Applied and Environmental Microbiology*, **54**, 2549–55.

Heitkamp, M. A., Freeman, J. P., Miller, D. W. & Cerniglia, C. E. (1988b). Pyrene degradation by a *Mycobacterium* sp.: identification of ring oxidation and ring fission products. *Applied and Environmental Microbiology*, **54**, 2556–65.

Hestbjerg, H., Willumsen, P. A., Christensen, M., Andersen, O. & Jacobsen, C. S. (2003). Bioaugmentation of tar-contaminated soils under field conditions using *Pleurotus ostreatus* refuse from commercial mushroom production. *Environmental Toxicology and Chemistry*, **22**, 692–8.

Hirano, S., Kitauchi, F., Haruki, M. *et al.* (2004). Isolation and characterization of *Xanthobacter polyaromaticivorans* sp. nov. 127 W that degrades polycyclic and heterocyclic aromatic compounds under extremely low oxygen conditions. *Bioscience, Biotechnology and Biochemistry*, **68**, 557–64.

Ho, Y., Jackson, M., Yang, Y., Mueller, J. G. & Pritchard, P. H. (2000). Characterization of fluoranthene- and pyrene-degrading bacteria isolated from PAH-contaminated soils and sediments. *Journal of Industrial Microbiology and Biotechnology*, **24**, 100–12.

Hu, Y., Ren, F., Zhou, P., Xia, M. & Liu, S. (2003). Degradation of pyrene and characterization of *Saccharothrix* sp. PYX-6 from the oligotrophic Tianchi Lake in Xinjiang Uygur Autonomous Region, China. *Chinese Science Bulletin*, **48**, 2210–5.

Huesemann, M. H., Hausmann, T. S. & Fortman, T. J. (2003). Assessment of bioavailability limitations during slurry biodegradation of petroleum hydrocarbons in aged soils. *Environmental Toxicology and Chemistry*, **22**, 2853–60.

Jeon, C. O., Park, W., Padmanabhan, P. *et al.* (2003). Discovery of a bacterium, with distinctive dioxygenase, that is responsible for *in situ* biodegradation in contaminated sediment. *Proceedings of the National Academy of Sciences of the United States of America*, **100**, 13 591–6.

Jeon, C. O., Park, W., Ghiorse, W. C. & Madsen, E. L. (2004). *Polaromonas naphthalenivorans* sp. nov., a naphthalene-degrading bacterium from naphthalene-contaminated sediment. *International Journal of Systematic and Evolutionary Microbiology*, **54**, 93–7.

Jiang, C., Alexander, R., Kagi, R. I. & Murray, A. P. (2000). Origin of perylene in ancient sediments and its geological significance. *Organic Geochemistry*, **31**, 1545–59.

Johnsen, A. R. & Karlson, U. (2004). Evaluation of bacterial strategies to promote the bioavailability of polycyclic aromatic hydrocarbons. *Applied Microbiology and Biotechnology*, **63**, 452–9.

Johnsen, A. R., Winding, A., Karlson, U. & Roslev, P. (2002). Linking of microorganisms to phenanthrene metabolism in soil by analysis of ^{13}C-labelled cell lipids. *Applied and Environmental Microbiology*, **68**, 6106–13.

Jones, K. D. & Tiller, C. L. (1999). Effect of solution chemistry on the extent of binding of phenanthrene by a soil humic acid: a comparison of dissolved and clay bound humic. *Environmental Science and Technology*, **33**, 580–7.

Joshi, M. M. & Lee, S. (1996). Effect of oxygen amendments and soil pH on bioremediation of industrially contaminated soils. *Energy Sources*, **18**, 233–42.

Juhasz, A. L. & Naidu, R. (2000). Bioremediation of high molecular weight polycyclic aromatic hydrocarbons: a review of the microbial degradation of benzo[*a*]pyrene. *International Biodeterioration and Biodegradation*, **45**, 57–88.

Juhasz, A. L., Stanley, G. A. & Britz, M. L. (2002). Metabolite repression inhibits degradation of benzo[*a*]pyrene and dibenz[*a, h*]anthracene by *Stenotrophomonas maltophilia* VUN 10,003. *Journal of Industrial Microbiology and Biotechnology*, **28**, 88–96.

Kalf, D. F., Crommentuijn, T. & van de Plassche, E. J. (1997). Environmental quality objectives for 10 polycyclic aromatic hydrocarbons (PAHs). *Ecotoxicology and Environmental Safety*, **36**, 89–97.

Kanaly, R. A. & Harayama, S. (2000). Biodegradation of high-molecular-weight polycyclic aromatic hydrocarbons by bacteria. *Journal of Bacteriology*, **182**, 2059–67.

Kanaly, R. A. & Watanabe, K. (2004). Multiple mechanisms contribute to the biodegradation of benzo[*a*]pyrene by petroleum-derived multicomponent nonaqueous-phase liquids. *Environmental Toxicology and Chemistry*, **23**, 850–6.

Kanaly, R. A., Bartha, R., Watanabe, K. & Harayama, S. (2000). Rapid mineralization of benzo[*a*]pyrene by a microbial consortium growing on diesel fuel. *Applied and Environmental Microbiology*, **66**, 4205–11.

Kazunga, C. & Aitken, M. D. (2000). Products from the incomplete metabolism of pyrene by polycyclic aromatic hydrocarbon-degrading bacteria. *Applied and Environmental Microbiology*, **66**, 1917–22.

Kazunga, C., Aitken, M. D., Gold, A. & Sangaiah, R. (2001). Fluoranthene-2,3- and -1,5-diones are novel products from the bacterial transformation of fluoranthene. *Environmental Science and Technology*, **35**, 917–22.

Kelley, I., Freeman, J. P. & Cerniglia, C. E. (1990). Identification of metabolites from degradation of naphthalene by a *Mycobacterium* sp. *Biodegradation*, **1**, 283–90.

Khan, A. A., Wang, R.-F., Cao, W.-W. *et al.* (2001). Molecular cloning, nucleotide sequence, and expression of genes encoding a polycyclic aromatic ring dioxygenase from *Mycobacterium* sp. strain PYR-1. *Applied and Environmental Microbiology*, **67**, 3577–85.

Koganti, A., Singh, R., Ma, B.-L. & Weyand, E. H. (2001). Comparative analysis of PAH:DNA adducts formed in lung of mice exposed to neat coal tar and soils contaminated with coal tar. *Environmental Science and Technology*, **35**, 2704–9.

Kosian, P. A., Makynen, E. A., Monson, P. D. *et al.* (1998). Application of toxicity-based fractionation techniques and structure-activity relationship models for the identification of phototoxic polycyclic aromatic hydrocarbons in sediment pore water. *Environmental Toxicology and Chemistry*, **17**, 1021–33.

Kotterman, M. J. J., Rietberg, H.-J., Hage, A. & Field, J. A. (1998). Polycyclic aromatic hydrocarbon oxidation by the white-rot fungus *Bjerkandera* sp. strain BOS55 in the presence of nonionic surfactants. *Biotechnology and Bioengineering*, **57**, 220–7.

Krivobok, S., Kuony, S., Meyer, C. *et al.* (2003). Identification of pyrene-induced proteins in *Mycobacterium* sp. strain 6PY1: evidence for two ring-hydroxylating dioxygenases. *Journal of Bacteriology*, **185**, 3828–41.

206   C. E. Cerniglia and J. B. Sutherland

Lahav, R., Fareleira, P., Nejidat, A. & Abielovich, A. (2002). The identification and characterization of osmotolerant yeast isolates from chemical wastewater evaporation ponds. *Microbial Ecology*, **43**, 388–96.

Landmeyer, J. E., Chapelle, F. H., Petkewich, M. D. & Bradley, P. M. (1998). Assessment of natural attenuation of aromatic hydrocarbons in groundwater near a former manufactured-gas plant, South Carolina, USA. *Environmental Geology*, **34**, 279–92.

Lau, K. L., Tsang, Y. Y. & Chiu, S. W. (2003). Use of spent mushroom compost to bioremediate PAH-contaminated samples. *Chemosphere*, **52**, 1539–46.

Lee, J., Kang, D., Lee, K.-H. *et al.* (2002). Influence of GSTM1 genotype on association between aromatic DNA adducts and urinary PAH metabolites in incineration workers. *Mutation Research*, **514**, 213–21.

Lee, K., Park, J.-W. & Ahn, I.-S. (2003). Effect of additional carbon source on naphthalene biodegradation by *Pseudomonas putida* G7. *Journal of Hazardous Materials*, **B105**, 157–67.

Lehto, K.-M., Puhakka, J. A. & Lemmetyinen, H. (2003). Biodegradation of selected UV-irradiated and non-irradiated polycyclic aromatic hydrocarbons (PAHs). *Biodegradation*, **14**, 249–63.

Leštan, D. & Lamar, R. T. (1996). Development of fungal inocula for bioaugmentation of contaminated soils. *Applied and Environmental Microbiology*, **62**, 2045–52.

Leys, N. M. E. J., Ryngaert, A., Bastiaens, L. *et al.* (2004). Occurrence and phylogenetic diversity of *Sphingomonas* strains in soils contaminated with polycyclic aromatic hydrocarbons. *Applied and Environmental Microbiology*, **70**, 1944–55.

Li, P., Sun, T., Stagnitti, F. *et al.* (2002). Field-scale bioremediation of soil contaminated with crude oil. *Environmental Engineering Science*, **19**, 277–89.

Lin, G.-H., Sauer, N. E. & Cutright, T. J. (1996). Environmental regulations: a brief overview of their applications to bioremediation. *International Biodeterioration and Biodegradation*, **38**, 1–8.

Liu, Y., Zhang, J. & Zhang, Z. (2004). Isolation and characterization of polycyclic aromatic hydrocarbons-degrading *Sphingomonas* sp. strain ZL5. *Biodegradation*, **15**, 205–12.

Lodovici, M., Akpan, V., Giovannini, L., Migliani, F. & Dolara, P. (1998). Benzo[a]pyrene diol-epoxide DNA adducts and levels of polycyclic aromatic hydrocarbons in autoptic samples from human lungs. *Chemico-Biological Interactions*, **116**, 199–212.

Lors, C., Mossmann, J. R. & Barbé, P. (2004). Phenotypic responses of the soil bacterial community to polycyclic aromatic hydrocarbon contamination in soils. *Polycyclic Aromatic Compounds*, **24**, 21–36.

Löser, C., Seidel, H., Zehnsdorf, A. & Hoffmann, P. (2000). Improvement of the bioavailability of hydrocarbons by applying nonionic surfactants during the microbial remediation of a sandy soil. *Acta Biotechnologica*, **20**, 99–118.

Lotfabad, S. K. & Gray, M. R. (2002). Kinetics of biodegradation of mixtures of polycyclic aromatic hydrocarbons. *Applied Microbiology and Biotechnology*, **60**, 361–5.

MacGillivray, A. R. & Shiaris, M. P. (1994). Relative role of eukaryotic and prokaryotic micro-organisms in phenanthrene transformation in coastal sediments. *Applied and Environmental Microbiology*, **60**, 1154–9.

Majcherczyk, A. & Johannes, C. (2000). Radical mediated indirect oxidation of a PEG-coupled polycyclic aromatic hydrocarbon (PAH) model compound by fungal laccase. *Biochimica et Biophysica Acta*, **1474**, 157–62.

Makkar, R. S. & Rockne, K. J. (2003). Comparison of synthetic surfactants and biosurfactants in enhancing biodegradation of polycyclic aromatic hydrocarbons. *Environmental Toxicology and Chemistry*, **22**, 2280–92.

Malachová, K. (1999). Using short-term mutagenicity tests for the evaluation of genotoxicity of contaminated soils. *Journal of Soil Contamination*, **8**, 667–80.

Maliszewska-Kordybach, B. & Smreczak, B. (2000). Ecotoxicological activity of soils polluted with polycyclic aromatic hydrocarbons (PAHs)–Effect on plants. *Environmental Technology*, **21**, 1099–110.

May, R., Schröder, P. & Sandermann, H. (1997). Ex-situ process for treating PAH-contaminated soil with *Phanerochaete chrysosporium*. *Environmental Science and Technology*, **31**, 2626–33.

Meckenstock, R. U., Safinowski, M. & Griebler, C. (2004). Anaerobic degradation of polycyclic aromatic hydrocarbons. *FEMS Microbiology Ecology*, **49**, 27–36.

Melcher, R. J., Apitz, S. E. & Hemmingsen, B. B. (2002). Impact of irradiation and polycyclic aromatic hydrocarbon spiking on microbial populations in marine sediment for future aging and biodegradability studies. *Applied and Environmental Microbiology*, **68**, 2858–68.

Mendonça, E. & Picado, A. (2002). Ecotoxicological monitoring of remediation in a coke oven soil. *Environmental Toxicology*, **17**, 74–9.

Meulenberg, R., Rijnaarts, H. H. M., Doddema, H. J. & Field, J. A. (1997). Partially oxidized polycyclic aromatic hydrocarbons show an increased bioavailability and biodegradability. *FEMS Microbiology Letters*, **152**, 45–9.

Michallet-Ferrier, P., Aït-Aïssa, S., Balaguer, P. *et al.* (2004). Assessment of estrogen (ER) and aryl hydrocarbon receptor (AhR) mediated activities in organic sediment extracts of the Detroit River, using in vitro bioassays based on human MELN and teleost PLHC-1 cell lines. *Journal of Great Lakes Research*, **30**, 82–92.

Miller, K. P. & Ramos, K. S. (2001). Impact of cellular metabolism on the biological effects of benzo[*a*]pyrene and related hydrocarbons. *Drug Metabolism Reviews*, **33**, 1–35.

Moody, J. D., Freeman, J. P., Doerge, D. R. & Cerniglia, C. E. (2001). Degradation of phenanthrene and anthracene by cell suspensions of *Mycobacterium* sp. strain PYR-1. *Applied and Environmental Microbiology*, **67**, 1476–83.

Moody, J. D., Freeman, J. P., Fu, P. P. & Cerniglia, C. E. (2004). Degradation of benzo[*a*]pyrene by *Mycobacterium vanbaalenii* PYR-1. *Applied and Environmental Microbiology*, **70**, 340–5.

Mougin, C. (2002). Bioremediation and phytoremediation of industrial PAH-polluted soils. *Polycyclic Aromatic Compounds*, **22**, 1011–43.

Mrozik, A., Piotrowska-Seget, Z. & Labużek, S. (2003). Bacterial degradation and bioremediation of polycyclic aromatic hydrocarbons. *Polish Journal of Environmental Studies*, **12**, 15–25.

Mueller, J. G., Cerniglia, C. E. & Pritchard, P. H. (1996). In *Bioremediation: Principles and Applications*, ed. R. L. Crawford & D. L. Crawford. Cambridge: Cambridge University Press, pp. 125–94.

Neumann, N. F. & Galvez, F. (2002). DNA microarrays and toxicogenomics: applications for ecotoxicology? *Biotechnology Advances*, **20**, 391–419.

Novotný, Č., Erbanová, P., Cajthaml, T. *et al.* (2000). *Irpex lacteus*, a white rot fungus applicable to water and soil bioremediation. *Applied Microbiology and Biotechnology*, **54**, 850–3.

Ohura, T., Amagai, T., Fusaya, M. & Matsushita, H. (2004). Polycyclic aromatic hydrocarbons in indoor and outdoor environments and factors affecting their concentrations. *Environmental Science and Technology*, **38**, 77–83.

Pan, F., Yang, Q., Zhang, Y., Zhang, S. & Yang, M. (2004). Biodegradation of polycyclic aromatic hydrocarbons by *Pichia anomala*. *Biotechnology Letters*, **26**, 803–6.

Peachey, R. B. J. (2003). Tributyltin and polycyclic aromatic hydrocarbon levels in Mobile Bay, Alabama: a review. *Marine Pollution Bulletin*, **46**, 1365–71.

Penning, T. M., Burczynski, M. E., Hung, C.-F. *et al.* (1999). Dihydrodiol dehydrogenases and polycyclic aromatic hydrocarbon activation: generation of reactive and redox active *o*-quinones. *Chemical Research in Toxicology*, **12**, 1–18.

Pickard, M. A., Roman, R., Tinoco, R. & Vázquez-Duhalt, R. (1999). Polycyclic aromatic hydrocarbon metabolism by white rot fungi and oxidation by *Coriolopsis gallica* UAMH 8260 laccase. *Applied and Environmental Microbiology*, **65**, 3805–9.

Pinto, L. J. & Moore, M. M. (2000). Release of polycyclic aromatic hydrocarbons from contaminated soils by surfactant and remediation of this effluent by *Penicillium* spp. *Environmental Toxicology and Chemistry*, **19**, 1741–8.

Pointing, S. B. (2001). Feasibility of bioremediation by white-rot fungi. *Applied Microbiology and Biotechnology*, **57**, 20–33.

Potin, O., Rafin, C. & Veignie, E. (2004). Bioremediation of an aged polycyclic aromatic hydrocarbons (PAHs)-contaminated soil by filamentous fungi isolated from the soil. *International Biodeterioration and Biodegradation*, **54**, 45–52.

Pozzoli, L., Gilardoni, S., Perrone, M. G. *et al.* (2004). Polycyclic aromatic hydrocarbons in the atmosphere: monitoring, sources, sinks and fate. I: Monitoring and sources. *Annali di Chimica*, **94**, 17–32.

Preuss, R., Angerer, J. & Drexler, H. (2003). Naphthalene – an environmental and occupational toxicant. *International Archives of Occupational and Environmental Health*, **76**, 556–76.

Rafin, C., Potin, O., Veignie, E., Lounès-Hadj Sahraoui, A. & Sancholle, M. (2000). Degradation of benzo[*a*]pyrene as sole carbon source by a non white rot fungus, *Fusarium solani*. *Polycyclic Aromatic Compounds*, **21**, 311–29.

Rama, R., Sigoillot, J.-C., Chaplain, V. *et al.* (2001). Inoculation of filamentous fungi in manufactured gas plant site soils and PAH transformation. *Polycyclic Aromatic Compounds*, **18**, 397–414.

Ramsay, J. A., Li, H., Brown, R. S. & Ramsay, B. A. (2003). Naphthalene and anthracene mineralization linked to oxygen, nitrate, Fe(III) and sulphate reduction in a mixed microbial population. *Biodegradation*, **14**, 321–9.

Rasmussen, G. & Olsen, R. A. (2004). Sorption and biological removal of creosote-contaminants from groundwater in soil/sand vegetated with orchard grass (*Dactylis glomerata*). *Advances in Environmental Research*, **8**, 313–27.

Ravelet, C., Grosset, C., Krivobok, S., Montuelle, B. & Alary, J. (2001). Pyrene degradation by two fungi in a freshwater sediment and evaluation of fungal biomass by ergosterol content. *Applied Microbiology and Biotechnology*, **56**, 803–8.

Renoux, A. Y., Millette, D., Tyagi, R. D. & Samson, R. (1999). Detoxification of fluorene, phenanthrene, carbazole and *p*-cresol in columns of aquifer sand as studied by the Microtox® assay. *Water Research*, **33**, 2045–52.

Rivera, L., Curto, M. J. C., Pais, P., Galceran, M. T. & Puignou, L. (1996). Solid-phase extraction for the selective isolation of polycyclic aromatic hydrocarbons, azaarenes and heterocyclic aromatic amines in charcoal-grilled meat. *Journal of Chromatography A*, **731**, 85–94.

Rockne, K. J., Chee-Sanford, J. C., Sanford, R. A. *et al.* (2000). Anaerobic naphthalene degradation by microbial pure cultures under nitrate-reducing conditions. *Applied and Environmental Microbiology*, **66**, 1595–601.

Rodrigues, J. L. M., Aiello, M. R., Urbance, J. W., Tsoi, T. V. & Tiedje, J. M. (2002). Use of both 16S rRNA and engineered functional genes with real-time PCR to quantify an engineered, PCB-degrading *Rhodococcus* in soil. *Journal of Microbiological Methods*, **51**, 181–9.

Romero, M. C., Cazau, M. C., Giorgieri, S. & Arambarri, A. M. (1998). Phenanthrene degradation by micro-organisms isolated from a contaminated stream. *Environmental Pollution*, **101**, 355–9.

Romero, M. C., Salvioli, M. L., Cazau, M. C. & Arambarri, A. M. (2002). Pyrene degradation by yeasts and filamentous fungi. *Environmental Pollution*, **117**, 159–63.

Rothermich, M. M., Hayes, L. A. & Lovley, D. R. (2002). Anaerobic, sulfate-dependent degradation of polycyclic aromatic hydrocarbons in petroleum-contaminated harbor sediment. *Environmental Science and Technology*, **36**, 4811–17.

Ruberto, L., Vazquez, S. C. & MacCormack, W. P. (2003). Effectiveness of the natural bacterial flora, biostimulation and bioaugmentation on the bioremediation of a hydrocarbon contaminated Antarctic soil. *International Biodeterioration and Biodegradation*, **52**, 115–25.

Saltiene, Z., Brukstiene, D. & Ruzgyte, A. (2002). Contamination of soil by polycyclic aromatic hydrocarbons in some urban areas. *Polycyclic Aromatic Compounds*, **22**, 23–35.

Samanta, S. K., Singh, O. V. & Jain, R. K. (2002). Polycyclic aromatic hydrocarbons: environmental pollution and bioremediation. *Trends in Biotechnology*, **20**, 243–8.

Santodonato, J. (1997). Review of the estrogenic and antiestrogenic activity of polycyclic aromatic hydrocarbons: relationship to carcinogenicity. *Chemosphere*, **34**, 835–48.

Saponaro, S., Bonomo, L., Petruzzelli, G., Romele, L. & Barbafieri, M. (2002). Polycyclic aromatic hydrocarbons (PAHs) slurry phase bioremediation of a manufacturing gas plant (MGP) site aged soil. *Water, Air, and Soil Pollution*, **135**, 219–36.

Saraswathy, A. & Hallberg, R. (2002). Degradation of pyrene by indigenous fungi from a former gasworks site. *FEMS Microbiology Letters*, **210**, 227–32.

Schneider, J., Grosser, R., Jayasimhulu, K., Xue, W. & Warshawsky, D. (1996). Degradation of pyrene, benz[*a*]anthracene, and benzo[*a*]pyrene by *Mycobacterium* sp. strain RJGII-135, isolated from a former coal gasification site. *Applied and Environmental Microbiology*, **62**, 13–19.

Selifonov, S. A., Chapman, P. J., Akkerman, S. B. *et al.* (1998). Use of ^{13}C nuclear magnetic resonance to assess fossil fuel biodegradation: fate of [1–^{13}C]acenaphthene in creosote polycyclic aromatic compound mixtures degraded by bacteria. *Applied and Environmental Microbiology*, **64**, 1447–53.

Šepič, E., Bricelj, M. & Leskovšek, H. (1998). Degradation of fluoranthene by *Pasteurella* sp. IFA and *Mycobacterium* sp. PYR-1: Isolation and identification of metabolites. *Journal of Applied Microbiology*, **85**, 746–54.

Shimada, T. & Fujii-Kuriyama, Y. (2004). Metabolic activation of polycyclic aromatic hydrocarbons to carcinogens by cytochromes P450 1A1 and 1B1. *Cancer Science*, **95**, 1–6.

Sho, M., Hamel, C. & Greer, C. W. (2004). Two distinct gene clusters encode pyrene degradation in *Mycobacterium* sp. strain S65. *FEMS Microbiology Ecology*, **48**, 209–20.

Shuttleworth, K. L. & Cerniglia, C. E. (1996). Bacterial degradation of low concentrations of phenanthrene and inhibition by naphthalene. *Microbial Ecology*, **31**, 305–17.

Sonnefeld, W. J., Zoller, W. H. & May, W. E. (1983). Dynamic coupled-column liquid chromatographic determination of ambient temperature vapor pressures of polynuclear aromatic hydrocarbons. *Analytical Chemistry*, **55**, 275–80.

Steffen, K. T., Hatakka, A. & Hofrichter, M. (2003). Degradation of benzo[*a*]pyrene by the litter-decomposing basidiomycete *Stropharia coronilla*: role of manganese peroxidase. *Applied and Environmental Microbiology*, **69**, 3957–64.

Straube, W. L., Jones-Meehan, J., Pritchard, P. H. & Jones, W. R. (1999). Bench-scale optimization of bioaugmentation strategies for treatment of soils contaminated with high molecular weight polyaromatic hydrocarbons. *Resources, Conservation and Recycling*, **27**, 27–37.

Straube, W. L., Nestler, C. C., Hansen, L. D. *et al.* (2003). Remediation of polyaromatic hydrocarbons (PAHs) through landfarming with biostimulation and bioaugmentation. *Acta Biotechnologica*, **23**, 179–96.

Sutherland, J. B. (2004). Degradation of hydrocarbons by yeasts and filamentous fungi. In *Fungal Biotechnology in Agricultural, Food, and Environmental Applications*, ed. D. K. Arora. New York: Marcel Dekker, pp. 443–55.

Sutherland, J. B., Freeman, J. P., Selby, A. L. *et al.* (1990). Stereoselective formation of a K-region dihydrodiol from phenanthrene by *Streptomyces flavovirens*. *Archives of Microbiology*, **154**, 260–6.

Sutherland, J. B., Selby, A. L., Freeman, J. P., Evans, F. E. & Cerniglia, C. E. (1991). Metabolism of phenanthrene by *Phanerochaete chrysosporium*. *Applied and Environmental Microbiology*, **57**, 3310–16.

Sutherland, J. B., Selby, A. L., Freeman, J. P. *et al.* (1992). Identification of xyloside conjugates formed from anthracene by *Rhizoctonia solani*. *Mycological Research*, **96**, 509–17.

Sutherland, J. B., Rafii, F., Khan, A. A. & Cerniglia, C. E. (1995). Mechanisms of polycyclic aromatic hydrocarbon degradation. In *Microbial Transformation and Degradation of Toxic Organic Chemicals*, ed. L. Y. Young & C. E. Cerniglia. New York: Wiley-Liss, pp. 269–306.

Sverdrup, L. E., Nielsen, T. & Krogh, P. H. (2002). Soil ecotoxicity of polycyclic aromatic hydrocarbons in relation to soil sorption, lipophilicity, and water solubility. *Environmental Science and Technology*, **36**, 2429–35.

Tiehm, A., Stieber, M., Werner, P. & Frimmel, F. H. (1997). Surfactant-enhanced mobilization and biodegradation of polycyclic aromatic hydrocarbons in manufactured gas plant soil. *Environmental Science and Technology*, **31**, 2570–76.

van der Oost, R., Beyer, J. & Vermeulen, N. P. E. (2003). Fish bioaccumulation and biomarkers in environmental risk assessment: a review. *Environmental Toxicology and Pharmacology*, **13**, 57–149.

van Herwijnen, R., Wattiau, P., Bastiaens, L. *et al.* (2003). Elucidation of the metabolic pathway of fluorene and cometabolic pathways of phenanthrene, fluoranthene, anthracene and dibenzothiophene by *Sphingomonas* sp. LB126. *Research in Microbiology*, **154**, 199–206.

Veignie, E., Rafin, C., Woisel, P., Lounès-Hadj Sahraoui, A. & Cazier, F. (2002). Metabolization of the polycyclic aromatic hydrocarbon benzo[*a*]pyrene by a non-white rot fungus (*Fusarium solani*) in a batch reactor. *Polycyclic Aromatic Compounds*, **22**, 87–97.

Veignie, E., Rafin, C., Woisel, P. & Cazier, F. (2004). Preliminary evidence of the role of hydrogen peroxide in the degradation of benzo[*a*]pyrene by a non-white rot fungus *Fusarium solani*. *Environmental Pollution*, **129**, 1–4.

Verdin, A., Lounès-Hadj Sahraoui, A. & Durand, R. (2004). Degradation of benzo[*a*]pyrene by mitosporic fungi and extracellular oxidative enzymes. *International Biodeterioration and Biodegradation*, **53**, 65–70.

Verdin, A., Lounès-Hadj Sahraoui, A., Newsam, R., Robinson, G. & Durand, R. (2005). Polycyclic aromatic hydrocarbons storage by *Fusarium solani* in intracellular lipid vesicles. *Environmental Pollution*, **133**, 283–91.

Vila, J., López, Z., Sabaté, J. *et al.* (2001). Identification of a novel metabolite in the degradation of pyrene by *Mycobacterium* sp. strain AP1: actions of the isolate on two- and three-ring polycyclic aromatic hydrocarbons. *Applied and Environmental Microbiology*, **67**, 5497–505.

Villeneuve, D. L., Khim, J. S., Kannan, K. & Giesy, J. P. (2002). Relative potencies of individual polycyclic aromatic hydrocarbons to induce dioxinlike and estrogenic responses in three cell lines. *Environmental Toxicology*, **17**, 128–37.

Volkering, F., Breure, A. M. & Rulkens, W. H. (1998). Microbiological aspects of surfactant use for biological soil remediation. *Biodegradation*, **8**, 401–17.

Watanabe, T., Katayama, S., Enoki, M., Honda, Y. & Kuwahara, M. (2000). Formation of acyl radical in lipid peroxidation of linoleic acid by manganese-dependent peroxidase from *Ceriporiopsis subvermispora* and *Bjerkandera adusta*. *European Journal of Biochemistry*, **267**, 4222–31.

White, P. A. (2002). The genotoxicity of priority polycyclic aromatic hydrocarbons in complex mixtures. *Mutation Research*, **515**, 85–98.

Wilcke, W. (2000). Polycyclic aromatic hydrocarbons (PAHs) in soil–a review. *Journal of Plant Nutrition and Soil Science*, **163**, 229–48.

Willumsen, P., Karlson, U., Stackebrandt, E. & Kroppenstedt, R. M. (2001). *Mycobacterium frederiksbergense* sp. nov., a novel polycyclic aromatic hydrocarbon-degrading *Mycobacterium* species. *International Journal of Systematic and Evolutionary Microbiology*, **51**, 1715–22.

Wrenn, B. A. & Venosa, A. D. (1996). Selective enumeration of aromatic and aliphatic hydrocarbon degrading bacteria by a most-probable-number procedure. *Canadian Journal of Microbiology*, **42**, 252–8.

Yang, J., Liu, X., Long, T. *et al.* (2003). Influence of nonionic surfactant on the solubilization and biodegradation of phenanthrene. *Journal of Environmental Sciences (China)*, **15**, 859–62.

Yu, H. (2002). Environmental carcinogenic polycyclic aromatic hydrocarbons: photochemistry and phototoxicity. *Journal of Environmental Science and Health C*, **20**, 149–83.

Zhang, H., Kallimanis, A., Koukkou, A. I. & Drainas, C. (2004). Isolation and characterization of novel bacteria degrading polycyclic aromatic hydrocarbons from polluted Greek soils. *Applied Microbiology and Biotechnology*, **65**, 124–31.

Zheng, Z. & Obbard, J. P. (2001). Effect of non-ionic surfactants on elimination of polycyclic aromatic hydrocarbons (PAHs) in soil-slurry by *Phanerochaete chrysosporium*. *Journal of Chemical Technology and Biotechnology*, **76**, 423–9.

Zheng, Z. & Obbard, J. P. (2002). Polycyclic aromatic hydrocarbon removal from soil by surfactant solubilization and *Phanerochaete chrysosporium* oxidation. *Journal of Environmental Quality*, **31**, 1842–7.

9

Biodegradation and biodeterioration of man-made polymeric materials

HRISTO A. SABEV, SARAH R. BARRATT, MALCOM
GREENHALGH, PAULINE S. HANDLEY AND
GEOFFREY D. ROBSON

Introduction

Man-made polymeric materials are ubiquitous in our everyday lives and have an enormous range of applications from man-made textiles to plastics, coatings, paints and additives. As a consequence, a vast array of man-made polymers accumulates in the environment and landfill waste sites where they cause considerable water and land pollution problems. Over the last few decades, plastics and plasticizers in particular, due to their wide production and distribution, have led to a large increased environmental burden (Bouwer, 1992). According to recent estimates, the annual production of plastics in the world exceeds more than 140 million tonnes per year (Shimao, 2001). Plastics possess a number of key characteristics including weight, inertness, flexibility and low production costs that make them widespread in many areas of human life. However, it is their inertness and durability, valuable during their use, that becomes a particular problem later during their disposal. Contrary to other synthetic chemicals and pesticides, synthetic polymers do not generally possess particular toxicological problems, unless supplied with protective agents such as biocides (Bentivegna & Piatkowski, 1998) or particular plasticizers, such as phthalates (Staples et al., 1997; Zeng et al., 2002). Plastics however contribute greatly to the amount of municipal solid waste (Palmisano & Pettigrew, 1992) and are an increasing problem due to improper disposal (Alexander, 1994). The UK alone consumed 4.7 million tonnes of plastics in 2002 with most of the material being used in packaging and the building/construction industries (Fig. 9.1).

While microbial biodegradation of synthetic polymers once they have been disposed has important environmental implications, the action of

Fungi in Biogeochemical Cycles, ed. G. M. Gadd. Published by Cambridge University Press. © British Mycological Society 2006.

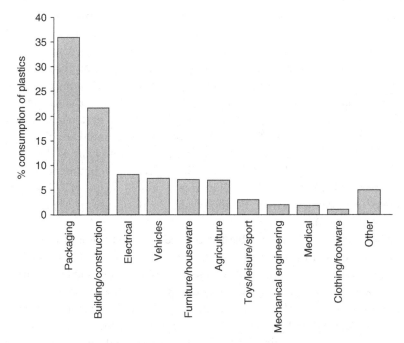

Fig. 9.1. Percentage UK consumption in 2002 of plastics by various sectors. Figures are based on data from the British Plastics Federation.

microorganisms can severely damage or limit the useful product life of many synthetic polymers whilst they are in use, causing vast industrial and medical problems (Eastwood, 1994; Morton & Surman, 1994; Flemming, 1998). Consequently, broad-spectrum biocides are often incorporated into polymer blends to inhibit microbial growth and so extend the lifetime of the product, particularly for products made from synthetic polymers that are known to be the most susceptible to microbial attack such as polyurethanes, plasticized polyvinyl chloride (pPVC) and polyesters (Gaylarde & Morton, 1999)

Susceptibility of man-made polymeric materials to deterioration and degradation

Biodeterioration refers to a loss in the physical integrity or defacement of a polymer whereas biodegradation refers to the chemical breakdown of a polymer or its additives, the products of which may (but not always) be used and incorporated into the cell mass (Palmisano & Pettigrew, 1992). The difference between the natural and man-made polymers is that, in contrast to the naturally produced biopolymers to which microbes have evolved hydrolases to utilize them as a nutrient source, many of the synthetic polymers are resistant due to the inability to catalyse the breakdown of the bonds in the backbone of the polymer (Albertsson & Karlsson, 1988; Andrady, 1994; Steinbuchel, 1996). However, certain polymers, such as polyester polyurethanes and those containing amide or peptide bonds are considered biodegradable to a variable degree (Steinbuchel, 1996; Nakajima-Kambe *et al.*, 1999). Where biodeterioration of a polymer occurs it can be due either to preferential attack by the microorganisms' enzymatic system towards the main chain of the polymer, for example polyurethane, or towards certain components of the polymeric blend, for example plasticizers present in pPVC which are added to increase polymer flexibility and so the range of applications (Andrady, 1994).

Many factors are responsible for the reduced useful life of a product made from a synthetic polymer. Polymers exposed outdoors for example are affected by physical factors such as sunlight, moisture and temperature (Grima *et al.*, 2000). These factors can promote microbial growth by conditioning the surface due to the deposition of organic molecules (Kumar & Anand, 1998). It has been shown that the structure of the composition of the main chain of the polymer, molecular weight and crystallinity are of crucial importance for its susceptibility to degradation (McNeill, 1991; Christensen *et al.*, 1995; Steinbuchel, 1996). The most inert and recalcitrant to biodegradation are the synthetic homo polymers, which contain C_2 units (C–C bonds) in their backbone structure, for example polyethylene, polypropylene, polystyrene, PVC and polytetrafluoroethylene (Fig. 9.2a). In contrast, the backbone of heterochain polymers contains additional elements other than carbon and can include for example oxygen, nitrogen and silicon (Fig. 9.2b). Examples of such polymers include polyethers (polyethylene glycol – PEG, polypropylene glycol – PPG), polyesters and polyamides, which contain a variety of bonds, potentially susceptible to enzymatic attack by esterases and amidases. Recent work with bacteria for example reported that *Acinetobacter*, *Pseudomonas* and *Flavobacterium* spp. were capable of degrading PEG, causing a reduction in the chain length of the substratum (Steinbuchel, 1996). The degree of hydrophilicity or hydrophobicity of the polymer also plays a role in its susceptibility to microbial degradation with biodegradability of the polymer increasing with increasing hydrophilicity (Kumar & Anand, 1998). In addition, hydrophobicity can also influence the degree of microbial attachment to the polymer surface. For example, a direct positive correlation was found between surface hydrophobicity and attachment for blastospores of *Aureobasidium pullulans* with more spores adhering the more hydrophobic the polymer (Webb *et al.*, 1999). The same study also demonstrated that plasticizers in pPVC caused increased attachment through electrostatic interactions. The molecular weight of the polymer has also been shown to strongly influence its susceptibility to biodegradation. For example, increasing the degree of polymerization of poly(caprolactone) was correlated with a decrease in the biodegradation rate by *A. pullulans* (Andrady, 1994).

Polymers differ in their chemical composition and content of plasticizers, additives, colorants and stabilizers. The leaching of compounds from polymers can promote microbial growth by providing a utilizable carbon source. It is interesting to note that a polymer surface, enriched in low-molecular-weight biodegradable components may allow more colonies of

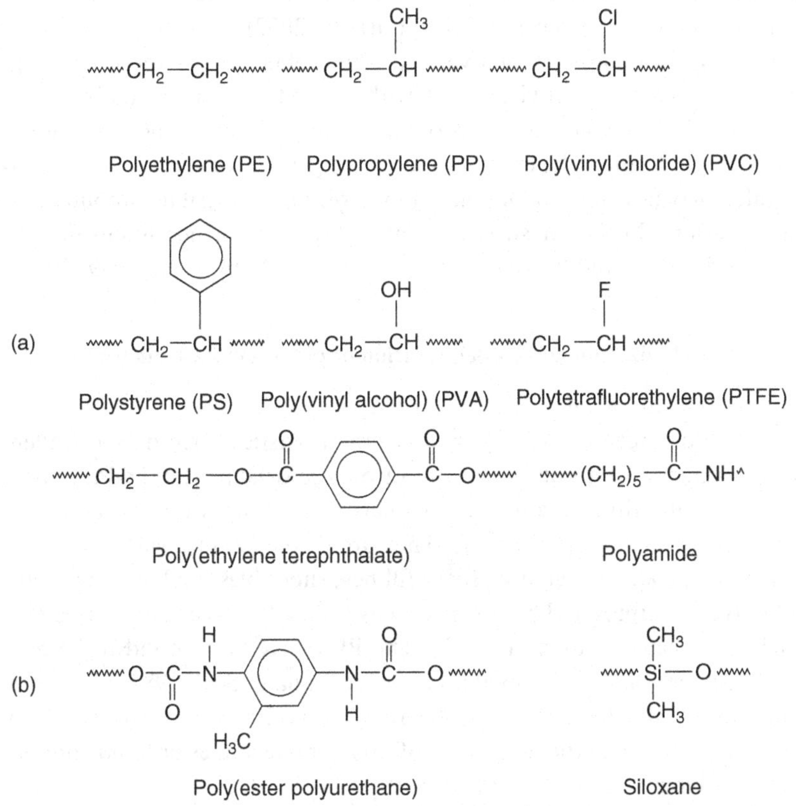

Fig. 9.2. Main types of (a) homopolymers and (b) heterochain synthetic polymers found in plastics (adapted from McNeill, 1991).

biodegrading microorganisms by 'conditioning' of the surface and selective enrichment of potential biodegradants (Kumar *et al.*, 1982). The environmental conditions, available light, moisture, deposited nutrients, minerals and organic molecules on the surface, also affect the level of polymer degradation (Gu, 2003). The transport of organic molecules to the surface of the polymer may promote better conditions for the first colonizing microorganisms and thereby increasing their chance of survival (Morton & Surman, 1994; Hazen & Glee, 1995). In one study, where pPVC was exposed to the air, the first colonizer to establish and multiply on the surface was *A. pullulans*, despite other degraders being found periodically on the surface (Webb *et al.*, 2000). Other degraders only became established after the initial colonization by *A. pullulans* and was suggested to be due to the organism's heavy melanization and ability to secrete the extracellular polysaccharide pullulan, which are known to help protect the cell from UV light and desiccation (Crang & Pechak, 1978; Andrews *et al.*, 1994; Turkovskii & Yurlova, 2002).

Clearly, the deposition of a variety of microbial genera and their adaptation to live on an environment with limited nutrient availability are important in the colonization process of the polymer surface and subsequent degradation. The capability of these genera for secretion of extracellular enzymes and production of low-molecular-weight metabolites aids in the establishment of successive microflora and hence intensifies the biodeterioration and biodegradation processes (Webb *et al.*, 1999, 2000).

Colonization and biodegradation of polyurethanes and pPVC

Polyurethanes

Polyurethanes (PUs) are a group of plastics derived as a condensation reaction of a polyisocyanate (aromatic or aliphatic) and a polyol, and make up a diverse family of polymers with a range of physical properties, structures and applications. They are employed in a vast number of consumer products including foam fillings, shoe soles, building materials, adhesives, coatings and paints as well as finding applications in medicine such as catheters (Howard, 2002). The PUs have been found to be relatively susceptible to microbial attack (Pathirana & Seal, 1984; Morton & Surman, 1994; Howard, 2002; Barratt *et al.*, 2003) and this has been attributed to enzymatic hydrolysis of urea, amide and ester bonds present in the polymer (Filip, 1978; Griffin, 1980, Howard, 2002).

Polyurethanes can be broadly divided into two groups, the first with repeating polyester units and the second with repeating polyether repeats.

The chemical heterogeneity of PU as a group is reflected in the different levels of susceptibility to microbial attack reported in the different formulations. In addition, a number of additives are also often added to influence the physical properties of the plastic including plasticizers that influence flexibility, lubricants, UV stabilizers/absorbers and flame retardants, all adding to the rich complexity of molecules that can be used as a microbial nutrient. Generally, polyester PU is far more susceptible to microbial degradation compared to polyether PU as ester linkages are more readily hydrolysed compared to ether linkages where little if any degradation occurs (Darby & Kaplan, 1968, Filip, 1978; Barratt *et al.*, 2003). As the backbone of the polymer is hydrolysed, biodegradation leads to a loss in tensile strength and dry weight of the polymer (Darby & Kaplan, 1968; Bentham, Morton & Allen, 1987; Kay *et al.*, 1991; Barratt *et al.*, 2003).

While many studies have focused on the degradation of PUs by bacteria (Pommer & Lorenz, 1985; Kay *et al.*, 1991; Nakajima-Kambe *et al.*, 1995; 1999; reviewed by Howard, 2002), numerous studies in the environment have shown that fungi are equally, if not more, important in PU degradation (Darby & Kaplan, 1968; Pathirana & Seal, 1984; Pommer & Lorenz, 1985; Stranger-Johannessen, 1985; Crabbe *et al.*, 1994; Baratt *et al.*, 2003). Of the fungi that have been isolated from PU, *Penicillium* and *Aspergillus* spp. have been the most frequently recovered, probably reflecting their broad distribution within the environment. Many of the PU-degrading fungi have been isolated from soil and using soil as a natural reservoir for fungi has been a highly successful method of recovering PU degraders (Pathirana & Seal, 1984; Bentham *et al.*, 1987; Owen *et al.*, 1996; Crabbe *et al.*, 1994; Barratt *et al.*, 2003). Table 9.1 summarizes the fungal species isolated from several studies and thought to be capable of the deterioration of PU.

While hydrolases, particularly esterases, have been implicated in being critical in PU degradation, there have been few studies to characterize these in the fungi. An enzyme with esterase activity capable of hydrolysing colloidal polyester PU (Impranil) was isolated from *Chaetomium globosum* and *Aspergillus terreus* (Boubendir, 1993) and Crabbe *et al.* (1994) purified a 28 Kd protein with polyester PU-degrading activity from *Curvularia senegalensis* that was stable at 100 °C. The enzyme was strongly inhibited by the addition of phenylmethylsulphonyl fluoride (PMSF), suggesting the enzyme responsible for PU hydrolysis could be a protease. In contrast, far more is known about the enzymes responsible for bacterial hydrolysis of polyester PU. Enzymes hydrolysing PU with esterase activity have been

Table 9.1. *Summary of fungal species isolated from polyurethanes*

Organism	Study
Aremonium spp.[a]	Stranger-Johannessen (1985)
Alternaria spp.[b]	Stranger-Johannessen (1985)
Alternaria alternata[c]	Pommer & Lorenz (1985)
Aspergillus sp.[a]	Stranger-Johannessen (1985)
Aspergillus fischeri[c]	Bentham *et al.* (1987)
Aspergillus flavus[c]	Pathirana & Seal (1984); Pommer & Lorenz (1985)
Aspergillus fumigatus[c]	Pathirana & Seal (1984)
Aspergillus niger[cd]	Filip, (1979); Subrahmanyam *et al.* (2001)
Fusarium tameri[c]	Subrahmanyam *et al.* (2001)
Aspergillus terreus[d]	Wales & Sager (1985)
Aspergillus ustus[c]	Bentham *et al.* (1987); Pommer & Lorenz (1985)
Aspergillus versicolor[c]	Bentham *et al.* (1987)
Aureobasidium pullulans[c]	Crabbe *et al.* (1994)
Chaetomium globosum[c]	Pathirana & Seal (1984)
Cladosporium sp.[c]	Crabbe *et al.* (1994)
Cladosporium herbarum[d]	Filip (1979)
Cryptococcus laurentii[d]	Cameron *et al.* (1987)
Curvularia senegalensis[c]	Crabbe *et al.* (1994)
Exophiala leanselmei[c]	Owen *et al.* (1996)
Fusarium sp.[c]	Pathirana & Seal (1984); Subrahmanyam *et al.* (2001)
Fusarium culmorum[c]	Bentham *et al.* (1987)
Fusarium oxysporum[c]	Pathirana & Seal (1984); Pommer & Lorenz (1985); Subrahmanyam *et al.* (2001)
Fusarium solani[c]	Bentham *et al.* (1987); Crabbe *et al.* (1994); Subrahmanyam *et al.* (2001)
Geomyces pannorum[c]	Barratt *et al.* (2003)
Gliocladium roseum[c]	Pathirana & Seal (1984)
Mucor sp.[c]	Pathirana & Seal (1984)
Nectria gliocladiodes[c]	Barratt *et al.* (2003)
Nitrospora spherical[c]	Pathirana & Seal (1984)
Paecillomyces variotti[c]	Pathirana & Seal (1984)
Penicillium spp.[c]	Pathirana & Seal (1984); Bentham *et al.* (1987)
Penicillium chrysogenum[c]	Pathirana & Seal (1984); Bentham *et al.* (1987)
Penicillium decumbens[c]	Pommer & Lorenz (1985)
Penicillium funiculosum[c]	Pommer & Lorenz (1985)
Penicillium notatum[c]	Pathirana & Seal (1984)
Penicillium ochrochloron[c]	Pommer & Lorenz (1985); Barratt *et al.* (2003)
Penicillium purpurogenum[c]	Pommer & Lorenz (1985)
Penicillium rugulosum[c]	Pommer & Lorenz (1985)
Penicillium variabile[c]	Pommer & Lorenz (1985)
Phoma spp.[b]	Stranger-Johannessen (1985)
Phoma fimenti[c]	Pommer & Lorenz (1985)
Rhizopus stolonifer[d]	Wales & Sager (1985)
Scopulariopsis brevicaulis[c]	Pathirana & Seal (1984)
Scopulariopsis fusca[c]	Pommer & Lorenz (1985)
Talaromyces spp.[c]	Pommer & Lorenz (1985)
Trichoderma spp.[c]	Pathirana & Seal (1984)
Trichoderma viride[c]	Pathirana & Seal (1984); Pommer & Lorenz (1985)

Table 9.1. (cont.)

Organism	Study
Ulocladium chartarum[b]	Griffin (1980)
Varicosporina rumulosa[b]	Stranger-Johannessen (1985)

[a] Marine piping,
[b] spoiled polyester PU,
[c] soil buried polyester PU,
[d] PU.

Table 9.2. *Summary of fungal species isolated from plasticized PVC*

Organism	Study
Alternaria spp.[a]	Hamilton (1983)
Alternaria alternata[b]	Webb *et al.* (2000)
Alternaria infectoria[b]	Webb *et al.* (2000)
Aspergillus sp.[a]	Hamilton (1983)
Aspergillus niger[b]	Webb *et al.* (2000)
Aspergillus ochraceus[b]	Webb *et al.* (2000)
Aureobasidium spp.[a]	Upsher & Roseblade (1984)
Aureobasidium pullulans[ab]	Upsher & Roseblade (1984); Webb *et al.* (2000)
Cladosporium spp.[a]	Hamilton (1983); Upsher & Roseblade (1984)
Cladosporium herbarum[b]	Webb *et al.* (2000)
Emericella nidulans[b]	Webb *et al.* (2000)
Epicoccum nigrum[b]	Webb *et al.* (2000)
Fomitopsis spp.[c]	Sabev (2004)
Fusarium spp.[a]	Upsher & Roseblade (1984)
Geomyces spp.[c]	Sabev (2004)
Klyveromyces spp.[b]	Webb *et al.* (2000)
Mucor fragilis[c]	Sabev (2004)
Neonectria ramularie[c]	Sabev (2004)
Paecilomyces spp.[a]	Hamilton (1983)
Penicillium canescens[c]	Sabev (2004)
Penicillium chrysogenum[c]	Sabev (2004)
Penicillium coprophilum[c]	Sabev (2004)
Penicillium expansum[c]	Sabev (2004)
Penicillium glabrum[b]	Webb *et al.* (2000)
Penicillium inflatum[c]	Sabev (2004)
Paecilomyces lilacinus[b]	Webb *et al.* (2000)
Penicillium swiecickii[c]	Sabev (2004)
Penicillium spp.[a]	Hamilton (1983)
Rhinocladiella spp.[a]	Upsher & Roseblade (1984)
Taphrina deformans[b]	Webb *et al.* (2000)

[a] Air exposed (tropical region),
[b] air exposed (temperate region),
[c] soil buried.

isolated from a number of bacteria including *Pseudomonas chlororaphis* (Howard *et al.*, 1999; Ruiz *et al.*, 1999a, b), *Pseudomonas fluorescens* (Howard & Blake, 1998; Ruiz & Howard, 1999), *Comamonas acidovarus* (Allen *et al.*, 1999; Akutsu *et al.*, 1998) and *Bacillus subtilis* (Rowe & Howard, 2002). Moreover, more than one PU-degrading enzyme may be produced at the same time; for example three enzymes with differing molecular weights have been isolated from *P. chlororaphis* (Ruiz *et al.*, 1999a, b). Two of these, *pueA* and *pueB* from *P. chlororaphis*, and *pulA* from *P. fluorescens* have been sequenced and appear to encode lipases (Stern & Howard, 2000; Howard *et al.*, 2001). However, it appears that some PU-degrading enzymes may also be proteases as some purified PU-degrading enzymes are strongly inhibited by PMSF or other protease inhibitors (Howard & Blake, 1998; Allen *et al.*, 1999) though none of this class has yet been cloned. Although the exact numbers and roles of the hydrolases involved in PU degradation have yet to be fully elucidated, it has been suggested that PU breakdown may be similar to that of other naturally occurring polymers such as cellulose and involve the concerted action of both exo (hydrolysing at the polymer ends) and endo (randomly hydrolysing internal bonds) esterases (Wales & Sager, 1988). In addition, it has been proposed that some of these enzymes may contain a PU substrate-binding domain enabling the enzyme to bind to the surface of solid PU (Howard, 2002) in a similar manner to that of other hydrolases that catalyse the degradation of insoluble natural polymers such as cellulases (Rabinovich *et al.*, 2002)

Plasticized polyvinylchloride (pPVC)

Polyvinyl chloride resin is composed of vinyl chloride monomeric units, polymerized in chains of different lengths according to the specific application (Rehm, 2002). Depending on the intended application, additives such as plasticizers, stabilizers, colorants and biocides might also be incorporated (Roberts & Davidson, 1986; Meier, 1990). Polyvinyl chloride has been extensively used for a broad range of applications due to its safety, effectiveness, manufacturing technology and cost. Sodium chloride and oil/gas are the basic raw sources for its production and the range of applications of PVC includes medical, industrial, household, automobile industry, furniture and appliances, piping industry, packaging materials, floor tiles, covers and wire insulations (Lorenz, 1990). While the PVC polymer itself is regarded as relatively recalcitrant (Griffin & Uribe, 1984; Roberts & Davidson, 1986; Andrady, 1994), plasticizers and other additives that are commonly added during manufacture to change the physical properties of the material are open to microbial attack (Tirpak, 1970; Webb *et al.*, 2000; Fig. 9.3). Primary plasticizers have the highest degree of compatibility with the PVC resin and are grouped into two classes – monomeric or polymeric. Modern monomeric plasticizers are synthetic organic chemicals with an ester base, such as adipates and phthalates (Fig. 9.2). The most common monomeric plasticizers include DEHP (bis 2-ethylhexyl phthalate, also known as dioctyl phthalate or DOP), DIDP (diisodecyl phthalate), DINP (diisononylphthalate), DOA(DEHA) bis-(2 ethylhexyl adipate), and TOTM (tri-octyl tri-mellitate). Azelates, laureates, oleates, sebacates and phosphoric acid derivatives are also commonly used (Roberts & Davidson, 1986; Cain 1992; Whitney, 1996; Gumargalieva *et al.*, 1999). The polymeric plasticizers include various molecular weight plasticizers polymerized from adipic and glutaric acids in combination with various glycols. Studies have been carried out on the microbial colonization and deterioration of plasticized PVC by bacteria (Booth *et al.*, 1968; Booth & Robb, 1968) and fungi

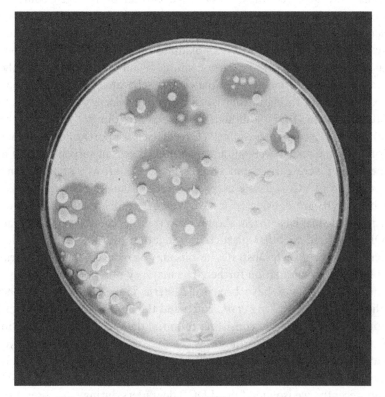

Fig. 9.3. Recovery of fungi from pPVC on plasticizer agar. Hydrolysis of the plasticizer can be seen as a zone of clearing around some colonies.

222   *H. A. Sabev* et al.

(Whitney, 1996; Webb *et al.*, 1999, 2000) where degradation of the additives, particularly the plasticizers, causes an increase in brittleness, loss of plasticity and fragmentation (Roberts & Davidson, 1986; Webb *et al.*, 2000).

When exposed to the outdoor environment, the major biodeterioration and biodegradation processes occur after the initial colonization of the substratum by microorganisms. The presence of different types of esters of adipic and phthalic acids are considered as important factors for the increased susceptibility of pPVC to fungal colonization (Andrady, 1994; Steinbuchel, 1996; Whitney, 1996; Webb *et al.*, 2000). Roberts and Davidson (1986) studied the degradation of pPVC and suggested the possibility that fatty acid derivatives from the additives could serve as a sole carbon source for fungi isolated from pPVC, which included representatives from the genera *Aspergillus, Cladosporium, Fusarium, Paecilomyces, Penicillium, Trichoderma and Verticillium*. Their study showed that these strains were able to utilize pPVC as a carbon or nitrogen source in suspension. In addition it was demonstrated that *Aspergillus* and *Paecilomyces* spp. were able to decompose the pPVC material without any additional nutrient supply. Epoxidized oil was found to be the only plasticizer to be utilized as a carbon source by all genera of the microbes tested. Plasticizers, including adipates and sebacates have been given a maximal growth rating of four by the American Society for Testing and Materials (Cain, 1992), although the researchers did not study if either phthalic acid, its derivates or the PVC resin itself supported fungal growth. Moreover, the results were obtained for fungal cells in suspension and no experiments were carried out to investigate the growth of fungi when attached to pPVC as a sole carbon source. Phthalate-based plasticizers such as DOP are considered less susceptible to degradation and fewer microorganisms have been shown to have the capability to degrade DOP compared to other plasticizers (Sugatt *et al.*, 1984). In addition, bioavailability of such substrates in soils may be reduced, thereby decreasing the rate of degradation further. The majority of reports on the degradation of phthalates have been by bacteria (Samsonova *et al.*, 1996; Valiente *et al.*, 1998; Zeng *et al.*, 2002) and the mode of bacterial degradation of DOP is known, consisting of initial de-esterification, formation of protocatechuic acid, ortho-cleavage and entry into the TCA cycle (Karpagam & Lalithakumari, 1999). Only a few studies however have reported on phthalate degradation by yeasts (Gartshore *et al.*, 2003). More recently, we isolated fungal DOP degraders during a long-term soil burial study that were identified as *Doratomyces* spp. (Sabev, 2004).

Among the fungi, the yeast-like Deuteromycete *A. pullulans* has been most frequently isolated from pPVC. Several studies have identified and demonstrated the degradative capabilities of this organism (Cooke, 1959; Webb *et al.*, 2000), which has been isolated from plant leaves (Pereira *et al.*, 2002), paints (Crang & Pechak, 1978), fruits, optic surfaces and oils (Cooke, 1959). The ability of the fungus to utilize long-chain fatty acids is an indication of substantial enzymatic activity, which is thought to be important for the degradation of plasticizers in pPVC (Webb *et al.*, 1999, 2000). Such activities may be one reason why this fungus is often isolated in association with other pioneer fungi such as *Alternaria* and *Cladosporium* and is often one of the primary colonizers in the process of microbial succession (Webb *et al.*, 2000). *Aureobasidium pullulans* produces an extracellular polysaccharide pullulan (α-glucan with polymaltotriose units, linked through α-1,6-bonds on the terminal glucose residues) and secretion of the polysaccharide may promote attachment, adhesion (Andrews *et al.*, 1994) and subsequent growth on different surfaces, including pPVC (Catley, 1971). *Aureobasidium pullulans* has also been reported to be able to grow on protocatechuic acid (Schoeman & Dickinson, 1996), suggesting that the fungus could potentially breakdown DOP, although this has not yet been reported.

Studies on the degradability of adipate-containing polymers by Uchida *et al.* (2000) reported a positive induction of lipase activity that was shown to be capable of breaking down a co-adipate ester. The authors proposed that due to the diversity of compounds incorporated into plastics, and the range of microorganisms capable of producing plastic-degrading enzymes, future studies should focus on clarifying the nature of the enzymes involved and their characteristics.

Colonization and biodegradation of other commercially important synthetic polymers

Polyesters

Polyester is a general term referring to any polymer where the monomers are linked by ester bonds and includes the biodegradable microbially derived polyhydroxyalkanoates, which, as they are naturally produced, are beyond the scope of this article (for a review see Kim & Rhee, 2003). Most synthetic polyesters in large-scale use are the aromatic poly(ethylene tetraphthalate) or poly(butylene tetraphthalate) polyesters as they have excellent material properties and are used in a wide range of applications including plastic containers, fibres for synthetic fabrics, films and sheets. However, studies with these aromatic polyesters indicate that they appear to be highly resistant to microbial degradation (Muller *et al.*, 2001) although *Aspergillus niger* has been reported to have been observed growing on some aromatic polyester surfaces (Huang & Byrne, 1980). By contrast, aliphatic polyesters including polyethylene adipate, polyethylene succinate, polypropylene adipate, polybutylene adipate, polybutylene succinate and polyhexylene carbonate are highly susceptible to microbial attack and degradation (Darby & Kaplan, 1968; Pranamuda *et al.*, 1995; Tansengo & Tokiwa, 1998; Jarerat & Tokiwa, 2001a). The enzyme responsible for the degradation of polyethylene adipate has been purified from a *Penicillium* strain and appeared to be a lipase with broad substrate specificity though it was not capable of degrading aromatic polyesters (Tokiwa & Suzuki, 1977). An extracellular enzyme causing non-specific hydrolysis of aliphatic polyesters has also been reported to be secreted by *Aspergillus fumigatus*, which showed its highest activity against polyethylene adipate, and again is most likely a lipase (Scherer *et al.*, 1999).

Poly (ε-caprolactone) is a linear polyester composed of a backbone of 6-hydroxyhexanoates that is known to be highly susceptible to microbial biodegradation in the environment. Most studies have highlighted the role of bacteria in the environmental degradation of this polymer (Suyama *et al.*, 1998), with few studies focusing on fungi (Tokiwa *et al.*, 1976; Benedict *et al.*, 1983; Sanchez *et al.*, 2000). Although little is known about the enzyme(s) responsible for degradation, lipases from *Rhizopus delemar* and *Rhizopus arrhizus* have been shown to readily degrade this polymer (Tokiwa & Suzuki, 1977; Tokiwa *et al.*, 1986), although there is some evidence that a cutinase from *Fusarium solani* is also capable of degrading poly (ε-caprolactone) (Murphy *et al.*, 1996).

Poly (L-lactide) is a linear polymer synthesized through the polymerization of lactic acid. Although poly (L-lactide) is open to microbial degradation, there have been relatively few reports with most degradation reported amongst some members of the actinomycetes (Tokiwa & Jarerat, 2004). Poly (L-lactide) is regarded as less susceptible compared to other synthetic polyesters (Suyama *et al.*, 1998). Few fungi have been isolated as poly (L-lactide) degraders with *Fusarium moniliforme* and *Penicillium roqueforti* proving to be relatively poor (Torres *et al.*, 1996). It appears that only one fungus to date, *Tritirachium album*, has significant degradative ability, which is greatly enhanced in the presence of gelatine, suggesting a protease may by responsible for poly (L-lactide) hydrolysis (Jarerat & Tokiwa, 2001b).

Polyamides

The polyamides are polymers that contain repeated amide groups within the polymer backbone. Commercially, two main polyamides are mainly produced, nylon (poly(hexamethylene adipamide)) and nylon 6 (polycaprolactam), and are used for the production of polyamide fibres and thermoplastics with a wide variety of uses. White-rot fungi, including *Phanerochaete chrysosporium*, have to date been the only microbes reported to be capable of causing significant breakdown of these materials under ligninolytic conditions and this appears to be catalysed by manganese peroxidase, which is involved in lignin degradation (Deguchi *et al.*, 1997; Deguchi *et al.*, 1998).

Polyvinyl alcohol

Polyvinyl alcohol is a vinyl polymer with a carbon–carbon backbone similar to other polymers such as polyethylene and polystyrene and is widely used as a water-soluble biodegradable polymer in the manufacture of delivery systems for fertilizers, pesticides and herbicides and is also used to manufacture containers and films. To date, only bacteria have been described as causing biodegradation of polyvinyl alcohols with some causing complete degradation of the polymer (see Shimao, 2001 for review).

Polyethylene, polypropylene and polystyrene

Polyethylenes, polypropylenes and polystyrenes are widely used in product packaging, engineering pipes and containers and their slow rate of deterioration in the environment pose serious environmental concerns (Kalb *et al.*, 1993). There is little evidence that microbes can cause significant degradation of these polymers and most of the degradation reported appears to be due to abiotic factors (Albertsson *et al.*, 1994). Most studies to date where microbial degradation has been observed have been with polyethylene blended with starch (Imam & Gould, 1990; Breslin & Swanson, 1993). Lugauskas *et al.* (2003) studied the colonization and deterioration of various polymeric materials over a period of up to nine years including polyethylene, polypropylene and polystyrene. However, these materials supported only limited fungal growth on the surface with no evidence of degradation. These polymers therefore, like unplasticized PVC, appear highly recalcitrant.

Assessing biodegradation of synthetic polymers

Generally, there is a difference between the theoretical and practically achieved level of biodegradation of any compound in contact with microorganisms in a given system (De Henau, 1998). In addition, the biodegradation of the compound of interest does not necessarily mean that complete degradation has been achieved. A number of different tests have been established as a means of assessing the susceptibility of a test polymer to microbial degradation and are outlined below.

Growth rating tests

Growth rate-based tests evaluate the ability of chosen test micro-organisms to grow and utilize components from the polymer blend as a sole source of carbon, nitrogen and energy. The tests consist of placing or embedding a test piece of plastic in a mineral-salts agar containing the microbe and examining the proliferation and growth above the material after a fixed length of incubation (Cain, 1992). However, the test suffers from a lack of quantification and the restricted number of fungal isolates used to assess the susceptibility of the material.

Agar plating methods and clearing zone production

Hydrolytic enzymes produced by attached microorganisms on the polymer surface are considered responsible for the degradation of polymer and/or additives and plasticizers in the polymeric blend. Therefore, the introduction of the test microorganism on an agar plate with incorporated compounds from the polymer blend and the subsequent production of a zone of clearing around the growing colonies is generally accepted as a positive sign of initial enzymatic degradation and is a widely used technique for screening microbial degraders and the degradation potential of microbial strains towards components of polymer formulations (Kumar *et al.*, 1982; Augusta *et al.*, 1993; Crabbe *et al.*, 1994; Webb *et al.*, 2000; Barratt *et al.*, 2003). Finally, Andrady (1994) suggested that for increased reliability in such tests a microbial consortium should be employed. The author pointed out that a complex biota is more likely to mimic the natural conditions and therefore the use of a single test microbe is only justified when more fundamental biochemical information is sought.

Viable count determination

While directly quantifying the numbers of microorganisms attached to the surface of a polymer by enumeration of colonies following incubation on a suitable medium has been widely used to indicate the amount of growth supported on the polymer surface (Christensen *et al.*, 1995), it has a number of severe limitations. The technique depends on a reliable and reproducible technique for detachment of the microbes by means of washing and/or scrapping of the polymer surface and the detachment procedure may be incomplete, cells may be destroyed, or cells may remain in clumps therefore resulting in an underestimate of the true number of attached microbes. In addition, this technique is problematic for filamentous fungi compared to unicellular organisms due to hyphal fragmentation and differences in the levels of sporulation between different species (Zak & Visser, 1996; Webb *et al.*, 2000).

Soil burial tests

Soil burial tests are important in the assessment of the degradation of a range of polymers by the corresponding microflora. They are performed either under field or laboratory conditions with the use of specific strains without the presence of any additional carbon source other than that supplied from the polymer (Whitney, 1996). There are certain drawbacks that should be taken into account, namely the lack of reproducibility due to the climatic conditions and the change of the microbial soil flora when *in situ* experiments are performed. However, *in situ* soil burial tests are often the most useful when the extent of biodegradability of a material is sought, especially when combined with tensile strength tests and microscopic examinations (Cain, 1992).

Tensile strength change and weight loss

Loss of plasticizers due to their utilization as a nutrient source or by physical leaching will increase the tensile strength of the pPVC and can be used as a measure of biodeterioration and biodegradation. Andrady (1994) demonstrated that the method is more sensitive in the early stages of the test because small changes in the molecular weight can greatly influence the mechanical properties of the plastic. Weight loss has also been reported to be an important parameter for determining pPVC degradation (Whitney, 1996) and thought to reflect weight loss due to microbial metabolism and utilization of the plastic as a carbon source for structural or energy needs.

Impedance measurements

Although the first reports demonstrating a change in medium conductivity due to bacterial growth appeared at the end of the nineteenth century, this technique did not start to receive particular attention until after the 1970s (Silley & Forsythe, 1996). Impedance is a measure of resistance to alternating current passing through conducting media. As a result of microbial metabolism, the conductance of the medium increases whilst the impedance decreases (Silley & Forsythe, 1996). This fact has been used for the design of impedance measurement systems for assessing microbial growth. Based on impedance measurement, ISO 14852 standard for biodegradability was proposed (Sawada, 1998). Direct impedance allows measurements of the net changes in impedance in the test medium at regular intervals. Any change, which exceeds a preliminary determined detection criteria value set up by the user, is considered to be due to microbial growth. The type of medium, its electrolyte concentration and released ions during the process can influence the result. However, the low detection limit of this method (<10 CFU ml^{-1}) (Silley & Forsythe, 1996) makes it very valuable for rapid determination of microbiological growth. The growth medium composition and particularly a high salt concentration may make the application of the direct impedance technique difficult.

The indirect impedance method overcomes these problems by monitoring microbial activity by measuring CO_2 released through respiration. This method has been applied and reported in many applications. The largest area of application is in the food industry, where indirect impedance is used to detect microbial activity. Silley and Forsythe (1996) showed the application of the technique for real-time observation of antibiotic resistance, biocide susceptibility, microbe–substratum interaction and colonization of inert materials. The method has been used for CO_2 release measurements of microbial flora grown on poly-ε-caprolactone (Starnecker & Menner, 1996a, b; Whitney, 1996). With the help of automated CO_2 analysis, a full carbon balance was performed and the rate of polymer decomposition assessed. However, Starnecker and Menner (1996a) pointed out that CO_2 evolution alone was not sufficient to estimate the extent of plastic degradation and, in addition, the slow growth and development of microorganisms on plastics may therefore result in insufficient CO_2 being generated to be detected.

Conclusions

The rise in the production and use of synthetic polymers worldwide was recently estimated at around 140 million tonnes per year, with most of these being petroleum-based and causing an increasing problem both as environmental pollutants and in the pressure they exert as landfill components. Biodegradation and biodeterioration of these materials can play an important role in microbial ecology and nutrient cycling in the environment. Compared to the importance of these materials in the environment, there have been relatively few studies on the colonization, microbial ecology and biochemical degradation of these polymers, particularly on the role played by the fungi. While many synthetic polymers are relatively recalcitrant or degraded only slowly, recent research has begun to focus on those that are completely decomposed in the environment to produce biodegradable plastics for use in products that most often cause problems in the environment such as food packaging (Krochta & De Muler-Johnston, 1997; Chandra & Rustgi, 1998; Pavlath & Robertson, 1999; Mohanty *et al.*, 2000; Gross & Kalra, 2002, Ishigaki *et al.*, 2004).

To date, our current understanding of the composition and interaction of fungi with polymer surfaces, their relationship with other microbes and the mechanisms underlying the complex process of polymer degradation is poorly developed. A far greater understanding of these processes will contribute toward the future development of polymer waste treatments and in the design and manufacture of novel biodegradable materials.

Acknowledgements

The authors would like to thank the British Foreign and Commonwealth Office and Central European University, Budapest for financial support to HAS and the BBSRC and Arch Biocides for CASE support to SRB.

References

Akutsu, Y., Nakajima-Kambe, T., Nomura, N. & Nakahara, T. (1998). Purification and properties of a polyester polyurethane-degrading enzyme from *Comamonas acidovorans* TB-35. *Applied and Environmental Microbiology*, **64**, 62–7.

Albertsson, A. C. & Karlsson, S. (1988). The 3 stages in degradation of polymers – polyethylene as a model substance. *Journal of Applied Polymer Science*, **35**, 1289–302.

Albertsson, A. C., Barenstedt, C. & Karlsson, S. (1994). Abiotic degradation products from enhanced environmentally degradable polyethylene. *Acta Polymerica*, **45**, 97–103.

Alexander, M. (1994). *Biodegradation and Bioremediation*. London: Academic Press.

Allen, A. B., Hilliard, N. P. & Howard, G. T. (1999). Purification and characterization of a soluble polyurethane degrading enzyme from *Comamonas acidovorans*. *International Biodeterioration and Biodegradation*, **43**, 37–41.

Andrady, A. L. (1994). Assessment of environmental biodegradation of synthetic-polymers. *Journal of Macromolecular Science*, **34**, 25–76.

Andrews, J. H., Harris, R. F., Spear, R. N., Gee, W. L. & Nordheim, E. V. (1994). Morphogenesis and adhesion of *Aureobasidium pullulans*. *Canadian Journal of Microbiology*, **40**, 6–17.

Augusta, J., Muller, R. J. & Widdecke, H. (1993). A rapid evaluation plate-test for the biodegradability of plastics. *Applied Microbiology and Biotechnology*, **39**, 673–8.

Barratt, S. R., Ennos, A. R., Greenhalgh, M., Robson, G. D. & Handley, P. S. (2003). Fungi are the predominant micro-organisms responsible for degradation of soil-buried polyester polyurethane over a range of soil water holding capacities. *Journal of Applied Microbiology*, **95**, 78–85.

Benedict, C. V., Cook, W. J., Jarrett, P. *et al.* (1983). Fungal degradation of polycaprolactones. *Journal of Applied Polymer Science*, **28**, 327–34.

Bentham, R. H., Morton, L. H. G. & Allen, N. G. (1987). Rapid assessment of the microbial deterioration of polyurethanes. *International Biodeterioration*, **23**, 377–86.

Bentivegna, C. S. & Piatkowski, T. (1998). Effects of tributyltin on medaka (*Oryzias latipes*) embryos at different stages of development. *Aquatic Toxicology*, **44**, 117–28.

Booth, G. H. & Robb, J. A. (1968). Bacterial degradation of plasticised PVC – Effect of some physical properties. *Journal of Applied Chemistry*, **18**, 194.

Booth, G. H., Cooper, A. W. & Robb, J. A. (1968). Bacterial degradation of plasticised PVC. *Journal of Applied Bacteriology*, **31**, 305–10.

Boubendir, A. (1993). Purification and biochemical evaluation of polyurethane degrading enzymes of fungal origin. *Dissertation Abstracts International*, **53**, 4632.

Bouwer, E. J. (1992). Bioremediation of organic contaminants in the subsurface. In *Environmental Microbiology*, ed. R. Mitchell. New York: Wiley-Liss, pp. 287–318.

Breslin, V. T. & Swanson, R. L. (1993). Deterioration of starch-plastic composites in the environment. *Journal of the Air and Waste Management Association*, **43**, 325–35.

Cain, R. B. (1992). Microbial degradation of synthetic polymers. In *Microbial Control of Pollution*, ed. J. C. Fry, G. M. Gadd, R. A. Herbert, C. W. Jones & I. A. Watson-Craik. Cambridge: Cambridge University Press, pp. 293–338.

Cameron, J. A., Bunch, C. L. & Huang, S. J. (1987). Microbial degradation of synthetic polymers. In *Biodeterioration*, 7th edn, ed. D. R. Houghton, R. N. Smith & H. O. W. Eggins. London: Elsevier Applied Science, pp. 553–61.

Catley, B. J. (1971). Role of pH and nitrogen limitation in the elaboration of the extracellular polysaccharide pullulan by *Pullularia pullulans*. *Applied Microbiology*, **22**, 650–4.

Chandra, R. & Rustgi, R. (1998). Biodegradable polymers. *Progress in Polymer Science*, **23**, 1273–335.

Christensen, G. D., Baldassarri, L. & Simpson, W. A. (1995). Methods for studying microbial colonisation of plastics. *Methods in Enzymology*, **253**, 477–500.

Cooke, W. B. (1959). An ecological life history of *Aureobasidium pullulans* (de Bary) Arnaud. *Mycopathologia et Mycologia Applicata*, **12**, 1–45.

Crabbe, J. R., Campbell, J. R., Thompson, L., Walz, S. L. & Schultz, W. W. (1994). Biodegradation of a colloidal ester-based polyurethane by soil fungi. *International Biodeterioration and Biodegradation*, **33**, 103–13.

Crang, R. E. & Pechak, D. G. (1978). *Aureobasidium pullulans*: fine structure and development. *Journal of Coatings Technology*, **50**, 36–42.

Darby, R. T. & Kaplan, A. M. (1968). Fungal susceptibility of polyurethanes. *Applied Microbiology*, **16**, 900–4.

De Henau, H. (1998). Biodegradation. In *Handbook of Ecotoxicology*, ed. P. Calow. London: Blackwell Science, pp. 355–77.

Deguchi, T., Kakezawa, M. & Nishida, T. (1997). Nylon biodegradation by lignin-degrading fungi. *Applied and Environmental Microbiology*, **63**, 329–31.

Deguchi, T., Kitaoka, Y., Kakezawa, M. & Nishida, T. (1998). Purification and characterization of a nylon-degrading enzyme. *Applied and Environmental Microbiology*, **64**, 1366–71.

Eastwood, I. M. (1994). Problems with biocides and biofilms. In *Bacterial Biofilms and their Control in Medicine and Industry*, eds. P. J. Wimpenny, W. Nichols, D. Stickler & H. Lappin-Scott. Cardiff: Bioline, pp. 169–72.

Filip, Z. (1978). Decomposition of polyurethane in garbage landfill leakage water and by soil microorganisms. *European Journal of Applied Microbiology and Biotechnology*, **5**, 225–31.

Filip, Z. (1979). Polyurethane as the sole nutrient source for *Aspergillus niger* and *Cladosporium herbarum*. *European Journal of Applied Microbiology*, **7**, 277–80.

Flemming, H. C. (1998). Relevance of biofilms for the biodeterioration of surfaces of polymeric materials. *Polymer Degradation and Stability*, **59**, 309–15.

Gartshore, J., Cooper, D. G. & Nicell, J. A. (2003). Biodegradation of plasticizers by *Rhodotorula rubra*. *Environmental Toxicology and Chemistry*, **22**, 1244–51.

Gaylarde, C. C. & Morton, L. H. G. (1999). Deteriogenic biofilms on buildings and their control: a review. *Biofouling*, **14**, 59–74.

Griffin, G. J. L. (1980). Synthetic-polymers and the living environment. *Pure and Applied Chemistry*, **52**, 399–407.

Griffin, G. J. L. & Uribe, M. (1984). Biodegradation of plasticised polyvinyl chloride. In *Biodeterioration 6*, Slough, UK: C. A. B. International, pp. 648–57.

Grima, S., Bellon-Maurel, V., Feuilloley, P. & Silvestre, F. (2000). Aerobic biodegradation of polymers in solid-state conditions: a review of environmental and physicochemical parameter settings in laboratory simulations. *Journal of Polymers and the Environment*, **8**, 183–95.

Gross, R. A. & Kalra, B. (2002). Biodegradable polymers for the environment. *Science*, **297**, 803–7.

Gu, J. D. (2003). Microbiological deterioration and degradation of synthetic polymeric materials: recent research advances. *International Biodeterioration and Biodegradation*, **52**, 69–91.

Gumargalieva, K. Z., Zaikov, G. E., Semenov, S. A. & Zhdanova, O. A. (1999). The influence of biodegradation on the loss of a plasticiser from poly(vinyl chloride). *Polymer Degradation and Stability*, **63**, 111–12.

Hamilton, N. F. (1983). Biodeterioration of flexible polyvinyl chloride films by fungal organisms. In *Biodeterioration 5*, ed. T. A. Oxley & S. Barry. Chichester: John Wiley & Sons, pp. 663–78.

Hazen, K. C. & Glee, P. M. (1995). Adhesion of fungi. Adhesion of microbial pathogens. *Methods In Enzymology*, **253**, 414–24.

Howard, G. T. (2002). Biodegradation of polyurethane: a review. *International Biodeterioration and Biodegradation*, **49**, 245–52.

Howard, G. T. & Blake, R. C. (1998). Growth of *Pseudomonas fluorescens* on a polyester-polyurethane and the purification and characterization of a polyurethanase-protease enzyme. *International Biodeterioration and Biodegradation*, **42**, 213–20.

Howard, G. T., Ruiz, C. & Newton, N. P. (1999). Growth of *Pseudomonas chlororaphis* on a polyester-polyurethane and the purification and characterization of a polyurethanase-esterase enzyme. *International Biodeterioration and Biodegradation*, **43**, 7–12.

Howard, G. T., Crother, B. & Vicknair, J. (2001). Cloning, nucleotide sequencing and characterization of a polyurethanase gene (pueB) from *Pseudomonas chlororaphis*. *International Biodeterioration and Biodegradation*, **47**, 141–9.

Huang, S. J. & Byrne, C. A. (1980). Biodegradable polymers: photolysis and fungal degradation of poly(arylene keto esters). *Journal of Applied Polymer Science*, **35**, 1951–60.

Imam, S. H. & Gould, J. M. (1990). Adhesion of an amylolytic arthrobacter sp to starch-containing plastic films. *Applied and Environmental Microbiology*, **56**, 872–76.

Ishigaki, T., Sugano, W., Nakanishi, A. *et al.* (2004). The degradability of biodegradable plastics in aerobic and anaerobic waste landfill model reactors. *Chemosphere*, **54**, 225–33.

232   *H. A. Sabev* et al.

Jarerat, A. & Tokiwa, Y. (2001a). Degradation of poly(tetramethylene succinate) by thermophilic actinomycetes. *Biotechnology Letters*, **23**, 647–51.
Jarerat, A. & Tokiwa, Y. (2001b). Degradation of poly(L-lactide) by a fungus. *Macromolecular Bioscience*, **1**, 136–40.
Kalb, P. D., Heiser, J. H. & Colombo, P. (1993). Long-term durability of polyethylene for encapsulation of low-level radioactive, hazardous, and mixed wastes. *ACS Symposium Series*, **518**, 439–49.
Karpagam, S. & Lalithakumari, D. (1999). Plasmid-mediated degradation of *o*- and *p*-phthalate by *Pseudomonas fluorescens*. *World Journal of Microbiology and Biotechnology*, **15**, 565–9.
Kay, M. J., Morton, L. H. G. & Prince, E. L. (1991). Bacterial-degradation of polyester polyurethane. *International Biodeterioration*, **27**, 205–22.
Kim, D. Y. & Rhee, Y. H. (2003). Biodegradation of microbial and synthetic polyesters by fungi. *Applied Microbiology and Biotechnology*, **61**, 300–8.
Krochta, J. M. & De Muler-Johnston, C. (1997). Edible and biodegradable polymer films: challenges and opportunities. *Food Technology*, **51**, 61–74.
Kumar, C. G. & Anand, S. K. (1998). Significance of microbial biofilms in food industry: a review. *International Journal of Food Microbiology*, **42**, 9–27.
Kumar, C. G., Kalpagam, V. & Nandi, U. S. (1982). Biodegradable polymers: prospects, problems, and progress. *Journal of Macromolecular Science*, **C22**, 225–60.
Lorenz, J. (1990). Biostabilizers, In *Plastics Additives Handbook*, ed. R. Gachter & H. Muller. Munich: Henser Publishers, pp. 791–809.
Lugauskas, A., Levinskaite, L. & Peciulyte, D. (2003). Micromycetes as deterioration agents of polymeric materials. *International Biodeterioration and Biodegradation*, **52**, 233–42.
McNeill, I. C. (1991). Fundamental aspects of polymer degradation. In *Polymers in Conservation*, ed. N. C. Allen, M. Edge & C. V. Horie. Cambridge: The Royal Society of Chemistry, pp. 15–17.
Meier, L. (1990). Plasticisers. In *Plastics Additives Handbook*, ed. R. Gachter & H. Muller. Munich: Henser Publishers, pp. 327–422.
Mohanty, A. K., Misra, M. & Hinrichsen, G. (2000). Biofibres, biodegradable polymers and biocomposites: an overview. *Macromolecular Materials and Engineering*, **276**, 1–24.
Morton, L. H. G. & Surman, S. B. (1994). Biofilms in biodeterioration – a review. *International Biodeterioration and Biodegradation*, **34**, 203–21.
Muller, R. J., Kleeberg, I. & Deckwer, W. D. (2001). Biodegradation of polyesters containing aromatic constituents. *Journal of Biotechnology*, **86**, 87–95.
Murphy, C. A., Cameron, J. A., Huang, S. J. & Vinopal, R. T. (1996). *Fusarium* polycaprolactone depolymerase is a cutinase. *Applied and Environmental Microbiology*, **62**, 456–60.
Nakajima-Kambe, T., Onuma, F., Akutsu, Y. & Nakahara, T. (1995). Determination of the polyester polyurethane breakdown products and distribution of the polyurethane degrading enzyme of *Comamonas acidovorans* strain TB-35. *Journal of Fermentation and Bioengineering*, **83**, 456–60.
Nakajima-Kambe, T., Shigeno-Akutsu, Y., Nomura, N., Onuma, F. & Nakahara, T. (1999). Microbial degradation of polyurethane, polyester polyurethanes and polyether polyurethanes. *Applied Microbiology and Biotechnology*, **51**, 134–40.
Owen, S., Otani, T., Masaoka, S. & Ohe, T. (1996). The biodegradation of low-molecular-weight urethane compounds by a strain of *Exophiala jeanselmei*. *Bioscience Biotechnology and Biochemistry*, **60**, 244–8.
Palmisano, A. C. & Pettigrew, C. A. (1992). Biodegradability of plastics. *Bioscience*, **42**, 680–5.

Pathirana, R. A. & Seal, K. J. (1984). Studies on polyurethane deteriorating fungi. 1. Isolation and characterization of the test fungi employed. *International Biodeterioration*, **20**, 163–8.

Pavlath, A. E. & Robertson, G. H. (1999). Biodegradable polymers vs. recycling: what are the possibilities? *Critical Reviews in Analytical Chemistry*, **29**, 231–41.

Pereira, P. T., de Carvalho, M. M., Girio, F. M., Roseiro, J. C. & Amaral-Collaco, M. T. (2002). Diversity of microfungi in the phylloplane of plants growing in a Mediterranean ecosystem. *Journal of Basic Microbiology*, **42**, 396–407.

Pommer, E. H. & Lorenz, G. (1985). The behaviour of polyester and polyether polyurethanes towards microorganisms. In *Biodeterioration and Biodegradation of Polymers 1*, ed. K. J. Seal. New York: Biodeterioration Society, pp. 77–86.

Pranamuda, H., Tokiwa, Y. & Tanaka, H. (1995). Microbial degradation of an aliphatic polyester with a high-melting point, poly(tetramethylene succinate). *Applied and Environmental Microbiology*, **61**, 1828–32.

Rabinovich, M. L., Melnik, M. S. & Boloboba, A. V. (2002). Microbial cellulases. *Applied Biochemistry and Microbiology*, **38**, 305–21.

Rehm, T. (2002). Polyvinyl chloride (PVC). *Kunststoffe – Plast Europe*, **92**, 22–3.

Roberts, W. T. & Davidson, P. M. (1986). Growth characteristics of selected fungi on polyvinyl chloride film. *Applied and Environmental Microbiology*, **51**, 673–6.

Rowe, L. & Howard, G. T. (2002). Growth of *Bacillus subtilis* on polyurethane and the purification and characterization of a polyurethanase-lipase enzyme. *International Biodeterioration and Biodegradation*, **50**, 33–40.

Ruiz, C. & Howard, G. T. (1999). Nucleotide sequencing of a polyurethanase gene (pulA) from *Pseudomonas fluorescens*. *International Biodeterioration and Biodegradation*, **44**, 127–31.

Ruiz, C., Hilliard, N. & Howard, G. T. (1999a). Growth of *Pseudomonas chlororaphis* on a polyester-polyurethane and the purification and characterization of a polyurethanase-esterase enzyme. *International Biodeterioration and Biodegradation*, **43**, 7–12.

Ruiz, C., Main, T., Hilliard, N. P. & Howard, G. T. (1999b). Purification and characterization of two polyurethanase enzymes from *Pseudomonas chlororaphis*. *International Biodeterioration and Biodegradation*, **43**, 43–7.

Sabev, H. A. (2004). Fungal biodeterioration and biodegradation of plasticised polyvinyl chloride in soil. Unpublished Ph.D. thesis, University of Manchester.

Samsonova, A. S., Aleshchenkova, Z. M., Syomochkina, N. F. & Baikova, S. V. (1996). Microbial decontamination of effluents from phthalate esters. In *Biodeterioration and Biodegradation. Papers of the 10th International Biodeterioration and Biodegradation Symposium, DECHEMA Monographs*, ed. W. Sand. Frankfurt: Schon & Wetzler, pp. 607–10.

Sanchez, J. G., Tsuchii, A. & Tokiwa, Y. (2000). Degradation of polycaprolactone at 50 °C by a thermotolerant *Aspergillus sp. Biotechnology Letters*, **22**, 849–53.

Sawada, H. (1998). ISO standard activities in standardization of biodegradability of plastics – development of methods and definitions. *Polymer Degradation and Stability*, **59**, 365–70.

Scherer, T. M., Fuller, R. C., Lenz, R. W. & Goodwin, S. (1999). Hydrolase activity of an extracellular depolymerase from *Aspergillus fumigatus* with bacterial and synthetic polyesters. *Polymer Degradation and Stability*, **64**, 267–75.

Schoeman, M. W. & Dickinson, D. J. (1996). *Aureobasidium pullulans* can utilize simple aromatic compounds as a sole source of carbon in liquid culture. *Letters in Applied Microbiology*, **22**, 129–31.

Shimao, M. (2001). Biodegradation of plastics. *Current Opinion in Biotechnology*, **12**, 242–7.

Silley, P. & Forsythe, S. (1996). Impedance microbiology – A rapid change for microbiologists. *Journal of Applied Bacteriology*, **80**, 233–43.

Staples, C. A., Peterson, D. R., Parkerton, T. F. & Adams, W. J. (1997). The environmental fate of phthalate esters: a literature review. *Chemosphere*, **35**, 667–749.

Starnecker, A. & Menner, M. (1996a). Assessment of biodegradability of plastics under simulated composting conditions in a laboratory test system. *International Biodeterioration and Biodegradation*, **37**, 85–92.

Starnecker, A. & Menner, M. (1996b). Kinetics of aerobic microbial degradation of aliphatic polyesters. In *Biodeterioration and Biodegradation. Papers of the 10th International Biodeterioration and Biodegradation Symposium, DECHEMA Monographs*, ed. W. Sand. Frankfurt: Schon & Wetzler, pp. 221–8.

Steinbuchel, A. (1996). Synthesis and production of biodegradable thermoplastics and elastomers: current state and outlook. *Kautschuk Gummi Kunststoffe*, **49**, 120–4.

Stern, R. V. & Howard, G. T. (2000). The polyester polyurethanase gene (pueA) from *Pseudomonas chlororaphis* encodes a lipase. *FEMS Microbiology Letters*, **185**, 163–8.

Stranger-Johannessen, M. (1985). Microbial degradation of polyurethane products in service. In *Biodeterioration and Biodegradation of Polymers 1*, ed. K. J. Seal. New York: Biodeterioration Society, pp. 264–7.

Subrahmanyam, S., Kodandapani, N., Ahamarshan, J. N. *et al.* (2001). Amperometric biochemical characterization of isolated fungal strains. *Electroanalysis*, **13**, 1454–8.

Sugatt, R. H., O'Grady, D. P., Banergee, S., Howard, P. H. & Gledhill, W. E. (1984). Shake flask biodegradation of 14 commercial phthalate esters. *Applied and Environmental Microbiology*, **47**, 601–6.

Suyama, T., Tokiwa, Y., Ouichanpagdee, P., Kanagawa, T. & Kamagata, Y. (1998). Phylogenetic affiliation of soil bacteria that degrade aliphatic polyesters available commercially as biodegradable plastics. *Applied and Environmental Microbiology*, **64**, 5008–11.

Tansengo, M. L. & Tokiwa, Y. (1998). Thermophilic microbial degradation of polyethylene succinate. *World Journal of Microbiology and Biotechnology*, **14**, 133–8.

Tirpak, G. (1970). Microbial degradation of plasticised PVC. *Society of Plastic Engineering Journal*, **26**, 511–20.

Tokiwa, Y. & Jarerat, A. (2004). Biodegradation of poly(L-lactide). *Biotechnology Letters*, **26**, 771–7.

Tokiwa, Y. & Suzuki, T. (1977). Microbial-degradation of polyesters 3. Purification and some properties of polyethylene adipate-degrading enzyme produced by *Penicillium sp* strain 14–3. *Agricultural and Biological Chemistry*, **41**, 265–74.

Tokiwa, Y., Ando, T. & Suzuki, T. (1976). Degradation of polycaprolactone by a fungus. *Journal of Fermentation Technology*, **54**, 603–8.

Tokiwa, Y., Suzuki, T. & Takeda, K. (1986). Hydrolysis of polyesters by *Rhizopus arrhizus* lipase. *Agricultural and Biological Chemistry*, **50**, 1323–5.

Torres, A., Li, S. M., Roussos, S. & Vert, M. (1996). Screening of microorganisms for biodegradation of poly(lactic acid) and lactic acid-containing polymers. *Applied and Environmental Microbiology*, **62**, 2393–7.

Turkovskii, I. I. & Yurlova, N. A. (2002). The photochemical and surface-active properties of melanins isolated from some black fungi. *Microbiology*, **71**, 410–16.

Uchida, H., Kambe, T. N., Akutsu, Y. S. *et al.* (2000). Properties of a bacterium which degrades solid poly(tetramethylene succinate)-co-adipate, a biodegradable plastic. *FEMS Microbiology Letters*, **189**, 25–9.

Upsher, F. J. & Roseblade, R. J. (1984). Assessment by tropical exposure of some fungicides in plasticized PVC. *International Biodeterioration*, **20**, 243–52.

Valiente, N., Lalot, T., Brigodiot, M. & Marechal, E. (1998). Enzymic hydrolysis of phthalic unit containing copolyesters as a potential tool for block length determination. *Polymer Degradation and Stability*, **61**, 409–15.

Wales, D. S. & Sager, B. F. (1985). Microbial degradation of synthetic polymers. In *Biodeterioration and Biodegradation of Plastics and Polymers 1*, ed. K. J. Seal. New York: Biodeterioration Society, pp. 44–8.

Wales, D. S. & Sager, B. F. (1988). Mechanistic aspects of polyurethane biodeterioration. In *Biodeterioration*, 7th edn, ed. D. R. Houghton, R. N. Smith & H. O. W. Eggins. London: Elsevier Applied Science, pp. 351–8.

Webb, J. S., Van der Mei, H. C., Nixon, M. *et al.* (1999). Plasticizers increase adhesion of the deteriogenic fungus *Aureobasidium pullulans* to polyvinyl chloride. *Applied and Environmental Microbiology*, **65**, 3575–81.

Webb, J. S., Nixon, M., Eastwood, I. M. *et al.* (2000). Fungal colonization and biodeterioration of plasticized polyvinyl chloride. *Applied and Environmental Microbiology*, **66**, 3194–200.

Whitney, P. J. (1996). A comparison of two methods for testing defined formulations of PVC for resistance to fungal colonization with two methods for the assessment of their biodegradation. *International Biodeterioration and Biodegradation*, **37**, 205–13.

Zak, J. C. & Visser, S. (1996). An appraisal of soil fungal biodiversity: the crossroads between taxonomic and functional biodiversity. *Biodiversity and Conservation*, **5**, 169–83.

Zeng, F., Cui, K. Y., Fu, J. M., Sheng, G. Y. & Yang, H. F. (2002). Biodegradability of di(2-ethylhexyl) phthalate by *Pseudomonas fluorescens* FS1. *Water Air and Soil Pollution*, **140**, 297–305.

10

Fungal dissolution and transformation of minerals: significance for nutrient and metal mobility

MARINA FOMINA, EUAN P. BURFORD
AND GEOFFREY M. GADD

Introduction

Fungi are chemoheterotrophic organisms, ubiquitous in subaerial and subsoil environments, and important as decomposers, animal and plant symbionts and pathogens, and spoilage organisms of natural and man-made materials (Gadd, 1993, 1999; Burford et al., 2003a). A fungal role in biogeochemical cycling of the elements (e.g. C, N, P, S, metals) is obvious and interlinked with the ability to adopt a variety of growth, metabolic and morphological strategies, their adaptive capabilities to environmental extremes and their symbiotic associations with animals, plants, algae and cyanobacteria (Burford et al., 2003a; Braissant et al., 2004; Gadd, 2004). Fungal polymorphism and reproduction by spores underpin successful colonization of different environments. Most fungi exhibit a filamentous growth habit, which provides an ability for adoption of either explorative or exploitative growth strategies, and the formation of linear organs of aggregated hyphae for protected fungal translocation (see Fomina et al., 2005b). Some fungi are polymorphic, occurring as both filamentous mycelium and unicellular yeasts or yeast-like cells, e.g. black meristematic or microcolonial fungi colonizing rocks (Sterflinger, 2000; Gorbushina et al., 2002, 2003). Fungi can also grow inside their own parental hyphae utilizing dead parts of the colony under the protection of parental cell walls (Gorbushina et al., 2003). The ability of fungi to translocate nutrients through the mycelial network is another important feature for exploring heterogeneous environments (Jacobs et al., 2002, 2004; Lindahl & Olsson, 2004).

The earliest fossil record of fungi in terrestrial ecosystems occurred during the Ordovician period (480 to 460 MYBP) (Heckman et al., 2001). Since that time fungi have been ubiquitous components of the microbial

Fungi in Biogeochemical Cycles, ed. G. M. Gadd. Published by Cambridge University Press. © British Mycological Society 2006.

communities of any terrestrial environment, including such hostile habitats as the Arctic, hot deserts and metal-rich and hypersaline soils (see Burford *et al.*, 2003a). The ability of many fungi to grow oligotrophically by scavenging nutrients from the air and rainwater helps them survive on stone and rock surfaces which are considered to be an inhospitable environment (Wainwright *et al.*, 1993). In addition, organic and inorganic residues on mineral surfaces or within cracks and fissures, waste products of other microorganisms, decaying plants and animals, dust particles, aerosols and animal faeces can also act as nutrient sources in the subaerial rock environment (Sterflinger, 2000). Inhabitants of subaerial surfaces include poikilotrophic fungi, which are able to deal with varying extremes in microclimatic conditions, e.g. irradiation, salinity, pH and water potential, and protect themselves by the presence of antioxidant protectors such as melanins and mycosporines in their cell walls and by embedding colonies in mucilaginous polysaccharide that often includes clay particles (Gorbushina *et al.*, 2003; Volkmann *et al.*, 2003). One of the most successful means for fungi to survive in the extreme subaerial environment is underpinned by their symbiotic associations with algae and cyanobacteria as lichens where the phototrophs provide a source of carbon and some protection from light and irradiation (Gorbushina *et al.*, 1993; Sterflinger, 2000). As will be further discussed in this chapter, fungi are able to weather a wide range of rocks (see Burford *et al.*, 2003a). In fact, bioweathering of basaltic outcrops by fungal communities in subpolar areas (Iceland) is believed to be chronologically the first process of weathering, which is followed by subsequent cryogenic processes (Etienne & Dupont, 2002).

The majority of fungi inhabit soil environments that are seemingly much more hospitable compared to bare rock surfaces. Fungal communities in soil are diverse and include free-living and symbiotic fungi, as well as plant and animal pathogens and unicellular yeasts. A particularly important fungal group in biogeochemical cycling of elements are mycorrhizal fungi that form symbiotic associations with plants, obtain carbon from the host plant and provide the plant hosts with phosphorus, nitrogen and other elements by increasing the nutrient absorbing area and by dissolution of weatherable minerals in the soil (Paris *et al.*, 1995; Hoffland *et al.*, 2002; Fomina *et al.*, 2004, 2005a). Because bacteria are prolific under both aerobic and anaerobic conditions and also play a significant role in the cycling of elements, previous geomicrobiological studies have generally focused on prokaryotes (Sterflinger, 2000; Burford *et al.*, 2003a). However, although fungi are chemoheterotrophic and predominantly aerobic, there is some evidence that several fungi can dissolve minerals and mobilize metals at higher pH values, and over a wider redox range, faster and more efficiently than bacteria (Gu *et al.*, 1998; Castro *et al.*, 2000; Burford *et al.*, 2003a).

Mechanisms of fungal weathering

Fungi grow in a specific microenvironment where the organisms, their exopolysaccharide slime, solid adsorbents and organic and inorganic surfaces all interact with each other. All the processes comprising fungal weathering of minerals such as dissolution, selective transport, diffusion and recrystallization of mobilized cations occur within that microenvironment (Burgstaller & Schinner, 1993; Banfield & Nealson, 1997; Fomina *et al.*, 2005b, c) (Fig. 10.1a, b). Two synergic actions by which fungi degrade mineral substrates are biomechanical and biochemical (Burford *et al.*, 2003a).

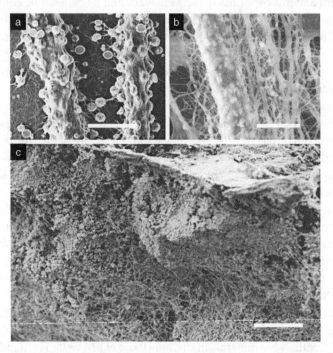

Fig. 10.1. (a, b) Environmental scanning electron microscope (ESEM) images of moolooite precipitation by *Beauveria caledonica* hyphae and cords on copper phosphate-containing ($\equiv$ 5 mM) agar medium: (a) high-vacuum mode ESEM image of air-dried Au/Pd-coated samples (b) wet mode ESEM image of hydrated samples revealing a thick mucilaginous sheath covering the mycelium and crystals (see Fomina *et al.*, 2005c). (c) High-vacuum mode ESEM image of extensive growth of fungi within a crack in a cleaved sample of limestone in the area of the weathering rind (Burford & Gadd, unpublished). Scale bars = (a) 200 μm; (b) 50 μm; (c) 100 μm.

Biomechanical weathering

Biomechanical weathering of minerals by fungi can be direct and indirect. Direct biomechanical degradation is associated with the invasive abilities of fungi (see Burford *et al.*, 2003a; Money, 2004). It occurs through extensive penetration by hyphae into decayed rocks and by burrowing into otherwise intact mineral matter along crystal planes, cleavage, cracks and grain boundaries in sandstone, calcitic and dolomitic rocks (Banfield *et al.*, 1999; Kumar & Kumar, 1999; Sterflinger, 2000). Fungal hyphae overcome mechanical resistance through the exertion of mechanical force that derives from osmotically generated turgor pressure within hyphae (Money, 2004). Biomechanical penetration into mineral matter is facilitated by such fungal features as thigmotropism and lubrication with mucilaginous slime, which may contain acidic and metal-chelating metabolites (see Burford *et al.*, 2003a). Thigmotropism or contact guidance is directed fungal growth towards grooves, ridges and pores in solid material that enables hyphae to explore and exploit weaknesses in mineral surfaces (Watts *et al.*, 1998; Fig. 10.1c). The process of fungal penetration into solids is also believed to be facilitated by synthesis of the protective black pigments, melanins, which increase the mechanical strength of cell walls, previously observed for fungal pathogens (Money & Howard, 1996). This is probably why many rock-dwelling fungi are melanized (see Burford *et al.*, 2003a): the significance of melanin in fungal protection from metal toxicity has long been established (Gadd, 1984, 1993).

Indirect biomechanical weathering is particularly associated with the action of hydrated mucilaginous slimes produced by fungi, which underpin fungal biofilm formation and attachment to solid surfaces. Due to shrinking and swelling of biofilms, mechanical pressure to the mineral unit can cause erosion or abrasion (Warscheid & Krumbein, 1994). All the biomechanical processes involved in fungal weathering are strongly connected with biochemical processes and these are believed to be more important in biogeochemical cycling (Kumar & Kumar, 1999; Burford *et al.*, 2003a).

Biochemical weathering

Processes involved in biochemical dissolution of minerals by fungi are often encompassed by the term 'heterotrophic leaching' (Burgstaller & Schinner, 1993: Gadd *et al.*, 2003). The four main mechanisms of solubilization are acidolysis, complexolysis, redoxolysis and metal accumulation by the fungal biomass (Burgstaller & Schinner, 1993). The primary fungal impact on mineral dissolution appears to result from acidolysis and complexolysis. Acidolysis (or proton-promoted dissolution) occurs because fungi acidify their microenvironment as a result of the excretion of protons and organic acids and the formation of carbonic acid resulting from respiratory CO_2 (Burgstaller & Schinner, 1993). Many fungi are able to excrete metabolites that are associated with complexolysis or ligand-promoted dissolution (Burford *et al.*, 2003a). These include carboxylic acids, amino acids, siderophores and phenolic compounds (Manley & Evans, 1986; Muller *et al.*, 1995; Gadd, 1999). Organic acid excretion by fungi is inter-

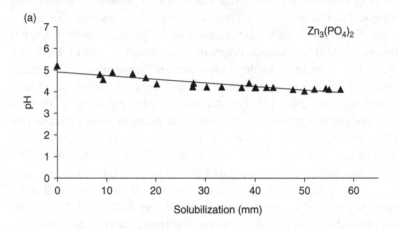

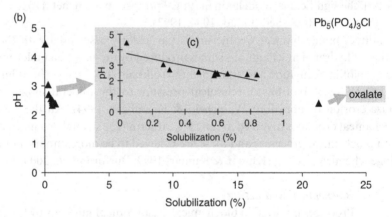

Fig. 10.2. Relationship between values of (a) pH of agar and zinc phosphate ($\equiv$ 5 mM) solubilization measured as clear halo diameters in the agar around colonies of mycorrhizal fungi (see Fomina *et al.*, 2005a); (b) pH of fungal culture filtrates and pyromorphite solubilization measured as the ratio (%) of free metal concentration in the liquid phase to that in the initial solid phase (initial (pyromorphite) $\equiv$ 0.5 mM); (c) as above but excluding oxalate over-excreting *Beauveria caledonica* (see Fomina *et al.*, 2004).

and intraspecific and can be strongly influenced by the presence of toxic metal minerals (Sayer & Gadd, 2001; Fomina *et al*., 2004, 2005c). Fungal-derived carboxylic acids with strong chelating properties (e.g. oxalic and citric) perform a strong aggressive attack on mineral surfaces (Gadd, 1999; Fomina *et al*., 2005c). Moreover, the production of organic acids provides another source of protons. Metal complexation depends on the concentration of anions and cations in the solution, pH and the stability constants of the various complexes (Devevre *et al*., 1996). The significance of proton- versus ligand-promoted dissolution may depend on the mineral, on the metabolism of the fungal cultures and conditions of growth including nutrient availability and carbon and nitrogen sources. It was shown that for the majority of tested ericoid mycorrhizal and ectomycorrhizal fungi grown in the presence of ammonium as a nitrogen source, the main mechanism of dissolution of toxic metal phosphates was acidolysis (Fomina *et al*., 2004, 2005a; Fig. 10.2a). However, if the fungus was capable of excreting large amounts of a strong chelator such as oxalate, the mechanism of dissolution switched to ligand-promoted and became much more efficient than acidolysis, as demonstrated for dissolution of pyromorphite (Fomina *et al*., 2004, 2005c; Fig. 10.2b, c). Metal accumulation by fungal biomass can also play a role in mineral solubilization where the mycelium functions as a sink for mobilized metal cations thereby increasing the efficiency and rate of dissolution (Gadd, 1993, 2000; Sterflinger, 2000).

Metal–fungal interactions

Metals that are essential for fungal growth and metabolism (e.g. Na, K, Cu, Zn, Co, Ca, Mg, Mn, Fe, etc) can all exert toxicity when present above certain threshold concentrations in bioavailable forms (Gadd, 1993). Metals that have no known biological function (e.g. Cd, Hg, Pb) can also be accumulated (Gadd, 1993). Toxicity is greatly affected by the physicochemical nature of the environment and the chemical behaviour of the particular metal species in question. Metals exert toxic effects in many ways, e.g. they can inhibit enzymes, displace or substitute for essential metal ions, cause disruption of membranes and interact with systems that normally protect against the harmful effects of free radicals (Gadd, 1993; Howlett & Avery, 1997). Toxic metals can inhibit growth and spore germination of fungi, affect reproduction and metabolism and reduce the ability of mycorrhizal fungi to colonize roots of host plants (see Fomina *et al*., 2005a). However, many fungi survive and grow in apparently metal-polluted locations and a variety of mechanisms, both active and incidental, contribute to tolerance. Several studies have reported a population shift from bacteria and streptomycetes to fungi in contaminated soils (Khan & Scullion, 2000; Chander *et al.*, 2001). Mechanisms of toxic metal tolerance in fungi include reduction of metal uptake and/or increased efflux, metal immobilization, e.g. cell wall adsorption, extracellular precipitation of secondary minerals, extracellular binding by polysaccharides and extra-cellular metabolites, intracellular sequestration, e.g. metallothioneins and phytochelatins, vacuolar localization (Gadd, 1993; Blaudez *et al.*, 2000; Perotto & Martino, 2001; Baldrian, 2003). Fungal ability to tolerate metals released from metal-containing minerals may be connected with the ability to dissolve those minerals (Fomina *et al.*, 2004, 2005a). For mycorrhizal fungi, it was found that more tolerant fungal strains solubilized toxic metal minerals more efficiently (Fomina *et al.*, 2004, 2005a).

Metal binding, accumulation and precipitation

Toxic metal species can be bound, accumulated and precipitated by fungi. Fungal biomass can function as a metal sink by metal biosorption (cell walls, pigments and extracellular polysaccharides), transport and intracellular accumulation and sequestration, and precipitation of metal compounds onto hyphae. Fungi can be highly efficient bioaccumulators of soluble and particulate forms of metals (e.g., Ni, Zn, Ag, Cu, Cd and Pb), especially from dilute external concentrations (Gadd, 1993, 2000, 2001; Baldrian, 2003). Metal binding can be an important passive process in both living and dead fungal biomass (Gadd, 1990, 1993; Sterflinger, 2000). Metal binding can be influenced by environmental pH, with binding decreasing at low pH for metals such as copper, zinc and cadmium (see Gadd, 1993). The cell wall has a key role in metal accumulation, comprising 38%–77% of copper uptake by wood-rotting fungi (Baldrian, 2003). The presence of melanin and chitin in fungal walls strongly influences the ability of fungi to act as biosorbents (Gadd & Mowll, 1985; Manoli *et al.*, 1997; Fomina & Gadd, 2002b). Gadd and Mowll (1985) reported that melanin-containing chlamydospores of *Aureobasidium pullulans* can absorb three times more copper than hyaline cells. For mycorrhizal fungi, metal-binding properties have been suggested as important factors in ameliorating effects of metal toxicity on the mycorrhizal symbiosis (Galli *et al.*, 1994; Vodnik *et al.*, 1998). Most studies on mycorrhizas indicate that extramatrical mycelium provides major metal-binding sites and that most metals are bound to cell wall components or, in interhyphal spaces, to extracellular polysaccharide (Denny & Wilkins, 1987; Jones & Hutchinson, 1988; Colpaert & Van Assche 1992, 1993; Turnau *et al.*,

1996; Van Tichelen *et al.*, 2001; Adriaensen *et al.*, 2003; Meharg, 2003). Radiotracer flux analyses of cadmium compartmentation in mycelium of the ectomycorrhizal fungus *Paxillus involutus* have shown that the cell wall-bound fraction contained 50% of accumulated cadmium, the cytoplasmic fraction contained 30% and the vacuolar compartment contained 20% of accumulated cadmium (Blaudez *et al.*, 2000). Accumulation of toxic metals by fungal cultures is inter- and intraspecific. It has been reported that lead uptake (^{210}Pb tracer) was higher in *Suillus bovinus* than for other ectomycorrhizal species (*Laccaria laccata, Lactarius piperatus, Pisolithus tinctorius* and *Amanita muscaria*) (Vodnik *et al.*, 1998). *Suillus* isolates were also observed to be efficient lead bioaccumulators when grown on media containing pyromorphite (Fomina *et al.*, 2004). *Suillus granulatus* was reported to accumulate all the cadmium solubilized from rock phosphate (Leyval & Joner, 2001). Investigation of the solubilization of zinc phosphate and pyromorphite by ericoid mycorrhizal and ectomycorrhizal fungi showed that a strain of *Thelephora terrestris* isolated from a metal-polluted environment accumulated the highest concentrations of zinc and lead in both axenic culture and mycorrhizal association (Fomina *et al.*, 2004). This ability of the same strain of *T. terrestris* to concentrate toxic metals could explain the successful protection of host pine (*Pinus sylvestris*) against copper toxicity despite the lack of extramatrical mycelium compared to *S. bovinus* (Van Tichelen *et al.*, 2001). However, such protection seems to be dependent on the fungal strain and its metal tolerance. For example, in a study of zinc toxicity in associations of ectomycorrhizal fungi with *P. sylvestris*, it was reported that another *T. terrestris* strain, much less tolerant to zinc, increased the zinc concentration in its host plant shoots compared to non-mycorrhizal plants and pines infected with other fungi (*Paxillus involutus, Laccaria laccata, S. bovinus, Scleroderma citrinum*) (Colpaert & Van Assche, 1992). It was demonstrated that, in general, zinc-tolerant isolates of ectomycorrhizal fungi accumulated significantly less zinc (including the sum of water-soluble and NaCl-extractable zinc) than non-tolerant strains (Fomina *et al.*, 2004) (Fig. 10.3). This could play a key role in metal tolerance and indicate some kind of 'avoidance' strategy, possibly by decreased uptake and/or enhanced efflux from cells (Gadd, 2000; Gadd & Sayer, 2000; Perotto & Martino, 2001; Meharg, 2003).

Mechanisms for metal immobilization include intracellular uptake with complexation to ligands such as sulphur-containing peptides (metallothioneins) and carboxylic acids (citrate, malate, oxalate) (Gadd, 1993; Sarret *et al.*, 1998, 2002; Fomina *et al.*, 2005c). Some fungi can also precipitate metals in amorphous and crystalline forms, e.g. oxalate salts and other secondary mycogenic minerals (Gadd, 1999; Burford *et al.*, 2003a). Powder X-ray diffraction (XRD) and extended X-ray absorption spectroscopy studies (EXAFS) of lichens grown in areas heavily contaminated with lead and zinc revealed that lead and zinc were coordinated by oxalates within biomass of *Diploschistes muscorum* whereas in *Xantorina parietina*, lead was complexed to carboxylic groups of a polyphenolic compound, parientinic acid, located in the fungal cell walls (Sarret *et al.*, 1998). Synchrotron-based X-ray microfluorescence and microEXAFS studies on zinc speciation in the zinc hyperaccumulating plant *Arabidopsis halleri* showed that zinc was distributed in zinc malate, zinc citrate and zinc phosphate (Sarret *et al.*, 2002). Because of the amorphous state or poor crystallization of such complexes and relatively low metal concentrations, determination of metal speciation in biological systems remains a challenging problem.

X-ray absorption spectroscopy has revealed that oxygen ligands play a major role in metal coordination within *Beauveria caledonica* mycelium, being carboxylic groups in copper phosphate-containing medium, and phosphate groups in pyromorphite-containing medium (Fomina *et al.*, 2005c).

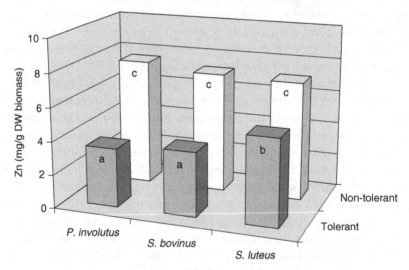

Fig. 10.3. Zinc accumulation by biomass of zinc-tolerant and non-tolerant strains of the ectomycorrhizal fungi *Paxillus involutus, Suillus bovinus* and *Suillus luteus* grown on liquid medium containing 1 mM zinc phosphate. The different letters denote significant differences at the 5% level using Fisher's LSD test (one-way ANOVA) (see Fomina *et al.*, 2004). DW denotes dry weight

Other experiments with axenic cultures of the ectomycorrhizal fungus *Rhizopogon rubescens* grown on media containing copper phosphate have shown that metal speciation within the mycelium may depend on the nitrogen source with mixed carboxylate/phosphate coordination of copper for ammonium and copper oxalate coordination for nitrate (Fomina & Gadd, unpublished; Fig. 10.4a, b). Further, our studies on *P. involutus/ P. sylvestris* ectomycorrhiza grown in mesocosms with zinc phosphate showed that ectomycorrhizal infection decreased the ratio of the most bioavailable zinc (water-soluble and salt-extractable) in the root biomass compared to non-mycorrhizal seedlings. However, the speciation of zinc within root biomass, as revealed by EXAFS, was similar in both ectomycorrhizal and non-mycorrhizal seedlings and showed octahedral coordination of zinc by oxygen-containing ligands fitting carboxylate coordination and, in part, phosphate coordination (Fomina & Gadd, unpublished; Fig. 10.4c).

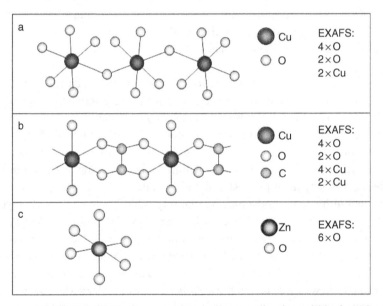

Fig. 10.4. Models for copper and zinc coordination within fungal and mycorrhizal biomass resulting from EXAFS analysis: (a, b) copper coordination by oxygen within the mycelium of *Rhizopogon rubescens* grown on medium containing copper phosphate ($\equiv$ 5 mM) (a) fitting a carboxylate/phosphate ligand model for ammonium-containing medium and (b) fitting an oxalate ligand model for nitrate-containing medium; (c) zinc coordination with six oxygens within *Paxillus involutus/Pinus sylvestris* ectomycorrhiza grown on zinc phosphate (0.5 g/mesocosm) suggesting binding by carboxylate and phosphate ligands (Fomina & Gadd, unpublished).

Formation of mycogenic minerals

Free-living and symbiotic fungi play an important role in mineral formation through precipitation of organic and inorganic secondary minerals and through nucleation and deposition of crystalline material on and within cell walls, e.g. oxalates and carbonates (Arnott, 1995; Ehrlich, 1998; Gadd, 1999; Urzi et al., 1999; Gorbushina et al., 2002; Burford et al., 2003a, b). The presence of calcified fungal filaments in limestone and calcareous soils has been widely reported (Bruand & Duval, 1999; Verrecchia, 2000; Burford et al., 2003a, b). Apart from calcium carbonate and oxalate, a variety of other mycogenic secondary minerals can be associated with fungal hyphae and lichen thalli, e.g. birnessite, desert varnish (MnO and FeO), ferrihydrite, iron gluconate, calcium formate, forsterite, goethite, halloysite, hydroserussite, todorokite, and moolooite (Grote & Krumbein, 1992; Hirsch et al., 1995; Verrecchia, 2000; Gorbushina et al., 2001, Arocena et al., 2003; Burford et al., 2003b). Precipitation, including crystallization, will immobilize metals and limit bioavailability, as well as perhaps lead to release of nutrients like sulphate and phosphate (Gadd, 2000).

Oxalates The most common form of oxalate in nature is calcium oxalate, which occurs as the dihydrate (weddelite) or the more stable monohydrate (whewellite) (Arnott, 1995; Burford et al., 2003a). Calcium oxalate crystals are associated with free-living, pathogenic and plant symbiotic fungi and are formed by reprecipitation of solubilized calcium as calcium oxalate (Arnott, 1995). The formation of calcium oxalate by fungi has a profound effect on geochemical processes in soils, acting as a reservoir for calcium but also influencing phosphate availability (Gadd, 1993; Gadd & Sayer, 2000). Fungi can also produce other metal oxalates with a variety of different metals and metal-bearing minerals (e.g. Cd, Co, Cu, Mn, Sr, Zn, Ni and Pb) (Sayer et al., 1997; 1999; Gadd, 2000; Magyarosy et al., 2002; Fomina et al., 2005c; Fig. 10.5a–c). The highly metal-tolerant fungus *B. caledonica* was able to solubilize cadmium, copper, lead and zinc minerals, converting them into oxalates in the local microenvironment but also in association with the mycelium (Fomina et al., 2005c). When grown on copper phosphate-containing medium, this fungus accumulated approximately 100 mg copper (g dry wt)$^{-1}$ with approximately 35% comprising copper oxalate (moolooite) crystals (Fig. 10.1a, b). The formation of metal oxalates may provide a mechanism whereby fungi can tolerate high concentrations of toxic metals. A similar mechanism may occur in lichens growing on copper sulphide-bearing rocks where precipitation of copper oxalate occurs within the thallus (Arnott, 1995; Easton, 1997). Many fungal species have the ability to produce and exude low-molecular-weight organic acids with strong chelating properties (Dutton & Evans, 1996; Gadd, 1999; Martino *et al.*, 2003). Oxalate excretion has been reported to be induced or enhanced by NO_3^- in contrast to NH_4^+, and also the presence of HCO_3^-, Ca^{2+} and some toxic metals (Cu, Al) or minerals (pyromorphite, zinc phosphate) (Lapeyrie *et al.*, 1987, 1991; Wenzel *et al.*, 1994; Gharieb & Gadd, 1999; Whitelaw *et al.*, 1999; Ahonen-Jonnarth *et al.*, 2000; Van Leerdam *et al.*, 2001; Arvieu *et al.*, 2003; Casarin *et al.*, 2003; Clausen & Green, 2003; Fomina *et al.*, 2004). Calcium oxalate crystals have commonly been associated with ectomycorrhizal fungi and ectomycorrhiza (Cromack *et al.*, 1979; Lapeyrie *et al.*, 1990). However, it has been

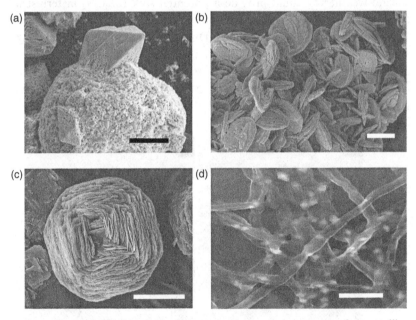

Fig. 10.5. Scanning electron microscope images of crystalline precipitation of toxic metals by fungi in medium and mycelium: (a) transformation of lead carbonate ($\equiv 15$ mM) into lead oxalate by *Beauveria caledonica* (see Fomina *et al.*, 2005c); (b) zinc oxalate crystals formed by ectomycorrhizal fungus *Rhizopogon roseolus* in axenic culture grown on zinc phosphate ($\equiv 5$ mM); (c) copper oxalate crystals formed by the ectomycorrhizal fungus *Rhizopogon rubescens* in axenic culture grown on copper phosphate ($\equiv 5$ mM); (d) zinc-containing crystals precipitated within extramatrical mycelium of *R. rubescens/Pinus sylvestris* ectomycorrhiza grown on zinc phosphate-containing ($\equiv 5$ mM) medium (Fomina & Gadd, unpublished). Scale bars = (a) 50 μm; (b, c) 20 μm; (d) 10 μm.

observed that oxalate-excreting strains of ectomycorrhizal fungi *R. rubescens, Rhizopogon roseolus* and *Suillus collinitus* are able to precipitate zinc oxalate when grown on NO_3^--containing medium in the presence of zinc phosphate, and copper oxalate (moolooite) in the presence of copper phosphate (Fomina & Gadd, unpublished) (Fig. 10.5b, c). Moreover, zinc-containing crystalline precipitates, presumably zinc oxalate, have been found within the extraradical mycelium of the ectomycorrhizal *R. rubescens/P. sylvestris* association grown in mesocosms with zinc phosphate (Fomina & Gadd, unpublished) (Fig. 10.5d).

Carbonates The precipitation of carbonates by microorganisms is widespread suggesting that microbial carbonate precipitation coupled with silicate weathering could provide a potential sink for CO_2 in terrestrial environments (Thompson & Ferris, 1990; Verrecchia *et al.* 1990; Rivadeneyra *et al.* 1993; Bruand & Duval, 1999; Folk & Chafetz, 2000; Fujita *et al.* 2000; Merz-Preiβ, 2000; Riding, 2000; von Knorre & Krumbein, 2000; Warren *et al.* 2001; Hammes & Verstraete, 2002). Manoli *et al.* (1997) demonstrated that chitin is a substrate on which calcite will readily nucleate and grow. Fungal filaments mineralized with calcite

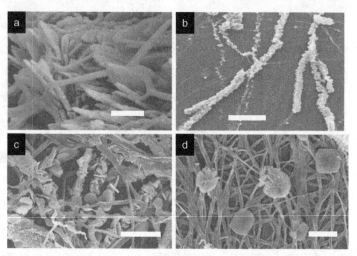

Fig. 10.6. Secondary carbonate precipitation: (a, b) calcite and whewellite precipitation by hyphae of (a) *Penicillium corylophilum* and (b) *Serpula himantioides* grown on limestone; (c, d) glushinskite ($Mg(C_2O_4).2H_2O$) and hydromagnesite ($Mg_5(CO_3)_4(OH)_2.4H_2O$) formed on mycelium of *Penicillium simplicissimum* grown on hydromagnesite and combinations of hydromagnesite with hematite, magnetite, cuprite and illite. Scale bars = (a) 2 μm; (b) 50 μm; (c, d) 20 μm. (Burford & Gadd, unpublished).

($CaCO_3$), together with whewellite (calcium oxalate monohydrate, $CaC_2O_4.H_2O$), have been reported in limestone and calcareous soils from a range of localities (Kahle, 1977; Klappa, 1979; Calvet, 1982; Callot *et al.*, 1985a, b; Verrecchia *et al.*, 1990; Monger & Adams, 1996; Bruand & Duval, 1999). In addition, near-surface limestones (calcretes), calcic and petrocalcic horizons in soils are often secondarily cemented and indurated with calcite and whewellite implying that fungi play a prominent role in the transformation and subsequent stabilization of limestone substrates (Verrecchia, 2000). It has been shown experimentally that fungi are able to reprecipitate secondary carbonates and mineralize hyphae with both calcite and whewellite or with glushinskite ($MgC_2O_4.2H_2O$) and hydromagnesite ($Mg_5(CO_3)_4(OH)_2.4H_2O$) (Burford *et al.*, 2003b) (Fig. 10.6). Calcite formation may occur through indirect processes via fungal excretion of oxalic acid and the precipitation of calcium oxalate (Verrecchia *et al.*, 1990; Gadd, 1999; Verrecchia, 2000). Oxalic acid excretion and the formation of calcium oxalate, results in dissolution of the internal pore walls of the limestone matrix so that the solution becomes enriched in free carbonate. During passage of the solution through the pore walls, calcium carbonate recrystallizes as a result of a decrease in CO_2, and this contributes to hardening of the material. Biodegradation of oxalate as a result of microbial activity can also lead to transformation into carbonate, resulting in precipitation of calcite in the pore interior, leading to closure of the pore system and hardening of the chalky parent material. Decomposition of fungal hyphae releases calcite crystals, which act as sites of further secondary calcite precipitation (Verrecchia, 2000).

Reduction of metal(loid)s

Many fungi can precipitate reduced forms of metals and metalloids (e.g. elemental silver, selenium, tellurium) within and around fungal cells (Gadd, 2004). The reductive ability of fungi is manifest by the appearance of black coloration of fungal colonies precipitating elemental silver or tellurium, and a red coloration for precipitation of elemental selenium (Gharieb *et al.*, 1999).

Fungi and silicates

Silicate and clay dissolution and transformation by fungi

The silicates are the largest class of minerals comprising 30% of all minerals and making up 90% of the Earth's crust (Ehrlich, 1996). Microorganisms including fungi play fundamental roles in the dissolution of silicate structures in the rock weathering process, in the genesis of clay minerals and soil and sediment formation (Banfield et al., 1999; Styriakova & Styriak, 2000; Bennett et al., 2001). Fungal dissolution of silicate minerals can occur biochemically by the excretion of acidic metabolites and chelators like oxalate and citrate (Webley et al., 1963; Cromack et al., 1979; de la Torre et al., 1993; Castro et al., 2000). Fungi (from *Aspergillus* and *Penicillium* genera) were demonstrated to leach metals from different silicates faster than bacteria (Castro et al., 2000; Burford et al., 2003a). In nature, the appearance of a secondary 2:1 clay mineral is a typical symptom of biogeochemically weathered silicate minerals in soils and rocks and has been observed for symbiotic fungi (lichens and ectomycorrhizas) (Barker & Banfield, 1996, 1998; Rodriguez Navarro et al., 1997; Arocena et al., 1999, 2003). The fungal partner is also reported to be involved in the formation of secondary silicates, e.g. opal and forsterite, in lichen thalli (Gorbushina et al., 2001). Following a series of cyclical biogeochemical transformations of silicon in nature, clay minerals can also be modified by fungi (Ehrlich, 1996; see Burford et al., 2003a). Fungi producing large amounts of oxalic, citric and gluconic acids (e.g. *Aspergillus niger, Penicillium frequentans, Hysterangium crassum*) were able to weather clay silicates and precipitate mobilized cations into organic salts such as calcium, magnesium and ferric oxalates and calcium citrates (Henderson & Duff, 1963; Cromack et al., 1979; de la Torre et al., 1993; Mandal et al., 2002).

Interactions between fungi and clay minerals

Clay minerals generally form the highest proportion of soil constituents and make a significant contribution to soil chemical and physical properties, e.g. ion-exchange capacity and buffering and adsorption powers (Burford et al., 2003a). Clay minerals can alter microbial growth and activity and, depending on conditions and the type of clay, may stimulate or inhibit fungal metabolism (Stotzky, 1966, 2000; Lee & Stotzky, 1999; Lotareva & Prozorov, 2000; Fomina & Gadd, 2002a). Stimulatory effects are thought to be a result of clays functioning as (1) pH buffers, (2) a source of inorganic nutrients through their cation-exchange capacity, (3) specific adsorbents of one or more metabolic inhibitors (as well as nutrients and growth stimulators) and (4) modifiers of the microbial microenvironment because of their surface area and adsorptive capacity that creates microzones with altered microbial growth conditions (Stotzky, 1966; Martin et al., 1976; Fletcher, 1987; Marshall, 1988). Clay minerals can affect the formation of fungal pellets and serve as a mineral

matrix for the colony (Fomina & Gadd, 2002a; Fig. 10.7). The presence of
clay within fungal microcolonies may influence the diffusion of nutrients
and oxygen and remove toxic metabolites (Fomina & Gadd, 2002a). It is
commonly assumed that clay minerals provide both physical and chemical
protection for fungi (Babich & Stotzky, 1977; McEldowney & Fletcher,
1986; Pereira *et al.*, 2000; Vettori *et al.*, 2000).

Fungal–clay mineral interactions also play an important role in soil
evolution, aggregation and stabilization (see Burford *et al.*, 2003a).
Fungi entangle soil particles in their hyphae forming stable microaggregates
and also take part in polysaccharide aggregation (Chantigny *et al.*, 1997;
Tisdall *et al.*, 1997; Puget *et al.*, 1999). Interactions between hyphae
and solid particles are subject to a complex of forces of both a physicochemical (electrostatic, ionic, hydrophobic effects etc.) and biological
nature (chemotaxis, production of specific enzymes, polysaccharides, lec-
tins and other adhesins, etc.) (Lunsdorf *et al.*, 2000). Direct biophysical
effects were shown in the microscale where, due to surface charge phenomena, fungal hyphae acted as nucleation zones and attracted and oriented
clay platelets resulting in narrow clay-lined channels (Ritz & Young,
2004). Interactions between clay minerals and fungi alter the adsorptive
properties of both clays and hyphae (Fomina & Gadd, 2002b). Under
certain conditions, sorption abilities of fungal–clay aggregates can be
decreased due to blocking and masking of binding sites or increased due
to modification of binding sites and emergence of new ones and this may
have further implications for the fate of toxic metals in soil (Fomina &
Gadd, 2002b).

Fig. 10.7. Scanning electron microscope images of fungal–clay interactions
within fungal microcolonies grown in the presence of 0.5% (w/v)
bentonite. (a) Fractured pellet of *Cladosporium cladosporioides* showing a
dense clay matrix in the central zone of microcolony (scale bar = 100 µm);
(b) surface hyphae of the *C. cladosporioides* microcolony covered with clay
particles (scale bar = 10 µm); (c) hyphae and clay matrix in the central zone
of a fractured pellet of *Humicola grisea* (scale bar = 10 µm) (see Fomina &
Gadd, 2002a)

Environmental biotechnology

Biogeochemical activity of fungi in the rhizosphere

Rocks and minerals represent a vast reservoir of elements, many of which are essential to life and must be released into bioavailable forms (Gadd, 2004). Free-living and symbiotic microbial populations associated with plant roots significantly alter the physicochemical characteristics of the rhizosphere, which may have significant consequences for the biogeochemical mobility of metals and associated elements (Wenzel *et al.*, 1994; Olsson & Wallander, 1998; Whitelaw *et al.*, 1999). In podzol E horizons under European coniferous forests, the weathering of hornblendes, feldspars and granitic bedrock has been attributed to exudation of oxalic, citric, succinic, formic and malic acids by saprotrophic and mycorrhizal hyphae. Oxalic acid concentrations in forest soil solution were correlated with hyphal length (van Hees *et al.*, 2003). Fungal hyphal tips produced micro- to millimolar concentrations of these acids that effectively dissolved calcium-rich plagioclase feldspars at rates of 0.3 to $30\,\mu m\ yr^{-1}$ (Jongmans *et al.*, 1997). The importance of mycorrhizas in plant phosphorus nutrition has been appreciated for a long time and their ability to dissolve and transform calcium-containing insoluble compounds and minerals (calcium phosphates, carbonate and sulphate) in pure culture and in mycorrhizal association has been widely studied (Callot *et al.*, 1985a, b; Lapeyrie *et al.*, 1990, 1991; Gharieb & Gadd, 1999). However, toxic metal mineral solubilization has received little attention. The ectomycorrhizal fungi *Suillus granulatus* and *Pisolithus tinctorius* can promote the release of cadmium and phosphorus from rock phosphate (Leyval & Joner, 2001) while the ericoid mycorrhizal fungus *Oidiodendron maius* can solubilize zinc oxide and phosphate (Martino *et al.*, 2003). Our experimental studies on ericoid mycorrhizal and ectomycorrhizal fungi showed that many tested cultures were able to solubilize zinc, cadmium, copper phosphates and lead chlorophosphate (pyromorphite), releasing phosphate and metals (Fomina *et al.*, 2004, 2005a). Both non-mycorrhizal Scots pines (*P. sylvestris*) and pines infected with the ectomycorrhizal fungus *P. involutus* were able to enhance zinc phosphate dissolution, withstand metal toxicity and acquire the mobilized phosphorus, increasing the phosphorus amount in shoots when zinc phosphate was present in the growth matrix (Fomina & Gadd, unpublished). In fact, trees have been suggested to respond to phosphorus limitation by increased exploitation of phosphorus-containing minerals by ectomycorrhizas (Hagerberg *et al.*, 2003). It can be concluded that soil biogeochemical activities of fungi have a significant impact on mineral nutrition of plants and other organisms (Fig. 10.8).

Deterioration of rock and building stone

Fungi are ubiquitous components of the microflora of all rocks and building stones. They have been reported from a wide range of rock types including limestone, soapstone, marble, granite, sandstone, andesite, basalt, gneiss, dolerite, amphibolite and quartz (Burford *et al.*, 2003a). Fungi often comprise a significant component of cryptoendolithic (i.e. actively penetrating the rock matrix to several millimetres in depth) and chasmolithic or endolithic (i.e. living in hollows, cracks and fissures within rocks) microbial communities inhabiting subaerial microenvironments (Staley *et al.*, 1982; Gorbushina *et al.*, 1993; Gerrath *et al.*, 1995; Kumar & Kumar, 1999; Sterflinger, 2000; Burford *et al.*, 2003b; Fig. 10.1c). They can exist in rocks as free-living fungi or as lichens. Among the most commonly reported fungal groups inhabiting exposed rock surfaces are

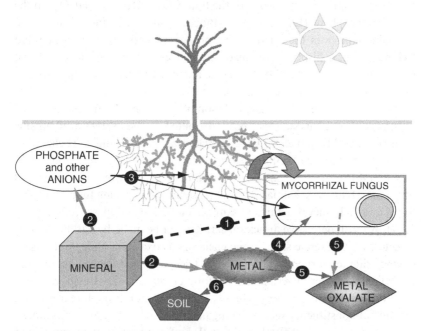

Fig. 10.8. Simple biogeochemical model for metal mineral transformations in the mycorhizosphere (the roles of the plant and other microorganisms contributing to the overall process are not shown). (1) Proton-promoted (proton pump, cation–anion antiport, organic anion efflux, dissociation of organic acids) and ligand-promoted (e.g. organic acids) dissolution of metal minerals. (2) Release of anionic (e.g. phosphate) nutrients and metal cations. (3) Nutrient uptake. (4) Intra- and extracellular sequestration of toxic metals: biosorption, transport, compartmentation, precipitation etc. (5) Immobilization of metals as oxalates. (6) Binding of soluble metal species to soil constituents, e.g. clay minerals, metal oxides, humic substances.

microcolonial fungi (or black yeasts or yeast-like black meristematic fungi) (Staley *et al.*, 1982; Gorbushina *et al.*, 1993; Wollenzien *et al.*, 1995; Sterflinger, 2000). The biofilms formed by fungi and other micro-organisms on and in rocks are believed to be major factors in rock decay, and involved in the formation of patinas, films, varnishes, crusts and stroma-tolites in rock substrates (Gorbushina & Krumbein, 2000). Alkaline (basic) rocks are generally more susceptible to fungal attack than acidic rocks (Eckhardt, 1985; Kumar & Kumar, 1999). In limestone, dolomite and marble, fungi are considered to be important agents of carbonate mineral deterioration through a combination of physical and chemical processes (Ehrlich, 1996; Kumar & Kumar, 1999). Carbonates (calcium and calcium/ magnesium carbonates) occur extensively as limestone and dolomite, cover-ing around 13% of the total land surface and serving as an important carbon reservoir comprising up to 80% of the total CO_2, CO_3^{2-} and HCO_3^- in the Earth's lithosphere (Ehrlich, 1996). Biochemical weathering of rocks can result in changes in the microtopography of minerals through pitting and etching, mineral displacement reactions and even complete dissolution of mineral grains (Ehrlich, 1998; Kumar & Kumar, 1999). Cavities in limestone provide a major habitat for fungi particularly in extreme environments like cold deserts (Ehrlich, 1996). Degradation of sandstones by fungi is also well documented and this has also been attributed to the excretion of organic acids (Gomes-Alarcon *et al.*, 1994; Hirsch *et al.*, 1995). Fungi are believed to contribute to the weathering of silicate-bearing rocks (e.g. mica and ortho-clase), and iron- and magnesium-bearing minerals (e.g. biotite, olivine and pyroxene) (Kumar & Kumar, 1999). Siderophore-producing fungi were able to pit and etch microfractures in samples of olivine and glass (Callot *et al.*, 1987). A most aggressive attack on silicates can be made by various hypho-mycetes, which excrete such strong chelators as oxalate and degrade basalt, olivine, dunite, serpentine, muscovite, feldspar, spodumene, kaolin and nepheline (see Sterflinger, 2000; Burford *et al.*, 2003a). Fungi can also deteriorate rock surfaces (e.g. siderite, rhodochrosite) through redox attack of mineral constituents such as manganese and iron (Grote & Krumbein, 1992). The oxidation of Fe(II) and Mn(II) by fungi leads to the formation of dark patinas on glass surfaces (Eckhardt, 1985). It appears that most types of natural and man-made rock, stone and other mineral materials can be susceptible to fungal deterioration.

Concrete biodeterioration in radioactive-waste disposal

Safe long-term storage of nuclear waste is of importance in pro-tecting the environment. Cement and concrete are used as barriers in all kinds of nuclear-waste repositories. Despite the theoretical service life of concrete reaching up to one million years, biological corrosion is an important factor to take into account. All types of building and ceramic materials, concrete and cement are deteriorated by microorganisms (Diercks *et al.* 1991; Gaylarde & Morton, 1999; Kikuchi & Sreekumari, 2002; Roberts *et al.*, 2002). In some environments, fungi may dominate the microflora and be the main cause of corrosion. The potential myco-corrosion problem for metal containers selected for storage of nuclear waste in terrestrial environments has been stressed (Geesey, 1993). Fungal attack on concrete can be strongly and mildly aggressive, caused by protons and organic acids and production of hydrophilic slimes leading to biochemical and biophysical/biomechanical deterioration (Sand & Bock, 1991a, b). Fungi include desiccant-resistant species and many can grow on traces of nutrients. *Fusarium* sp., *Penicillium* sp. and *Hormoconis* sp. can degrade hydrocarbon-based lubricants and produce organic acids that cause local-ized corrosion of post-tension structures used in buildings, bridges and nuclear power plants (Little & Staehle, 2001). Many studies have indicated that fungi play an important role in the deterioration of concrete (Perfettini *et al.* 1991; Gu *et al.*, 1998; Nica *et al.*, 2000). Fungal degrada-tion proceeded more rapidly than bacterium-mediated degradation with complexolysis suggested as the main mechanism of calcium mobilization (Gu *et al.*, 1998). Several fungi exhibit very high levels of radiation resis-tance and can survive and colonize concrete barriers even under the severe radioactive contamination that occurred after the Chernobyl accident in 1986 (Zhdanova *et al.*, 2000). The high radiation selection pressure inside Reactor No. 4 of the Chernobyl nuclear power plant (radiation $\alpha = 500$ Bq cm^{-2}, $\beta = 20\,000$ Bq cm^{-2}, $\gamma = 700$ mR h^{-1}) led to genetic adaptation of fungal strains inhabiting the concrete surfaces (Mironenko *et al.*, 2000). The most frequently isolated microfungi in the first years after the accident and later in the habitats with severe radiation were melanized species of *Alternaria, Cladosporium* and *Aureobasidium* (Zhdanova *et al.*, 2000). It was also shown that microfungi from the genera *Aspergillus, Alternaria* and *Cladosporium* were able to colonize the samples of the concrete used as the radioactive-waste barrier and leached iron, aluminium, silicon and calcium, and reprecipited silicon and calcium oxalate in their microenviron-ment (Fomina *et al.*, 2005d; Fig. 10.9).

Bioremediation

Fungal involvement in element cycling at local and global scales has important implications for living organisms, plant production and human health. There is an increasing risk from toxic metals in natural, industrial and agricultural soils. The ability of fungi to solubilize, immobilize and transform toxic metals and metalloids provides the potential to treat contaminated land and substrates (Thomson-Eagle & Frankenberger, 1992; Gadd, 2000, 2001, 2004; Hochella, 2002). Solubilization provides a route for removal of metals from industrial wastes and by-products, low-grade ores and metal-bearing minerals that is relevant to bioremediation of soil matrices and solid wastes, and metal recovery and recycling (Burgstaller & Schinner, 1993; Gadd, 2000; Gadd & Sayer, 2000; Brandl, 2001). Immobilization processes enable metals to be transformed into chemically more inert forms (Gadd, 2000). Fungi and their by-products have received considerable previous attention as biosorbent materials for metals and radionuclides (Gadd, 1990).

Increasing attention has been given to the potential of mycorrhizal associations for reforestation and clean-up of metal-contaminated areas (Van der Lelie *et al.* 2001). Mycorrhizas may enhance metal phytoextraction by host plants by increasing plant biomass and increasing metal

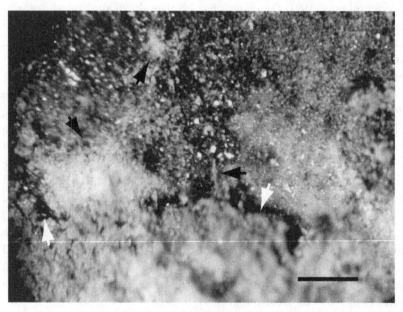

Fig. 10.9. Mycodeterioration of barrier concrete by *Aspergillus niger* in a microcosm: cotton-wool-like microcolonies of fungus (black arrows) arising after one year on a sample of barrier concrete and associated with microcracks on the concrete surface (white arrows) (Fomina, unpublished). Scale bar = 1 mm.

accumulation by the plants. However, the potential impact of mycorrhizal fungi on bioremediation may be conditional and dependent on the metal tolerance of fungal strains, their mycorrhizal status as well as the nutritional status of contaminated soils.

Conclusions

Fungi are ubiquitous members of subaerial and subsoil environments and often become a dominant group in metal-rich or metal-polluted habitats. Their ability for oligotrophic growth, their explorative filamentous growth habit, flexible polymorphic growth strategies and resistance to extreme environmental factors including metal toxicity, irradiation and desiccation make them successful colonizers of rock surfaces and other metal-rich habitats. Fungal weathering of rocks and minerals occurs through biomechanical and biochemical attack. The main mechanisms of biochemical weathering are proton-promoted, ligand-promoted and metal accumulation ('sinking') within fungal biomass, and fungal impact on dissolution depends on species, mineral types and conditions. Despite proton-promoted mechanisms being the most common mechanism for the majority of fungi, ligand-promoted dissolution of minerals by strong chelators such as oxalic and citric acids may provide a more efficient attack. One of the paramount reasons for fungal success in biogeochemical processes in soil and on rock surfaces is their ability to form symbiotic associations with photosynthetic plants, algae and cyanobacteria (mycorrhizas and lichens), which makes them responsible for major transformations and redistribution of inorganic nutrients, e.g. metals and phosphate, as well as carbon flow. Fungi are also major agents of the biodeterioration of stone, wood, plaster, cement, concrete and other building materials, and are important components of rock-inhabiting microbial communities with significant roles in mineral dissolution and secondary mineral formation.

Acknowledgements

The authors' research described within was funded by the BBSRC/BIRE programme (94/BRE13640), and CCLRC Daresbury SRS (SRS Grant 40107), which is gratefully acknowledged. EPB gratefully acknowledges receipt of a NERC postgraduate research studentship. We thank Professor Andy Meharg and Dr David Genney (University of Aberdeen, Scotland), Dr Elena Martino (University of Torino, Italy), Dr Derek Mitchell (University College Dublin, Ireland), Dr Jan Colpaert and Ms Kristin Adriaensen (Limburgs University Centre, Belgium), Dr Claude Plassard (INRA, France) and Dr Håkan Wallander (Lund

258   *M. Fomina* et al.

University, Sweden) for the provision of fungal strains. We are very grateful to Dr J. Charnock, Dr Lorrie Murphy and Dr Bob Bilsborrow (Stations 7.1 and 16.5, CCLRC Daresbury SRS, UK) for their help with X-ray absorption spectroscopy scanning and analysis, and also to Mr Martin Kierans (Centre for High Resolution Imaging and Processing (CHIPs), School of Life Sciences, University of Dundee, Scotland) for assistance with scanning electron microscopy.

References

Adriaensen, K., Van Der Leile, D., Van Laere, A., Vangronsveld, J. & Colpaert, J. V. (2003). A zinc-adapted fungus protects pines from zinc stress. *New Phytologist*, **161**, 549–55.
Ahonen-Jonnarth, U., van Hees, P. A. W., Lundström, U. S. & Finlay, R. D. (2000). Production of organic acids by mycorrhizal and non-mycorrhizal *Pinus sylvestris* L. seedlings exposed to elevated concentrations of aluminium and heavy metals. *New Phytologist*, **146**, 557–67.
Arnott, H. J. (1995). Calcium oxalate in fungi. In *Calcium Oxalate in Biological Systems*, ed. S. R. Khan. Boca Raton: CRC Press, pp. 73–111.
Arocena, J. M., Glowa, K. R., Massicotte, H. B. & Lavkulich, L. (1999). Chemical and mineral composition of ectomycorrhizosphere soils of subalpine fir (*Abies lasiocarpa* (Hook.) Nutt.) in the AE horizon of a Luvisol. *Canadian Journal of Soil Sciences*, **79**, 25–35.
Arocena, J. M., Zhu, L. P. & Hall, K. (2003). Mineral accumulations induced by biological activity on granitic rocks in Qinghai Plateau, China. *Earth Surface Processes and Landforms*, **28**, 1429–37.
Arvieu, J. C., Leprince, F., & Plassard, C. (2003). Release of oxalate and protons by ectomycorrhizal fungi in response to P-deficiency and calcium carbonate in nutrient solution. *Annals of Forest Sciences*, **60**, 815–21.
Babich, H. & Stotzky, G. (1977). Reduction in the toxicity of cadmium to microorganisms by clay minerals. *Applied and Environmental Microbiology*, **33**, 696–705.
Baldrian, P. (2003). Interaction of heavy metals with white-rot fungi. *Enzyme and Microbial Technology*, **32**, 78–91.
Banfield, J. F. & Nealson, K. H. (eds.) (1997). *Geomicrobiology: Interactions between Microbes and Minerals*. Washington, DC: Mineralogical Society of America.
Banfield, J. F., Barker, W. W., Welch, S. A. & Taunton, A. (1999). Biological impact on mineral dissolution: application of the lichen model to understanding mineral weathering in the rhizosphere. *Proceedings of the National Academy of Sciences of the United States of America*, **96**, 3404–11.
Barker, W. W. & Banfield, J. F. (1996). Biologically versus inorganically mediated weathering reactions: relationships between minerals and extracellular microbial polymers in lithobiotic communities. *Chemical Geology*, **132**, 55–69.
Barker, W. W. & Banfield, J. F. (1998). Zones of chemical and physical interaction at interfaces between microbial communities and minerals: a model. *Geomicrobiology Journal*, **15**, 223–44.
Bennett, P. C., Rogers, J. R. & Choi, W. J. (2001). Silicates, silicate weathering, and microbial ecology. *Geomicrobiology Journal*, **18**, 3–19.

Blaudez, D., Botton, B. & Chalot, M. (2000). Cadmium uptake and subcellular compartmentation in the ectomycorrhizal fungus *Paxillus involutus*. *Microbiology*, **146**, 1109–17.

Braissant, O., Cailleau, G., Aragno, M. & Verrecchia, E. P. (2004). Biologically induced mineralization in the tree *Milicia excelsa* (Moraceae): its causes and consequences to the environment. *Geobiology*, **2**, 59–66.

Brandl, H. (2001). Heterotrophic leaching. In *Fungi in Bioremediation*, ed. G. M. Gadd. Cambridge: Cambridge University Press, pp. 383–423.

Bruand, A. & Duval, O. (1999). Calcified fungal filaments in the petrocalcic horizon of Eutrochrepts in Beauce, France. *Soil Science Society of America Journal*, **63**, 164–9.

Burford, E. P., Fomina, M. & Gadd, G. M. (2003a). Fungal involvement in bioweathering and biotransformation of rocks and minerals. *Mineralogical Magazine*, **67**, 1127–55.

Burford, E. P., Kierans, M. & Gadd, G. M. (2003b). Geomycology: fungi in mineral substrata. *Mycologist*, **17**, 98–107.

Burgstaller, W. & Schinner, F. (1993). Leaching of metals with fungi. *Journal of Biotechnology*, **27**, 91–116.

Callot, G., Guyon, A. & Mousain, D. (1985a). Inter-relation entre aiguilles de calcite et hyphes mycéliens. *Agronomie*, **5**, 209–16.

Callot, G., Mousain, D. & Plassard, C. (1985b). Concentration of calcium carbonate on the walls of fungal hyphae. *Agronomie*, **5**, 143–50.

Callot, G., Maurette, M., Pottier, L. & Dubois, A. (1987). Biogenic etching of microfractures in amorphous and crystalline silicates. *Nature*, **328**, 147–9.

Calvet, F. (1982). Constructive micrite envelope developed in vadose continental environment in pleistocene eoliantes of Mallorca (Spain). *Acta Geologica Hispanica*, **17**, 169–78.

Casarin, V., Plassard, C., Souche, G. & Arvieu, J.-C. (2003). Quantification of oxalate ions and protons released by ectomycorrhizal fungi in rhizosphere soil. *Agronomie*, **23**, 461–9.

Castro, I. M., Fietto, J. L. R., Vieira, R. X. *et al.* (2000). Bioleaching of zinc and nickel from silicates using *Aspergillus niger* cultures. *Hydrometallurgy*, **57**, 39–49.

Chander, K., Dyckmans, J., Joergensen, R. G., Meyer, B. & Raubuch, M. (2001). Different sources of heavy metals and their long-term effects on soil microbial properties. *Biology and Fertility of Soils*, **34**, 241–7.

Chantigny, M. H., Angers, D. A., Prevost, D., Vezina, L. P. & Chalifour, F. P. (1997). Soil aggregation and fungal and bacterial biomass under annual and perennial cropping systems. *Soil Science Society of America Journal*, **61**, 262–7.

Clausen, C. A. & Green, F. (2003). Oxalic acid overproduction by copper-tolerant brown-rot basidiomycetes on southern yellow pine treated with copper-based preservatives. *International Biodeterioration and Biodegradation*, **51**, 139–44.

Colpaert, J. V. & Van Assche, J. A. (1992). Zinc toxicity in ectomycorrhizal *Pinus sylvestris*. *Plant and Soil*, **143**, 201–11.

Colpaert, J. V. & Van Assche, J. A. (1993). The effect of cadmium on ectomycorrhizal *Pinus sylvestris* L. *New Phytologist*, **123**, 325–33.

Cromack, K. Jr, Solkins, P., Grausten, W. C. *et al.* (1979). Calcium oxalate accumulation and soil weathering in mats of the hypogeous fungus *Hysterangium crassum*. *Soil Biology and Biochemistry*, **11**, 463–8.

de la Torre, M. A., Gomez-Alarcon, G. Vizcaino C. & Garcia M. T. (1993) Biochemical mechanisms of stone alteration carried out by filamentous fungi living on monuments. *Biogeochemistry*, **19**, 129–47.

Denny, H. J. & Wilkins, D. A. (1987). Zinc tolerance in *Betula ssp*. IV. The mechanism of ectomycorrhizal amelioration of zinc toxicity. *New Phytologist*, **106**, 545–53.

Devevre, O., Garbaye, J. & Botton, B. (1996). Release of complexing organic acids by rhizosphere fungi as a factor in Norway Spruce yellowing in acidic soils. *Mycological Research*, **100**, 1367–74.

Diercks, M., Sand, W. & Bock, E. (1991). Microbial corrosion of concrete. *Experientia*, **47**, 514–16.

Dutton, M. V. & Evans, C. S. (1996). Oxalate production by fungi: its role in pathogenicity and ecology in the soil environment. *Canadian Journal of Microbiology*, **42**, 881–95.

Easton, R. M. (1997). Lichen-rock-mineral interactions: an overview. In *Biological–Mineralogical Interactions*, Vol. 21, eds. J. M. McIntosh & L. A. Groat. Mineralogical Association of Canada Short Course Series, Ottawa, Ontario, Canada, pp. 209–39.

Eckhardt, F. E. W. (1985). Solubilisation, transport, and deposition of mineral cations by microorganisms – efficient rock-weathering agents. In *The Chemistry of Weathering*, ed. J. Drever, Nato Asi Ser C, **149**, 161–73.

Ehrlich, H. L. (1996). *Geomicrobiology*. New York: Marcel Dekker.

Ehrlich, H. L. (1998). Geomicrobiology: its significance for geology. *Earth-Science Reviews*, **45**, 45–60.

Etienne, S. & Dupont, J. (2002). Fungal weathering of basaltic rocks in a cold oceanic environment (Iceland): comparison between experimental and field observations *Earth Surface Processes and Landforms*, **27**, 737–48.

Fletcher, M. (1987). How do bacteria attach to solid surfaces? *Microbiological Sciences*, **4**, 133–6.

Folk, R. L. & Chafetz, H. S. (2000). Bacterially induced microscale and nanoscale carbonate precipitates. In *Microbial Sediments*, ed. R. E. Riding & S. M. Awramik. Berlin: Springer-Verlag, pp. 41–9.

Fomina, M. & Gadd, G. M. (2002a). Influence of clay minerals on the morphology of fungal pellets. *Mycological Research*, **106**, 107–17.

Fomina, M. & Gadd, G. M. (2002b). Metal sorption by biomass of melanin-producing fungi grown in clay-containing medium. *Journal of Chemical Technology and Biotechnology*, **78**, 23–34.

Fomina, M. A., Alexander, I. J., Hillier, S. & Gadd, G. M. (2004). Zinc phosphate and pyromorphite solubilization by soil plant-symbiotic fungi. *Geomicrobiological Journal*, **21**, 351–66.

Fomina, M. A., Alexander, I. J., Colpaert, J. V. & Gadd, G. M. (2005a). Solubilization of toxic metal minerals and metal tolerance of mycorrhizal fungi. *Soil Biology and Biochemistry*, **37**, 851–66.

Fomina, M., Burford, E. P. & Gadd, G. M. (2005b) Toxic metals and fungal communities. In *The Fungal Community. Its Organization and Role in the Ecosystem*, ed. J. Dighton, J. F. White & P. Oudemans. Boca Raton: CRC Press, pp. 733–58.

Fomina, M., Hillier, S., Charnock, J. M. *et al.* (2005c). Role of oxalic acid overexcretion in toxic metal mineral transformations by *Beauveria caledonica*. *Applied and Environmental Microbiology*, **71**, 371–81.

Fomina, M. A., Olishevskaya, S. V., Kadoshnikov, V. M., Zlobenko, B. P. & Podgorsky, V. S. (2005d). Concrete colonization and destruction by mitosporic fungi in model experiment. *Mikrobiologichny Zhurnal*, **67**, 97–106 (in Russian with English summary).

Fujita, Y., Ferris, F. G., Lawson, D. R., Colswell, F. S. & Smith, R. W. (2000). Calcium carbonate precipitation by ureolytic subsurface bacteria. *Geomicrobiology Journal*, **17**, 305–18.

Gadd, G. M. (1984). Effect of copper on *Aureobasidium pullulans* in solid medium: adaptation not necessary for tolerant behaviour. *Transactions of the British Mycological Society*, **82**, 546–9.

Gadd, G. M. (1990). Fungi and yeasts for metal accumulation. In *Microbial Mineral Recovery*, ed. H. L. Ehrlich & C. Brierley. New York: McGraw-Hill, pp. 249–275.

Gadd, G. M. (1993). Interactions of fungi with toxic metals. *New Phytologist*, **124**, 25–60.

Gadd, G. M. (1999). Fungal production of citric and oxalic acid: importance in metal speciation, physiology and biogeochemical processes. *Advances in Microbial Physiology*, **41**, 47–92.

Gadd, G. M. (2000). Bioremedial potential of microbial mechanisms of metal mobilization and immobilization. *Current Opinion in Biotechnology*, **11**, 271–9.

Gadd, G. M. (2001). Metal transformations. In *Fungi in Bioremediation*, ed. G. M. Gadd. Cambridge: Cambridge University Press, pp. 359–82.

Gadd, G. M. (2004). Mycotransformation of organic and inorganic substrates. *Mycologist*, **18**, 60–70.

Gadd, G. M. & Mowll, J. L. (1985). Copper uptake by yeast-like cells, hyphae and chlamydospores of *Aureobasidium pullulans*. *Experimental Mycology*, **9**, 230–40.

Gadd, G. M. & Sayer, J. A. (2000). Fungal transformations of metals and metalloids. In *Environmental Microbe-Metal Interactions*, ed. D. R. Lovley. Washington, DC: American Society for Microbiology, pp. 237–56.

Gadd, G. M., Burford, E. P. & Fomina, M. (2003). Biogeochemical activities of microorganisms in mineral transformations: consequences for metal and nutrient mobility. *Journal of Microbiology and Biotechnology*, **13**, 323–31.

Galli, U., Schuepp, H. & Brunold, C. (1994). Heavy metal binding by mycorrhizal fungi. *Physiologia Plantarum*, **92**, 364–8.

Gaylarde, C. C. & Morton, L. H. G. (1999). Deteriogenic biofilms on buildings and their control: a review. *Biofouling*, **14**, 59–74.

Geesey, G. (1993). *A Review of the Potential for Microbially Influenced Corrosion of High-Level Nuclear Waste Containers*. San Antonio, TX: Nuclear Regulatory Commission.

Gerrath, J. F., Gerrath, J. A. & Larson, D. W. (1995). A preliminary account of endolithic algae of limestone cliffs of the Niagara Escarpment. *Canadian Journal of Botany*, **73**, 788–93.

Gharieb, M. M. & Gadd, G. M. (1999). Influence of nitrogen source on the solubilization of natural gypsum. *Mycological Research*, **103**, 473–81.

Gharieb, M. M., Kierans, M. & Gadd, G. M. (1999). Transformation and tolerance of tellurite by filamentous fungi: accumulation, reduction and volatilization. *Mycological Research*, **103**, 299–305.

Gomes-Alarcon, G., Munor, M. L. & Flores, M. (1994). Excretion of organic acids by fungal strains isolated from decayed sandstones. *International Biodeterioration and Biodegradation*, **34**, 169–80.

Gorbushina, A. A. & Krumbein, W. E. (2000). Subaerial microbial mats and their effects on soil and rock. In *Microbial Sediments*, ed. R. E. Riding & S. M. Awramik. Berlin: Springer-Verlag, pp. 161–9.

Gorbushina, A. A., Krumbein, W. E., Hamann, R. *et al.* (1993). On the role of black fungi in colour change and biodeterioration of antique marbles. *Geomicrobiology Journal*, **11**, 205–21.

Gorbushina, A. A., Boettcher, M., Brumsack, H. J., Krumbein, W. E. & Vendrell-Saz, M. (2001). Biogenic forsterite and opal as a product of biodeterioration and lichen stromatolite formation in table mountain systems (tepuis) of Venezuela. *Geomicrobiology Journal*, **18**, 117–32.

Gorbushina, A. A., Krumbein, W. E. & Volkmann, M. (2002). Rock surfaces as life indicators: new ways to demonstrate life and traces of former life. *Astrobiology*, **2**, 203–13.

Gorbushina, A. A., Whitehead, K., Dornieden, T. *et al.* (2003). Black fungal colonies as units of survival: hyphal mycosporines synthesized by rock-dwelling microcolonial fungi. *Canadian Journal of Botany*, **81**, 131–8.

262   *M. Fomina* et al.

Grote, G. & Krumbein, W. E. (1992). Microbial precipitation of manganese by bacteria and fungi from desert rock and rock varnish. *Geomicrobiology Journal*, **10**, 49–57.

Gu, J. D., Ford, T. E., Berke, N. S. & Mitchell, R. (1998). Biodeterioration of concrete by the fungus *Fusarium*. *International Biodeterioration and Biodegradation*, **41**, 101–9.

Hagerberg, D., Thelin, G. & Wallander, H. (2003). The production of ectomycorrhizal mycelium in forests: relation between forest nutrient status and local mineral sources. *Plant and Soil*, **252**, 279–90.

Hammes, F. & Verstraete, W. (2002). Key roles of pH, and calcium metabolism in microbial carbonate precipitation. *Reviews in Environmental Science and Biotechnology*, **1**, 3–7.

Heckman, D. S., Geiser, D. M., Eidell, B. R. *et al.* (2001). Molecular evidence for the early colonisation of land by fungi and plants. *Science*, **293**, 1129–33.

Henderson, M. E. K. & Duff, R. B. (1963). The release of metallic and silicate ions from minerals, rocks and soils by fungal activity. *Journal of Soil Science*, **14**, 236–46.

Hirsch, P., Eckhardt, F. E. W. & Palmer, R. J. Jr. (1995). Fungi active in weathering rock and stone monuments. *Canadian Journal of Botany*, **73**, 1384–90.

Hochella, M. F. (2002). Sustaining Earth: thoughts on the present and future roles in mineralogy in environmental science. *Mineralogical Magazine*, **66**, 627–52.

Hoffland, E., Giesler, R., Jongmans, T. & van Breemen, N. (2002). Increasing feldspar tunneling by fungi across a north Sweden podzol chronosequence. *Ecosystems*, **5**, 11–22.

Howlett, N. G. & Avery, S. V. (1997). Relationship between cadmium sensitivity and degree of plasma membrane fatty acid unsaturation in *Saccharomyces cerevisiae*. *Applied Microbiology and Biotechnology*, **48**, 539–45.

Jacobs, H., Boswell, G. P., Ritz, K., Davidson, F. A. & Gadd, G. M. (2002). Solubilization of calcium phosphate as a consequence of carbon translocation by *Rhizoctonia solani*. *FEMS Microbiology Ecology*, **40**, 65–71.

Jacobs, H., Boswell, G. P., Scrimgeour, C. M. *et al.* (2004). Translocation of carbon by *Rhizoctonia solani* in nutritionally-heterogeneous environments. *Mycological Research*, **108**, 453–62.

Jones, M. D. & Hutchinson, T. C. (1988). Nickel toxicity in mycorrhizal birch seedlings infected with *Lactarius rufus* or *Scleroderma flavidum*. II Uptake of nickel, calcium, magnesium, phosphorus and iron. *New Phytologist*, **108**, 461–70.

Jongmans, A. G., van Breemen, N., Lungstrom, U. *et al.* (1997). Rock-eating fungi. *Nature*, **389**, 682–3.

Kahle, C. F. (1977). Origin of subaerial Holocene calcareous crusts: role of algae, fungi and sparmicristisation. *Sedimentology*, **24**, 413–35.

Khan, M. & Scullion, J. (2000). Effect of soil on microbial responses to metal contamination. *Environmental Pollution*, **110**, 115–25.

Kikuchi, Y. & Sreekumari, K. R. (2002). Microbially influenced corrosion and biodeterioration of structural metals. *Journal of the Iron and Steel Institute of Japan*, **88**, 620–8.

Klappa, C. F. (1979). Calcified filaments in quaternary calcretes: organo-mineral interactions in the subaerial vadose environment. *Journal of Sedimentary Petrology*, **49**, 955–68.

Kumar, R. & Kumar, A. V. (1999). *Biodeterioration of Stone in Tropical Environments: An Overview*. The J. Paul Getty Trust, USA.

Lapeyrie, F., Chilvers, G. A. & Bhem, C. A. (1987). Oxalic acid synthesis by the mycorrhizal fungus *Paxillus involutus* (Batsch.ex fr.). *New Phytologist*, **106**, 139–46.

Lapeyrie, F., Picatto, C., Gerard J. & Dexheimer, J. (1990). TEM study of intracellular and extracellular calcium oxalate accumulation by ectomycorrhizal fungi in pure culture or in association with *Eucalyptus* seedlings. *Symbiosis*, **9**, 163–6.

Lapeyrie, F., Ranger, J. & Vairelles, D. (1991). Phosphate-solubilizing activity of ectomycorrhizal fungi *in vitro*. *Canadian Journal of Botany*, **69**, 342–6.

Lee, G.-H. & Stotzky, G. (1999). Transformation and survival of donor, recipient, and transformants of *Bacillus subtilis* in vitro and in soil. *Soil Biology and Biochemistry*, **31**, 1499–508.

Leyval, C. & Joner, E. J. (2001). Bioavailability of heavy metals in the mycorrhizosphere. In *Trace Elements in the Rhizosphere*, ed. G. R. Gobran, W. W. Wenzel & E. Lombi. Boca Raton: CRC Press, pp. 165–85.

Lindahl, B. D. & Olsson, S. (2004). Fungal translocation – creating and responding to environmental heterogeneity. *Mycologist*, **18**, 79–88.

Little, B. & Staehle, R. (2001). Fungal influenced corrosion in post-tension structures. *The Electrochemical Society Interface*, Winter 2001, 44–8.

Lotareva, O. V. & Prozorov, A. A. (2000). Effect of the clay minerals montmorillonite and kaolinite on the generic transformation of competent *Bacillus subtilis* cells. *Microbiology*, **69**, 571–4.

Lunsdorf, H., Erb, R. W., Abraham, W. R. & Timmis, K. N. (2000). 'Clay hutches': a novel interaction between bacteria and clay minerals. *Environmental Microbiology*, **2**, 161–8.

McEldowney, S. & Fletcher, M. (1986). Effect of growth conditions and surface characteristics of aquatic bacteria on their attachment to solid surfaces. *Journal of General Microbiology*, **132**, 513–23.

Magyarosy, A., Laidlaw, R. D., Kilaas, R. *et al.* (2002). Nickel accumulation and nickel oxalate precipitation by *Aspergillus niger*. *Applied Microbiology and Biotechnology*, **59**, 382–8.

Mandal, S. K., Roy, A. & Banerjee, P. C. (2002). Iron leaching from china clay by fungal strains. *Transactions of the Indian Institute of Metals*, **55**, 1–7.

Manley, E. & Evans, L. (1986). Dissolution of feldspars by low-molecular-weight aliphatic and aromatic acids. *Soil Science*, **141**, 106–12.

Manoli, F., Koutsopoulos, E. & Dalas, E. (1997). Crystallization of calcite on chitin. *Journal of Crystal Growth*, **182**, 116–24.

Marshall, K. C. (1988). Adhesion and growth of bacteria at surfaces in oligotrophic habitats. *Canadian Journal of Microbiology*, **34**, 593–606.

Martin, J. P., Filip, Z. & Haider, K. (1976). Effect of montmorillonite and humate on growth and metabolic activity of some actinomyces. *Soil Biology and Biochemistry*, **8**, 409–13.

Martino, E., Perotto, S., Parsons, R. & Gadd, G. M. (2003). Solubilization of insoluble inorganic zinc compounds by ericoid mycorrhizal fungi derived from heavy metal polluted sites. *Soil Biology and Biochemistry*, **35**, 133–41.

Meharg, A. A. (2003). The mechanistic basis of interactions between mycorrhizal associations and toxic metal cations. *Mycological Research*, **107**, 1253–65.

Merz-Preiβ, M. (2000). Calcification in cyanobacteria. In *Microbial Sediments*, ed. R. E. Riding & S. M. Awramik. Berlin: Springer-Verlag, pp. 51–5.

Mironenko, N. V., Alekhina, I. A., Zhdanova, N. N. & Bulat, S. A. (2000). Intraspecific variation in gamma-radiation resistance and genomic structure in the filamentous fungus *Alternaria alternata*: a case study of strains inhabiting Chernobyl Reactor No. 4. *Ecotoxicology and Environmental Safety*, **45**, 177–87.

Money, N. P. (2004). The fungal dining habit – a biomechanical perspective. *Mycologist*, **18**, 71–6.

Money, N. P. & Howard, R. J. (1996). Confirmation of a link between fungal pigmentation, turgor pressure, and pathogenicity using a new method of turgor measurement. *Fungal Genetics and Biology*, **20**, 217–27.

264   *M. Fomina* et al.

Monger, C. H. & Adams, H. P. (1996). Micromorphology of calcite-silica deposits, Yucca Mountain, Nevada. *Soil Science Society of America Journal,* **60**, 519–30.

Muller, B., Burgstaller, W., Strasser, H., Zanella, A. & Schinner, F. (1995). Leaching of zinc from an industrial filter dust with *Penicillium, Pseudomonas and Corynebacterium*: citric acid is the leaching agent rather than amino acids. *Journal of Industrial Microbiology,* **14**, 208–12.

Nica, D., Davis, J. L., Kirby, L., Zuo, G. & Roberts, D. J. (2000). Isolation and characterization of microorganisms involved in the biodeterioration of concrete in sewers. *International Biodeterioration and Biodegradation,* **46**, 61–8.

Olsson, P. A. & Wallander, H. (1998). Interactions between ectomycorrhizal fungi and the bacterial community in soils amended with various primary minerals. *FEMS Microbiology Ecology,* **27**, 195–205.

Paris, F., Bonnaud, P., Ranger, J. & Lapeyrie, F. (1995). *In vitro* weathering of phlogopite by ectomycorrhizal fungi I. Effect of K^+ and Mg^{2+} deficiency on phyllosilicate evolution. *Plant Soil,* **177**, 191–201.

Pereira, M. O., Vieira, M. J. & Melo, L. F. (2000). The effect of clay particles on the efficacy of a biocide. *Water Science and Technology,* **41**, 61–4.

Perfettini, J. V., Revertegat, E. & Langomazino, N. (1991). Evaluation of cement degradation by the metabolic activities of two fungal strains. *Experientia,* **47**, 527–33.

Perotto, S. & Martino, E. (2001). Molecular and cellular mechanisms of heavy metal tolerance in mycorrhizal fungi: what perspectives for bioremediation? *Minerva Biotechnologica,* **13**, 55–63.

Puget, P., Angers, D. A. & Chenu, C. (1999). Nature of carbohydrates associated with water-stable aggregates of two cultivated soils. *Soil Biology and Biochemistry,* **31**, 55–63.

Riding, R. (2000). Microbial carbonates: the geological record of calcified bacterial-algal mats and biofilms. *Sedimentology,* **47**, 179–214.

Ritz, K. & Young, I. M. (2004). Interaction between soil structure and fungi. *Mycologist,* **18**, 52–9.

Rivadeneyra, M. A., Delgado, R., Delgado, G. *et al.* (1993). Precipitation of carbonates by *Bacillus* sp. isolated from saline soils. *Geomicrobiology Journal,* **11**, 175–84.

Roberts, D. J., Nica, D., Zuo, G. & Davis, J. L. (2002). Quantifying microbially induced deterioration of concrete: initial studies. *International Biodeterioration and Biodegradation,* **49**, 227–34.

Rodriguez Navarro, C., Sebastian, E. & Rodriguez Gallego, M. (1997). An urban model for dolomite precipitation: authigenic dolomite on weathered building stones. *Sedimentary Geology,* **109**, 1–11.

Sand, W. & Bock, E. (1991a). Biodeterioration of mineral materials by microorganisms – Biogenic sulphuric and nitric-acid corrosion of concrete and natural stone. *Geomicrobiology Journal,* **9**, 129–38.

Sand, W. & Bock, E. (1991b). Biodeterioration of ceramic materials by biogenic acids. *International Biodeterioration,* **27**, 175–83.

Sarret, G., Manceau, A., Cuny, D. *et al.* (1998). Mechanisms of lichen resistance to metallic pollution. *Environmental Science and Technology,* **32**, 3325–30.

Sarret, G., Saumitou-Laprade, P., Bert, V. *et al.* (2002). Forms of zinc accumulated in the hyperaccumulator *Arabidopsis halleri. Plant Physiology,* **130**, 1815–26.

Sayer, J. A. & Gadd, G. M. (2001). Binding of cobalt and zinc by organic acids and culture filtrates of *Aspergillus niger* grown in the absence or presence of insoluble cobalt or zinc phosphate. *Mycological Research,* **105**, 1261–7.

Sayer, J. A., Kierans, M. & Gadd, G. M. (1997). Solubilization of some naturally occurring metal-bearing minerals, limescale and lead phosphate by *Aspergillus niger*. *FEMS Microbiology Letters*, **154**, 29–35.

Sayer, J. A., Cotter-Howells, J. D., Watson, C., Hillier, S. & Gadd, G. M. (1999). Lead mineral transformation by fungi. *Current Biology*, **9**, 691–4.

Staley, J. T., Palmer, F. & Adams, J. B. (1982). Microcolonial fungi: common inhabitants on desert rocks. *Science*, **215**, 1093–5.

Sterflinger, K. (2000). Fungi as geologic agents. *Geomicrobiology Journal*, **17**, 97–124.

Stotzky, G. (1966). Influence of clay minerals on microorganisms-II. Effect of various clay species, homoionic clays, and other particles on bacteria. *Canadian Journal of Microbiology*, **12**, 831–48.

Stotzky, G. (2000). Persistence and biological activity in soil of insecticidal proteins from *Bacillus thuringiensis* and of bacterial DNA bound on clays and humic acids. *Journal of Environmental Quality*, **29**, 691–705.

Styriakova, I. & Styriak, I. (2000). Iron removal from kaolins by bacterial leaching. *Ceramics-Silikaty*, **44**, 135–41.

Thompson, J. B. & Ferris, F. G. (1990). Cyanobacterial precipitation of gypsum, calcite and magnesite from natural alkaline lake water. *Geology*, **18**, 995–8.

Thomson-Eagle, E. T. & Frankenberger, W. T. (1992). Bioremediation of soils contaminated with selenium. In *Advances in Soil Science*, ed. R. Lal & B. A. Stewart. New York: Springer-Verlag, pp. 261–309.

Tisdall, J. M., Smith, S. E. & Rengasamy, P. (1997). Aggregation of soil by fungal hyphae. *Australian Journal of Soil Research*, **35**, 55–60.

Turnau, K., Kottke, I. & Dexheimer, J. (1996). Toxic element filtering in *Rhizopogon roseolus/ Pinus sylvestris* mycorrhizas collected from calamine dumps. *Mycological Research*, **100**, 16–22.

Urzi, C., Garcia-Valles, M. T., Vendrell, M. & Pernice, A. (1999). Biomineralization processes of the rock surfaces observed in field and in laboratory. *Geomicrobiology Journal*, **16**, 39–54.

Van der Lelie, D., Schwitzguebel, J. P., Glass, D. J., Vangronsveld, J. & Baker, A. (2001). Assessing phytoremediation's progress in the United States and Europe. *Environmental Science and Technology*, **35**, 446A–52A.

van Hees, P. A. V., Goldbold, D. L., Jentschke, G. & Jones, D. L. (2003). Impact of ectomycorrhizas on the concentration and biodegradation of simple organic acids in a forest soil. *European Journal of Soil Science*, **54**, 697–706.

Van Leerdam, D. M., Williams, P. A. & Cairney, J. W. G. (2001). Phosphate-solubilizing abilities of ericoid mycorrhizal endophytes of *Woollsia pungens* (Epacridaceae). *Australian Journal of Botany*, **49**, 75–80.

Van Tichelen, K. K., Colpaert, J. V. & Vangronsveld, J. (2001). Ectomycorrhizal protection of *Pinus sylvestris* against copper toxicity. *New Phytologist*, **150**, 203–13.

Verrecchia, E. P. (2000). Fungi and sediments. In *Microbial Sediments*, ed. R. E. Riding & S. M. Awramik. Berlin: Springer-Verlag, pp. 69–75.

Verrecchia, E. P., Dumont, J. L. & Rolko, K. E. (1990). Do fungi building limestones exist in semi-arid regions? *Naturwissenschaften*, **77**, 584–6.

Vettori, C., Gallori, E. & Stotzky, G. (2000). Clay minerals protect bacteriophage PBS1 of *Bacillus subtilis* against inactivation and loss of transducing ability by UV radiation. *Canadian Journal of Microbiology*, **46**, 770–3.

Vodnik, D., Byrne, A. R. & Gogala, N. (1998). The uptake and transport of lead in some ectomycorrhizal fungi in culture. *Mycological Research*, **102**, 953–8.

Volkmann, M., Whitehead, K., Rutters, H., Rullkotter, J. & Gorbushina, A. A. (2003). Mycosporine-glutamicol-glucoside: a natural UV-absorbing secondary metabolite of rock-inhabiting microcolonial fungi. *Rapid Communications in Mass Spectrometry*, **17**, 897–902.

von Knorre, H. & Krumbein, W. E. (2000). Bacterial calcification. In *Microbial Sediments*, ed. R. E. Riding & S. M. Awramik. Berlin: Springer-Verlag, pp. 25–31.

Wainwright, M., Tasnee, A. A. & Barakah, F. (1993). A review of the role of oligotrophic microorganisms in biodeterioration. *International Biodeterioration and Biodegradation*, **31**, 1–13.

Warren, L. A., Maurice, P. A., Parmer, N. & Ferris, F. G. (2001). Microbially mediated calcium carbonate precipitation: implications for interpreting calcite precipitation and for solid-phase capture of inorganic contaminants. *Geomicrobiology Journal*, **18**, 93–115.

Warscheid, T. & Krumbein, W. E. (1994). Biodeterioration processes on inorganic materials and means of countermeasures. *Materials and Corrosion*, **45**, 105–13.

Watts, H. J., Very, A. A., Perera, T. H. S., Davies, J. M. & Gow, N. A. R. (1998). Thigmotropism and stretch-activated channels in the pathogenic fungus *Candida albicans*. *Microbiology*, **144**, 689–95.

Webley, D. M., Henderson, M. E. F. & Taylor, I. F. (1963). The microbiology of rocks and weathered stones. *Journal of Soil Science*, **14**, 102–12.

Wenzel, C. L., Ashford, A. E. & Summerell, B. A. (1994). Phosphate-solubilizing bacteria associated with proteoid roots of seedlings of warratah [*Telopea speciosissima* (Sm.) R. Br.]. *New Phytologist*, **128**, 487–96.

Whitelaw, M. A., Harden, T. J. & Helyar, K. R. (1999). Phosphate solubilization in solution culture by the soil fungus *Penicillium radicum*. *Soil Biology and Biochemistry*, **31**, 655–65.

Wollenzien, U., de Hoog, G. S., Krumbein, W. E. & Urzi, C. (1995). On the isolation of microcolonial fungi occurring on and in marble and other calcareous rocks. *Science of the Total Environment*, **167**, 287–94.

Zhdanova, N. N., Zakharchenko, V. A., Vember, V. V. & Nakonechnaya, L. T. (2000). Fungi from Chernobyl: mycobiota of the inner regions of the containment structures of the damaged nuclear reactor. *Mycological Research*, **104**, 1421–26.

11
Fungal activities in subaerial rock-inhabiting microbial communities

ANNA A. GORBUSHINA

Introduction

There were times on our planet when the barren dryness of uninhabited continents sharply contrasted with the densely populated sea. The continental lithosphere was then essentially represented by rock surfaces of different types. Sedimentary rocks were rare, if not absent. As rock materials became exposed to the subaerial environment at the Earth's surface, they encountered a whole range of environmental challenges such as temperature fluctuations, water, unbuffered cosmic and solar irradiation and atmospheric gases and solids instead of dissolved species. These influences resulted in rocks undergoing alterations in material properties leading to erosion and breakdown into ever-smaller particles and constituent minerals, formation of sandy sediments, and mineral soils (Ehrlich, 1996). Primordial terrestrial environments can therefore be visualized as a freshly exposed and only slightly physically pre-weathered rock surface.

However, physical and chemical changes in rock-forming minerals even during the very first stages of the terrestrial evolution were accompanied by an initially slow but steady establishment and spread of living organisms. Life started to colonize rock surfaces during the Archean. The first settlers were undoubtedly biofilms and later mature microbial mats not unlike modern desert or intertidal stromatolithic systems (Costerton & Stoodley, 2003). Environmental and geochemical settings of these ancient subaerial habitats were probably very similar to the conditions of present-day deserts. Rock surface environments were then, and remain now, exceptionally hostile with respect to all conditions necessary for the maintenance of living systems. They challenge new settlers with a variety of important stress factors including desiccation, temperature fluctuations, nutrient shortage and intensive irradiation. Unlike the buffering and

Fungi in Biogeochemical Cycles, ed. G. M. Gadd. Published by Cambridge University Press. © British Mycological Society 2006.

highly protective environment of any water body these influences on cells and cell envelopes are direct and, in addition, may change rapidly and, sometimes more adversely, sporadically.

However, many organisms have specialized on living in this primordial niche. Highly specialized microbial communities, now usually termed subaerial biofilms, represent a group of ubiquitous settlers on sediment and rock surfaces in a diversity of environments including the coldest and hottest, as well as the physically and chemically most adverse conditions measurable on this planet (Gorbushina & Krumbein, 1999; 2000). The firm establishment of a microbial biofilm community on and in terrestrial rock surfaces represents a starting point for the development of all land ecosystems. Especially important is the crucial role of subaerial microbial communities in acceleration of rock weathering (Brehm *et al.*, 2005), leading to soil formation and vegetation development (Gromov, 1957; Ehrlich, 1996; Chertov *et al.*, 2004). Associations of microscopic organisms were the first settlers on inhospitable primordial land surfaces at times when the dominant life forms on our planet were biofilms and later mature microbial mats. In the earliest periods these chiefly consisted of prokaryotic organisms like cyanobacteria and archaea. Later on eukaryotic forms were established (Margulis, 1970; 1982) by mixing of bacterial and archaeal sets of genes (Rivera & Lake, 2004). Early eukaryotes on terrestrial surfaces must have included algae, fungi and protoctists. All three major microscopic groups of eukaryotes were basically fitter for stress exposure than the less protected prokaryotes. These specialized genetic constructs and chimera probably expanded their influence in an explosive way and achieved dominance in all land environments, preparing the ground for more sophisticated approaches such as fungi-supported lichens and later plants (McMenamin & McMenamin, 1994). Present-day rock surfaces are inhabited by chemolithotrophic, chemoorganotrophic and photoautotrophic bacteria, algae, fungi and lichens. These interact with each other and form a complex biofilm community. Free-living and symbiotic fungi are important settlers on all rock surfaces (Staley *et al.*, 1982; Braams, 1992; May *et al.*, 1993; Sterflinger, 2000). The aerobic metabolism of fungi is perfectly suitable for colonization of the interface between an oxygen-enriched atmosphere and the lithosphere. Due to their hyphal and/ or yeast-like growth mode and the remarkable stress tolerance of certain ecological groups, free-living fungi are the most persistent components of subaerial rock biofilms. Their interactions with other subaerial biofilm settlers, as well as with the mineral substrate are multifaceted and to a large extent are neither well understood nor recognized. Additionally, as fungi have significantly contributed to the evolution of higher plants on land (McMenamin & McMenamin, 1994; Taylor *et al.*, 1995), they might well have contributed to the first steps in rock colonization in the early stages of evolution on Earth. Therefore, among the rock-inhabiting eukaryotes we shall concentrate our attention on the diverse and omnipresent group of rock-inhabiting micromycetes.

In this article, the subaerial rock surface environment, an ancient and inhospitable, but nevertheless densely colonized ecological niche, will be examined. Attention is directed to the establishment, interactions and biogeomorphogenetic impact of microscopic free-living fungi in selected rock environments. The view that microscopic fungal life creates geological evidence, and the important biological role fungi play in the subaerial rock-inhabiting communities is a main focus of discussion despite the fact that geologists, including micropaleontologists, have only rarely dealt with this group.

Atmosphere-exposed rock as a biofilm substrate: environmental prerequisites and biofilm matrix

The present-day bare rock surfaces in their environmental settings are very similar to terrestrial environments as they were present in the Archean. Several environmental factors strongly influence settlement, growth and development of microorganisms on bare rock surfaces. Microbial communities on atmosphere-exposed rock surfaces may be considered as an example of a subaerial biofilm. Temperature, humidity, light and short-wave radiation intensity and organic and inorganic nutrient deposition vary from one extreme to the other in this unsheltered habitat. Exposure to the atmosphere is very important as bare rock surfaces experience diurnal, annual and more irregular fluctuations of temperature and humidity, as well as intensive irradiation (e.g. sunspot bursts, Milankovic-cycles, cosmic clouds, etc.). These peculiar habitats allow for settlement and development of only highly specialized organisms able to adapt to these broad-spectrum alterations.

Microbiota dwelling on rock surfaces metabolize according to these fluctuations and are forced to develop defensive reactions to the presence of assorted stress factors. Protective reactions to external conditions primarily involve biofilm formation. In nature, microorganisms rarely grow as single-species colonies and the majority exist as more or less heterogeneous aggregations (Stoodley *et al.*, 2002; Costerton & Stoodley, 2003). The study of a rock-inhabiting biofilm therefore addresses a complex microbial ecosystem. These interorganismal associations, unified by a connecting extracellular matrix and capable of surviving adverse and fluctuating environmental conditions, were first termed subaerial biofilms by Gorbushina and Krumbein (2000). This interacting community is in general more protected from physical stress than would be a single organism. Rock inhabitants have to withstand diverse physical stresses (desiccation and UV radiation being among the most prominent), but probably rarely experience any kind of biological competition or antibiosis. This was stated for lithobiontic Antarctic communities (Friedmann, 1993) and is probably correct for all rock-inhabiting biofilms. In the hostile rock environment the metabolic costs of survival are very high (Nienow & Friedmann, 1993), and antibiosis is a rare phenomenon (Friedmann & Ocampo-Friedmann, 1984). Growth in a subaerial biofilm therefore confers a protective measure, allowing for an increased survival of any individual cell in this habitat. Propagation and the spread of specialized cells or cell units, often a result of complex sexual patterns, is much too energetically expensive. Wind-blown small cell aggregates or microcolonies may settle elsewhere and continue growth when good conditions for metabolic activity occur.

Most subaerial microbial communities on a rock surface are embedded in an extracellular polymer matrix. However, in the subaerial environment, the extracellular polymeric substances (EPS) are highly specialized for maintenance of life in the presence of a minimum amount of water. The EPS possess a high potential for water retention and favour desiccation tolerance (Hill *et al.*, 1994; Potts, 1999). Synthesis of EPS is an important survival adaptation of a rock surface inhabitant and is present also in rock-inhabiting fungi. Various fungal species are known to be active slime producers and mucilage production by rock-inhabiting fungi *in situ* and in vitro is illustrated in Fig. 11.1. The importance of cellular exudates in subaerial biofilms cannot be overestimated. Interactions with the substratum involve adhesion and chemical interaction caused by specific compounds present in the extracellular mucilage. The colloidal aggregates of acidic polysaccharides and proteins form a physical contact between the cells and the inorganic substrate (Leppard, 1986), but most important is chemical dissolution of the substrate by microbial EPS (Welch & Ullman, 1993; Welch & Vandevivere, 1994; Barker *et al.*, 1997; Banfield *et al.*, 1999). The EPS often contains soluble organic ligands and other complex-forming agents that can influence chemically sensitive substrates. Protons provided by low-molecular-weight organic acids and/or EPS locally dissolve the rocks in the nearest vicinity of hyphae (Avakian *et al.*, 1981; Jongmans *et al.*, 1997; Gadd, 1999; Burford *et al.*, 2003).

Ions are transferred from the substrate into solution and then into ion–organic complexes stimulating a new dissolution process of the bulk rock material (Warscheid & Krumbein, 1996). Organic EPS components already in the first stage of colonization slowly change the substrate including the pH of the surface (chemical vulnerability) and grain cohesion (physical vulnerability) as documented by Dornieden *et al.* (1997).

Chemical reactions with the mineral substrate in a homeostatic environment are possible only because of the presence of EPS and form the basis of biogeochemical impact on the rock surface. The presence of EPS also changes such material surface properties as hydrophobicity. Fungal hyphae or colonies serve as condensation and water-collecting points, where water is not only quickly absorbed, but retained for longer periods than on a neighbouring patch of the rock surface. Free-living and symbiotic epi- and endolithic fungi are capable of using atmospheric water vapour and the condensed water appearing during diurnal temperature cycles. The EPS-conditioned rock substrate plays an important role in collecting and storing this water. This allows chemical reactions and processes of dissolution (wear-down) to occur for considerably longer periods and brings about evident traces of the microbial presence like biopitting patterns (Krumbein 1969; Brehm *et al.*, 2005).

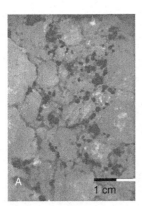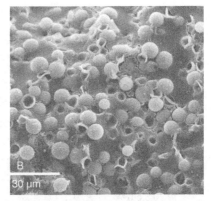

Fig. 11.1. Fungal microcolonies and associated extracellular polymeric substances (EPS). (A) On the surface of marble in Greece (Athens). The EPS network around black fungal colonies is stained red (on the photo – light grey) by periodic acid – Schiff's reagent stain (PAS). (B) Cryo-scanning electron microscope image of a fungal microcolony with its inherent mucilage. This colony of strain A49 was cultured in subaerial conditions with some additional supply of nutrients from the substrate.

Rock-inhabiting fungi often demonstrate a yeast-like growth pattern (Fig. 11.2). Most of them are heavily melanized (Gorbushina *et al.*, 1993). This pigmentation not only changes the aesthetic appearance of stone monuments upon advanced biofilm development (Diakumaku *et al.*, 1995), but also supplies specific mechanical properties to the growing and penetrating fungal hyphae (Dornieden *et al.*, 1997; Sterflinger & Krumbein, 1997). The capacity of melanized fungal structures to create a considerable growth pressure (Bechinger *et al.*, 1999) is supported by the properties of the melanin cell wall layer (Butler & Day, 1998; Butler *et al.*, 2001). Differential heat transfer and irreversible thermal expansion of rock particles is another indirect physical or mechanical factor (Dornieden *et al.*, 1997). Deeper penetration into the substrate represents yet another mode of adaptation to the formation of networks and biofilms in the surface layers of rock substrates. The original crystals will be separated from each other at first by EPS and then by growing hyphae. Because of this and the previously mentioned heat-transfer patterns, deeper cracks and crevices with considerably more stable microclimatic conditions will be created and successfully inhabited. Such penetration increases the contact surface of fungal hyphae and their metabolites with the substrate, amplifying their geochemical influence. Development of an interconnected

0.5 cm

Fig. 11.2. Rock-inhabiting fungi often demonstrate a specific yeast-like growth pattern. Image shows a fungal colony, isolated from a marble surface in Crimea (Ukraine). The whole structure resembles a cauliflower and is intensively pigmented by melanin. The colony is very compact and mechanically hard and thus capable of producing intense pressure on the surrounding material.

fungal biofilm is shown in Fig. 11.3. The advance of this complex and three-dimensional structure on the marble surface took only three months to develop under laboratory conditions. A longer subaerial exposure will undoubtedly bring about further development of three-dimensional fungal biofilms and networks (biodictyon *sensu* Krumbein *et al.*, 2003).

Growth expansion into deeper rock layers also helps to protect the cells from the harmful influence of UV radiation (Friedmann & Ocampo-Friedmann, 1984; Friedmann *et al.*, 1987; Nienow & Friedmann, 1993; Cockell & Knowland, 1999) and causes even more distinct and articulated interactions with the mineral material. Additionally, protection from UV radiation is enhanced through the synthesis of a diverse spectrum of protective pigments like carotenoids (Cockell & Knowland, 1999; Sterflinger *et al.*, 1999), mycosporines (Volkmann *et al.*, 2003) and melanins (Gorbushina *et al.*, 1993; Diakumaku *et al.*, 1995).

The presence of EPS in subaerial biofilms (Fig. 11.1) is also an important prerequisite for inter-organismal interactions and communication. The EPS layer enables other microorganisms to join the biofilm by direct adhesion and might also stimulate growth of accompanying bacteria (Fig. 11.4), as well as promote contact with photobionts (Gorbushina *et al.*, 2005).

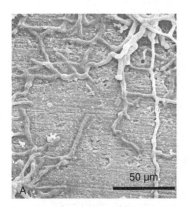 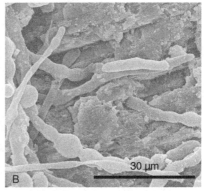

Fig. 11.3. Fungal colonies of *Coniosporium uncinatum*, forming a network on a marble block surface. The formation of interconnecting hyphae and hyphal bundles seemingly represent an adaptation to survive under the hostile conditions on the mineral surface. Interconnected fungal biofilm often demonstrates the development of a complex and three-dimensional structure on the mineral surface (A, B), which frequently results in deeper penetration into the rock substrate (B). This elaborated network has been developed on the marble surface after only three months of laboratory cultivation.

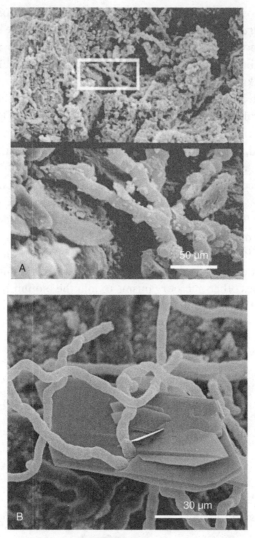

Fig. 11.4. The development of the fungal network on the marble surface (Crimea, Ukraine). As demonstrated by scanning electron microscopy, these hyphae are accompanied by intensive bacterial growth on their surface (A) and thus probably are the pioneer organisms of subaerial biofilms. (B) oxalate crystal formation was observed only when the fungal colonies were cultivated subaerially or in sterile sand. Coverage of the colony with podzol soil blocked the biomineralization process (data not shown) probably by metabolic products of soil micro-organisms.

In spite of an obviously hostile situation, the solid/air interface environment has also several advantages. For instance, the presence of the earliest life is nowadays more frequently associated with the stimulatory effect of surfaces than with the presence of liquid water, which may dilute out nutrients. It is assumed that solid surfaces were extremely important for the origin of life on early Earth (Reysenbach & Cady, 2001). Subaerial microbial biofilms inhabit rocks and take advantage of the interface between different media, where a turbulent atmospheric phase is responsible for a constantly changing movement and supply of materials. Living at the interface between the atmosphere and the rock substrate also has advantages in that particulate and volatile matter accumulate passively from the atmosphere. Such deposition of particulate matter is actively enhanced by subaerial biofilms as they increase the stickiness and absorptive characteristics of the substrate by EPS deposition. This polymer matrix transforms rock surfaces into highly adhesive aeolian catchment surfaces, which accumulate dust and other air-borne material and significantly increase 'soiling' of the biofilm overgrown surface.

The strong survival potential of the rock biofilm indicates that the permanent presence of water is not the most essential attribute for the evolution and spread of life (Reysenbach & Cady, 2001; Costerton & Stoodley, 2003). The existence of a microbial biofilm on the rock surface is more determined by the interactions of the organisms with the mineral substrate. Lack of water even over several years is tolerated by rock biofilms. However sporadic the supply might be in subaerial conditions, some water from rain, snow, ice, dew, or fog is always present in terrestrial environments. However, for chemoorganotroph life forms the mineral substrate harbours additional difficulties as it is, or almost immediately becomes, deficient in organic matter, as nutrients and energy resources reach this habitat mainly from the atmosphere as particulates and volatile matter. This difficulty is partially overcome by the presence of EPS, which significantly increase the residence time of air-borne particles on any rock surface.

Environmental conditions related to rock surface exposure have profound effects on biofilm development, as metabolic activity and growth are directly connected to the availability of water, energy sources and nutrients, as well as to conditions of temperature and irradiation. Another important factor for establishment of subaerial biofilms is the resistance of the supporting substrate to environmentally and biologically influenced disintegration and dissolution (wear-down). Rapidly weathering rock surfaces (e.g. porous sandstone in an intertidal coastal zone) show little or sparse biofilm development, whereas hard rocks (basalt, quartzite, dense limestone, siliceous chert) have the longest time of contact with the atmosphere and thus exhibit the most prominent microbial biofilms as well as other traces of subaerial exposure and growth. On a hard and weathering-resistant support (granite, basalt, quartzite) the weathering front is stable for longer time periods and accumulates multiple layers of biofilm and atmospheric deposits (Krumbein, 1969). In some cases, such developments may lead to the formation of superimposed mineralized deposits termed microstromatolites or desert stromatolites not unlike the microscopic layers of desert varnish (Allen, 1978; Dorn, 1984; Broecker & Liu, 2001). In this case the accumulation of mainly fungal-derived terrestrial deposits replaces biological wear-down (Gorbushina & Krumbein, 2005).

Diversity and biological interactions of subaerial rock-inhabiting fungi

Many different organisms inhabit rock surfaces worldwide and, by covering grains and intergranular spaces, form a global biofilm (Fig. 11.1, 11.5). Under the already mentioned complex gaseous/solid microclimatic conditions the whole structure of the biofilm is different from those encountered in any subaquatic environment. Although specialized extracellular biofilm matrices are observed on the rock surface (Fig. 11.1A), the whole biofilm system under the atmosphere is based on completely different principles of survival. A permanently flowing and diluting liquid represents totally different conditions from the permanently desiccating breeze from the atmosphere. The diversity and abundance of fungi on subaerial rock substrates thus depends on the structure and

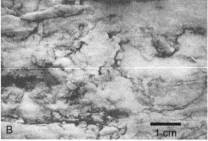

Fig. 11.5. Many different organisms inhabit rock surfaces and by covering the grains and intergranular spaces form a subaerial biofilm. In the deeper layers phototrophic organisms are situated (A), while the surface is mainly inhabited by stress-tolerant dark-pigmented fungi (B).

chemistry of the substrate as well as on the complex climatic conditions of exposure to the atmosphere.

Climate influences the seasonal, as well as the general persistent distribution of fungi on rock surfaces. At present rock surfaces in even the most harsh conditions as the Antarctic (Friedmann, 1982) and hot deserts (Staley *et al.*, 1982) and in moderate climatic conditions as in Central Europe (Krumbein, 1969; Hirsch *et al.*, 1995) are inhabited by specialized microbial communities, which have a considerable geochemical impact on the substrate. Fungi are predominantly strictly aerobic and therefore primarily occupy the upper portion of the lithosphere. This makes them ideal biofilm producers on subaerial rock surfaces. Fungi dominate in all temperate and polar soils that are well aerated and oligotrophic (Kjoller & Struwe, 1982). It can be expected that a high diversity of these organisms is also found on rock surfaces. The mycelial growth habit along with a capability of some species to grow dimorphically – changing from hyphal to yeast-like growth – significantly contribute to the dominance of these organisms in the habitats in question. The growth mode of fungi definitely supplies them with a whole set of advantages in colonizing solid surfaces. However, mineral substrates are poor in nutrients and are exposed to desiccation, irradiation and temperature stress and thus refractory to rapid mycelial growth. Because of these reasons some fungal groups have abandoned their branched hyphal elongation pattern, which appears more appropriate for penetration and utilization of nutritious substrates. Some autochthonous rock-inhabiting fungi often form microcolonies – structures that enhance their survival and persistence (Palmer *et al.*, 1987; Sterflinger & Krumbein, 1995; Gorbushina *et al.*, 2003). These microcolonies present a special form of growth typical for subaerial biofilms (Staley *et al.*, 1982). Probably, the relationship of a microcolony to the normal hyphal colony is analogous to the difference between a planktonic and a sessile cell aggregate in the aqueous environment (Costerton & Stoodley, 2003). This microcolonial yeast-like growth form is brought about by the adaptation to growth in unfavourable environmental conditions and on solid substrates far removed from softer substrates that enable hyphal embedding and spreading.

Subaerial rock communities metabolize under conditions of limited water availability and high solar and cosmic irradiation levels and can be found even on desert rocks and at high altitudes. In these places, they have found an ideal environment that allows for a stressful but less competitive existence. In more favourable conditions, these communities are quickly succeeded by more developed but less stress-tolerant symbiotic lichen or higher plant communities (Chertov *et al.*, 2004). Extremely arid desert or pseudo-desert conditions on the rock surface usually result in the dominance of xerotolerant organisms such as free-living micromycetes (Staley *et al.*, 1982; Braams, 1992; Grote & Krumbein, 1992; Hirsch *et al*, 1995; Sterflinger, 2000). Rock-inhabiting fungi can be split into two ecological groups: (1) hyphomycetes of soil and epiphytic origin (de Leo *et al.*, 1996) and (2) black (melanized) microscopic fungi forming compact microcolonies and known to be typical inhabitants of rock surfaces (Gorbushina *et al.*, 1993; Diakumaku *et al.*, 1995; Sterflinger *et al.*, 1999). Due to their characteristic restricted-growth pattern the latter group of black rock-inhabiting fungi are also termed microcolonial fungi (MCF, Staley *et al.*, 1982). Typical adaptational features of MCF are melanized cell walls (Gorbushina *et al.*, 1993) and optimal surface to volume ratio both for their spherical yeast-like cells as well as for the compact microcolonies (Wollenzien *et al.*, 1995). These characteristics are important for MCF survival in challenging epilithic desert-type conditions and make MCF the most stress-tolerant and persistent inhabitants of subaerial rock biofilms on desert rocks (Staley *et al.*, 1982), or on any other rock surfaces with prolonged subaerial exposure (Gorbushina, 1998).

The biogeochemical and evolutionary importance of subaerial fungal communities has been largely underestimated. Terrestrial fungi are commonly portrayed as being connected to degradation and transformation of organic substances. However, this common opinion has been significantly changed since numerous examples of fungal interactions with inorganic materials have been recently reported (e.g. Gadd & Mowll, 1985; Diakumaku *et al.*, 1995; Gadd, 1999; Sterflinger, 2000; Burford *et al.*, 2003; Fomina *et al.*, 2004).

Subaerial biofilms and geochemical change in the rocky substrate by organic metabolite secretion

Interactions between subaerial microorganisms and their organic products (which may be termed biomarkers in marine sediments and sedimentary rocks) with the mineral substrate are noteworthy for terrestrial geological evidence. Here, the formation and longevity of microbially mediated complex organic molecules in terrestrial subaerial environments will be discussed. The most important components are pigments that stay geochemically stable for long periods of time. Organic, highly complex and polymerized pigments may act as geological tracers of an extant microbial terrestrial community. The production of dark and irradiation-protective organic pigments is closely related to metal accumulation (Gadd & Mowll,

1985; Gadd & de Rome, 1988; Purvis *et al.*, 2004) and deposition on all rock surfaces worldwide (Krumbein, 2003). Metal oxidation as well as metal enrichment is often connected to the presence of melanin and other polymeric pigments (Krumbein, 1969; Gadd, 1999). In warm dry land areas, where it is thought that high temperatures and low moisture levels promote thicker and darker desert varnish accumulation on hard rock types, a combination of organic dark pigments with heavy metal oxides may often be observed. Desert varnish forms very slowly, $\sim$10 000 years for a heavy and thick varnish (Elvidge, 1982), and becomes thicker and darker with increasing age. Krumbein (2003) has demonstrated the evolution of fresh varnish layers in the Negev desert over a period of 30 years. Freshly carved rock surfaces were exposed at intervals of 3–7 years and the speed of organic pigment and iron/manganese oxide accumulation was observed (Krumbein & Jens 1981; Krumbein, 2003). Neo-mineral formation by microcolonial fungi was mainly achieved in collaboration with metal-oxidizing bacteria. The most prominent mineralization potential is probably therefore a synergistic effect of the bacterial and fungal communities of rock biofilms. It still remains difficult to determine the relative biogeochemical importance of fungi in comparison to bacteria, because in these communities the overall activity of a biofilm is difficult to separate from the activity of separate members of the ecosystem. For instance, calcium carbonate production in soils is based on joint activities of oxalate- and citrate-producing fungi and oxalate- and citrate-utilizing bacteria (Braissant & Verrecchia, 2002). Calcification in microbial mats is initiated within a polymer biofilm of a mixed cyanobacterial–chemotrophic bacterial origin that embeds microbial communities (Dupraz *et al.*, 2004). The mineralization activity of such fungal/bacterial communities is frequently determined by a combination of low-molecular-weight organic acid-(LMWOA) producing fungi and bacteria that utilize these LMWOA, resulting in calcite deposition. Here, the activity of terrestrial cyanobacteria also has to be mentioned. Heterotrophic fungi settle on rocks, derive organics from air-borne particles, excrete oxalic acid and produce large amounts of oxalates (mostly with calcium, although magnesium, iron and manganese oxalates have also been reported). Even if the majority of persistent rock-inhabiting fungi (microcolonial fungi, for instance) have not been shown to increase weathering through the production of low-molecular-weight organic acids this potential can be inferred from their general metabolic features. However, their mere presence helps to increase local CO_2 concentrations and decrease the pH of the rock surface in the contact zone. Subsequent evolution of complex associations with algae, cyanobacteria and chemotrophic bacteria enhance the weathering potential of subaerial rock communities. Sometimes the fungi get entrapped in their own mineral products and produce formations, which may be interpreted as boring or drilling structures or as cavities left after the death and disintegration of the hyphae. In many cases the mineralization process is unspecific and related only vaguely to the extracellular physical/chemical conditioning of the immediate environment of the microbial cells.

In pure cultures minerals are often formed only under subaerial and loose-aerated sand exposure but not in an aqueous environment. In several experiments crystal formation was observed only when the fungal colonies were cultivated subaerially or in sterile sand (Fig. 11.4B, Gorbushina, unpublished data). Coverage of the colony with podzol soil blocked the mineralization process, probably by metabolic products of soil microorganisms (presumably the same oxalate-utilizing bacteria mentioned before). However, other soil components (organic and inorganic) as well as a slightly different chemical environment could also have influenced this. In addition, the presence of EPS mediates multiple biological functions such as UV-light and desiccation protection (Hill *et al.*, 1994; Cockell & Knowland, 1999; Potts, 1999; Costerton & Stoodley, 2003), which may interfere with mineralization. The complex relationships and communication strategies between diverse pro- and eukaryotic organisms as well as photo- and chemotrophs during establishment and development of a subaerial biofilm are not well understood at present. One of the most successful strategies for growth in primitive ecosystems may have been the combination of different metabolic strategies. Such combinations of approaches in stressed situations have to include the absence of antibiosis, which is absent from rock-inhabiting communities (Nienow & Friedmann, 1993).

Rock-inhabiting communities are obviously illuminated during daytime and subaerial communities must be able to protect themselves from the harmful influence of sunlight and cosmic radiation. There are advantages and drawbacks in this situation. On the one hand, the abundant sunlight would allow for the development of a metabolically integrated community with cooperating photosynthetic and chemoorganotrophic organisms, which could result in a well-balanced structure with a successful energy budget. Subaerial epi- and endolithic biofilms create and maintain biologically modified environments where mineral solubility and dissolution rates are significantly altered. However, in complex subaerial biofilms, free-living fungi frequently maintain mutually beneficial contacts with bacteria and phototrophic organisms. Fungal influence on solid rocks is manifest through (1) a series of excreted secondary metabolic products (low-molecular-weight organic acids and organic polymers including diverse pigments and antioxidants) and (2) biologically induced formation of new minerals (carbonates, oxalates, oxides, phosphates and silicates or silica). Further the geomicrobiological and biogeochemical role and modification of fungal metabolic products on rock surfaces will be documented by a complex system of accumulation and degradation processes for organic and inorganic compounds that alter not only the appearance of rock surfaces but also light reflectance and adsorption properties (Gorbushina *et al.*, 2002).

Biogeomorphological impact of fungi

Biological processes have an important bearing on the geomorphology of all air-exposed rock and soil surfaces (Viles, 1984; Viles & Naylor, 2002). Ultimately the biogeomorphogenetic potential of subaerial biofilms and the potential of creating interesting and unusual mineral assemblages will be included into the geobiology of these communities (Gorbushina *et al.*, 2002). The survival potential of symbiotic (e.g. lichen) biofilms as compared to biofilms composed of free-living organisms (e.g. cyanobacteria, bacteria, algae, fungi) must also be taken into account. Biofilms composed of meristematically growing black MCF or 'black yeasts' are weathering agents creating patina or rock varnish, biopitting and rock exfoliation on a geological scale in space and time. Fungal influence on solid rocks is manifest through mechanical impact on grain cohesion through turgor pressure. These processes lead to the formation of a patina that protects the organisms and the rock they live on, but also can lead to heavy damage and destruction of architectural units, mural paintings and rock art (petroglyphs).

As a kind of a (bio)geomorphological summary, it can be stated that numerous investigations strongly suggest that the weathering of minerals can be accelerated by the growth of some fungal and lichen species. The effects of fungi on their mineral substrates can be attributed to both physical and chemical processes. Physical effects are manifest by mechanical disruption of rocks caused by hyphal penetration, and the swelling action of the organic and inorganic salts originating from fungal metabolism. Fungi also have a significant impact on the chemical weathering of rocks by the excretion of various organic acids, particularly oxalic acid, which can effectively dissolve minerals and chelate metallic cations. As a result of the weathering induced by subaerial biofilms, many rock-forming minerals exhibit extensive surface corrosion. The precipitation of poorly ordered iron oxides and amorphous alumino-silica gels, the neoformation of crystalline metal oxalates and secondary clay minerals have been frequently identified in a variety of rocks colonized by fungi. In general, the principles of the geological impact of fungi can be stated as follows: (1) Mn(II) ions are microbially transformed into manganese oxides; (2) this process involves at least one if not several metalloenzymes; (3) laccase activity is proved for the responsible enzyme (e.g. in *Acremonium* sp.); (4) Ascomycetes are ubiquitous in natural environments and many of them possess laccase activity; (5) for rock-inhabiting fungi, manganese and iron oxidation has been demonstrated many times (Grote & Krumbein, 1992).

As mineral soil horizons are the result of intense biological activity in the litter zone and the upper soil, the zonation of weathered rock transforming into a soil profile is very much the result of biological growth on its surface (Banfield *et al.*, 1999). Geochemical zonation of subaerial rock surfaces takes place in accordance with the distance from direct atmospheric influences and then turns into pedological processes (iron and manganese encrustations in the B horizons). On the other hand directly atmosphere-exposed environments, e.g. desert varnish (brown to almost black deposits with a laminar structure composed of melanin pigments combined with iron and manganese oxides in a dispersed clay matrix) are found across the world on rock surfaces under arid conditions where it is thought that high temperatures and low moisture levels promote thicker and darker desert varnish accumulations. Desert varnish forms very slowly (Elvidge, 1982), and becomes thicker and darker with increasing age. Even rocks deficient in manganese, iron and trace elements usually found within desert varnish are equally capable of developing a desert varnish coating, which suggests an external source of these components. Wind-blown dust has been proposed as an external source, with microbiological rather than physico-chemical mechanisms being considered as the dominant processes driving the selective fixation of desert varnish components from atmospherically introduced material. Another important factor for establishment of accumulated biogenic patination is the resistance of the supporting substrate to wear-down. Hard and weathering-resistant rock surfaces (chert, basalt, quartzite, dense limestone) have the longest time of contact with the atmosphere and thus exhibit the most prominent traces of former and extant subaerial microbial biofilms as well as multiple layers of atmospheric deposits (Krumbein, 1969).

Complex (bio)geomorphological rules and principles, which are not well understood, determine when and where soils are formed, and this drastically modifies the interplay between bare rock and the atmosphere in terrestrial environments. Whether soils or a permanent decrease of rock surface or rock patinations will be generated and maintained depends on so many different climatic and biological factors that a conclusive treatment seems to be difficult at this moment. It suffices to state that in biogeochemistry and geomorphology, the marine environment is seemingly much better documented than the much more important and fractally dominant terrestrial/continental surface of planet Earth.

Conclusions

For organisms prevailing on desert or desert-like rock surfaces, long-term survival is of higher importance then any sporulation/propagation strategy. This may apply to all environments with extreme or extremely fluctuating physical/chemical conditions. Rock-inhabiting microbial communities are a living system tuned to the conditions of this specific niche environment. Earth was inhabited by diverse, small organisms for a long period of time (Schopf, 1999), and these microorganisms and their assemblages are still playing important roles in the biogeochemistry of our planet (Krumbein *et al.*, 2003). Terrestrial environments especially are inhabited by complex multispecies associations, which are interconnected both functionally and metabolically. Subaerial microbial communities have been active and abundant in practically all geologic eras. They present a complex interactive system with cooperative communication between rock biofilm community members, the atmospheric environment and the rock (mineral) substrate.

- A concise picture of how life prospers on a subaerial rock surface may retrospectively explain how primary terrestrial environments were conquered with the help of biogeochemical adaptations. Adaptation mechanisms of modern rock inhabitants may provide important clues for understanding the processes of early land colonization.
- It is probable, that fungi evolved as the most stress-tolerant surface inhabitants and thus made this habitat available for succeeding and accompanying organisms.
- External protection for less tolerant photobionts and bacteria – a new niche for sensitive but useful neighbours – finally resulted in close symbiotic cooperation. Experimental evidence of in vitro interactions between MCF and isolated lichen photobionts (Gorbushina *et al.*, 2005) confirm the hypothesis that the colonization of land by eukaryotes was facilitated by a symbiosis between a photosynthesizing organism and a fungus (Heckmann

et al., 2001). It should be pointed out that (1) subaerial biofilm communities cannot be considered as 'primitive' because of the very long history of their development and high degree of specialized organization; and (2) free-living fungal, bacterial and algal biofilms are important predecessors of lichen communities, which evolved much later.

- Rock biofilms may also host the remnants of previous lichen symbioses that have lost one of their symbionts (e.g. fungi of lichens).
- Microbial colonization of bare rock surfaces represents the initial stage of all land (terrestrial) successions.

Revelation of the mechanisms used by rock-inhabiting microscopic life to establish itself on rock surfaces can answer a broad spectrum of fundamental questions. Although no direct analogies can be drawn between fossil organisms and their modern counterparts, understanding the rules and laws of rock-inhabiting community development on modern Earth could elucidate the principles of land colonization in the far past.

Acknowledgements

The author expresses her deep gratitude to Geoffrey M. Gadd for his editorial help, and thanks Wolfgang Krumbein for his unfailing support and stimulating discussions. The financial support of an EU grant BIODAM (EVK4-CT-2002–000098 'Inhibitors of biofilm damage on mineral materials') and a DFG-grant Go 897/2–2 is gratefully acknowledged.

References

Allen, C. C. (1978). Desert varnish of the Sonoran Desert – optical and electron probe microanalysis. *Journal of Geology*, **86**, 743–52.

Avakian, Z. A., Karavaiko, G. I., Mel'nikova, E. O., Krutsko, V. S. & Ostroushko, I. I. (1981). Role of microscopic fungi in the process of weathering of pegmatite deposit rocks and minerals. *Mikrobiologiya*, **50**, 156–62.

Banfield, J. F., Barker, W. W., Welch, S. A. & Taunton, A. (1999). Biological impact on mineral dissolution: application of the lichen model to understanding mineral weathering in the rhizosphere. *Proceedings of the National Academy of Sciences of the United States of America*, **96**, 3403–11.

Barker, W. W., Welch, S. A. & Banfield, J. F. (1997). Biogeochemical weathering of silicate minerals. *Reviews in Mineralogy*, **35**, 391–428.

Bechinger, C., Giebel, K. F., Schnell, M. *et al.* (1999). Optical measurements of invasive forces exerted by appressoria of a plant pathogenic fungus. *Science*, **285**, 1896–9.

Braams, J. (1992). Ecological studies on the fungal microflora inhabiting historic sandstone monuments. In *Geomicrobiology*, Oldenburg: Oldenburg University, pp. 128.

Braissant, O. & Verrecchia, E. P. (2002). Microbial biscuits of vaterite in Lake Issyk-Kul (Republic of Kyrgyzstan) – Discussion. *Journal of Sedimentary Research*, **72**, 944–6.

Brehm, U., Gorbushina, A. A. & Mottershead, D. (2005). The role of micro-organisms and biofilms in the breakdown and dissolution of quartz and glass. *Paleogeography, Paleoclimatology, Paleobiology*, **219**, 117–29. (Published online 6 January, 2005.)

Broecker, W. S. & Liu, T. (2001). Rock varnish: recorder of desert wetness. *GSA Today*, **11**, 4–10.

Burford, E. P., Fomina, M. & Gadd, G. M. (2003). Fungal involvement in bioweathering and biotransformation of rocks and minerals. *Mineralogical Magazine*, **67**, 1127–55.

Butler, M. J. & Day, A. W. (1998). Fungal melanins: a review. *Canadian Journal of Botany*, **44**, 1115–36.

Butler, M. J., Day, A. W., Henson, J. M. & Money, N. P. (2001). Pathogenic properties of fungal melanins. *Mycologia*, **93**, 1–8.

Chertov, O., Gorbushina, A. & Deventer, B. (2004). A model for microcolonial fungi growth on rock surfaces. *Ecological Modelling*, **177**, 415–26.

Cockell, C. S. & Knowland, J. (1999). Ultraviolet radiation screening compounds. *Biological Reviews of the Cambridge Philosophical Society*, **74**, 311–45.

Costerton, J. W. & Stoodley, P. (2003). Microbial biofilms: protective niches in ancient and modern microbiology. In *Fossil and Recent Biofilms: A Natural History of Life on Earth*, ed. W. E. Krumbein, D. M. Paterson & G. A. Zavarzin, Dordrecht: Kluwer, pp. xv–xxi.

de Leo, F., Criseo, G. & Urzi, C. (1996). Impact of surrounding vegetation and soil on the colonization of marble statues by dematiaceous fungi. In *Proceedings of 8th International Congress on Deterioration and Conservation of Stone*, ed. J. Riederer. Berlin: Moeller pp. 625–30.

Diakumaku, E., Gorbushina, A. A., Krumbein, W. E., Panina, L. & Soukharjevski, S. (1995). Black fungi in marble and limestones – an aesthetical, chemical and physical problem for the conservation of monuments. *Science of the Total Environment*, **167**, 295–304.

Dorn, R. I. (1984). Cause and implications of rock varnish microchemical laminations. *Nature*, **310**, 767–70.

Dornieden, T., Gorbushina, A. & Krumbein, W. E. (1997). Änderungen der physikalischen Eigenschaften von Marmor durch Pilzbewuchs. *International Journal for Restoration of Buildings and Monuments*, **3**, 441–56.

Dupraz, C., Visscher, P. T., Baumgartner, L. K. & Reid, R. P. (2004). Microbe-mineral interactions: early carbonate precipitation in a hypersaline lake (Eleuthera Island, Bahamas). *Sedimentology*, **51**, 745–65.

Ehrlich, H. L. (1996). *Geomicrobiology*. New York: Marcel Dekker.

Elvidge, C. D. (1982). Reexamination of the rate of desert varnish formation reported south of Barston, California. *Earth Surface Processes and Landforms*, **7**, 345–8.

Fomina, M., Alexander, I. J., Hillier, S. & Gadd, G. M. (2004). Zinc phosphate and pyromorphite solubilization by soil plant-symbiotic fungi. *Geomicrobiology Journal*, **21**, 351–66.

Friedmann, E. I. (1982). Endolithic microorganisms in the Antarctic cold desert. *Science*, **215**, 1045–1053.

Friedmann, E. I. (1993). *Antarctic Microbiology*. New York: Wiley.

Friedmann, E. I. & Ocampo-Friedmann, R. (1984). Endolithic microorganisms in extremely dry environments: analysis of a lithobiontic microbial habitat. In *Current*

Perspectives in Microbial Ecology, ed. M. J. Klug & C. A. Reddy. Washington, DC: American Society for Microbiology, pp. 177–85.

Friedmann, E. I., McKay, C. P. & Nienow, J. A. (1987). The cryptoendolithic microbial environment in the Ross desert of Antarctica – satellite-transmitted continuous nanoclimate data, 1984 to 1986. *Polar Biology*, 7, 273–87.

Gadd, G. M. (1999). Fungal production of citric and oxalic acid: importance in metal speciation, physiology and biogeochemical processes. *Advances in Microbial Physiology*, 41, 47–92.

Gadd, G. M. & Mowll, J. L. (1985). Copper uptake by yeast-like cells, hyphae and chlamydospores of *Aureobasidium pullulans*. *Experimental Mycology*, 9, 230–40.

Gadd, G. M. & de Rome, L. (1988). Biosorption of copper by fungal melanins. *Applied Microbiology and Biotechnology*, 29, 610–17.

Gorbushina, A. A. (1998) Biodiversity of rock dwelling poikilotroph fungal communities with decreasing nutrient content of the habitat. In *Proceedings of VIth International Mycological Congress*, p. 140. Jerusalem.

Gorbushina, A. A. & Krumbein, W. E. (1999). Poikilotrophic response of micro-organisms to shifting alkalinity, salinity, temperature and water potential. In *Microbiology and Biogeochemistry of Hypersaline Environments*, ed. A. Oren. London: CRC Press, pp. 75–86.

Gorbushina, A. A. & Krumbein, W. E. (2000). Subaerial microbial mats and their effects on soil and rock. In *Microbial Sediments*, ed. R. E. Riding & S. M. Awramik. Berlin: Springer-Verlag, pp. 161–70.

Gorbushina, A. A. & Krumbein, W. E. (2005). Role of microorgansims in wear down of rocks and minerals. In *Soil Biology*, Vol. 3. *Microorganisms in Soils: Role in Genesis and Functions*, ed. F. Buscot & A. Varma. Berlin: Springer-Verlag, pp. 59–84.

Gorbushina, A. A., Krumbein, W. E., Hamann, C. H. *et al.* (1993). Role of black fungi in color change and biodeterioration of antique marbles. *Geomicrobiology Journal*, 11, 205–21.

Gorbushina, A. A., Krumbein, W. E. & Volkmann, M. (2002). Rock surfaces as life indicators: new ways to demonstrate life and traces of former life. *Astrobiology*, 2, 203–13.

Gorbushina, A. A., Whitehead, K., Dornieden, Th. *et al.* (2003). Black fungal colonies as units of survival: hyphal mycosporines synthesized by rock dwelling microcolonial fungi. *Canadian Journal of Botany*, 81 (2), 131–8.

Gorbushina, A. A., Beck, A. & Schulte, A. (2005). Microcolonial rock inhabiting fungi and lichen photobionts: evidence for mutualistic interactions. *Mycological Research*, 109 (in press).

Gromov, B. V. (1957). Microflora of rock substrates and primitive soils from northern areas of the USSR. *Mikrobiologiya*, 26, 52–9.

Grote, G. & Krumbein, W. E. (1992). Microbial precipitation of manganese by bacteria and fungi from desert rock and rock varnish. *Geomicrobiology Journal*, 10, 49–57.

Heckmann, D. S., Geiser, D. M., Eidell, B. R. *et al.* (2001). Molecular evidence for the early colonization of land by fungi and plants. *Science*, 293, 1129–33.

Hill, D. R., Peat, A. & Potts, M. (1994). Biochemistry and structure of the glycan secreted by desiccation-tolerant *Nostoc commune* (Cyanobacteria). *Protoplasma*, 182, 126–48.

Hirsch, P., Eckhardt, F. E. W. & Palmer, R. J. J. (1995). Fungi active in weathering of rock and stone monuments. *Canadian Journal of Botany*, 73, 1384–90.

Jongmans, A. G., van Breemen, N., Lundstrom, U. *et al.* (1997). Rock-eating fungi. *Nature*, 389, 682–3.

Kjoller, A. & Struwe, S. (1982). Microfungi in ecosystems – fungal occurrence and activity in litter and soil. *Oikos*, 39, 391–422.

Krumbein, W. E. (1969). Über den Einfluss der Mikroflora auf die exogene Dynamik (Verwitterung und Krustenbildung). *Geologische Rundschau*, 58, 333–63.

Krumbein, W. E. (2003). Patina and cultural heritage – a geomicrobiologist's perspective. In *Proceedings of the Vth EC Conference Cultural Heritage Research: a Pan-European Challenge*, ed. R. Kozlowski. Cracow: EC and ISC, pp. 39–47.

Krumbein, W. E. & Jens, K. (1981) Biogenic rock varnish of the Negev desert (Israel): an ecological study of iron and manganese transfer by cyanobacteria and fungi. *Oecologia*, **50**, 25–38.

Krumbein, W. E., Brehm, U., Gerdes, G. *et al.* (2003). Biofilm, biodictyon, biomat, microbialites, ooolites, stromatolites, geophysiology, global mechanisms, parahistology. In *Fossil and Recent Biofilms: A Natural History of Life on Earth*, ed. W. E. Krumbein, D. M. Paterson & G. A. Zavarzin. Kluwer Academic Publishers, Dordrecht, pp. 1–27.

Leppard, G. G. (1986). The fibrillar matrix component of lacustrine biofilms. *Water Research*, **20**, 697–702.

McMenamin, M. A. S. & McMenamin, D. L. S. (1994). *Hypersea:Life on Land*. New York: Columbia University Press.

Margulis, L. (1970). *Origin of the Eukaryotic Cells*. New Haven: Yale University Press.

Margulis, L. (1982). *Early Life*. Boston: Science Books International, Inc.

May, E., Lewis, F. J., Pereira, S. *et al.* (1993). Microbial deterioration of building stone – a review. *Biodeterioration Abstracts*, **7**, 109–23.

Nienow, J. A. & Friedmann, E. I. (1993). Terrestrial lithophytic (rock) communities. In *Antarctic Microbiology*, ed. E. I. Friedmann. New York: Wiley, pp. 342–412.

Palmer, F. E., Emery, J. & Staley, J. T. (1987). Survival and growth of microcolonial fungi as affected by temperature and humidity. *New Phytologist*, **107**, 155–62.

Potts, M. (1999). Mechanisms of desiccation tolerance in cyanobacteria. *European Journal of Phycology*, **34**, 319–28.

Purvis, O. W., Bailey, E. H., McLean, J., Kasama, T. & Williamson, B. J. (2004). Uranium biosorption by the lichen *Trapelia involuta* at a uranium mine. *Geomicrobiology Journal*, **21** (3), 159–67.

Reysenbach, A. L. & Cady, S. L. (2001). Microbiology of ancient and modern hydrothermal systems. *Trends in Microbiology*, **9**, 79–86.

Rivera, M. C. & Lake, J. A. (2004). The ring of life provides evidence for a genome fusion origin of eukaryotes. *Nature*, **431**, 152–5.

Schopf, J. W. (1999). *Cradle of Life*. Princeton, New Jersey: Princeton University Press.

Staley, J. T., Palmer, F. & Adams, J. B. (1982). Microcolonial fungi: common inhabitants on desert rocks? *Science*, **215**, 1093–5.

Sterflinger, K. (2000). Fungi as geologic agents. *Geomicrobiology Journal*, **17**, 97–124.

Sterflinger, K. & Krumbein, W. E. (1995). Multiple stress factors affecting growth of rock-inhabiting black fungi. *Botanica Acta*, **108**, 490–6.

Sterflinger, K. & Krumbein, W. E. (1997). Dematiaceous fungi as a major agent for biopitting on Mediterranean marbles and limestones. *Geomicrobiology Journal*, **14**, 219–22.

Sterflinger, K., Krumbein, W. E., Lellau, T. & Rullkötter, J. (1999). Microbially mediated orange patination of rock surfaces. *Ancient Biomolecules*, **3**, 51–65.

Stoodley, P., Sauer, D. G., Davies, D. G. & Costerton, J. W. (2002). Biofilms as complex differentiated communities. *Annual Review of Microbiology*, **56**, 187–209.

Taylor, T. N., Remy, W., Hass, H. & Kerp, H. (1995). Fossil arbuscular mycorrhizae from the Early Devonian. *Mycologia*, **87**, 560–73.

Viles, H. A. (1984). Biokarst – review and prospect. *Progress in Physical Geography*, **8** (4), 523–42.

Viles, H. A. & Naylor, L. A. (2002). Biogeomorphology – Editorial. *Geomorphology*, **47** (1), 1–2.

Volkmann, M., Whitehead, K., Rütters, H., Rullkötter, J. & Gorbushina, A. A. (2003). Mycosporine-glutamicol-glucoside: a natural UV-absorbing secondary metabolite of rock-inhabiting microcolonial fungi. *Rapid Communications in Mass Spectrometry*, **17**, 897–902.

Warscheid, T. & Krumbein, W. E. (1996). Biodeterioration of inorganic non-metallic materials – general aspects and selected cases. In *Microbially Induced Corrosion of Materials*, ed. H. Heintz, W. Sand & H. C. Flemming. Berlin: Springer-Verlag, pp. 273–95.

Welch, S. A. & Ullman, W. J. (1993). The effect of organic-acids on plagioclase dissolution rates and stochiometry. *Geochimica et Cosmochimica Acta*, **57**, 2725–36.

Welch, S. A. & Vandevivere, P. (1994). Effect of microbial and other naturally occurring polymers on mineral dissolution. *Geomicrobiology Journal*, **12**, 227–38.

Wollenzien, U., De Hoog, G. S., Krumbein, W. E. & Urzi, C. (1995). On the isolation of microcolonial fungi occurring on and in marble and other calcareous rocks. *Science of the Total Environment*, **167**, 287–94.

12

The oxalate–carbonate pathway in soil carbon storage: the role of fungi and oxalotrophic bacteria

ERIC P. VERRECCHIA, OLIVIER BRAISSANT
AND GUILLAUME CAILLEAU

Introduction

Although fungi are generally disregarded in the biogeochemical literature, they undoubtedly constitute crucial biogeochemical factors in many elemental cycles. This fact, combined with their abundance in the soil, warrants greater detailed study into their geoecological impact. The network formed by fungal filaments can represent 10 000 km of thread-like mycelia in 1 m^2 of fertile soil. Their mass is evaluated at 3500 kg ha^{-1} at a depth of 20 cm in an average continental soil, i.e. taking into account all the different terrestrial environments on Earth (Gobat et al., 2004). In comparison, bacteria and algae would represent 1500 and 10–1000 kg ha^{-1} respectively, in the same virtual average soil. Fungi are not only biologically important as saprophytes in the recycling of organic matter, but also play a geological role by excreting notable amounts of organic acids, among which oxalic acid is particularly important (Gadd, 1999), contributing to continental weathering as well as to mineral neogenesis (Verrecchia & Dumont, 1996; Verrecchia, 2000; Burford et al., 2003 a, b).

The first fossil fungi have been identified in rocks dated from the Ordovician, i.e. 460 to 455 Ma ago (Redecker et al., 2000). However, molecular clock estimates for the evolution of fungi have suggested a Late Precambrian (600 Ma) colonization on land (Berbee & Taylor 2000). Recent molecular studies, based on protein sequence analysis, indicate that fungi were present on continents 1 billion years ago and possibly affected (together with plants) the evolution of Earth's atmosphere and climate since 700 Ma (Heckman et al., 2001). Therefore, if fungi have been present on the Earth's surface for such a long time, producing large amounts of oxalic acid able to precipitate as metal oxalates, why is there no evidence of oxalate accumulation in paleosols?

Fungi in Biogeochemical Cycles, ed. G. M. Gadd. Published by Cambridge University Press. © British Mycological Society 2006.

The question is also valid for present-day soils. The aim of this chapter is to demonstrate (1) that plants and fungi can produce high amounts of calcium oxalate polymorphs (weddellite, $CaC_2O_4.2H_2O$ and whewellite, $CaC_2O_4.H_2O$) by various processes, and (2) that oxalotrophic bacteria must have used these abundant oxalate crystals as carbon, electron and energy sources by oxidizing them into calcium carbonate ($CaCO_3$), a common polymorphic mineral (in the form of calcite, vaterite or aragonite) found in soils and surficial sediments. Thermodynamic approaches will show that spontaneous oxidation of oxalate is impossible and therefore necessitates a biomediated process, leading to secondary formation of calcium carbonate and a pH increase in the soil solution. In addition, it can be demonstrated that the oxalate–carbonate pathway constitutes a carbon sink, as the carbon source is organic and not an inherited lithogenic mineral carbon (e.g. fossil limestones).

The oxalate pool

Plants, fungi and oxalate

There are several ways by which plants modify the surrounding soil to their advantage. They may be biotic, like the accumulation of plant-protecting rhizobacteria increasing the soil's ability to fight against root disease, or the acclimation of mycorrhizal fungi. There may also be abiotic factors, like the precipitation of toxic cations and minerals. Accumulation of calcite in carbonate-poor soils may improve the soil structure and function. There is increasing evidence that the calcium oxalate cycle is a major pathway in calcite biomineralization. This was shown with different oxalate-accumulating plants, such as iroko trees (Braissant *et al.*, 2004) and Cactaceae (Garvie, 2003). Nevertheless, a question is still pending: how and why do plants form oxalate crystals? In plant metabolism, oxalic acid is produced in varying amounts, depending on plant taxon and external conditions. This oxalic acid may be either accumulated in the vacuole, or precipitated in the form of insoluble calcium oxalate crystals in the cell, called crystal idioblasts (Fig. 12.1). Calcium oxalate crystal formation in plants appears to play a central role in a variety of important functions, including tissue calcium regulation, protection from herbivores, and metal detoxification (Nakata, 2003). It seems that ascorbic acid is the primary precursor for oxalate biosynthesis. The ascorbic acid can be synthesized directly within the calcium oxalate crystal-accumulating cell (Nakata, 2003). Obviously, calcium oxalate accumulated in the aerial parts of the plant can eventually be transferred to the litter after plant death, during organic matter oxidation.

As an example, iroko trees (*Milicia excelsa*, Moraceae), which are trees from the tropical African forest, can accumulate large amounts of oxalate crystals in their trunk as well as their roots (Cailleau *et al.*, 2005). In the wood, calcium oxalate occupies some idioblasts, forming euhedral crystals (Fig. 12.2A), often occurring as chains inside neighbouring cells. In roots, calcium oxalate forms a mineral network of crystals between the inner root and the thin cuticle constituting the outer part of the root (Braissant *et al.*, 2004). When fungi decay parts of dead roots or rotting wood (Fig. 12.2B, C), they free oxalate crystals inside the soil or litter (Fig. 12.2D), increasing their proportion outside the living tissues. This plant oxalate pool is a widespread and abundant carbon source for oxalate consumers.

Contrary to plants, whose excretion of oxalate ions is generally considered as negligible, fungi (mycorrhizal as well as saprophytic) are essentially

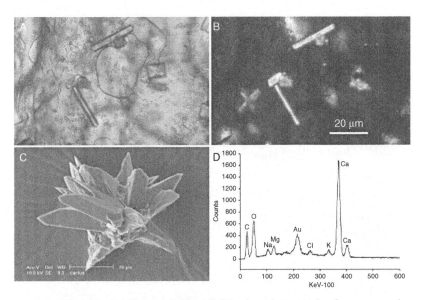

Fig. 12.1. (A) Plane-polarized light microphotograph of a cross-section inside a *Cereus* sp. succulent stem: prism and styloids of calcium oxalate are visible as crystal idioblasts. Scale bar shown on image B. (B) The same view in cross-polarized light showing the mono-crystalline nature of the biominerals. (C) Scanning electron micrograph showing the structure of a drusic agglomerate of calcium oxalate inside an *Opuntia* sp. (D) Electron dispersive energy spectrum for druse: elements found are mainly carbon, oxygen and calcium (for the calcium oxalate), magnesium, potassium, sodium and chlorine as accessory salts, and traces of phosphorus. The gold peak is due to sample coating. The identification of the mineral (calcium oxalate monohydrate or COM) was confirmed by X-ray diffraction analysis.

292   *E. P. Verrecchia* et al.

oxalate excreters. Metal oxalates, essentially calcium oxalates (Lapeyrie *et al.*, 1990; Gadd, 1999; Tait *et al.*, 1999), then crystallize at the mycelium surface or in the nearby soil, mainly as monohydrate (COM – calcium oxalate monohydrate or whewellite) and dihydrate (COD – calcium oxalate dihydrate or weddellite; Fig. 12.3A, B). Each of them belongs to a different crystallographic system, monoclinic and tetragonal, respectively (see synthesis in Verrecchia *et al.*, 1993). An extensive review of oxalate biosynthesis in fungi has been produced by Gadd (1999). Oxalic acid production by fungi appears to depend on whether glucose or citrate is used as the carbon source (Fig. 12.3C). When glucose is used, oxalate is produced through oxidation of glucose to pyruvate (Verrecchia, 1990; Wolschek & Kubicek, 1999):

$$C_6H_{12}O_6 + O_2 \rightarrow 2CH_3COCOOH + 2H_2O \qquad (12.1)$$

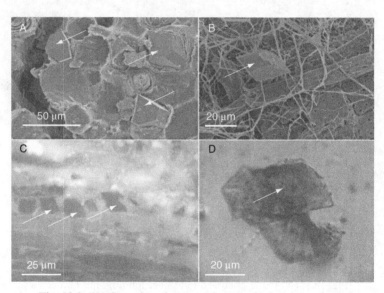

Fig. 12.2. (A) Scanning electron micrograph showing euhedral crystals (arrows) of COD (calcium oxalate dihydrate or weddellite): the other part of the image is mainly cellulose. (B) COD crystal observed with scanning electron microscope. The crystal (arrow) is surrounded by fungi decaying iroko wood whose cellulose fibrous structure is still observable. (C) Ultraviolet-fluorescent light microphotograph of a cross-section inside decaying wood. Between the wood fibres, a chain of dark euhedral crystals of COD has almost been freed in the medium. (D) View with optical binoculars of free crystals of COD inside a soil aggregate. The shape (habitus) of the crystal shown with the arrow is exactly the same as the one of the crystals shown in B.

Then, carboxylation of pyruvate yields oxaloacetate:

$$2CH_3COCOOH + 2CO_2 \rightarrow 2HOOC.CH_2CO.COOH \quad (12.2)$$

The hydrolysis of oxaloacetate allows the formation of oxalate and acetate:

$$2HOOC.CH_2CO.COOH + 2H_2O \rightarrow 2(COOH)_2 + 2CH_3COOH \quad (12.3)$$

Oxalic acid reacts with calcium, forming calcium oxalate crystals, either COM or COD. Although calcium oxalate crystals are easily observed

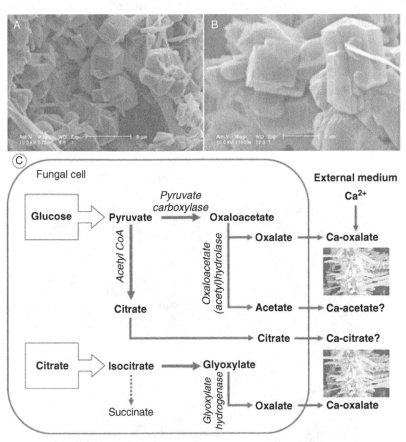

Fig. 12.3. (A) Scanning electron micrograph showing euhedral crystals of COD (calcium oxalate dihydrate or weddellite) associated with fungal filaments in decaying iroko wood (Ivory Coast). Scale bar $= 5\,\mu m$. (B) Detail of COD crystal shown in A. The COD crystals belong to the tetragonal system. Scale bar $= 2\,\mu m$. (C) Diagram of various pathways for calcium oxalate production by fungi. Biosynthesis of calcium oxalate is often accompanied by other low-molecular-weight organic acids, such as citrate or acetate.

associated with fungi, calcium acetate or calcium citrate crystals have never been identified. On the one hand, it is true that they constitute extremely soluble salts. But, on the other hand, organic substrates such as acetate or citrate can easily be used as carbon sources by bacteria. The culture of oxalatrophic bacteria (e.g. *Ralstonia eutropha* and *Xanthobacter autotrophicus*) for a few days on acetate-rich and citrate-rich media results in the production of calcite (calcium carbonate) crystals at the expense of the low-molecular-weight organic acids (Fig. 12.4; Braissant *et al.*, 2002). Therefore, these by-products can also be a source of secondary calcium carbonate in soils, even if they are highly soluble, and probably rapidly leached. In contrast, calcium oxalate crystals precipitate and may consti- tute almost 25% of soil hyphae and rhizomorphs' dry weight in some ecosystems (Cromack *et al.*, 1977). When citrate is the source, oxalate is produced through the isocitrate and glyoxylate cycle involving glyoxylate hydrogenase (Dutton *et al.*, 1993).

The accumulation of oxalate crystals by fungi has also unexpected consequences: calcium oxalate can be disseminated inside the soil and

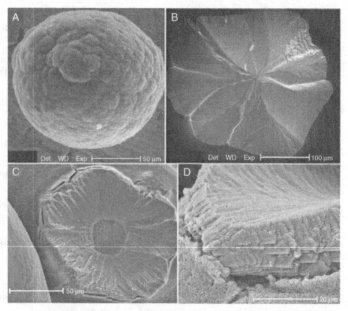

Fig. 12.4. Scanning electron micrograph showing calcium carbonate crystals found (A) associated with a *Xanthobacter autotrophicus* culture on an acetate-rich medium, (B) on the same medium with *Ralstonia eutropha*, (C) with the same bacterium on a citrate-rich medium and (D) on the same medium with *Xanthobacter autotrophicus*.

the litter through oribatid mites. Such mites feed on calcium oxalate crystals produced by the fungi and then reprecipitate the mineral in their cuticular hardened parts (Norton & Behan-Pelletier, 1991). They are thought to process a significant portion of the calcium pool in some ecosystems (Gist & Crossley, 1975). In conclusion, calcium oxalate is a common product of the biosphere–lithosphere interface, it can be found in numerous environments and it results from biomineralization under the control of organisms. So why does it not accumulate in surficial environments and soils?

Calcium oxalate stability

In order to solve the problem of missing calcium oxalate accumulation, the first question to ask is whether or not the mineral is stable, i.e. is calcium oxalate able to spontaneously oxidize when in contact with the atmosphere? If this is the case, the explanation is simple: all the oxalate produced is rapidly oxidized as CO_2, and therefore can be neither accumulated in the surficial environment nor in the fossil record. This assumes that the transformation of oxalate into CO_2 must be complete and rapid in normal conditions, i.e. at 25 °C (298.15 K) and a pressure of 1 atm. This complete oxalate oxidation in solution is given by the following reaction:

$$C_2O_4{}^{2-} + 1/2\,O_2 + 2H^+ \rightleftharpoons 2CO_2 + H_2O \tag{12.4}$$

This equation can be divided into two redox equations:

$$C_2O_4{}^{2-} \rightleftharpoons 2CO_2 + 2e^- \tag{12.5}$$

$$1/2\,O_2 + 2H^+ + 2e^- \rightleftharpoons H_2O \tag{12.6}$$

For each of the redox couples, the potential can be calculated using the Nernst equation. This equation correlates Gibb's free energy, known as ΔG, and the electromotive force provided by an oxido-reduction reaction (such a reaction acts as a galvanic cell). Given the following equation due to a chemical reaction:

$$aA + bB \rightleftharpoons cC + dD \tag{12.7}$$

reaction coefficient Q is calculated using the following ratio:

$$Q = \frac{[C]^c\,[D]^d}{[A]^a\,[B]^b} \tag{12.8}$$

At equilibrium in the solution, $Q = K_{eq}$, K_{eq} being the equilibrium constant associated with the reaction. Gibb's law is expressed as:

$$\Delta G = \Delta G^0 + RT \ln Q \text{ and } \Delta G = -n\mathcal{F}\Delta E \qquad (12.9)$$

This last equation gives the relationship between Gibb's free energy and the electromotive force, ΔE. Consequently, by expressing ΔG and ΔG^0 (standard free energy) in terms of ΔE and ΔE^0 (standard electrode potential), the following equation is obtained:

$$-n\mathcal{F}\Delta E = -n\mathcal{F}\Delta E^0 + RT \ln Q \Leftrightarrow$$

$$\Delta E = \Delta E^0 - \frac{RT}{n\mathcal{F}} \ln \frac{[C]^c [D]^d}{[A]^a [B]^b} \qquad (12.10)$$

Equation (12.10) is known as the Nernst equation, where R is the gas constant ($8.314 \text{ J mol}^{-1} \text{ K}^{-1}$), T the temperature in Kelvin, Q the reaction quotient as defined in Eq. (12.8), $\mathcal{F}$ the Faraday constant ($9.65 \times 10^4 \text{ C mol}^{-1}$) and n being the number of electrons involved during the oxido-reduction reaction. By using log (to the base 10) instead of the natural logarithm, replacing the variables with their numerical values and fixing the temperature at 25 °C (298 K), the Nernst equation becomes:

$$\Delta E = \Delta E^0 - \frac{0.059}{n} \log \frac{[C]^c [D]^d}{[A]^a [B]^b} \qquad (12.11)$$

Therefore, by using Eq. (12.11), the potential E_1 for Eq. (12.5) is given by:

$$E_1 = -0.49 + \frac{0.059}{2} \log \frac{(pCO_2)^2}{|C_2O_4{}^{2-}|} \qquad (12.12)$$

where $E_0 = -0.49$ is obtained via experimental measurement. For Eq. (12.6), the Nernst equation is written as:

$$E_2 = E_{water}^0 + \frac{0.059}{2} \log \left((pO_2)^{\frac{1}{2}} |H^+|^2 \right) \Longleftrightarrow$$

$$E_2 = 1.23 - \frac{0.059}{2} \times 2 \times \underbrace{(-\log|H^+|)}_{pH} + \frac{0.059}{2} \log \sqrt{(pO_2)} \Longleftrightarrow$$

$$E_2 = 1.23 - 0.059 \times pH + \frac{0.059}{2} \log \sqrt{(pO_2)} \qquad (12.13)$$

At equilibrium in the solution, $E_2 = E_1$. By combining Eqs. (12.12) and (12.13) and by replacing variables with numerical values, such as $pO_2 = 0.2095 \text{ atm}$, $pCO_2 = 3.3 \times 10^{-4} \text{ atm}$, the equilibrium condition yields:

$$-0.49 + \frac{0.059}{2}\log\frac{(pCO_2)^2}{|C_2O_4{}^{2-}|} = 1.23 - 0.059 \times pH + \frac{0.059}{2}\log\sqrt{(pO_2)} \Leftrightarrow$$

$$\underbrace{\frac{-1.72 \times 2}{0.059}}_{-58.305} + 2 \times pH + \underbrace{\log\frac{(pCO_2)^2}{\sqrt{(pO_2)}}}_{-6.623} = \log|C_2O_4{}^{2-}| \Leftrightarrow$$

$$- 64.928 + 2 \times pH = \log|C_2O_4{}^{2-}| \tag{12.14}$$

By varying the pH between 7 and 14 in Eq. (12.14), the concentration of oxalate ions is:

$$7 < pH < 14 \Rightarrow 1.18 \times 10^{-51} < |C_2O_4{}^{2-}| < 1.18 \times 10^{-37}$$

From these calculations, it appears that there is effectively no $C_2O_4{}^{2-}$ in solution. But in fact, this reaction occurs in the solution at an almost infinitely low rate under normal conditions of pressure and temperature. Indeed, many oxalate salts have a low solubility index. However, this point cannot explain why a solution of potassium oxalate, which is soluble, will not oxidize spontaneously into CO_2. The problem is linked to a lack of *activation energy*. Such an oxidation needs a certain initial amount of energy to occur. If a solution of calcium oxalate is left in contact with the atmosphere, it will never be oxidized into CO_{2aq}, $HCO_3{}^-$, $CO_3{}^{2-}$ and Ca^{2+}. Nevertheless, if oxalatrophic bacteria are in contact with the solution, they will provide energy to start the oxidation and feed on the oxalate carbon source.

In conclusion, any metal oxalate can be considered as a compound in a metastable equilibrium. The only sub-spontaneous diagenetic evolution of calcium oxalate is a possible transformation of weddellite into whewellite by dehydration (Frey-Wyssling, 1981; Verrecchia et al., 1993). Consequently, activation energy has to be provided for a complete oxidation of COM or COD, and life is the best and the most obvious candidate. Therefore, biogenic activity could be the key explanation for the absence of oxalate in paleosols as well as in the geological sedimentary record.

Oxalate oxidation

Oxalotrophic bacteria
The fate of oxalate in natural systems remains unclear, although oxalate catabolism by bacteria is a well-recorded phenomenon (Tamer & Aragno, 1980; Allison et al., 1995). Oxalate is used as an energy, electron

and carbon source by bacteria belonging to diverse taxonomical groups (Tamer & Aragno, 1980; Jenni *et al.*, 1987, 1988; Sahin, 2003). These bacteria may occupy different oxalate-containing niches, like the rhizosphere (Knutson *et al.*, 1980), the litter, or the gut of soil animals (Cromack *et al.*, 1977). It seems that, after oxalate is transported into the bacterial cell, its fate is determined by either the glyoxylate or formate pathway (Fig. 12.5). If oxalate is reduced into glyoxylate by the enzyme oxalyl-CoA reductase, biosynthesis can take one of two routes: the glycolate pathway (Allison *et al.*, 1995) and a variant of the serine pathway (Sahin, 2003). The glycolate pathway is common in soil oxalotrophic bacteria such as *Ralstonia oxalatica* or *Ralstonia eutropha*. The serine pathway is mainly used by oxalate-consuming pink-pigmented facultatively methylotrophic bacteria, which contain L-serine glyoxylate aminotransferase and hydroxypyruvate reductase, key enzymes of the serine pathway (Sahin, 2003). If, however, oxalate is transformed into formate through oxalate decarboxylation, the formate is then used for cell energy and finally results in the production of CO_2 (Dijkhuizen *et al.*, 1977). Carbon dioxide is usually transported and excreted as bicarbonate ions ($HCO_3{}^-$) through the bacterial membrane. Its combination with calcium ions leads to precipitation of calcium carbonate and concomitantly generates a proton motive force (Fig. 12.5). Therefore, oxalotrophic bacteria perform a complete oxidation of oxalate to CO_2, leading to bicarbonate ion excretion, and finally carbonate deposition. But are soil conditions favourable for calcium carbonate storage?

The fact that protons are used as a motive force can contribute to an increase in the pH of the soil solution, due to the H^+ uptake by bacteria. Moreover, the pH can also be increased due to the fact that reactions involve the transformation of oxalic acid into carbonic acid, i.e. of a strong ($pK_1 = 1.25$, $pK_2 = 4.27$) to a weak ($pK_1 = 6.35$, $pK_2 = 10.33$) acid (Braissant *et al.*, 2002). This alkalinization facilitates precipitation of calcium carbonate ($CaCO_3$). By modifying a general equation for oxalate metabolism by oxalatrophic bacteria (Harder *et al.*, 1974), the following balanced equation can be proposed:

$$1000CaC_2O_4.nH_2O + 372O_2 + 32NH_4{}^+ \rightarrow$$
$$32C_4H_8O_2N^+{}_{biomass} + 936CaCO_3 + 64Ca(OH)_{2aq} + \cdots$$
$$\cdots + (1000.(n-2) + 1872)H_2O + 936CO_2 \qquad (12.15)$$

It is easy to see in this balanced equation that 93.6% of the organic carbon is transformed into mineral carbon (as precipitated $CaCO_3$ in addition to

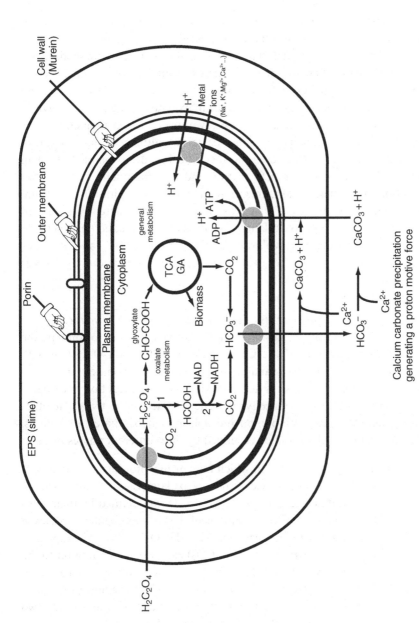

Fig. 12.5. The two main processes identified that lead to the precipitation of calcium carbonate from an oxalate source in oxalotrophic bacteria, the formate and the glyoxylate pathways. TCA, tricarboxylic acid cycle; GA, glyoxylic acid cycle; EPS, exopolysaccharides; 1, oxalate decarboxylation into formate; 2, formate dehydrogenation; ● permease/transporters.

CO_2, the latter being dispersed in the environment and eventually degassed to the atmosphere), whereas the other part of the carbon (i.e. only 6.4%) is used to increase the biomass. These figures demonstrate the far-reaching impact of such processes on organic carbon mineralization and sequestration.

Finally, it must be noted that oxalate oxidation by oxalate oxidase is a different means of degrading oxalate. By this mechanism, oxalate is completely oxidized to CO_2 with concomitant production of H_2O_2. It has no trophic significance and its importance in oxalate cycling is not known.

All these results are related to observations and measurements in the field as well as from oxalotrophic bacterial cultures in the laboratory (Braissant *et al.*, 2004). There is no theoretical model available to explain the oxalate–carbonate transformation and its consequences on the soil solution properties, i.e. alkalinization facilitating precipitation of calcium carbonate ($CaCO_3$) and CO_2 release into the atmosphere. This kind of model should be able to explain the process of oxido-reduction reactions, pH regulation and the evolution of the various phase concentrations involved in the system, i.e. oxalate, carbonate, water and CO_2. This is the aim of the next section.

A theoretical analysis of oxalate–carbonate transformation

The objective of this theoretical approach is to try to compare the equilibrium diagrams of the $CaCO_3$–CaC_2O_4–CO_2–H_2O system with the biogeochemical data available on the oxalate–carbonate transformation. The initial hypothesis is that the oxido-reduction reaction of oxalate–carbonate occurs biochemically, i.e. due to bacterial activity. In other words, activation energy is present and high enough to initiate the reactions. The theoretical system studied is defined by three different phases present:

1. The aqueous solution, which may or may not be associated with the two mineral species, calcium oxalate (CaC_2O_4) and calcium carbonate ($CaCO_3$). These two species are defined by their solubility products. Concentrations of the various chemical species in solution (including carbonates – HCO_3^-, CO_3^{2-}; oxalates – $C_2O_4^{2-}$, $C_2O_4H^-$; Ca^{2+} and CO_2) can vary depending on to the outside environment. Therefore, the system is considered as open.
2. Inside the system, the acido–basic equilibria are always reached. Nevertheless, solutions can be supersaturated regarding the two mineral species, calcite ($CaCO_3$) and calcium oxalate (CaC_2O_4).
3. Gas exchange is assumed to be fast enough to reach equilibrium between $CO_{2(dissolved)}$ and $CO_{2(gas)}$.

The fact that the system is open is justified because atmospheric oxygen is needed for the transformation (oxidation) of oxalate into carbonate by bacteria. In addition, the gas exchange with the atmosphere must be fast enough to meet the condition given in point 3. The system is described in Fig. 12.6. This system is supposed to evolve reversibly, i.e. the system is in permanent equilibrium. Therefore, at each moment, there is chemical equilibrium in the solution between the carbonate species (carbonates, bicarbonates, dissolved CO_2) and the two oxalate species (oxalate salt and oxalic acid). Nevertheless, equilibrium between solution and solids (i.e. minerals) could not be reached when the solution is supersaturated regarding one or the other of the minerals. Concerning the exchange kinetics between dissolved species inside the system and the exterior environment (Fig. 12.6), it is considered as extremely slow because equilibria between dissolved species inside the system are assumed to be continuously reached. Consequently, transformation of oxalate into carbonate is described as an exchange between dissolved species, the amount of oxalate and carbonate being variable. In other words, the transformation of oxalate into carbonate is an input of the model and considered to occur outside this theoretical system, i.e. through oxalotrophic bacterial activity (Fig. 12.6). Species' activities inside the system will be considered as equal to concentrations to simplify calculations. Except for H^+, the only cation present in the system is Ca^{2+}. This point is also only a means to simplify the calculations, but is true in laboratory bacterial cultures. The anions are the carbonate and the oxalate species, as well as OH^-. Therefore, the alkalinity can be defined as:

$$|Alc| = 2|Ca^{2+}| \tag{12.16}$$

The various limits of the variables in the system are given as follows. The values taken for the total quantity of oxalate $A = |C_2O_4{}^{2-}| + |C_2O_4H^-|$ are $0 < A < 10^{-1}$ mol l^{-1}. The concentration of calcium ions in the model is $10^{-6} < |Ca^{2+}| < 10^{-1}$ mol l^{-1}. The partial pressure of CO_2 will vary between 0 and 1 atm. Finally, the pH ranges between 4 and 11. The equilibrium constants at 1 atm and 25 °C (298 K) are the following:

$$k = \frac{(C_2O_4{}^{2-})(H^+)}{(C_2O_4H^-)} = 10^{-4.3} \tag{12.17}$$

$$K_1 = \frac{(CO_3H^-)(H^+)}{(CO_2)} = 10^{-6.4} \tag{12.18}$$

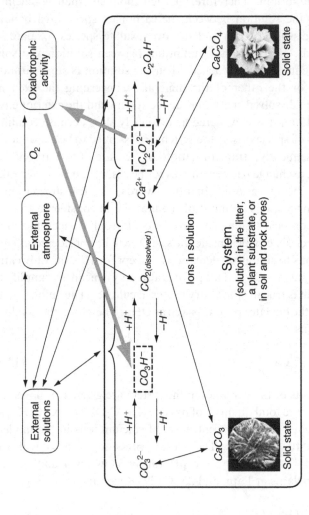

Fig. 12.6. Sketch of the system upon which is based a theoretical model of oxalate transformation into carbonate through bacterial oxalotrophic activity. The system is defined as open and therefore, exchanges are possible with the external environment.

$$K_2 = \frac{(CO_3{}^{2-})(H^+)}{(CO_3H^-)} = 10^{-10.2} \tag{12.19}$$

$$K_w = (H^+)(OH^-) = 10^{-14} \tag{12.20}$$

The mineral solid phases (calcium carbonate and calcium oxalate) are characterized by their solubility products:

$$\Pi_1 = (Ca^{2+})(CO_3{}^{2-}) = 10^{-8.3} \tag{12.21}$$

$$\Pi_2 = (Ca^{2+})(C_2O_4{}^{2-}) = 10^{-8.7} \tag{12.22}$$

The model is built using equations that follow two laws. The mass action law is applied to acido–basic reactions of carbonate and oxalate species with water. The equilibrium between the CO_2 concentrations in the gas and the solution is described by Henry's law:

$$|CO_3H_2| \text{ or } |CO_2| = k_H . pCO_2 \tag{12.23}$$

Finally, the electro-neutrality of the solution is given by:

$$|C_2O_4H^-| + 2|C_2O_4{}^{2-}| + |CO_3H^-| + 2|CO_3{}^{2-}| + |OH^-|$$
$$= 2|Ca^{2+}| + |H^+| \tag{12.24}$$

By applying these laws to the data given in Eqs. (12.18) to (12.24), a general equation describing the relationship between the different variables of the system can be found. The key equation describing the model is expressed as:

$$2|Ca^{2+}| = -|H^+| + \frac{K_w}{|H^+|} + \cdots$$

$$\cdots + \left(|C_2O_4{}^{2-}| + |C_2O_4H^-| \right) \left\{ 1 + \frac{1}{1 + \frac{|H^+|}{k}} \right\}$$

$$+ K_1 k_H \frac{pCO_2}{|H^+|} \left\{ 1 + 2\frac{K_2}{|H^+|} \right\} \tag{12.25}$$

In a two-dimensional plot, only three variables can be put in relationship to one another. Therefore, in Eq. (12.25), three variables have to be chosen among $|H^+| = 10^{-pH}$, $|C_2O_4H^-| + |C_2O_4{}^{2-}|$, $|Ca^{2+}|$ and pCO_2. For example, curves can be drawn for a variety of pH and concentrations of $|Ca^{2+}|$. In this example, curves represent the function $pH = f(\log(|Ca^{2+}|), pCO_2)$ (Fig. 12.7). The next step consists of calculating the saturation curves related to the two mineral species: calcium oxalate and calcium carbonate.

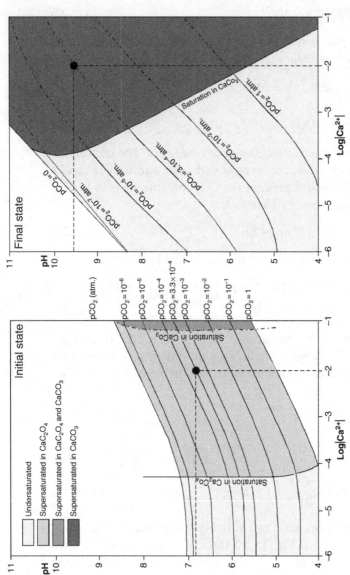

Fig. 12.7. Curves of the function pH = f(log| Ca^{2+} |, pCO_2) obtained using the model given in Fig. 12.6. Two states are described: the initial state, in which oxalate is present and begins to be oxidized by oxalotrophic bacteria, and the final state, showing the evolution of pH and calcium concentration after total disappearance of oxalate. The dotted lines refer to observed conditions. Measurements are in agreement with the model. All the oxalate (intersection inside the area related to oxalate supersaturation in the initial state) has been transformed into carbonate (intersection inside the area corresponding to calcite supersaturation in the final state) accompanied by an increase in pH compared with the initial state.

If we consider that the oxalate concentration is known: the saturation curve (i.e. solution in the presence of solid calcium oxalate) is represented by the following equation obtained by using Eqs. (12.25), (12.22) (the solubility product), (12.18) (the value of k) and the concentration in oxalate species ($|C_2O_4H^-| + |C_2O_4^{2-}|$):

$$2|Ca^{2+}| = -|H^+| + \frac{\Pi_2}{|Ca^{2+}|}\left(2 + \frac{|H^+|}{k}\right)$$
$$+ K_1 k_H \frac{pCO_2}{|H^+|}\left(1 + \frac{2K_2}{|H^+|}\right) + \frac{K_w}{|H^+|} \qquad (12.26)$$

The pCO_2 values are given by the intersection of the saturation curve with the isobar. The result is given in Fig. 12.7 when the only source of calcium is considered to be the oxalates, i.e. $|Ca^{2+}| = |C_2O_4H^-| + |C_2O_4^{2-}|$. The same approach can be used to calculate the saturation curve regarding calcite. By combining Eqs. (12.19), (12.23) and (12.21), the following relationship is obtained:

$$|Ca^{2+}| = \frac{\Pi_1}{k_H K_1 K_2} \cdot \frac{|H^+|^2}{pCO_2} \Rightarrow pH$$
$$= -\frac{1}{2}\log(|Ca^{2+}|) - \frac{1}{2}\log(pCO_2) + 4.86 \qquad (12.27)$$

Once again, the result given in Fig. 12.7 assumes that oxalates are the only source of calcium. To check this hypothesis, the model can be compared with an in vitro culture of oxalotrophic bacteria in the presence of calcium oxalate as the sole source of carbon, energy and calcium. Such an experiment has been conducted by Braissant *et al.* (2004). They show that, in a liquid medium, calcium oxalate consumption is followed by an increase in pH of the solution, accompanied by precipitation of calcium carbonate. During the experiment, the initial pH of the solution was 6.9. At the end of the experiment, i.e. after total consumption of oxalate, the final pH was 9.6. The concentration of $|Ca^{2+}|$ does not change during the experiment: it can be used as a probe during the transformation of calcium oxalate into calcium carbonate. The initial concentration of calcium is around 10^{-2} mol l^{-1}, i.e. $\log(|Ca^{2+}|) = -2$. Fig. 12.7 (initial state) shows that, for such a concentration of calcium in the presence of calcium oxalate crystals and at a $pCO_2 = 3.10^{-4}$ atm, the theoretical pH should be between 6.8 and 6.9, a range corroborated by the measurements. At the end of the experiment (Fig. 12.7, final state), there is no more oxalate. Calcium carbonate can precipitate because the intersection of the $pCO_2 = 3.10^{-4}$ atm curve and a concentration of calcium around 10^{-2} mol l^{-1} is situated

305 306   E. P. Verrecchia et al.

inside the area of supersaturation regarding $CaCO_3$. In addition, the pH should be between 9.5 and 9.6, which is also consistent with the experiment.

In conclusion, such a model is convenient to get an idea of the calcium oxalate concentration, CO_2 pressure and conditions for potential precipitation of secondary calcium carbonate through oxalotrophic bacterial activity. It demonstrates that as long as calcium is available and oxalotrophic bacteria are present, transformation of oxalate into carbonate can occur under normal conditions found in soils and surficial sediments. Therefore, an oxalate–carbonate cycle, or at least pathway, must exist at the surface of continents (Verrecchia & Dumont, 1996), explaining the absence of calcium oxalate accumulation in soils and the fossil record.

The oxalate–carbonate pathway

A synthetic coupled cycle of calcium and carbon through the oxalate–carbonate pathway is shown in Fig. 12.8. Atmospheric CO_2 is fixed by the plants through photosynthesis to produce biomass. Inside the plant, oxalate crystals form. In addition, fungal mycelium may also accumulate oxalate. Mainly in the form of calcium oxalate (COM or COD), this carbon pool is used by oxalotrophic bacteria as a carbon, energy and electron source. The transformation of oxalate can occur in the soil

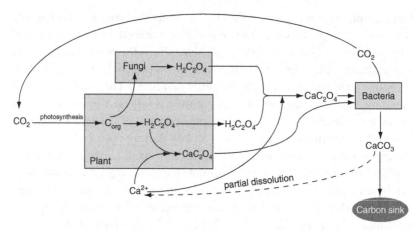

Fig. 12.8. Simplified sketch showing main relationships inside the coupled calcium and carbon cycles of the oxalate–carbonate pathway in a hypothetical ecosystem. Plants and fungi are oxalate producers. Oxalotrophic bacteria (in the soil or animal guts) use oxalate as carbon, energy and electron sources, leading to CO_2 and calcium carbonate production. Calcium carbonate can accumulate inside the soils. Because the carbon of the carbonate originates from organic carbon, its fossilization in the soil constitutes a carbon sink.

solution inside the pores, or in soil animal guts. The oxalate oxidation results in CO_2 production and calcium carbonate precipitation. Released CO_2 can be used for photosynthesis, forming a loop in the cycle. Calcium carbonate precipitation is enhanced by increasing the pH during oxalotrophy, as observed by Jayasuriya (1955) and Braissant *et al.* (2002) and demonstrated in this chapter (see pp. 305–6). Now, if we suppose that all the oxalate available in the soil is oxidized in order to measure the potential occurrence of a carbon sink, the overall reactions can be summarized by the following two equations:

$$CaC_2O_4.nH_2O + 1/2O_2 \rightarrow Ca^{2+} + 2CO_2 + (n-1)H_2O + 2OH^- \quad (12.28)$$

$$Ca^{2+} + 2OH^- + CO_2 \rightarrow CaCO_3 + H_2O \quad (12.29)$$

the first one describing the oxidation, the second taking advantage of alkalization to enhance calcium carbonate precipitation as stipulated above. Therefore, the final balance is the following:

$$CaC_2O_4.nH_2O + 1/2O_2 \rightarrow CaCO_3 + CO_2 + (n)H_2O \quad (12.30)$$

The initial two moles of organic carbon are both transformed into inorganic carbon. One mole is precipitated as a mineral, with carbon being trapped inside the calcium carbonate crystals, i.e. stored on a long-term time scale. The other mole can be released into the atmosphere and reused for phototrophy. Therefore, the oxalate–carbonate pathway constitutes a true carbon sink because one out of two moles of organic carbon is stored in a mineral state with a long residence time, whereas the other one returns to the atmospheric reservoir.

Conclusions

Consequences of the oxalate–carbonate cycle have been investigated in different environments. In surficial formations, this cycle is believed to participate in the genesis of calcrete (indurated carbonate soils) in semi-arid environments (Verrecchia 1990; Verrecchia & Dumont 1996). In temperate forests, it is related to mineral nutrient cycles including calcium, and more generally, other elements such as iron, aluminium and phosphorus (Cromack *et al.* 1977). The possible oxidation of calcium oxalate pools by oxalotrophic bacteria probably means that a large amount of the secondary calcium carbonate found in soils and surficial sediments should originate in the transformation of low-molecular-weight organic acids (including citrate and acetate) into carbonates. Calcium carbonate (as various $CaCO_3$ polymorphs) may then accumulate, modifying the soil conditions. Theoretical studies have demonstrated that soil solutions can be alkalinized enough to enhance calcium carbonate accumulation and storage. Therefore, when associated with plant biomineralization, fungi and bacteria contribute to long-term carbon storage by potentially transforming half of the organic carbon from oxalate (or low-molecular-weight organic acids) into mineral carbon (carbonate), which has a much longer residence time in soils than organic substrates (Cailleau *et al.*, 2004). The other half is released as CO_2 into the atmosphere. In conclusion, because oxalate salts are organic in origin, the oxalate–carbonate pathway represents a potentially major carbon sink (Cailleau *et al.*, 2005) and probably acts as a regulator of atmospheric pCO_2.

Acknowledgements
The authors would like to thank Professor G. M. Gadd for his invitation to contribute and his never-ending patience, Professor M. Aragno for his indispensable scientific input and Dr M. Dadras for his help operating the SEM. This work is supported by the Swiss National Science Foundation, grant no FNS-205320–101564.

References

Allison, M., Daniel, S. L. & Cornick, N. A. (1995). Oxalate-degrading bacteria. In *Calcium Oxalate in Biological Systems*, ed. S. R. Khan. Boca Raton: CRC Press, pp. 131–68.

Berbee, M. L. & Taylor, J. W. (2000). Fungal molecular evolution: gene trees and geologic time. In *The Mycota*, Vol. 7B. *Systematics and Evolution*, ed. D. J. McLaughlin, E. G. McLaughlin & P. A. Lemke. Berlin: Springer–Verlag, pp. 229–46.

Braissant, O., Verrecchia, E. P. & Aragno, M. (2002). Is the contribution of bacteria to terrestrial carbon budget greatly underestimated? *Naturwissenschaften*, **89**, 366–70.

Braissant, O., Cailleau, G., Aragno, M. & Verrecchia, E. P. (2004). Biologically induced mineralization in the tree *Milicia excelsa* (Moraceae): its causes and consequences to the environment. *Geobiology*, **2**, 59–66.

Burford, E., Fomina, M. & Gadd, G. M. (2003a). Fungal involvement in bioweathering and biotransformation of rock and minerals. *Mineralogical Magazine*, **67**, 1127–55.

Burford, E., Kierans, M. & Gadd, G. M. (2003b). Geomycology: fungi in mineral substrata. *Mycologist*, **17**, 98–107.

Cailleau, G., Braissant, O. & Verrecchia, E. P. (2004). Biomineralization in plants as long-term carbon sink. *Naturwissenschaften*, **91**, 191–4.

Cailleau, G., Braissant, O., Dupraz, C., Aragno, M. & Verrecchia, E. P. (2005). Biologically induced accumulations of $CaCO_3$ in orthox soils of Biga, Ivory Coast. *Catena*, **59**, 1–17.

Cromack, K., Sollins, P., Todd, R. L. *et al.* (1977). The role of oxalic acid and bicarbonate in calcium cycling by fungi and bacteria: some possible implications for soil animals. *Ecological Bulletin*, **25**, 246–52.

Dijkhuizen, L., Wiermsa. M. & Harder. W. (1977). Energy production and growth of *Pseudomonas oxalaticus* OX1 on oxalate and formate. *Archives in Microbiology*, **115**, 229–36.

Dutton, M. V., Kathiara, N., Gallacher, I. M. & Evans, C. S. (1993). Oxalate production by basidiomycetes, including the white-rot species *Coriolus versicolor* and *Phanerochaete chrysosporium*. *Applied Microbiology and Biotechnology*, **39**, 5–10.

Frey-Wyssling, A. (1981). Crystallography of the two hydrates of crystalline calcium oxalate in plants. *American Journal of Botany*, **68**, 130–41.

Gadd, G. M. (1999). Fungal production of citric and oxalic acid: importance in metal speciation, physiology and biogeochemical processes. *Advances in Microbial Physiology*, **41**, 47–92.

Garvie, L. A. J. (2003). Decay-induced biomineralization of the saguaro cactus (*Carnegiea gigenta*). *American Mineralogist*, **88**, 1879–88.

Gist, C. S. & Crossley, D. A., Jr. (1975). A model of mineral cycling for an arthropod foodweb in a southeastern hardwood forest litter community. In *Mineral Cycling in Southeastern Ecosystems*, ed. F. G. Howell & M. H. Smith. Washington DC: ERDA Symposium Series, Conference 740513, pp. 84–106.

Gobat, J.-M., Aragno, M. & Matthey, W. (2004). *The Living Soil – Fundamentals of Soil Science and Soil Biology*. Enfield: Science Publishers.

Harder, W., Wiermsa, M. & Groen, L. (1974). Transport of substrate and energetics of growth of *Pseudomonas oxalaticus* during growth on formate or oxalate in continuous culture. *Journal of General Microbiology*, **81**, ii–iii.

Heckman, D. S., Geiser, D. M., Eidell, B. R. *et al.* (2001). Molecular evidence for early colonization of land by fungi and plants. *Nature*, **293**, 1129–33.

Jayasuriya, G. C. N. (1955). The isolation and characteristics of an oxalate-decomposing organism. *Journal of General Microbiology*, **12**, 419–28.

Jenni, B., Aragno, M. & Wiegel, J. K. W. (1987). Numerical analysis and DNA-DNA hybridization studies on *Xanthobacter* and emendation of *Xanthobacter flavus*. *Systematic and Applied Microbiology*, **9**, 247–53.

Jenni, B., Realini, L., Aragno, M. & Tamer, A. U. (1988). Taxonomy of non H$_2$-lithotrophic, oxalate-oxidizing bacteria related to *Alcaligenes eutrophus*. *Systematic and Applied Microbiology*, **10**, 126–30.

Knutson, D. M., Hutchins, A. S. & Cromack, K. J. (1980). The association of calcium oxalate-utilizing *Streptomyces* with conifer ectomycorrhizae. *Antonie van Leeuwenhoek*, **46**, 611–19.

Lapeyrie, F., Picatto, C., Gerard, J. & Dexheimer, J. (1990). TEM study of intracellular and extracellular calcium oxalate accumulation of ectomycorrhizal fungi in pure culture or in association with *Eucalyptus* seedlings. *Symbiosis*, **9**, 163–6.

Nakata, P. A. (2003). Advances in our understanding of calcium oxalate crystal formation and function in plants. *Plant Science*, **164**, 901–9.

Norton, R. A. & Behan-Pelletier, V. M. (1991). Calcium carbonate and calcium oxalate as cuticular hardening agents in oribatid mites (Acari: Oribatida). *Canadian Journal of Zoology*, **69**, 1504–11.

Redecker, D., Kodner, R. & Graham, L. E. (2000). Glomalean fungi from the Ordovician. *Science*, **289**, 1920–1.

Sahin, N. (2003). Oxalotrophic bacteria. *Research in Microbiology*, **154**, 399–407.

Tait, K., Sayer, J. A., Gharieb, M. M. & Gadd, G. M. (1999). Fungal production of calcium oxalate in leaf litter microcosms. *Soil Biology and Biochemistry*, **31**, 1189–92.

Tamer, A. U. & Aragno, M. (1980). Isolement, caractérisation et essai d'identification de bactéries capables d'utiliser l'oxalate comme seule source de carbone et d'énergie. *Bulletin de la Société Neuchâteloise de Sciences Naturelles*, **103**, 91–104.

Verrecchia, E. P. (1990). Litho-diagenetic implications of the calcium oxalate-carbonate biogeochemical cycle in semiarid calcretes, Nazareth, Israel. *Geomicrobiology Journal*, **8**, 87–99.

Verrecchia, E. P. (2000). Fungi and sediments. In *Microbial Sediments*, ed. R. Riding & S. M. Awramik. New York: Springer-Verlag, pp. 68–75.

Verrecchia, E. P. & Dumont, J.-L. (1996). A biogeochemical model for chalk alteration by fungi in semiarid environments. *Biogeochemistry*, **35**, 447–70.

Verrecchia, E. P., Dumont, J.-L. & Verrecchia, K. E. (1993). Role of calcium oxalate biomineralization by fungi in the formation of calcretes: a case study from Nazareth, Israel. *Journal of Sedimentary Petrology*, **63**, 1000–6.

Wolscheck, M. F. & Kubicek, C. P. (1999). Biochemistry of citric acid accumulation by *Aspergilus niger*. In *Citric Acid Biotechnology*, ed. B. Kristiansen, M. Mattey & J. Linden. London: Taylor and Francis, pp. 11–31.

13

Mineral tunnelling by fungi

MARK SMITS

Introduction

This chapter reviews the distribution, mechanism and impact of mineral tunnelling by soil ectomycorrhizal fungi (EMF). Most trees in boreal forests live in close relation with EMF (Smith & Read, 1997). These EMF mediate nutrient uptake; they form an extension of the tree roots. In turn they obtain carbohydrates from the tree. Over the years ectomycorrhizal (EM) research has a strong focus on nutrient acquisition by EMF from organic sources (Read, 1991). In boreal forest systems, however, minerals could also be an important nutrient source, especially for calcium, potassium and phosphorus (Likens et al., 1994, 1998; Blum et al., 2002). Recent developments in EM research suppose a role for EMF in mobilizing nutrients from minerals (see Wallander, Chapter 14, this volume).

In 1997, Jongmans et al. described small tunnel-like features in feldspar and hornblende grains from Swedish forest soils. These tunnels have the shape of fungal hyphae: a constant width between 3 and 10 μm, smooth borders and a rounded end. In that way they differ from other weathering phenomena such as etch pitches and cracks (Fig. 13.1). In some tunnels hyphae were found. Jongmans et al. (1997) postulated that EMF created these tunnels by mineral dissolution through the exudation of low-molecular-weight organic compounds and subsequent removal of the weathering products. The weathering products like calcium, magnesium and potassium are supposed to be transported to the tree roots. In this way the host tree has direct access to mineral-bound nutrients, bypassing the bulk soil solution (van Breemen et al., 2000a; Landeweert et al., 2001).

Where are tunnels found?

An extensive survey of thin sections of soils from all over the world, revealed that tunnelled mineral grains occur almost exclusively in

Fungi in Biogeochemical Cycles, ed. G. M. Gadd. Published by Cambridge University Press. © British Mycological Society 2006.

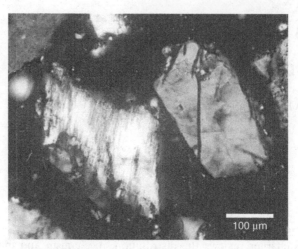

Fig. 13.1. Thin section micrograph in cross-polarized visible light of two feldspar grains in the upper cm of the mineral soil of a podzol formed in a 5400-year-old dune in northern Michigan. The left grain shows signs of chemical weathering, the right grain is tunnelled.

podzols in temperate and boreal zones (Hoffland, unpublished data). Tunnels were not observed in very young soils (<1000 years) and in very old soils (>100 000 years). Apparently, tunnel formation is either a slow process, or it requires a certain level of soil formation. The reason why tunnels are not found in very old soils is probably that in these soils almost all weatherable minerals are gone.

In podzols, tunnels are concentrated in the upper few centimetres of the mineral soil (Hoffland *et al.*, 2002), but tunnels can occasionally be found up to 30 cm in depth (Smits *et al.*, 2005). They occur mainly in feldspar grains, but have also been found in hornblendes.

Jongmans *et al.* (1997) were not the first to report fungal tunnels in minerals. Hyphae from lichens living on dolmens in Portugal penetrate the granite rock and form 'canals' in all granite minerals (Romao & Rattazzi, 1996). Also, tunnels have been found in ancient teeth (Soggnaes, 1950; Werelds, 1962; Poole & Tratman, 1978; Bell *et al.*, 1991; Dye *et al.*, 1995) and in ancient bones (Hacket, 1976; Piepenbrink, 1986). In both teeth and bones the tunnels have a smaller diameter ($\pm 2\,\mu m$) than found in forest soil minerals (3–10 μm). From tunnelled bones, several fungal species have been isolated: *Stachybotrys cyclindrosporum, Doratomyces stemonitis, Pythium* sp. and *Rhizoctonia* sp. (Piepenbrink, 1986). Werelds (1962) and Poole and Tratman (1978) suggested that filamentous bacteria from within the actinomycetes also form tunnels in ancient teeth. The focus in this chapter will be on tunnelling in soil mineral grains.

How are tunnels formed?

Tunnel formation is restricted to the upper few centimetres of the mineral soil, which suggests a biotic origin. The tunnel morphology suggests that these tunnels are formed by fungal activity, although there is no direct evidence yet. Fungal exudation of small organic compounds like oxalic acid at the hyphal tips, and simultaneous removal of weathering products would increase the tunnel length. The widespread ability to produce and exude small organic compounds, like oxalic acid, by fungi (Gadd, 1999) supports the involvement of fungi in tunnel formation.

Fungi, because of their hyphal growth form, are able to transport products over a long distance in the soil, and in that way can cope with a heterogeneous soil environment (Davidson, 1998; Davidson & Olsson, 2000). Cracks and tunnels in mineral grains can be possible niches for fungi as long as they can obtain their carbon from elsewhere. This is in contrast to bacteria, which depend on local carbon sources. Therefore it is unlikely that bacteria are responsible for the tunnel formation, although a combined effect of bacteria and fungi is possible.

Scanning electron microscope observations show that fungal growth follows the mineral surface and individual hyphae enter cracks and holes in the mineral surface (van Breemen *et al.*, 2000a, b). However, less than 1% of the observed tunnels were seen to be occupied by fungal hyphae (Jongmans *et al.*, 1997; van Breemen *et al.*, 2000a). The latter is possibly due to the fact that in unstained thin sections only melanin-containing hyphae are visible.

The hypothesis that EMF are specifically involved in tunnel formation (Jongmans *et al.*, 1997; Landeweert *et al.*, 2001) is supported by several observations. Firstly, Hoffland *et al.* (2003) found a positive linear relationship between tunnel length and EM density. Secondly, tunnel formation seems restricted to podzols in the boreal and temperate zone, where the dominant trees species are mainly EM.

If EMF are responsible for creating these tunnels, they must be active in the mineral soil. There are only few studies on the vertical distribution of EMF in the soil. Most studies are based on morphological identification of EM root tips ('morphotyping') (Rambold & Agerer, 1997). The studies were based on field sampling of EM root tips in the organic and mineral soil layers (Egli, 1981; Goodman & Trofymow, 1998; Fransson *et al.*, 2000; Rosling *et al.*, 2003; Tedersoo *et al.*, 2003) or on bait seedlings grown on organic and mineral soil samples (Danielsson & Visser, 1989; Heinonsalo *et al.*, 2001, 2004). In all studies, EM root tips are found within the mineral soil; all studies suggest large differences in the EM community between the organic and mineral soil layers, although variation was high. Sequencing techniques reveal that the EM community was highly variable at a 5 cm scale (Tedersoo *et al.*, 2003). Molecular identification on three-dimensional distribution of EM root tips in a *Larix kaempferi* stand showed no clear vertical distribution pattern (Zhou & Hogetsu, 2002).

While the mycorrhizal root tip is the most striking part of the mycorrhizal symbiosis, interactions of the fungus with the soil takes place at the hyphal tips. The vertical hyphal distribution could give a clue about which fungi are possibly involved in the tunnelling process. Unfortunately, morphological determination does not work for hyphae. Characterizations of the EM community, based on DNA extracted from the soil (Dickie *et al.*, 2002; Landeweert *et al.*, 2003) revealed that this community changes with depth in a North Swedish podzol profile, both with regard to EM root tips (Rosling *et al.*, 2003) and extramatrical hyphae (Landeweert *et al.*, 2003). Analysis of DNA also revealed that EM basidiomycetes dominate in both the organic and the mineral soil layers (Landeweert *et al.*, 2003). Mesh-bag experiments also indicated that in both the organic and the mineral soil layers EMF dominated over saprotrophic fungi (Wallander *et al.*, 2004).

Ecological impact

Podzolization process

Podzols are the main soil type in boreal forests. They are characterized by a dark-coloured organic surface (O) horizon over a white bleached, eluvial (E) horizon, abruptly underlain by a black-to-brown accumulation (B) horizon. Tunnelled minerals are abundant in the E horizon, but only few tunnelled minerals occur in the underlying B horizon. This suggests a link to podzolization via weathering and transport of weathering products (van Breemen *et al.*, 2000b). The next paragraphs discuss the impact of tunnelling on weathering and transport.

Velocity of tunnel formation

Dissolution of feldspars and hornblende is a slow process. If tunnels are formed by dissolution processes, laboratory experiments are not a promising tool to study the rate of tunnel formation. In that case, soil chronosequences can be of high value (Huggett, 1998). Tunnel formation has been studied in two chronosequences.

The first study was done in Mid Sweden, near Umeå. The Mid-Sweden chronosequence was formed through glacial rebound after the Pleistocene ice sheet melted, resulting in a series of soils ranging from 0 to about 9000 years old (Hoffland *et al.*, 2002). The parent material contains mainly quartz and about 20%–30% feldspars. The second study was done in northern Michigan, where the soil chronosequence consists of a series of sand dunes parallel to the northern coast of Lake Michigan. The dunes are formed at intervals of about 70 years, due to glacial rebound and oscillations of the Lake Michigan water level. Their age ranges from 0 to 5400 years (Delcourt *et al.*, 1996; Petty *et al.*, 1996). The dunes have been built up with sediments from Lake Michigan, which contains mainly quartz, with only 10%–15% feldspars (Smits *et al.*, 2005). Soils of both chronosequences are vegetated with coniferous forest, and their climates are similar.

Tunnelling was quantified in the upper 2 cm of the mineral soil, using thin sections. In the thin sections, the fraction of tunnelled feldspars was counted. Tunnels appeared after a lag phase of 1500 to 2000 years (Fig. 13.2);

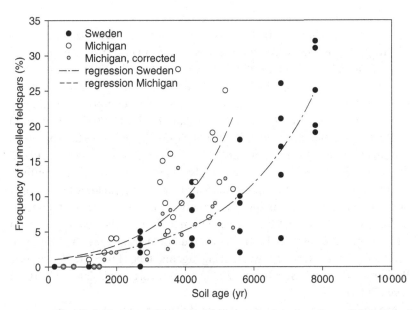

Fig. 13.2. Fraction of tunnelled feldspars in the upper 2 cm of the mineral soil against soil age in two chronosequences. Data of the Michigan chronosequence is also corrected for the difference in feldspar content between the Mid-Sweden and the Michigan chronosequence, to get a measure for fraction of tunnelled minerals (grey dots) (modified from Smits *et al.*, 2005).

at that time the most easily weatherable minerals such as biotite and hornblende had disappeared (Hoffland *et al.*, 2002; Smits *et al.*, 2005). In both chronosequences the fraction of tunnelled feldspars against soil age followed an exponential or sigmoid function. The increase in the fraction of tunnelled feldspars was higher in the Michigan chronosequence than in the Sweden chronosequence, and this difference is statistically significant.

In Michigan, feldspar content, as a fraction of total mineral content, is approximately half of that in the Mid-Sweden chronosequence. In contrast, the fraction of tunnelled feldspars in the Michigan chronosequence is twice that of the Mid-Sweden chronosequence. That could indicate that tunnelling is driven by nutrient demand. But, if the data are corrected for the difference in feldspar content, to get the fraction of tunnelled minerals, both curves match (Fig. 13.2). This suggests that feldspar tunnelling activity is independent of feldspar abundance, and could be determined by EM density. Indeed, a positive correlation was found between the frequency of tunnelled feldspars and EM density (Hoffland *et al.*, 2003).

Weathering due to tunnelling

In this chapter, the focus is on weathering of feldspars, alumino-silicate minerals, which are the most abundant mineral species in the earth crust (Banfield & Hamers, 1997). Feldspars contain aluminium and silicon, which are arranged in a tetrahedral structure, with other cations in the voids of this structure. The common feldspars have compositions ranging between albite ($NaAlSi_3O_8$) and K-feldspar ($KAlSi_3O_8$) (alkali feldspars) and between albite and anorthite ($CaAl_2Si_2O_8$) (plagioclase feldspars).

Feldspar weathering is a much-studied subject (see White, 1995), because of the proposed key role of feldspars in (1) neutralizing the effects of anthropogenic acid precipitation (Sverdrup, 1996), (2) long-term supply of calcium and potassium and (3) regulation of atmospheric CO_2 concentrations over geological time scales (Berner, 1995). The role of biota on feldspar weathering has been recognized for a long time. The original emphasis on bacteria as weathering agents has shifted to fungi (see Barker *et al.*, 1997; Sterflinger, 2000; Landeweert *et al.*, 2001; Hoffland *et al.*, 2004).

Measuring the contribution of fungi to mineral weathering in the field has often been hampered by the impossibility to distinguish fungal weathering from other types of weathering. Fungal tunnels, however, can be identified visually and be distinguished from other weathering phenomena (Hoffland *et al.*, 2002) (see Fig. 13.1).

In the Michigan chronosequence, the contribution of tunnelling to the total weathering was determined for different stages of soil formation (Smits *et al.*, 2005). Again, the focus was on the upper 2 cm of the mineral soil. The contribution of tunnelling to total weathering was expressed as the total tunnel volume divided by the total volume dissolved in time. Total tunnel volume was determined on thin sections, using image analysis software. The contribution of tunnelling to total weathering was found to be below 0.5%. Though the variation is high, the difference between the two feldspar species is clear. Tunnelling in the more easily weatherable oligoclase contributes more to its total weathering than in the more persistent microcline (Smits *et al.*, 2005).

Quantitatively, tunnelling seems to play a minor role in weathering and fluxes of calcium and potassium (Smits *et al.*, 2005). However, tunnelling is only a part of total fungal weathering. Hyphae not only occur inside tunnels, but also cover mineral surfaces. By exudation of organic compounds, hyphae attached to the mineral surfaces also contribute to fungal weathering. It is possible that surface weathering is quantitatively more important than tunnelling, but tunnelling may reflect trends of the total impact of fungi on weathering. With that perspective, the difference in tunnelling between oligoclase and microcline can be of interest.

Ectomycorrhizal weathering

Several in vitro studies illustrate the weathering capacities of EMF (Ochs *et al.*, 1993; Paris *et al.*, 1995a, b, 1996; Ochs, 1996; Chang & Li, 1998; Welch *et al.*, 2002; Glowa *et al.*, 2003). Pot experiments, where the fungi grow in symbiosis with the host tree, also show evidence of EM-induced mineral weathering (Leyval *et al.*, 1990; Wallander *et al.*, 1997; Wallander & Wickman, 1999; Wallander, 2000). Few attempts have been made to determine the impact of EMF on mineral weathering in the field (Blum *et al.*, 2002; Sverdrup *et al.*, 2002; Wallander *et al.*, 2002, 2003; Hagerberg *et al.*, 2003) (see also Wallander, Chapter 14 this volume).

Because of the difficulties of interpreting field data in complex systems with low controllability and high heterogeneity, modelling offers a way to quantify fungal weathering in the field. Sverdrup *et al.* (2002) used such an approach, and concluded that EM weathering is quantitatively unimportant. They based their efforts on a theoretical framework, specified in the PROFILE weathering model (Sverdrup, 1990), with weathering kinetics based on the transition state theory (see Lasaga & KirkPatrick, 1981). Weathering rate is determined by five different weathering reactions: proton reaction, water reaction, hydroxide reaction, CO_2 reaction and an organic acid reaction. In general, the reaction rate increases with higher temperature, but every reaction shows its own specific dependency on temperature. The rates of the proton and water reactions increase faster with temperature than those of the organic acid and CO_2 reactions. Field studies of soils in temperature gradients show a temperature dependency of the total weathering rate in the range of that of water and proton reactions (Sverdrup *et al.*, 2002). On that basis it was concluded that water and proton reactions are the dominant weathering reactions. Further, modelling results are in line with mass balance data for a soil profile in Sweden (Sverdrup *et al.*, 2002). The calculated data accounts for 98% of the measured weathering data. Still, according to the model, 15% of total weathering is due to the organic acid reaction and 20% is due to the CO_2 reaction. It must be noted that organic acids not only have a direct effect on weathering through the organic acid reaction, but also via complexation of aluminium and base cations. These cations have an inhibiting effect on the water and proton reactions (Chou & Wollast, 1985; Gautier *et al.*, 1994; Oelkers *et al.*, 1994), which is also built into the PROFILE model. This means that organic acids and CO_2 do have a bigger effect on the weathering rate than is reflected by their weathering reactions. Sverdrup *et al.* (2002) stated that EM weathering is only a marginal process because the PROFILE model does explain 98% of the total weathering, but they also recognize that EMF could act via the processes already described in the model. According to predictions made by the PROFILE model, biota play an important role in weathering, since 35% of total weathering is due to the organic acid and the CO_2 reactions. The question is how much EMF contribute to both reactions. A tree girdling experiment in Sweden shows that at least 52% of total soil respiration in a boreal forest is directly linked to photosynthesis (Högberg *et al.*, 2001). Apparently (EM) roots and their extending hyphae exude high amounts of CO_2 and easily degradable organic compounds, like organic acids, and in that way, account for a high contribution to 'biotic' weathering reactions.

A more direct approach to study EM weathering is the use of tracer elements as used by Blum *et al.* (2002). They examined calcium uptake by different tree species in the Hubbert Brook Experimental Forest in north-eastern USA. They used the fact that the different calcium sources (atmospheric deposition, calcium silicate weathering and apatite weathering) have different Ca/Sr ratios. In the ecosystem, strontium is believed to behave in a similar fashion as calcium (Åberg *et al.*, 1990). Atmospheric deposition, calcium silicate minerals and soil water have similar Ca/Sr ratios (120–300), but apatite has a much higher Ca/Sr ratio (2200–2920).

The tree species differed in their Ca/Sr ratio in the leaves. American beech (*Fagus grandifolia* Ehrh), yellow birch (*Betula alleghaniensis* Britton), sugar maple (*Acer saccharum* Marsh.) and toothed wood fern (*Dryopteris spinulosa* Mull.) all had Ca/Sr ratios similar to the soil solution, but balsam fir (*Abies balsamea* (L.) Mill.) and red spruce (*Picea rubens Sarg.*) had Ca/Sr ratios more similar to apatite (Fig. 13.3). Blum *et al.* (2002) concluded that the Ca/Sr ratios in the different tree species reflected their different calcium sources, where fir and spruce derived more than 90% of their calcium from apatite, whereas the other species largely depended on other calcium sources. They hypothesized that fir and spruce reached the apatite-derived calcium via EMF. Although this approach seems rather straightforward, some of the assumptions made in the interpretation seem questionable. In these forests, external ecosystem inputs of calcium are very small compared to fluxes by recycling via litter (Likens *et al.*, 1998, Dijkstra & Smits, 2002). Furthermore, fractionation between calcium and strontium may take place: relatively more calcium than strontium is transported from roots to leaves, at least under acidic conditions (Poszwa *et al.*, 2000, Watmough & Dillon, 2003). The combined effect of calcium recycling and fractionation could explain the high Ca/Sr ratios in fir and spruce, without taking into account different calcium inputs (Smits, unpublished data). The strikingly higher Ca/Sr ratios in the two coniferous EM trees, spruce and fir, than in the broadleaf EM trees, beech and birch (Fig. 13.3), may well be due to differences in physiology or nutrient cycling, including

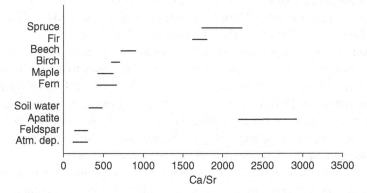

Fig. 13.3. The range of Ca/Sr ratios (w/w) as measured in the leaves of five different tree species and a fern, in different minerals and soil water in a mixed forest ecosystem of north-eastern USA. (Modified from Blum *et al.*, 2002.) Atm. dep., atmospheric depression.

litter decomposition, between coniferous and deciduous trees rather than pointing to a role of EMF in apatite-derived calcium uptake.

Fate of weathering products

The absence of secondary mineral precipitation in fungal tunnels indicates that the weathering products are removed from the tunnel interior, either by diffusion or fungal transport. Removal by diffusion is likely because it is probably faster than feldspar weathering. If the tunnel is occupied by a fungus, the diffusion rate would be depressed, so that fungal transport could become important.

When feldspars dissolve, the constituents (Si, Al and Na, K or Ca) are released in ionic form. Aluminium is of particular interest because of its phytotoxic properties (Kinraide, 1991). Furthermore, aluminium binds to organic matter (Ritchie, 1995), and in that way decreases organic matter dissolution and decomposition (Mulder et al., 2001). Aluminium binding on cation exchange sites competes with other cations, resulting in reduced storage of calcium and magnesium in the rooting zone (Lawrence et al., 1995).

The traditional view of aluminium dynamics in podzol systems is that aluminium is released by mineral weathering in the E horizon (upper mineral soil layer) and is translocated further down into the soil into the B horizon (Ugolini et al., 1987). However, large amounts of aluminium are found in the O horizon of podzols (Lawrence et al., 1995; Rustad & Cronan, 1995; Giesler et al., 1998, 2000). Budgets studies indicate a significant upward translocation of aluminium from the E horizon to the O horizon (Rustad & Cronan, 1995; Giesler et al., 2000). A shift from mixed forest to pure Norway spruce forest induced a strong increase in aluminium content of the O horizon within 12 years (Brandtberg & Simonsson, 2003). Mechanisms proposed to explain this are (1) transfer of weathering products by hyphae from the minerals to mycorrhizal roots in the O horizon, followed by the release of aluminium that is not taken up by the plants (Van Breemen et al., 2000b), (2) root transport (Vogt et al., 1987; Rustad & Cronan, 1988), (3) diffusion under water-saturated conditions during snow melt and (4) mixing of mineral soil with the O horizon and subsequently weathering (Rustad & Cronan, 1995).

The first prerequisite for a possible role of EMF in aluminium dynamics is that they are able to transport reasonable amounts of aluminium. Some maintain that fungi cannot transport aluminium at all (Brandes et al., 1998). Mushrooms, however, have aluminium contents from 20 to 330 µg g^{-1} (dry weight) (Leski et al., 1995; Müller et al., 1997; Rudawska & Leski,

1998; Falandysz *et al.*, 2001). It is unlikely that all the aluminium originates from atmospheric deposition, so aluminium must be transported to
the fruit body. In vitro tests of EMF on two-compartment Petri dishes
revealed that at least some EMF are able to transport aluminium (Smits,
unpublished data).

In some aluminium-tolerant plants, aluminium is complexed by carboxylates like citrate and oxalate (Ma *et al.*, 1997, 1998; Ma & Hiradate, 2000).
Besides carboxylates, phosphate groups are the strongest Al^{3+} binders in
biological systems (Martin *et al.*, 1994). In EMF, phosphorus is stored in
the form of polyphosphates (polyP), linear condensed polymers of inorganic phosphate, which can, after hydrolysis, be transferred to the host
plant (Bücking & Heyser, 1999). They are found in the vacuoles in the form
of granules (Bücking & Heyser, 1999). Small soluble polyP chains, mainly
associated with potassium (Bücking & Heyser, 1999) are thought to be
transported via a network of spherical vacuoles (Shepherd *et al.*, 1993a, b;
Ashford *et al.*, 1994, 1999). A possible transport mechanism is that aluminium is complexed with small, soluble polyP chains and, indeed, aluminium has been found to be associated with polyP in both soluble (Martin
et al., 1994) and precipitated form (Väre, 1990).

A second prerequisite for a role of EMF in aluminium transport from
the E to the O horizon is that hyphae growing in the E horizon are attached
to EM root tips in the O horizon. Work on the vertical distribution of
hyphae and EM root tips in a Swedish podzol profile showed that several
EM species live in both the O horizon and the E horizon (Landeweert *et al.*,
2003, Rosling *et al.*, 2003). For *Dermocybe*, *Russula*, *Piloderma* and
Cortinarius, the clones found in the extracted DNA from the E horizon
match with DNA from EM root tips in the O horizon of the same soil
profile. That suggests that at least some EM root tips from the O horizon
could have extramatrical hyphae growing into the mineral soil.

Conclusions
Fungal tunnelling of mineral grains appears to be quantitatively
unimportant in weathering, yet this phenomenon is still of interest, firstly
because of the exclusive access to possible nutrients like phosphorus in the
form of apatite inclusions in feldspar grains (van Breemen *et al.*, 2000a;
Blum *et al.*, 2002) and secondly because the only part of total mineral
weathering by fungi that so far can easily be quantified is tunnelling.
Modelling is another way but needs to be interpreted with great care.
New developments in crystal dissolution modelling (e.g. Lasaga &
Luttge, 2003) have consolidated part of the theoretical background of

322   *M. Smits*

the PROFILE model. However, besides model development, more information is needed on the data input side, for instance the fungal impact on the mineral surface chemistry and physics. Another means is using tracer elements to unravel the different nutrient sources. Nutrient cycling and different behaviour between elements complicate this approach, as described in this chapter. The use of multiple tracer elements and specific experimental designs could increase the power of this approach.

Tracer elements can also be of great use in studying nutrient transport. In the case of aluminium, gallium may be considered an appropriate tracer. However, gallium behaves rather differently to aluminium (Shiller & Frilot, 1996; Bénézeth *et al.*, 1997), and so caution is needed when using it as a tracer for aluminium. The use of the long-lived isotope ^{26}Al, which can be measured with high sensitivity by accelerated mass spectrometry (Faarinen *et al.*, 2001; Faarinen, 2003) may be a feasible alternative in further investigations of the role of fungi in aluminium dynamics in podzols.

Acknowledgements
This work was supported by the Netherlands Organization of Scientific Research (NWO). Ellis Hoffland, Nico van Breemen and Håkan Wallander are kindly acknowledged for reviewing this chapter.

References
Åberg, G., Jacks, G., Wickman, T. & Hamilton, P. J. (1990). Strontium isotopes in trees as an indicator for calcium availability. *Catena*, 17, 1–11.
Ashford, A. E., Ryde, S. & Barrow, K. D. (1994). Demonstration of a short chain polyphosphate in *Pisolithus tinctorius* and the implications for phosphorus transport. *New Phytologist*, 126, 239–47.
Ashford, A. E., Vesk, P. A., Orlovich, D. A., Markovina, A. L. & Allaway, W. G. (1999). Dispersed polyphosphate in fungal vacuoles in *Eucalyptus pilularis/Pisolithus tinctorius* ectomycorrhizas. *Fungal Genetics and Biology*, 28, 21–33.
Banfield, J. F. & Hamers, R. J. (1997). Processes at minerals and surfaces with relevance to microorganisms and prebiotic synthesis. In *Geomicrobiology: Interactions between Microbes and Minerals*, Vol. 35, ed. J. F. Banfield & K. H. Nealson. Washington DC: Mineralogical Society of America, pp. 81–122.
Barker, W. W., Welch, S. A. & Banfield, J. F. (1997). Biogeochemical weathering of silicate minerals. In *Geomicrobiology: Interactions between Microbes and Minerals*, Vol. 35, ed. J. F. Banfield & K. H. Nealson. Washington DC: Mineralogical Society of America, pp. 391–428.
Bell, L. S., Boyde, A. & Jones, S. J. (1991). Diagenetic alteration to teeth *In situ* illustrated by backscattered electron imaging. *Scanning*, 13, 173–83.
Bénézeth, P., Diakonov, I. I., Pokrovski, G. S. *et al.* (1997). Gallium speciation in aqueous solution. Experimental study and modelling: Part 2. Solubility of a-GaOOH in

acidic solutions from 150 to 250 °C and hydrolysis constants of gallium (III) to 300 °C. *Geochimica et Cosmochimica Acta*, **61**, 1345–57.

Berner, R. A. (1995). Chemical weathering and its effect on atmospheric CO_2 and climate. In *Chemical Weathering Rates in Silicate Minerals*, ed. A. F. White & S. L. Brantley. Washington DC: Mineralogical Society of America, pp. 565–83.

Blum, J. D., Klaue, A., Nezat, C. A. et al. (2002). Mycorrhizal weathering of apatite as an important calcium source in base-poor forest ecosystems. *Nature*, **417**, 729–31.

Brandes, B., Godbold, D. L., Kuhn, A. J. & Jentschke, G. (1998). Nitrogen and phosphorus acquisition by the mycelium of the ectomycorrhizal fungus *Paxillus involutus* and its effect on host nutrition. *New Phytologist*, **140**, 735–43.

Brandtberg, P. O. & Simonsson, M. (2003). Aluminum and iron chemistry in the O horizon changed by a shift in tree species composition. *Biogeochemistry*, **63**, 207–28.

Bücking, H. & Heyser, W. (1999). Elemental composition and function of polyphosphates in ectomycorrhizal fungi – an X-ray microanalytical study. *Mycological Research*, **103**, 31–9.

Chang, T. T. & Li, C. Y. (1998). Weathering of limestone, marble, and calcium phosphate by ectomycorrhizal fungi and associated microorganisms. *Taiwan Journal of Forest Science*, **13**, 85–90.

Chou, L. & Wollast, R. (1985). Steady-state kinetics and dissolution mechanisms of albite. *American Journal of Science*, **285**, 963–93.

Danielsson, R. M. & Visser, S. (1989). Effects of forest soil acidification on ectomycorrhizal and vesicular-arbuscular mycorrhizal development. *New Phytologist*, **112**, 41–7.

Davidson, F. A. (1998). Modelling the qualitative response of fungal mycelia to heterogeneous environments. *Journal of Theoretical Biology*, **195**, 281–92.

Davidson, F. A. & Olsson, S. (2000). Translocation induced outgrowth of fungi in nutrient-free environments. *Journal of Theoretical Biology*, **205**, 73–84.

Delcourt, P. A., Petty, W. H. & Delcourt, H. R. (1996). Late-Holocene formation of Lake Michigan beach ridges correlated with a 70-yr oscillation in global climate. *Quaternary Research*, **45**, 321–6.

Dickie, I. A., Xu, B. & Koide, R. T. (2002). Vertical niche differentiation of ectomycorrhizal hyphae in soil as shown by T-RFLP analysis. *New Phytologist*, **156**, 527–35.

Dijkstra, F. & Smits, M. (2002). Tree species effects on calcium cycling: the role of calcium uptake in deep soil. *Ecosystems*, **5**, 385–98.

Dye, T. J., Lucy, D. & Pollard, A. M. (1995). The occurrence and implications of postmortem pink teeth in forensic and archaeological cases. *International Journal of Osteoarchaeology*, **5**, 339–48.

Egli, S. (1981). Mycorrhizae and their vertical distribution in oak stands. *Schweizerische Zeitschrift für Forstwesen*, **132**, 345–53.

Faarinen, M. (2003). Development of the Lund AMS facility for detection of [26]Al – with applications in plant ecology. Unpublished Ph.D. thesis, Lund University, Lund.

Faarinen, M., Magnusson, C. E., Hellborg, R. et al. (2001). Al-26 investigations at the AMS-laboratory in Lund. *Journal of Inorganic Biochemistry*, **87**, 57–61.

Falandysz, J., Szymczyk, K., Ichihashi, H. et al. (2001). ICP/MS and ICP/AES elemental analysis (38 elements) of edible wild mushrooms growing in Poland. *Food Additives and Contaminants*, **18**, 503–13.

Fransson, P. M. A., Taylor, A. F. S. & Finlay, R. D. (2000). Effects of continuous optimal fertilization on belowground ectomycorrhizal community structure in a Norway spruce forest. *Tree Physiology*, **20**, 599–606.

Gadd, G. M. (1999). Fungal production of citric and oxalic acid: importance in metal speciation, physiology and biogeochemical processes. *Advances in Microbial Physiology*, **41**, 47–92.

324   *M. Smits*

Gautier, J. M., Oelkers, E. H. & Schott, J. (1994). Experimental study of K-feldspar dissolution rates as a function of chemical affinity at 150 °C and pH-9. *Geochimica et Cosmochimica Acta*, **58**, 4549–60.

Giesler, R., Högberg, M. & Högberg, P. (1998). Soil chemistry and plants in Fennoscandian boreal forest as exemplified by a local gradient. *Ecology*, **79**, 119–37.

Giesler, R., Ilvesniemi, H., Nyberg, L. *et al.* (2000). Distribution and mobilization of Al, Fe and Si in three podzolic soil profiles in relation to the humus layer. *Geoderma*, **94**, 249–63.

Glowa, K. R., Arocena, J. M. & Massicotte, H. B. (2003). Extraction of potassium and/or magnesium from selected soil minerals by *Piloderma*. *Geomicrobiology Journal*, **20**, 99–111.

Goodman, D. M. & Trofymow, J. A. (1998). Distribution of ectomycorrhizas in microhabitats in mature and old-growth stands of Douglas-fir on southeastern Vancouver island. *Soil Biology and Biochemistry*, **30**, 2127–38.

Hacket, C. (1976). Microscopical focal destruction (tunnels) in exhumed human bone. *Medicine Science and the Law*, **21**, 243–64.

Hagerberg, D., Thelin, G. & Wallander, H. (2003). The production of ectomycorrhizal mycelium in forests: relation between forest nutrient status and local mineral sources. *Plant and Soil*, **252**, 279–90.

Heinonsalo, J., Jorgensen, K. S. & Sen, R. (2001). Microcosm-based analyses of Scots pine seedling growth, ectomycorrhizal fungal community structure and bacterial carbon utilization profiles in boreal forest humus and underlying illuvial mineral horizons. *FEMS Microbiology Ecology*, **36**, 73–84.

Heinonsalo, J., Hurme, K. R. & Sen, R. (2004). Recent C[14]-labelled assimilate allocation to Scots pine seedling root and mycorrhizosphere compartments developed on reconstructed podzol humus, E- and B-mineral horizons. *Plant and Soil*, **259**, 111–21.

Hoffland, E., Giesler, R., Jongmans, T. & van Breemen, N. (2002). Increasing feldspar tunneling by fungi across a North Sweden podzol chronosequence. *Ecosystems*, **5**, 11–22.

Hoffland, E., Giesler, R., Jongmans, A. G. & van Breemen, N. (2003). Feldspar tunneling by fungi along natural productivity gradients. *Ecosystems*, **6**, 739–46.

Hoffland, E., Kuyper T, W., Wallander, H. *et al.* (2004). The role of fungi in weathering. *Frontiers in Ecology*, **2**, 258–64.

Högberg, P., Nordgren, A., Buchmann, N. *et al.* (2001). Large-scale forest girdling shows that current photosynthesis drives soil respiration. *Nature*, **411**, 789–92.

Huggett, R. J. (1998). Soil chronosequences, soil development, and soil evolution: a critical review. *Catena*, **32**, 155–72.

Jongmans, A. G., van Breemen, N., Lundström, U. S. *et al.* (1997). Rock-eating fungi. *Nature*, **389**, 682–3.

Kinraide, T. B. (1991). Identity of the rhizotoxic aluminum species. *Plant and Soil*, **134**, 167–78.

Landeweert, R., Hoffland, E., Finlay, R. D., Kuyper, T. W. & van Breemen, N. (2001). Linking plants to rocks: ectomycorrhizal fungi mobilize nutrients from minerals. *Trends in Ecology and Evolution*, **16**, 248–54.

Landeweert, R., Leeflang, P., Kuyper, T. W. *et al.* (2003). Molecular identification of ectomycorrhizal mycelium in soil horizons. *Applied and Environmental Microbiology*, **69**, 327–33.

Lasaga, A. & KirkPatrick, R. (1981) *Kinetics of Geochemical Processes*. Washington, DC: Mineralogical Society of America.

Lasaga, A. C. & Luttge, A. (2003). A model for crystal dissolution. *European Journal of Mineralogy*, **15**, 603–15.

Lawrence, G. B., David, M. B. & Shortle, W. C. (1995). A new mechanism for calcium loss in forest-floor soils. *Nature*, **378**, 162–5.

Leski, T., Rudawska, M. & Kieliszewska-Rokicka, B. (1995). Intraspecific aluminium response in *Suillus luteus* (L.) S. F.Gray., an ectomycorrhizal symbiont of Scots Pine. *Acta Societatis Botanicorum Poloniae*, **64**, 97–105.

Leyval, C., Laheurte, F., Belgy, G. & Berthelin, J. (1990). Weathering of micas in the rhizospheres of maize, pine and beech seedlings influenced by mycorrhizal and bacterial inoculation. *Symbiosis*, **9**, 105–9.

Likens, G. E., Driscoll, C. T., Buso, D. C. *et al.* (1994). The biogeochemistry of potassium in the Hubbard Brook Experimental Forest, New Hampshire. *Biogeochemistry*, **25**, 61–125.

Likens, G. E., Driscoll, C. T., Buso, D. C. *et al.* (1998). The biogeochemistry of calcium at Hubbard Brook. *Biogeochemistry*, **41**, 89–173.

Ma, J. F. & Hiradate, S. (2000). Form of aluminium for uptake and translocation in buckwheat (*Fagopyrum esculentum* Moench). *Planta*, **211**, 355–60.

Ma, J. F., Hiradate, S., Nomoto, K., Iwashita, T. & Matsumoto, H. (1997). Internal detoxification mechanism of Al in hydrangea – identification of Al form in the leaves. *Plant Physiology*, **113**, 1033–9.

Ma, J. F., Hiradate, S. & Matsumoto, H. (1998). High aluminum resistance in buckwheat – II. Oxalic acid detoxifies aluminum internally. *Plant Physiology*, **117**, 753–9.

Martin, F., Rubini, P., Côté, R. & Kottke, I. (1994). Aluminium polyphosphate complexes in the mycorrhizal basidiomycete *Laccaria bicolor*: A [27]Al-nuclear magnetic resonance study. *Planta*, **194**, 241–6.

Mulder, J., De Wit, H. A., Boonen, H. W. J. & Bakken, L. R. (2001). Increased levels of aluminium in forest soils: effects on the stores of soil organic carbon. *Water Air and Soil Pollution*, **130**, 989–94.

Müller, M., Anke, M. & Illing-Günther, H. (1997). Aluminium in wild mushrooms and cultivated *Agaricus bisporus*. *Zeitschrift für Lebensmitteluntersuchung und Forschung A*, **205**, 242–7.

Ochs, M. (1996). Influence of humified and non-humified natural organic compounds on mineral dissolution. *Chemical Geology*, **132**, 119–24.

Ochs, M., Brunner, I., Strumm, W. & Cosovic, B. (1993). Effects of root exudates and humic substances on weathering kinetics. *Water Air and Soil Pollution*, **68**, 213–29.

Oelkers, E. H., Schott, J. & Devidal, J. L. (1994). The effect of aluminum, pH, and chemical affinity on the rates of aluminosilicate dissolution reactions. *Geochimica et Cosmochimica Acta*, **58**, 2011–24.

Paris, F., Bonnaud, P., Ranger, J. & Lapeyrie, F. (1995a). *In vitro* weathering of phlogopite by ectomycorrhizal fungi: I. Effect of K^+ and Mg^{2+} deficiency on phyllosilicate evolution. *Plant and Soil*, **177**, 191–201.

Paris, F., Bonnaud, P., Ranger, J., Robert, M. & Lapeyrie, F. (1995b). Weathering of ammonium- or calcium-saturated 2:1 phyllosilicates by ectomycorrhizal fungi in vitro. *Soil Biology and Biochemistry*, **27**, 1237–44.

Paris, F., Botton, B. & Lapeyrie, F. (1996). *In vitro* weathering of phlogopite by ectomycorrhizal fungi. II. Effect of K^+ and Mg^+ deficiency and N sources on accumulation of oxalate and H^+. *Plant and Soil*, **179**, 141–50.

Petty, W. H., Delcourt, P. A. & Delcourt, H. R. (1996). Holocene lake-level fluctuations and beach-ridge development along the northern shore of Lake Michigan, USA. *Journal of Paleolimnology*, **15**, 147–69.

Piepenbrink, H. (1986). Two examples of biogenous dead bone decomposition and their consequences for taphonomic interpretation. *Journal of Archaeological Science*, **13**, 417–30.

326   *M. Smits*

Poole, D. F. G. & Tratman, E. K. (1978). Postmortem changes in human-teeth from late upper Paleolithic-Mesolithic occupants of an English limestone cave. *Archives of Oral Biology*, 23, 1115–20.

Poszwa, A., Dambrine, E., Pollier, B. & Atteia, O. (2000). A comparison between Ca and Sr cycling in forest ecosystems. *Plant and Soil*, 225, 299–310.

Rambold, G. & Agerer, R. (1997). DEEMY – The concept of a characterization and determination system for ectomycorrhizae. *Mycorrhiza*, 7, 113–16.

Read, D. J. (1991). Mycorrhizas in ecosystems. *Experientia*, 47, 376–91.

Ritchie, G. S. P. (1995). Soluble aluminium in soils: principles and practicalities. *Plant and Soil*, 171, 17–27.

Romao, P. M. S. & Rattazzi, A. (1996). Biodeterioration on megalithic monuments. Study of lichens' colonization on Tapadao and Zambujeiro dolmens (Southern Portugal). *International Biodeterioration and Biodegradation*, 37, 23–35.

Rosling, A., Landeweert, R., Lindahl, B. *et al.* (2003). Vertical distribution of ectomycorrhizal root tips in a podzol soil profile. *New Phytologist*, 159, 775–83.

Rudawska, M. & Leski, T. (1998). Aluminium tolerance of different *Paxillus involutus* Fr. strains originating from polluted and nonpolluted sites. *Acta Societatis Botanicorum Poloniae*, 67, 115–22.

Rustad, L. E. & Cronan, C. S. (1988). Element loss and retention during litter decay in a red spruce stand in Maine. *Canadian Journal of Forest Research*, 18, 947–53.

Rustad, L. E. & Cronan, C. S. (1995). Biogeochemical controls on aluminum chemistry in the O horizon of a red spruce (*Picea rubens* Sarg.) stand in central Maine, USA. *Biogeochemistry*, 29, 107–29.

Shepherd, V. A., Orlovich, D. A. & Ashford, A. E. (1993a). Cell-to-cell transport via motile tubules in growing hyphae of a fungus. *Journal of Cell Science*, 105, 1173–8.

Shepherd, V. A., Orlovich, D. A. & Ashford, A. E. (1993b). A dynamic continuum of pleiomorphic tubules and vacuoles in growing hyphae of a fungus. *Journal of Cell Science*, 104, 495–507.

Shiller, A. M. & Frilot, D. M. (1996). The geochemistry of gallium relative to aluminum in Californian streams. *Geochimica et Cosmochimica Acta*, 60, 1323–8.

Smith, S. E. & Read, D. J. (1997) *Mycorrhizal Symbiosis*. San Diego: Academic Press.

Smits, M., Hoffland, E. & van Breemen, N. (2005). Contribution of mineral tunneling to total feldspar weathering. *Geoderma*, 125, 59–69.

Soggnaes, R. (1950). Histological studies of ancient and recent teeth with special regard to differential diagnosis between intra-vitam and post mortem characteristics. *American Journal of Physical Anthropology*, 8, 269–70.

Sterflinger, K. (2000). Fungi as geologic agents. *Geomicrobiology Journal*, 17, 97–124.

Sverdrup, H. (1990). *Kinetics of Base Cation Release due to Chemical Weathering*. Lund: Lund University Press.

Sverdrup, H. (1996). Geochemistry, the key to understanding environmental chemistry. *Science of the Total Environment*, 183, 67–87.

Sverdrup, H., Hagen-Thorn, A., Holmqvist, J. *et al.* (2002). Biogeochemical processes and mechanisms. In *Developing Principles and Models for Sustainable Forestry in Sweden*, ed. H. Sverdrup & I. Stjernquist. Dordrecht: Kluwer Academic Publishers, pp. 91–196.

Tedersoo, L., Koljalg, U., Hallenberg, N. & Larsson, K. H. (2003). Fine scale distribution of ectomycorrhizal fungi and roots across substrate layers including coarse woody debris in a mixed forest. *New Phytologist*, 159, 153–65.

Ugolini, F. C., Dahlgren, R., Righi, D. & Chauvel, A. (1987). The mechanism of podzolization as revealed by soil solution studies. In *Podzols et Podzolisation, Proceedings, Symposium, 10–11 April 1986*. Poitiers: CNRS, pp. 195–203.

van Breemen, N., Finlay, R. D., Lundström, U. S. *et al.* (2000a). Mycorrhizal weathering: a true case of mineral plant nutrition? *Biogeochemistry*, **49**, 53–67.

van Breemen, N., Lundstrom, U. S. & Jongmans, A. G. (2000b). Do plants drive podzolization via rock-eating mycorrhizal fungi? *Geoderma*, **94**, 163–71.

Väre, H. (1990). Aluminum polyphosphate in the ectomycorrhizal fungus *Suillus variegatus* (Fr.) O. Kunze as revealed by energy dispersive spectrometry. *New Phytologist*, **116**, 663–8.

Vogt, K. A., Dahlgren, R. A., Ugolini, F. C. *et al.* (1987). Aluminium, Fe, Ca, Mg, K, M, Cu, Zn and P in above- and belowground biomass. II. Pools and circulation in a subalpine *Abies amabilis* stand. *Biogeochemistry*, **4**, 295–311.

Wallander, H. (2000). Uptake of P from apatite by *Pinus sylvestris* seedlings colonised by different ectomycorrhizal fungi. *Plant and Soil*, **218**, 249–56.

Wallander, H. & Wickman, T. (1999). Biotite and microcline as potassium sources in ectomycorrhizal and non-mycorrhizal *Pinus sylvestris* seedlings. *Mycorrhiza*, **9**, 25–32.

Wallander, H., Wickman, T. & Jacks, G. (1997). Apatite as a P source in mycorrhizal and non-mycorrhizal *Pinus sylvestris* seedlings. *Plant and Soil*, **196**, 123–31.

Wallander, H., Johansson, L. & Pallon, J. (2002). PIXE analysis to estimate the elemental composition of ectomycorrhizal rhizomorphs grown in contact with different minerals in forest soil. *FEMS Microbiology Ecology*, **39**, 147–56.

Wallander, H., Mahmood, S., Hagerberg, D., Johansson, L. & Pallon, J. (2003). Elemental composition of ectomycorrhizal mycelia identified by PCR-RFLP analysis and grown in contact with apatite or wood ash in forest soil. *FEMS Microbiology Ecology*, **44**, 57–65.

Wallander, H., Goransson, H. & Rosengren, U. (2004). Production, standing biomass and natural abundance of N[15] and C[13] in ectomycorrhizal mycelia collected at different soil depths in two forest types. *Oecologia*, **139**, 89–97.

Watmough, S. A. & Dillon, P. J. (2003). Ecology – mycorrhizal weathering in base-poor forests. *Nature*, **423**, 823–4.

Welch, S. A., Taunton, A. E. & Banfield, J. F. (2002). Effect of microorganisms and microbial metabolites on apatite dissolution. *Geomicrobiology Journal*, **19**, 343–67.

Werelds, R. (1962). Nouvelles observations sur les dégradations post-mortem de la dentine et du cément des dents inhumées. *Bulletin du Groupment Internationale pour la Recherche Scientifique en Stomatologie*, **5**, 559–91.

White, A. F. (1995). Chemical weathering rates of silicate minerals in soils. In *Chemical Weathering Rates of Silicate Minerals*, ed. A. F. White & S. L. Brantley. Washington, DC: Mineralogical Society of America, pp. 407–61.

Zhou, Z. H. & Hogetsu, T. (2002). Subterranean community structure of ectomycorrhizal fungi under *Suillus grevillei* sporocarps in a *Larix kaempferi* forest. *New Phytologist*, **154**, 529–39.

14

Mineral dissolution by ectomycorrhizal fungi

HÅKAN WALLANDER

Introduction

Ectomycorrhizal fungi (EMF) form symbioses with forest trees. These fungi are mainly basidiomycetes and ascomycetes and they probably evolved from saprotrophic fungi when organic matter began to accumulate in certain soils 200 million years ago (Cairney, 2000). The tree host benefits from the symbiosis through improved nutrient acquisition since the fungus explores the soil efficiently for nutrients (especially N and P) in return for host carbon (Smith & Read, 1997). Forest trees with ectomycorrhiza usually dominate in acidic soils with thick litter layers, with the associated EMF forming extensive external mycelia with a high capacity to take up nutrients from the soil. The fungi form mantles around the root tips with large storage capacities (Read, 1991). The trees invest large amounts of carbon in the ectomycorrhizal (EM) symbionts, especially under nutrient-poor conditions, with up to 20% of photosynthetic assimilates allocated to the mycorrhizal symbionts (Finlay & Söderström, 1992). This large carbon source for EMF gives them an advantage over non-symbiotic microorganisms in the soil ecosystem since these are usually carbon-limited (Aldén et al., 2001). The importance of current photosynthate for soil processes was recently demonstrated by Högberg et al. (2001), who girdled trees in northern Sweden and found a rapid reduction in soil respiration of more than 50% within 1–2 months. More precise calculations the second year after girdling demonstrated that 65% of total respiration was contributed by EM roots, their associated EMF and other rhizosphere microorganisms (Bhupinderpal-Singh et al., 2003). Furthermore, it has been shown that EM roots are directly involved in production of dissolved organic carbon (DOC) in forest soils (Griffiths et al., 1994; Högberg & Högberg, 2002) and low-molecular-weight organic

Fungi in Biogeochemical Cycles, ed. G. M. Gadd. Published by Cambridge University Press. © British Mycological Society 2006.

acids constitute a significant part of this DOC (van Hees *et al.*, 2000). In addition, the flux through this pool is extremely rapid, which suggests that it contributes substantially to the total CO_2 efflux from soil (van Hees *et al.*, 2005). The large carbon flux to the soil ecosystem via EM hyphae thus includes substantial production of carbonic acid and organic acids, making them potentially important in dissolution of soil minerals (see reviews by Landeweert *et al.*, 2001 and Hoffland *et al.*, 2004).

The more than 6000 known EM species (Smith & Read, 1997) use a wide range of growth strategies in the soil. Although the density of EM root tips is highest in the topmost organic horizon (Stober *et al.*, 2001), recent investigations with minirhizotrons and more thorough sampling approaches have shown that many EM root tips are also present in the mineral soil (Majdi *et al.*, 2001, Rosling *et al.*, 2003a). Mineral particles in the E and C horizons are usually covered by fungal hyphae (van Breemen *et al.*, 2000). Furthermore, DNA extractions of soil samples from different soil depths have demonstrated that EM hyphae are present throughout the soil profile, even in the C horizon (Landeweert *et al.*, 2003). In fact, studies by Rosling *et al.* (2003a) and Landeweert *et al.* (2003) at Svartberget Forest Research Station in northern Sweden demonstrated that many EM species were only present in the mineral soil (e.g. *Suillus* sp.). Many taxa inhabiting the mineral soil, such as *Suillus* sp., also have a high weathering capacity in laboratory studies (e.g. Wallander, 2000a). Ectomycorrhizal hyphae have also been found inside rocks in semi-arid chaparral ecosystems in California, with EM hyphae colonizing weathered granitic bedrock underlying the topsoil (Egerton-Warburton *et al.*, 2003). The authors suggested that EM hyphae were important for extracting water stored in the bedrock during summer drought. The different environmental conditions prevailing at different soil depths may partially explain the broad diversity of EMF found in many boreal forest soils (Bruns, 1995).

The possible involvement of EMF in weathering of soil minerals was recognized by Hintikka and Näykki (1967), who found that the formation of eluvial horizons in podzols was strongly promoted under fungal colonies of *Hydnellum ferrugineum* (Fig 14.1). The importance of EMF for the podzolization processes is further discussed by Lundström *et al.* (2000).

Many forests in the northern hemisphere are normally nitrogen-limited (Tamm, 1991), but due to the increased anthropogenic deposition of nitrogen together with increased leaching of base cations by acid rain, other mineral nutrients such as potassium, phosphorus, magnesium and calcium may become limiting in some forest stands (reviewed by Thelin, 2000). Modelling of weathering rates and mass balance calculations in southern

Fig 14.1. Development of the eluvial horizon in a podzol profile of a Scots pine forest in central Sweden (Kroksbo). The soil sample to the right was collected under a dense mycelial mat of *Hydnellum ferrugineum* while the soil sample to the left was collected at a distance of 20 cm from the *Hydnellum* mat where no dense fungal mat was present.

Sweden indicate that potassium will be an element in short supply in the future (Barkman & Sverdrup, 1996). To understand how future forests will respond to nutrient imbalances caused by anthropogenic activities it is essential to understand the influence of EMF on mineral dissolution.

In the present chapter, I will discuss the potential of EMF to influence weathering rates of soil minerals, and review experiments where the effects of EMF on dissolution of minerals have been investigated in axenic cultures, pot cultures and field experiments. I will then discuss the influence of the nutrient status of the trees on weathering induced by EMF and, finally, to what extent this process may influence the overall cycling of elements in forest ecosystems.

Potential role of EMF in mineral dissolution

Weathering of soil minerals is usually faster in forested landscapes than in areas covered by sparse field layer vegetation. For example, silicon concentrations increased in stream water with an increased presence of forests along rivers in northern Sweden (Humborg *et al.*, 2002). Augusto *et al.* (2000) demonstrated that coniferous trees stimulated weathering more efficiently than deciduous trees when feldspar was experimentally inserted in the soil and incubated for up to nine years, presumably because of greater soil acidification by coniferous trees than by deciduous trees. Furthermore, in large monitored sandbox ecosystems, Bormann *et al.* (1998) demonstrated that planted pine trees (*Pinus resinosa*) dramatically increased the release of calcium and magnesium from primary minerals.

Minerals are dissolved through a number of chemical reactions in the soil. The main causes of mineral dissolution are reaction with protons, hydrolysis with water, reaction with organic acid ligands and reaction with CO_2 (Sverdrup *et al.*, 2002). Trees and their EM symbionts can influence dissolution of minerals through most of these mechanisms. Fungi exude organic anions and protons through the hyphal tips during growth (Gadd, 1999). Ectomycorrhizal fungi usually exhibit higher biomass production on ammonium than on nitrate as nitrogen source and since uptake of ammonium and cations is followed by efflux of protons this leads to subsequent acidification of the substrate, thereby accelerating mineral dissolution. Soil pH is also lowered by the production of CO_2 by EM roots and their associated microorganisms. Carbon dioxide dissolves in the soil water to form carbonic acid. Ectomycorrhizal fungi can also produce organic acids, which, apart from reducing the pH, also function as chelating agents, which can form complexes with metals released through weathering. This chelation lowers the activity of free cations in the soil solution, which further promotes mineral weathering. Organic acid production by EMF has been demonstrated in axenic cultures (Arvieu *et al.*, 2003) and in vivo when EMF are growing with spruce seedling (Ahonen-Jonnarth *et al.*, 2000). Casarin *et al.* (2003) demonstrated that oxalate production was stimulated by addition of $CaCO_3$ but not by phosphorus deficiency, suggesting that oxalate production might be a way to decrease concentrations of dissolved calcium in calcareous soil. The increased proton efflux often associated with oxalate exudation may, however, still increase mobilization of phosphorus in calcareous soils (Arvieu *et al.*, 2003). Colonization of EMF has increased concentrations of organic acids in the soil solution in several studies (Wallander, 2000a, b; Casarin *et al.*, 2003; van Hees *et al.*, 2003), although it is unclear whether these organic acids originate directly from the mycorrhizal fungi or from other associated microorganisms. Certain genera of EMF such as *Rhizopogon* (Ahonen-Jonnarth *et al.*, 2000; Casarin *et al.*, 2003) and *Suillus* (Ahonen-Jonnarth *et al.*, 2000; Wallander, 2000a) seem to be especially active in production of oxalic acid.

Some researchers have suggested that the estimated concentrations of organic acids in the soil solution are too low to promote significant

weathering (Drever & Stillings, 1997). Concentrations of oxalate, suppo-
sedly one of the most active organic acids in weathering, are usually less
than 10 μm (van Hees et al., 2003). Bulk estimates of concentrations of
organic acids are, however, not very useful for predicting the influence on
mineral weathering since concentrations will probably vary spatially, with
higher concentrations around root tips and hyphal tips where production
is supposed to be high (Sun et al., 1999; Jones et al., 2003). Organic acids
have a very high turnover rate in soil since they are an attractive substrate
for microbes (van Hees et al., 2003). As a consequence, any effects on
dissolution of minerals are likely to be near hyphal tips and root tips.

Production of organic acids by EMF has also been suggested to result in
penetration of mineral particles by EM hyphae (Jongmans et al., 1997),
although recent estimates have found that tunnel-shaped structures within
mineral grains contribute very little (less than 1%) of the total weathering
in the soil (Smits et al., 2005; see Smits, Chapter 13 this volume).

Although it seems clear that EMF have the potential to influence
dissolution of soil minerals it is difficult to estimate how much this contributes to the total weathering of soil minerals. It is also not known to
what extent the EM contribution to total weathering depends on the
nutritional status of the forest. It is possible that trees respond to nutrient
deficiencies by allocating resources below ground that accelerate mineral
weathering. To discuss this possibility further we need to know how forest
trees and their associated EMF respond to nutrient deficiencies.

Response of EMF to nutrient conditions of their host trees

Most EMF are adapted to nitrogen-limiting conditions and in
forests with high nitrogen input, growth of EMF is usually inhibited. Other
effects of increased nitrogen inputs include reduced production of fruit
bodies (Termorshuizen & Schaffers, 1991; Brandrud, 1995; Wiklund et al.,
1995; Lilleskov et al., 2001), reduced diversity of EMF on root tips (Kårén &
Nylund, 1997; Wallenda & Kottke, 1998; Fransson et al., 2000; Jonsson
et al., 2000; Taylor et al., 2000; Peter et al., 2001; Lilleskov et al., 2002;
Erland & Taylor, 2002), and, in laboratory experiments, reduced growth
of the external mycelium from the root tips (Wallander & Nylund, 1992).
Species sensitivity to nitrogen additions differs among EMF (Arnebrant,
1994). By using ingrowth mesh bags to quantify the production of EM
mycelia in the field, Nilsson and Wallander (2003) demonstrated that
growth of external mycelia was reduced by about 50% in field experiments
with the addition of $100 \, kg \, N \, ha^{-1} \, yr^{-1}$ as $(NH_4)_2SO_4$ for 10 years in the
Skogaby experimental forest in southern Sweden. Elevated nitrogen inputs

to forests may thereby reduce the potential for EMF to influence weathering rates since less carbon is allocated to the EMF under these conditions.

The nutrient status of trees strongly influences allocation between above-ground and below-ground parts. In general, nutrient deficiencies should result in increased allocation to roots to facilitate better nutrient uptake, but this is not always the case. Ericsson (1995) demonstrated in laboratory experiments that deficiencies of nitrogen and phosphorus increased the allocation of dry matter to the roots as expected but potassium and magnesium deficiencies decreased allocation to roots in birch seedlings. This was suggested to result from reduced carbon fixation under potassium and magnesium deficiency leading to a shortage of carbohydrates, and an accompanying reduced allocation of carbon to below-ground parts. Similar results have been found for carbon allocation to EMF in laboratory systems. Wallander and Nylund (1992) and Ekblad *et al.* (1995) found increased allocation of dry matter to the EM symbionts under phosphorus deficiency whereas potassium deficiency resulted in reduced allocation of dry matter to EM symbionts. These findings suggest that the potential to counteract phosphorus deficiency through increased weathering of phosphorus-containing minerals by EMF is much higher than the potential to counteract potassium deficiency by increased weathering of potassium-containing minerals.

Examples of experiments on the influence of EMF on weathering

Axenic cultures

Ectomycorrhizal fungi and other fungi have dissolved minerals such as tri-calcium phosphate and calcium carbonate in Petri dishes when the minerals have been added as a fine powder (Lapeyrie *et al.*, 1991; Chang & Li, 1998; Mahmood *et al.*, 2001). Because the dissolution of these minerals is usually mediated through acidification of the substrate, Rosling *et al.* (2003b) recently developed a method to quantify substrate acidification in these systems. By using this method, the authors found significant species–mineral interactions regulating substrate acidification.

In axenic cultures, some EMF can get access to non-exchangeable potassium and calcium reserves in micas such as phlogopite and vermiculite (Paris *et al.*, 1995a, b, 1996). Application of limiting amounts of potassium and magnesium in the nutrient media increased the release of potassium from phlogopite, and seemed to be accompanied by increased exudation of oxalic acid by the EMF (Paris *et al.*, 1995a, 1996). Exposure of the phlogopite to one of the tested EMF (*Paxillus involutus*) led to irreversible transformation of the phlogopite toward vermiculitic structures, indicating that the fungus strongly influenced mineral weathering. It seems clear from the above studies that many EMF can potentially dissolve minerals. This activity seems to depend on the amount of carbon and energy available to the fungus. As mentioned above, EMF have an advantage over other microorganisms in this respect since they can obtain carbon and energy from their host plants. However, given the difficulties in extrapolating results from axenic cultures to field situations, it is also desirable to examine the influence of plant–fungal symbioses on mineral dissolution in culture studies.

Pot and microcosm cultures

Ectomycorrhizal plants and non-mycorrhizal control plants have been grown in pots and microcosms by many researchers to investigate the role of EM colonization on dissolution of minerals. In such experiments it is usually not possible to separate effects of the fungus itself and effects of other microorganisms in the soil that might be influenced by the presence of the EM fungus.

In a pot culture experiment, beech grown in association with *Laccaria laccata* showed increased uptake of phosphorus from rock phosphate and increased uptake of potassium and magnesium from added mica (phlogopite) compared to non-colonized seedlings (Leyval & Berthelin, 1989). In a later study, Leyval and Berthelin (1991) found that *Laccaria laccata* increased potassium loss from phlogopite by increasing the exchangeable surface area while dual inoculation with *Agrobacterium* sp. further increased potassium loss by acidifying the substrates. Several more recent experiments have also shown that phosphate minerals such as apatite are usually dissolved more rapidly when an EM fungus is present (Wallander et al., 1997, Wallander, 2000a, Casarin et al., 2003) and that potassium release from micas increases under the influence of EMF (Wallander & Wickman, 1999, Wallander, 2000b). The EM species or community of EM species colonizing the root systems influence how efficient the process is. In a recent rhizobox study, Casarin et al. (2004) showed that an EM species with high oxalic acid production (*Rhizopogon roseolus*) was more efficient than a species with low oxalate production (*Hebeloma cylindrosporum*) in using phosphorus from poorly soluble sources. Addition of $CaCO_3$ suppressed the positive effect of *R. roseolus* on phosphorus uptake, suggesting that one important function of oxalate is to sequester excess calcium in calcareous soils. Genera with high oxalate production such as *Suillus*

(Ahonen-Jonnarth *et al.*, 2000) appear to be more efficient than taxa with low oxalate production in dissolving minerals containing phosphorus such as apatite (Wallander, 2000a)

In a recent microcosm study using a reconstructed podzol, Heinonsalo *et al.* (2004) estimated the allocation of ^{14}C-labelled carbon to mycorrhizal roots in different soil horizons. They found similar allocation patterns in organic, eluvial and illuvial soil horizons. Interestingly, large amounts of carbon were allocated to *Suillus*-type mycorrhizas in the illuvial (B) horizon, indicating that *Suillus*-type mycorrhizas are active in this soil layer where weatherable minerals are present in high amounts. Presumably some of this labelled carbon is oxalic acid exuded in the mycorrhizosphere. In another microcosm study, Rosling *et al.* (2004) showed stronger proliferation of external EM mycelium of *Hebeloma crustuliniforme* on patches of feldspar than on patches of quartz, suggesting that the fungus used nutrients from the feldspar.

Spruce seedlings and their associated EM fungus (*P. involutus*) significantly affected weathering of E horizon soil added in pot cultures (van Hees *et al.*, 2004). Mobilization of silicon increased 50% in the presence of the plant alone and increased about 100% when the mycorrhizal fungus was also present. The plant alone did not affect aluminium mobilization but the EM fungus increased aluminium mobilization by more than 100%. In contrast, addition of humic acid did not affect silicon and aluminium mobilization.

Although EMF appear important in increasing dissolution of minerals, EM biomass on roots is usually a poor predictor of mineral dissolution since many EM species may not be active in weathering processes (Wallander *et al.*, 2004).

Field studies

By comparing Ca/Sr ratios in foliage and in different potential calcium sources, Blum *et al.* (2002) suggested that EM trees used calcium derived from apatite present in the B horizon while trees colonized by arbuscular mycorrhizal fungi (AMF) could not access this mineral source. The reason may be that AMF do not produce organic acids such as oxalic acid, which are known to dissolve apatite. Using Ca/Sr ratios to estimate calcium sources has, however, been criticized by Watmough and Dillon (2003), who argue that Ca/Sr ratios can vary substantially in plant parts and cannot therefore be used to identify the source of calcium in trees.

It is difficult to directly study the influence of EMF on dissolution of minerals in the field. Several studies on EM fungal mat communities in

Oregon, USA have, however, shown that concentrations of oxalic acid can be very high in such communities. Oxalic acid and phosphate concentrations were highly correlated, suggesting that oxalate increases the weathering and solubility of phosphate in these soils (Griffiths et al., 1994). Many soil-inhabiting fungi (including EMF) are covered by calcium oxalate crystals (Cromack et al., 1979) and large amounts of calcium in EM hyphae growing in soil have been detected by scanning electron microscopy equipped with energy dispersive systems (SEM-EDS) (Arocena et al., 1999) and particle-induced X-ray emission (PIXE), especially when the fungus has been growing in contact with a calcium source such as apatite or wood ash (Wallander et al., 2003). It is likely that these calcium oxalate crystals could be residues left after dissolving calcium-rich minerals in the soil to obtain phosphorus.

Arocena et al. (1999) and Arocena and Glowa (2000) investigated the soil solution and mineral composition of soil collected near EM roots in subalpine fir forests in Canada. The authors found that mica and chlorite had been transformed to 2:1 expandable clays to a larger extent in ectomycorrhizosphere soil than in non-ectomycorrhizosphere soil. Mineral degradation increased the availability of K^+, Ca^{2+} and Mg^{2+} in the soil solution. It was suggested that organic acid production and uptake of potassium and magnesium had led to dissolution of the minerals. Subsequent laboratory experiments confirmed that one of the EMF tested in the field (*Piloderma*) could extract potassium and magnesium from different minerals (Glowa et al., 2003). *Piloderma* is known to produce oxalic acid in high amounts and its hyphae are usually covered in calcium-rich encrustations that are composed primarily of calcium oxalate (Arocena et al., 2001).

Wallander et al. (2001) studied the mycorrhizal response to adding minerals to ingrowth bags where only fungal hyphae, but not roots, could enter. Growth of EM mycelia in Norway spruce forests in southern Sweden with different potassium status or phosphorus status did not vary significantly but adding apatite in phosphorus-poor forests resulted in increased growth of EM mycelia and increased production of EM root tips around ingrowth mesh bags. The strong stimulatory influence of apatite on formation of mycorrhizal root tips suggests that the tree responded to the phosphorus source in the soil by increasing the amount of carbon allocated to this microsite, which in turn increased the production of mycorrhizal root tips and external mycelia. A significant part of the carbon allocated to the EMF was probably used to produce oxalic acid to facilitate solubilization of the apatite.

Rhizomorphs growing in contact with apatite had elevated concentrations of calcium, indicating that calcite ($CaCO_3$) or calcium oxalate could accumulate in fungal tissue as a result of the interaction between the EM fungus and the apatite (Wallander *et al.*, 2003). The dissolution of the apatite was quantified by estimating the flux of rare earth elements from the apatite to the mycorrhizal roots. These elements are in high concentration in the apatite and, although transported by EMF, are not transferred to the host tissue (Vesk *et al.*, 2000). The EMF strongly influenced the dissolution of apatite, with the EM-induced dissolution increasing by 40% in the phosphorus-deficient soil compared to the phosphorus-sufficient soil. This finding suggests that EMF can compensate for phosphorus deficiency in Norway spruce stands by increasing the dissolution of apatite. Forests with a larger range of phosphorus status need to be studied to confirm these results. Hagerberg (2003) calculated the extent to which this process could compensate for nutrient losses induced by whole-tree harvesting but found that EM-induced weathering could compensate for at most only half of the phosphorus lost from whole tree harvesting in Norway spruce forests in southern Sweden.

In contrast to the results found with apatite, growth of EM mycelia was not affected by adding the potassium-containing mineral biotite to the ingrowth mesh bags in any of the forests (Hagerberg *et al.*, 2003). Furthermore, the elemental composition of rhizomorphs growing in contact with biotite did not differ from rhizomorphs growing in contact with pure quartz sand (Wallander *et al.*, 2003). However, the biotite is naturally enriched in [87]Sr and minute amounts of [87]Sr originating from the biotite was detected in root tips on the outside of biotite-amended mesh bags (Hagerberg, 2003). Ectomycorrhizal fungi may thus influence the dissolution of biotite although the effects appear to be much smaller than the effects on dissolution of apatite.

Conclusions

The present chapter has described the presence of EMF where soil minerals occur and their formation of intimate contact with soil minerals. It seems clear that EMF can dissolve at least some of these minerals when exposed to them in laboratory conditions. Furthermore, EMF obtain large amounts of carbon from their host trees and convert much of it to organic and carbonic acids, making them potentially important in dissolving minerals in natural ecosystems. The main questions for future research are to quantify the contribution of EMF to total weathering of minerals in the soil and to determine whether this contribution will change as a result of more intensively managed forest ecosystems and increased anthropogenic depletion of nutrients. We need to know this to predict whether nutrients need to be added to counteract nutrient losses due to harvest and leaching or if the system itself can counteract these effects through more intensive weathering rates induced by EMF. It is likely that certain EMF are adapted to nutrient-poor conditions and have evolved traits to use phosphorus sources efficiently. Phosphorus deficiency is common in many soils since phosphate is immobile and many soils have a large phosphorus-fixing capacity. In contrast, potassium is much more mobile in the soil and potassium deficiency is probably a rare phenomenon in most ecosystems. Specific traits to enhance access to potassium sources are therefore less likely to have evolved among EMF.

Acknowledgements

This work was supported by the Swedish National Energy Administration and the Swedish Research Council for Environment, Agricultural Sciences and Spatial Planning (grants to H. Wallander). Thanks to Erik Hobbie and Roger Finlay for valuable comments on the manuscript.

References

Ahonen-Jonnarth, U., van Hees, P. A. W., Lundström, U. & Finlay, R. (2000). Production of organic acids by mycorrhizal and non-mycorrhizal *Pinus sylvestris* seedlings exposed to elevated concentrations of aluminium and heavy metals. *New Phytologist*, 146, 557–67.

Aldén, L., Demoling, F. & Bååth, E. (2001). Rapid method of determining factors limiting bacterial growth in soil. *Applied and Environmental Microbiology*, 67, 1830–8.

Arnebrant, K. (1994). Nitrogen amendments reduce the growth of extramatrical mycelium. *Mycorrhiza*, 5, 7–15.

Arocena, J. M. & Glowa, K. R. (2000). Mineral weathering in ectomycorrhizosphere of subalpine fir (*Abies lasiocarpa* (Hook.) Nutt.) as revealed by soil solution composition. *Forest Ecology and Management*, 133, 61–70.

Arocena, J. M., Glowa, K. R., Massicotte, H. B. & Lavkulich, L. (1999). Chemical and mineral composition of ectomycorrhizosphere soils of subalpine fir (*Abies lasiocarpa* (Hook) Nutt.) in the Ae horizon of a Luvisol. *Canadian Journal of Soil Science*, 79, 25–35.

Arocena, J. M., Glowa, K. R. & Massicotte, H. B. (2001). Calcium-rich hypha encrustations on *Piloderma*. *Mycorrhiza*, 10, 209–15.

Arvieu, J.-C., Leprince, F. & Plassard, C. (2003). Release of oxalate and protons by ectomycorrhizal fungi in response to P-deficiency and calcium carbonate in nutrient solution. *Annals of Forest Science*, 60, 815–21.

Augusto, L., Turpault, M.-P. & Ranger, J. (2000). Impact of forest tree species on feldspar weathering rates. *Geoderma*, 96, 215–37.

Barkman, A. & Sverdrup, H. (1996). *Critical loads of acidity and nutrient imbalance for forest ecosystems in Skåne*, Report 1. Department of Chemical Technology II, Lund University, Lund, Sweden. 64 p.

Bhupinderpal-Singh, Nordgren, A., Ottoson Löfvenius, M. *et al.* (2003). Tree root and soil heterotrophic respiration as revealed by girdling of boreal Scots pine forest: extending observations beyond the first year. *Plant Cell and Environment*, **26**, 1287–96.

Blum, J. D., Klaue, A., Nezat, C. A. *et al.* (2002). Mycorrhizal weathering of apatite as an important calcium source in base-poor forest ecosystems. *Nature*, **417**, 729–31.

Bormann, B. T., Wang, D., Borrmann, F. H. *et al.* (1998). Rapid plant-induced weathering in an aggrading experimental ecosystem. *Biogeochemistry*, **43**, 129–55.

Brandrud, T. E. (1995). The effects of experimental N addition on the ectomycorrhizal flora in an oligotrophic spruce forest at Gårdsjön, Sweden. *Forest Ecology and Management*, **71**, 111–22.

Bruns, T. D. (1995). Thoughts on the processes that maintain local species diversity of ectomycorrhizal fungi. *Plant and Soil*, **170**, 63–73.

Cairney, J. W. G. (2000). Evolution of mycorrhiza systems. *Naturwissenschaften*, **87**, 467–75.

Casarin, V., Plassard, C., Souche, G. & Arvieu, J.-C. (2003). Quantification of oxalate ions and protons released by ectomycorrhizal fungi in rhizosphere soil. *Agronomie*, **23**, 461–9.

Casarin, V., Plassard, C., Hinsinger, P., Arvieu, J.-C. (2004). Quantification of ectomycorrhizal fungal effects on the bioavailability and mobilization of soil P in the rhizosphere of *Pinus pinaster*. *New Phytologist*, **163**, 177–85.

Chang, T. T. & Li, C. Y. (1998). Weathering of limestone, marble and calcium phosphate by ectomycorrhizal fungi and associated microorganisms. *Taiwan Journal of Forest Science*, **13**, 85–90.

Cromack, K., Sollins, P., Graustein, W. *et al.* (1979). Calcium oxalate accumulation and soil weathering in mats of the hypogeous fungus *Hysterangium crassum*. *Soil Biology and Biochemistry*, **11**, 463–8.

Drever, J. I. & Stillings, L. L. (1997). The role of organic acids in mineral weathering. *Colloids and Surfaces*, **120**, 167–81.

Egerton-Warburton, L. M., Graham, R. C. & Hubbert, K. R. (2003). Spatial variability in mycorrhizal hyphae and nutrient and water availability in a soil-weathered bedrock profile. *Plant and Soil*, **249**, 331–42.

Ekblad, A., Wallander, H., Carlsson, R. & Huss-Danell, K. (1995). Fungal biomass in roots and extramatrical mycelium in relation to macronutrients and plant biomass of ectomycorrhizal *Pinus sylvestris* and *Alnus incana*. *New Phytologist*, **131**, 443–51.

Ericsson, T. (1995). Growth and shoot:root ratio of seedlings in relation to nutrient availability. *Plant and Soil*, **168–169**, 205–14.

Erland, S. & Taylor, A. F. S. (2002). Diversity of ecto-mycorrhizal fungal communities in relation to the abiotic environment. In *Mycorrhizal Ecology, Ecological Studies 157*, ed. M. G. A. van der Heijden & I. Sanders. Berlin: Springer-Verlag, pp. 163–224.

Finlay, R. & Söderström, B. (1992). Mycorrhiza and carbon flow to the soil. In *Mycorrhizal Functioning*, ed. M. F. Allen. New York: Chapman and Hall, pp. 134–60.

Fransson, P. M. A., Taylor, A. F. S. & Finlay, R. D. (2000). Effects of continuous optimal fertilization upon a Norway spruce ectomycorrhizal community. *Tree Physiology*, **20**, 599–606.

Gadd, G. M. (1999). Fungal production of citric and oxalic acid: importance in metal speciation, physiology and biogeochemical processes. *Advances in Microbial Physiology*, **41**, 48–91.

Glowa, K. R., Arocena, J. M. & Massicotte, H. B. (2003). Extraction of potassium and/or magnesium from selected minerals by *Piloderma*. *Geomicrobiology Journal*, **20**, 99–111.

Griffiths, R. P., Baham, J. E. & Caldwell, B. A. (1994). Soil solution chemistry of ectomycorrhizal mats in forest soil. *Soil Biology and Biochemistry*, **26**, 331–7.

Hagerberg, D. (2003). The growth of external ectomycorrhizal mycelia in the field in relation to host nutrient status and local addition of mineral sources. Unpublished Ph.D. thesis, Lund University, Lund, Sweden.

Hagerberg, D., Thelin, G. & Wallander, H. (2003). The production of ectomycorrhizal mycelium in forests: relation between forest nutrient status and local mineral sources. *Plant and Soil*, **252**, 279–90.

Heinonsalo, J., Hurme, K.-R. & Sen, R. (2004). Recent ^{14}C-labelled assimilate allocation to Scots pine seedling root and mycorrhizosphere compartments developed on reconstructed podzol humus, E- and N-mineral horizons. *Plant and Soil*, **259**, 111–21.

Hintikka, V. & Näykki, O. (1967). Notes on the effects of the fungus *Hydnellum ferrugineum* (FK.) Karst. on forest soil and vegetation. *Communicationes Instituti Forestalis Fenniae*, **62**, 1–23.

Hoffland, E., Kuyper, T. W., Wallander, H. et al. (2004). The role of fungi in weathering. *Frontiers in Ecology and the Environment*, **2**, 258–64.

Högberg, M. N. & Högberg, P. (2002). Extramatrical ectomycorrhizal mycelium contributes one-third of microbial biomass and produces, together with associated roots, half the dissolved organic carbon in a forest soil. *New Phytologist*, **154**, 791–5.

Högberg, P., Nordgren, A., Buchmann, N. et al. (2001). Large-scale forest girdling shows that current photosynthesis drives soil respiration. *Nature*, **411**, 789–92.

Humborg, C., Blomqvist, S., Avsan, E., Bergensund, Y. & Smedberg, E. (2002). Hydrological alterations with river damming in northern Sweden: implications for weathering and river biogeochemistry. *Global Biogeochemical Cycles*, **16**, 1–13.

Jones, D. L., Dennis, P. G., Owen, A. G. & van Hees, P. A. W. (2003). Organic acid behaviour in soils–misconceptions and knowledge gaps. *Plant and Soil*, **248**, 31–41.

Jongmans, A. G., van Breemen, N., Lundström, U. et al. (1997). Rock-eating fungi. *Nature*, **389**, 682–3.

Jonsson, L., Dahlberg, A. & Brandrud, T.-E. (2000). Spatiotemporal distribution of an ectomycorrhizal community in an oligotrophic Swedish *Picea abies* forest subjected to nitrogen addition: above- and below-ground views. *Forest Ecology and Management*, **132**, 143–56.

Kårén, O. & Nylund, J.-E. (1997). Effects of ammonium sulphate on the community structure and biomass of ectomycorrhizal fungi in a Norway spruce stand in southwestern Sweden. *Canadian Journal of Botany*, **75**, 1628–42.

Landeweert, R., Hoffland, E., Finlay, R. & van Breemen, N. (2001). Linking plants to rocks: ectomycorrhizal fungi mobilize nutrients from minerals. *Trends in Ecology and Evolution*, **16**, 248–54.

Landeweert, R., Leeflang, P., Kuyper, T. W. et al. (2003). Molecular identification of ectomycorrhizal mycelium in soil horizons. *Applied and Environmental Microbiology*, **69**, 327–53.

Lapeyrie, F., Ranger, J. & Vairelles, D. (1991). Phosphate-solubilizing activity of ectomycorrhizal fungi in vitro. *Canadian Journal of Botany*, **69**, 342–6.

Leyval, C. & Berthelin, J. (1989). Interactions between *Laccaria laccata*, *Agrobacterium radiobacter* and beech roots: influence on P, K, Mg and Fe mobilization from minerals and plant growth. *Plant and Soil*, **117**, 103–20.

Leyval, C. & Berthelin, J. (1991). Weathering of a mica by roots and rhizospheric microorganisms of pine. *Soil Science Society of America Journal*, **55**, 1009–16.

Lilleskov, E., Fahey, T. J. & Lovett, G. M. (2001). Ectomycorrhizal fungal aboveground community change over an atmospheric nitrogen deposition gradient. *Ecological Applications*, **11**, 397–410.

Lilleskov, E. A., Fahey, T. J., Horton, T. R. & Lovett, G. M. (2002). Belowground ectomycorrhizal fungal community change over a nitrogen deposition gradient in Alaska. *Ecology*, **83**, 104–15.

Lundström, U., van Breemen, N., Bain, D. C. *et al.* (2000). Advances in understanding the podzolization process resulting from a multidisciplinary study of three coniferous forest soils in the Nordic countries. *Geoderma*, **94**, 335–53.

Mahmood, S., Finlay, R. D., Erland, S. & Wallander, H. (2001). Solubilization and colonisation of wood ash by ectomycorrhizal fungi isolated from a wood ash fertilised spruce forest. *FEMS Microbiology Ecology*, **35**, 151–61.

Majdi, H., Damm, E. & Nylund, J.-E. (2001). Longevity of mycorrhizal roots depend on branching order and nutrient availability. *New Phytologist*, **150**, 195–202.

Nilsson, L. O. & Wallander, H. (2003). The production of external mycelium by ectomycorrhizal fungi in a Norway spruce forest was reduced in response to nitrogen fertilization. *New Phytologist*, **158**, 409–16.

Paris, F., Bonnaud, P., Ranger, J. & Lapeyrie, F. (1995a). In vitro weathering of phlogopite by ectomycorrhizal fungi. I. Effect of K^+ and Mg^{2+} deficiency on phyllosilicate evolution. *Plant and Soil*, **177**, 191–205.

Paris, F., Bonnaud, P., Ranger, J., Robert, M. & Lapeyrie, F. (1995b). Weathering of ammonium or calcium saturated 2:1 phyllosilicates by ectomycorrhizal fungi *in vitro*. *Soil Biology and Biochemistry*, **27**, 1237–44.

Paris, F., Botton, B. & Lapeyrie, F. (1996). *In vitro* weathering of phlogopite by ectomycorrhizal fungi II. Effect of K^+ and Mg^{2+} deficiency and N source on accumulation of oxalate and H^+. *Plant and Soil*, **179**, 141–50.

Peter, M., Ayer, F. & Egli, S. (2001). Nitrogen addition in a Norway spruce stand altered macromycete sporocarp production and below-ground ectomycorrhizal species composition. *New Phytologist*, **149**, 311–25.

Read, D. J. (1991). Mycorrhizas in ecosystems. *Experientia*, **47**, 376–91.

Rosling, A., Landeweert, R., Lindahl, B. D. *et al.* (2003a). Vertical distribution of ectomycorrhizal fungal taxa in a podzol profile. *New Phytologist*, **159**, 775–83.

Rosling, A., Lindahl, B. D., Taylor, A. F. S. & Finlay, R. D. (2003b). Mycelial growth and substrate acidification of ectomycorrhizal fungi in response to different minerals. *FEMS Microbiology Ecology*, **47**, 31–7.

Rosling, A., Lindahl, B. D. & Finlay, R. D. (2004). Carbon allocation to ectomycorrhizal roots and mycelium colonising different mineral substrates. *New Phytologist*, **162**, 795–802.

Smith, S. E. & Read, D. J. (1997). *Mycorrhizal Symbiosis*. San Diego: Academic Press.

Smits, M. M., Hoffland, E., Jongmans, A. G. & van Breemen, N. (2005). Contribution of feldspar tunneling by fungi in weathering. *Geoderma*, **125**, 59–69.

Stober, C., George, E. & Persson, H. (2001). Root growth and response to nitrogen. In *Carbon and Nitrogen Cycling in European Forest Ecosystems, Ecological Studies 142*, ed. E.-D. Schulze. Berlin: Springer-Verlag, pp. 99–119.

Sun, Y. P., Unestam, T., Lucas, S. D. *et al.* (1999). Exudation-reabsorption in a mycorrhizal fungus, the dynamic interface for interaction with soil and soil microorganisms. *Mycorrhiza*, **9**, 137–44.

Sverdrup, H., Hagen-Thorn, A., Holmqvist, J. *et al.* (2002). Biogeochemical processes and mechanisms. In *Developing Principles for Sustainable Forestry in Southern Sweden*, ed. H. Sverdrup & I. Stjernquist. Dordrecht: Kluwer Academic Publishers, pp. 91–196.

342   *H. Wallander*

Tamm, C.-O. (1991). *Nitrogen in Terrestrial Ecosystems. Questions of Productivity*. Berlin: Springer-Verlag.

Taylor, A. F. S., Martin, F. & Read, D. J. (2000). Fungal diversity in ectomycorrhizal communities of Norway spruce [*Picea abies* (l.) Karst.] and beech (*Fagus sylvatica* L.) along north-south transects in Europe. In *Carbon and Nitrogen Cycling in European Forest Ecosystems, Ecological Studies 142*, ed. E.-D. Schulze. Berlin: Springer-Verlag, pp. 343–65.

Termorshuizen, A. J. & Schaffers, A. (1991). The decline of carpophores of ectomycorrhizal fungi in stands of *Pinus sylvestris L.* in The Netherlands: possible causes. *Nova Hedwigia*, **53**, 267–89.

Thelin, G. (2000). Nutrient imbalance in Norway spruce. Unpublished Ph.D. thesis, Lund University, Lund, Sweden.

van Breemen, N., Finlay, R. D., Lundström, U. *et al.* (2000). Mycorrhizal weathering: a true case of mineral nutrition? *Biogeochemistry*, **49**, 53–67.

van Hees, P. A. W., Lundström, U. & Giesler, R. (2000). Low molecular weight organic acids and their Al-complexes in soil solution – composition, distribution and seasonal variation in three podzolized soils. *Geoderma*, **94**, 173–200.

van Hees, P. A. W., Jones, D. L., Jentschke, G. & Godbold, D. L. (2003). Mobilization of aluminium, iron and silicon by *Picea abies* and ectomycorrhizas in a forest soil. *European Journal of Soil Science*, **55**, 101–11.

van Hees, P. A. W., Godbold, D. L., Jentschke, G. & Jones, D. (2004). Impact of ectomycorrhizas on the concentration and biodegradation of simple organic acids in a forest soil. *European Journal of Soil Science*, **54**, 697–706.

van Hees, P. A. W., Jones, D., Finlay, R., Godbold, D. L. & Lundström, U. S. (2005). The carbon we do not see–the impact of low molecular weight compounds on carbon dynamics and respiration in forest soils: a review. *Soil Biology and Biochemistry*, **37**, 1–13.

Vesk, P. A., Ashford, A. E., Markovina, A.-L. & Allaway, W. G. (2000). Apoplasmic barriers and their significance in the exodermis and sheath of *Eucalyptus pilularis–Pisolithus tinctorius* ectomycorrhizas. *New Phytologist*, **145**, 333–46.

Wallander, H. (2000a). Uptake of P from apatite by *Pinus sylvestris* seedlings colonised by different ectomycorrhizal fungi. *Plant and Soil*, **218**, 249–56.

Wallander, H. (2000b). Use of strontium isotopes and foliar K content to estimate weathering of biotite induced by pine seedlings colonised by ectomycorrhizal fungi from two different soils. *Plant and Soil*, **222**, 215–29.

Wallander, H. & Nylund, J.-E. (1992). Effects of excess nitrogen and phosphorus starvation on extramatrical mycelium in Scots pine seedlings. *New Phytologist*, **120**, 495–503.

Wallander, H. & Wickman, T. (1999). Biotite and microcline as a K source in mycorrhizal and non-mycorrhizal *Pinus sylvestris* seedlings. *Mycorrhiza*, **9**, 25–32.

Wallander, H., Wickman, T. & Jacks, G. (1997). Apatite as a P source in mycorrhizal and non-mycorrhizal *Pinus sylvestris* seedlings. *Plant and Soil*, **196**, 123–31.

Wallander, H., Nilsson, L. O., Hagerberg, D. & Bååth, E. (2001). Estimation of the biomass and production of external mycelium of ectomycorrhizal fungi in the field. *New Phytologist*, **151**, 751–60.

Wallander, H., Mahmood, S., Hagerberg, D., Johansson, L. & Pallon, J. (2003). Elemental composition of ectomycorrhizal mycelia identified with PCR/RFLP and grown in contact with apatite or wood ash in forest soil. *FEMS Microbiology Ecology*, **44**, 57–65.

Wallander, H., Fossum, A., Rosengren, U. & Jones, H. (2004). Ectomycorrhizal colonisation and uptake of P from apatite by *Pinus sylvestris* seedlings growing in forest soil with and without wood ash amendment. *Mycorrhiza*, **15**, 143–8.

Wallenda, T. & Kottke, I. (1998). Nitrogen deposition and ectomycorrhizas. *New Phytologist*, **139**, 169–87.

Watmough, S. A. & Dillon, P. J. (2003). Ecology-mycorrhizal weathering in base-poor forests. *Nature*, **423**, 823–4.

Wiklund, K., Nilsson, L. O. & Jacobsson, S. (1995). Effect of irrigation, fertilization, and drought on basidioma production in a Norway spruce stand. *Canadian Journal of Botany*, **73**, 200–8.

15

Lichen biogeochemistry

JOHNSON R. HAAS
AND O. WILLIAM PURVIS

Introduction

This volume focuses primarily on the influence of free-living fungi in biogeochemistry. Lichens, fungi that exist in facultative or obligate symbiosis with one or more photosynthesizing partners, also play an important role in many biogeochemical processes. Pioneer colonizers of fresh rock outcrops, lichens were possibly one of the first life forms to occupy Earth's land surfaces. The unique lichen symbiosis formed between the fungal partner (mycobiont) and the photosynthesizing partner, an alga or cyanobacterium (photobiont), enables lichens to grow in all surface terrestrial environments. These include extreme environments where no other multicellular vegetation can survive, such as the dry Antarctic valleys (Nash, 1996). An estimated 6% of the Earth's land surface is covered by lichen-dominated vegetation.

Globally, lichens play an important biogeochemical role in the retention and distribution of nutrient (e.g. C, N) and trace elements (e.g. Knops et al., 1991; Garty et al., 1995), in soil formation processes (Ascaso et al., 1976; Jones, 1988) and in rock weathering (Hallbauer & Jahns, 1977; Wilson et al., 1981; Wessels & Schoeman, 1988; McCarroll & Viles, 1995; Barker et al., 1997; Lee & Parsons, 1999). Lichens tend to accumulate trace elements such as lead, copper and other heavy metals of environmental concern (see below), including radionuclides (Yliruokanen, 1975; Nieboer & Richardson, 1981; Beckett et al., 1982; Boileau et al., 1982, 1985a, b; Richardson et al., 1985; Fahselt et al., 1995; Haas et al., 1998; McLean et al., 1998; Jacquiot & Daillant, 1999; Purvis et al., 2004). Many lichen species form metal–organic biominerals. Like certain bryophytes, metallophyte lichens thrive on metal-rich substrates where high metal concentrations prevent colonization by vascular plants (Purvis & Halls, 1996). Lichens grow on anthropogenic metal-rich surfaces, near metal smelters, on mine spoil and directly on ore minerals (Table 15.1).

Fungi in Biogeochemical Cycles, ed. G. M. Gadd. Published by Cambridge University Press. © British Mycological Society 2006.

Table 15.1. *Some metallophyte lichens observed growing directly on minerals, mineralized substrates and metal structures*

Lichens	Substrate	Formula	Country	Reference
Scoliciosporum chlorococcum	Weathered aluminium sheeting	Al (probably a weathered oxide surface)	UK	Brightman & Seaward (1977)
Psilolechia leprosa	Uncharacterized secondary copper minerals/beneath lightening conductors on churches	Cu (uncharacterized weathered surfaces)	Sweden, UK	Purvis (1996)
Acarospora rugulosa	Atacamite	$Cu_2Cl(OH)_3$	Sweden	Purvis (1984)
Acarospora rugulosa	Brochantite	$Cu_4SO_4(OH)_6$	Sweden	Purvis (1984)
Lecidea inops	Malachite	$Cu_2(CO_3)(OH)_2$	England	Purvis & Halls (1996)
Lecanora galactiniza and *Hypotrachyna lecanoracea*	Malachite	$Cu_2(CO_3)(OH)_2$	S Africa	Wild (1968)
Lecidea inops	Azurite	$Cu_3(CO_3)_2(OH)_2$	UK	Purvis & James (1985)
Acarospora sinopica	Chalcopyrite	$CuFeS_2$	Scandinavia	*
Lecanora cascadensis (= *L. garovaglii*)	Bornite	Cu_5FeS_4	USA	Czehura (1977)
Acarospora smaragdula	Chalcopyrite	$CuFeS_2$	UK	*
Trapelia involuta	Metatorbernite	$Cu(UO_2)_2(PO_4)_2.8H_2O$	UK	Purvis *et al.* (2004)
Trapelia involuta	Metazeunerite	$Cu(UO_2)_2(AsO_4)_2.8H_2O$	UK	Purvis *et al.* (2004)
Trapelia involuta	Autunite	$Ca(UO_2)_2(PO_4)_2.10H_2O$	UK	Purvis *et al.* (2004)
Trapelia involuta	Beudantite	$PbFe_3(AsO_4, PO_4)(SO_4)(OH)_6$	UK	Purvis *et al.* (2004)
Candelariella vitellina, *Trapelia coarctata*	Uraninite	UO_2	UK	*

Table 15.1. (cont.)

Lichens	Substrate	Formula	Country	Reference
Acarospora, Caloplaca citrina, Candelariella vitellina, Candelariella aurella, Lecanora dispersa, Physcia caesia, Rinodina subexigua, Scoliciosporum umbrinum, Stereocaulon pileatum, Xanthoria parietina	On lead structures	Pb (probably a weathered oxide surface)	UK, France	Brightman & Seaward (1977); Hickmott (1980)
Gylaideopsis sp.	Anglesite	$PbSO_4$		*
Gylaideopsis sp.	Galena	PbS		*
Trapelia sp.	Pyromorphite	$Pb_5(PO_4)_3Cl$		*
Trapelia involuta	Coronadite	$Pb(Mn^{4+}, Mn^{2+})_8O_{16}$	UK	Kasama *et al.* (2001); Purvis & Halls (1996)
e.g. *Xanthoria, Physcia, Caloplaca, Candelariella, Rhizocarpon, Aspicilia* spp.; *Bacidia saxenii; Acarosporion* community	Iron	Fe (or hydrated iron oxides 'rust')	UK	Brightman & Seaward (1977); Gilbert (1990); Purvis & Halls (1996)
Acarospora sinopica	Aresenopyrite	FeAsS	UK	*
Acarorpora sinopica	Pyrite	FeS_2	UK, Scandinavia	*

Taxon	Substrate	Composition	Location	Reference
Acarosporion sinopicae community				
Trapelia involuta	Goethite	$HFeO_2$	Europe	Purvis & Halls (1996)
	Haematite	Fe_2O_3	UK, Scandinavia	Kasama *et al.* (2001); Purvis & Halls (1996)
Lecidea sp.	Cinnabar	HgS	Romania	*
Vezdaea leprosa	Zinc	Zn (no reports directly on Zn; some species occur in drip zone beneath galvanized structures	Germany, UK	Purvis (1996)
Trapelia sp.	Sphalerite	$(Zn, Fe)S$	UK	*
Acarospora smaragdula	Timber	Wood preserved with Chromated Copper Arsenate (CCA), a chemical mixture consisting of three pesticidal compounds (arsenic, chromium and copper)	UK	Purvis *et al.* (1985)
Lecanora vinetorum, *Acarospora anomala*	Vine supports	Bordeaux fungicide applied containing $CuSO_4.3Cu(OH)_2. 3CaSO_4$	Austria	Poelt & Huneck (1968)

*collected by O. W. Purvis (Natural History Museum)

Metal uptake and retention by lichens have broad implications for subjects ranging from environmental monitoring to global geochemical cycles. For example, metal complexation from substrate minerals by rock-growing (saxicolous) lichens can profoundly affect the rate of mineral weathering, which impacts the rate of lithogenic nutrient release, soil formation and carbon drawdown from the atmosphere. Adsorption and biomineralization of metals by lichens provide a robust method of tracking environmental pollution. Understanding these phenomena requires that the influence of lichens on geochemical processes be clarified and quantified. In this chapter we briefly review key aspects of lichen biogeochemistry, emphasizing the role of lichens in accumulating metals, the impact of bioaccumulation on macroscopic processes, and the chemical mechanisms governing these phenomena.

Metal accumulation by lichens

Much literature focuses on the tendency of lichens to bioaccumulate relatively high micronutrient and non-nutrient element concentrations. The reviews of Nieboer *et al.* (1978), Tyler (1989), Brown (1991), Easton (1994), Richardson (1995), Wilson (1995), Purvis (1996), Branquinho (2001), Garty (2001) and Bargagli and Mikhailova (2002) provide a comprehensive summary, which will only be summarized here. Additional references are included in Jacquiot and Daillant (1997). The capacity of lichens to bioaccumulate metals is due both to their structure and chemical coordination reactions involving lichen biomass components and dissolved cations (Fig. 15.1). Foliose and fruticose lichens with complex branching or corrugated morphologies provide a relatively high surface to volume ratio. Many lichens have large intercellular medullary spaces where particulates may accumulate. Unlike vascular plants, lichens lack a protective outer cuticle and absorb both nutrients and many pollutants over much of their surface. The role of the substrate as a source of nutrients has yet to be established (Fig. 15.2).

Lichens can accumulate metals from dry or wet deposition (via rainfall, snowfall, aerosols, through-fall or surface flow), from dissolution of trapped particulates, or by substrate weathering (Fig. 15.1). Physical trapping of particulates is thought to be an important accumulation mechanism. Very high element concentrations recorded (>1000 ppm) usually result from particle entrapment. Particles ranging in size from macroscopic (>1 mm) dust grains to microscopic (<1 μm) aerosol particles derived from natural or anthropogenic sources, such as quarries, volcanoes, mines, smelters, vehicle exhaust, industrial releases and

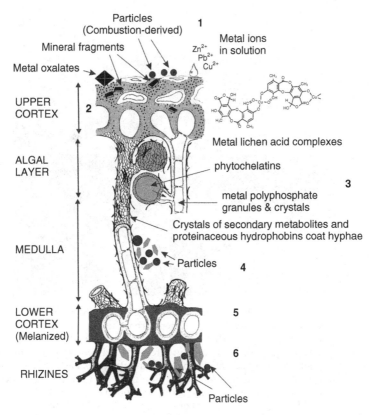

Fig. 15.1. Diagrammatic representation of possible metal uptake mechanisms and metal localization in a foliose macrolichen adapted from Honegger (1997) and Purvis (2000) (not to scale). (1) Accumulation of particles (mineral and combustion-derived) in dry and wet deposition, metal ion sorption via soluble phases, particulate entrapment in intercellular spaces. (2) Particulates are not necessarily inert and may be solubilized by acid precipitation and/or lichen-derived organic acids leading to metal sorption to e.g. extracellular hydrophilic β-glucans secreted by mycobiont cortical cells with negatively charged anionic sites (e.g. Sarret *et al.*, 1998); extracellular metal oxalate and metal lichen acid-complex formation (Cu-norstictic acid shown) (Purvis *et al.*, 1987, 1990; Takani *et al.*, 2002). (3) Intracellular phytochelatin, thiol peptides containing metal-chelating sulphydryl groups of cysteine help protect photobionts from metal toxicity (Pawlik-Skowronska *et al.*, 2002). (4) Particles (mineral, combustion-derived and biogenic, e.g. oxalates) trapped in intercellular spaces of fungal hyphae and/or attached to medullary hyphae coated with hydrophobic mycobiont-derived secondary metabolites that may further act as sites for metal complexation (Purvis *et al.*, 1987, 1990). Also coated by water-repellant proteins, 'hydrophobins', which protect thalli from intracellular penetration of heavy metal-containing solutions (Pawlik-Skowronska *et al.*, 2002). (5) Lower cortex containing extracellular hydrophilic β-glucans

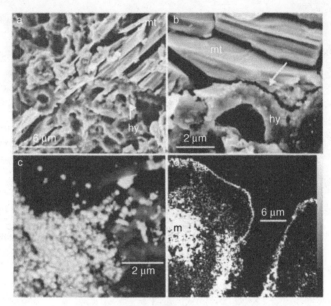

Fig. 15.2. (a, b), Scanning electron microscope (SEM) images of *Trapelia involuta* sampled growing on metatorbernite, $Cu(UO_2)_2(PO_4)_2.8H_2O$ (mt) from South Terras Mine (see McLean *et al.*, 1998; Purvis *et al.*, 2004). (a) Hyphae (hy) were observed in direct contact penetrating along cleavage plane and may derive phosphate as a nutrient as has been demonstrated for many non-lichenized fungi (Fomina *et al.*, Chapter 10, this volume). (b) Close-up of hyphae (hy) showing mineral dissolution (arrowed); (c, d) SEM images of *Acarospora smaragdula* subject to extreme metal particulate fall-out and sulphur dioxide pollution near Zlatna smelter, Romania (see Purvis *et al.*, 2000). (c) Microbotryoidal granules of Pb/S/Sn phases visible down to 20 nm which may be primary in origin or secondarily mobilized within the lichen. (d) X-ray map for Pb mainly localized on the surface and medulla (m).

Fig. 15.1. (cont.)
secreted by cortical cells which may bind metals as in (2). Melanins are often present, which may sorb metals (U, Cu and Fe) in melanized tissues in fruiting bodies (not shown) (McLean *et al.*, 1998, Purvis *et al.*, 2004). (6) Particulates (mineral or combustion-derived) may become trapped by rhizines through surface flow; sorbed to cell walls (Goyal & Seaward, 1982) and in extracellular melanized regions (as in the outer wall of fruiting bodies, McLean *et al.*, 1998, Purvis *et al.*, 2004). Rhizines occupy by far the bulk of the thallus in section and may extend to several millimetres, hyphae to several centimetres in the substrate. Particles and metals may also be removed from thalli by a variety of processes.

transboundary pollution can be efficiently captured. Particles may be trapped by fall-out onto the upper surfaces of the thallus, or by adhesion onto lower cortical surfaces, including rhizines attached to the substrate. How particles reach intercellular thalline spaces is unclear although ultra-structural surface features are likely to be important (Bargagli & Mikhailova, 2002). In several studies a direct relationship between thallus metal content and concentrations of dry particles on filters has been found (e.g. Saeki *et al.*, 1977). Lacking roots they do not sample either deep soil horizons or groundwater. Long-lived, slow-growing and perennial in nature, lichens continuously accumulate material over long periods and are therefore an excellent medium for pollution biomonitoring (Nimis & Purvis, 2002). Numerous recent studies demonstrate their utility (e.g. Sensen & Richardson, 2002; Cicek & Koparal, 2003; Loppi & Corsini, 2003; Loppi & Pirintsos, 2003; Nash *et al.*, 2003; Loppi *et al.*, 2004).

Particles trapped by lichen thalli are not necessarily inert, and may be dissolved by rain, by acid delivery in rainfall, or by lichen organic acid secretion. In many cases it is difficult to determine if metal uptake occurred originally by sorption, or by *in situ* dissolution of trapped particles, followed by sorption of the dissolved constituents. Numerous studies have examined the accumulation of lead in lichens along roadsides or in urban regions (Garty, 2001) both as total biomass lead and as easily exchangeable (adsorbed) lead cations. Lead is normally emitted from automotive exhaust in the form of metal bromide salt aerosols, which are soluble in water. Trapping of these particles by lichens will lead to their eventual dissolution and adsorption by lichen biomass. In contrast, metal sulphide particles (e.g. galena PbS, sphalerite ZnS) or native elements (Ni, Cu) derived from mining and smelting operations, or metal oxide dusts (e.g. goethite FeOOH) are less soluble, and will be retained intact within the thallus for longer periods. Uptake and retention of cations directly by adsorption from rainfall would, in the long term, be indistinguishable from dissolution of retained particles.

Lichens and mineral weathering

Rocks colonized by lichens weather faster than uncolonized rock surfaces, as a direct result of metal chelation by lichen substances and organic acids (Fig. 15.3). Fomina *et al.* and Smits (Chapters 10 and 13, this volume) provide comprehensive discussions of the effects of fungi on mineral weathering processes, and therefore only a brief summary of the effects of lichens on this phenomenon will be presented here. Essentially, lichen colonization can influence the rate of silicate weathering through both physical and chemical processes, including mechanical disruption by hyphae, expansion and contraction of infiltrating biomass during wet and dry periods, secretion of organic acids – particularly oxalic acid, promoting mineral dissolution and metal chelation, and increased local concentrations of CO_2 that promote carbonic acid dissolution. The significance of this topic is evidenced by the numerous review articles focused on lichen-induced weathering (e.g. Jones, 1988; Easton, 1994; Barker & Banfield, 1996; Adamo & Violante, 2000; Chen *et al.*, 2000).

The effects of lichens on rock weathering are difficult to quantify because most lichens grow slowly, and because many lichens are difficult to transplant. Field studies have demonstrated that lichen colonization

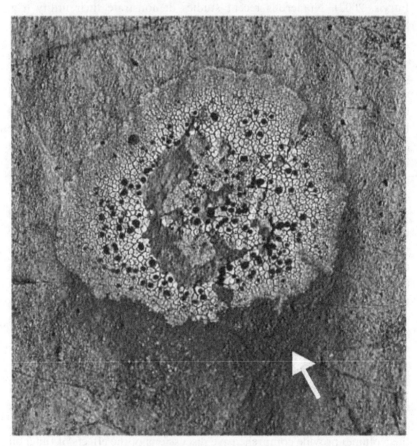

Fig. 15.3. Crustose lichen *Lecidea lactea* growing on dunite, a rock composed of the silicate olivine ((Mg, $Fe)_2SiO_4$). Note dark stain (arrowed) below lichen representing amorphous silica coated in iron produced by lichen weathering (Purvis, 2000).

accelerates net silicate weathering rates over background (abiotic) rates. In a classic study of field weathering rates, Berner and Cochran (1998) quantitatively examined the extent of weathering of bare, lichen-colonized and plant-colonized basalt, using Hawaiian lava flows of known age (~3400 years). The physical results of weathering were examined using scanning electron microscopy (SEM) and energy dispersive spectroscopy (EDS). Overall weathering rates were calculated based on the extent of secondary porosity development. Rocks were found to have weathered predominantly by congruent dissolution of primary silicates, leaving no aluminosilicate or ferruginous secondary mineral residue. Vascular plant-colonized basalts showed the highest overall weathering rates, averaging to approximately $100 \, \mathrm{g\,m^{-2}}$ of land area per year. In contrast, lichen-dominated rock surfaces exhibited weathering rates two orders of magnitude lower, around $1 \, \mathrm{g\,m^{-2}}$ of land area per year. Although slow in comparison to weathering by plant roots, lichen-induced weathering rates exceeded background weathering rates on bare basalt by an order of magnitude. McCarroll and Viles (1995) measured historical weathering rates of lichen-colonized moraine sediment in Norway, and found that relative to uncolonized surfaces, endolithic lichen development resulted in an increase over background rates of at least $3.56 \, \mathrm{g\,m^{-2}}$ of land area per year. Lee and Parsons (1999) estimated a comparable penetration rate for *Rhizocarpon geographicum* into granite of at least 0.002–$0.003 \, \mathrm{mm\,yr^{-1}}$, based on SEM and EDS analyses of field samples of known exposure age. Absolute weathering rates vary according to substrate rock type, annual precipitation, temperature and individual species. However, field data are consistent with lichens increasing silicate weathering rates by at least an order of magnitude.

Fungal corrosion of silicate minerals can result in complete dissolution (Berner & Cochran, 1998) of primary phases, or alteration to secondary clays or aluminosilicates (Ascaso & Wierzchos, 1995; Barker & Banfield, 1996; Lee & Parsons, 1999; Asta *et al.*, 2001). Berner and Cochran (1998) show evidence of the former, where in one SEM image fungal hyphae penetrate through low-calcium basaltic glass to completely dissolve a higher-calcium plagioclase phenocryst. Barker and Banfield (1996) show energy filtered transmission electron microscopy (EFTEM) images of the lichen–rock interface between crustose lichen *Porpidia albocaerulescens* and an amphibole syenite. The results illustrate an upper zone of physical substrate disruption from hyphae infiltration, and a lower zone of micro- to nanocrystalline calcium-, potassium- and iron-rich clay minerals (e.g. smectite) and Fe(III) oxyhydroxides (e.g. goethite) in a matrix of

354   *J. R. Haas and O. W. Purvis*

mucopolysaccharide gel exuded by the infiltrating hyphae, and extending down to ~10 mm (Barker & Banfield, 1996). Adamo *et al.* (1997) examined the interface between leucite-bearing rock from Mt Vesuvius and *Stereocaulon vesuvianum*, using TEM, X-ray diffraction (XRD) and SEM. Their results also show accumulation of ferric oxyhydroxides beneath the thalli. Wilson *et al.* (1981) report XRD, SEM and EDS results that show leaching of cations from chrysotile serpentinite beneath *Lecanora atra*, yielding residual amorphous silica.

The production of carboxylic organic acids by lichens appears to be especially significant in promoting silicate chemical weathering, particularly the release of oxalic acid (Wilson *et al.*, 1981; Jones, 1988; Wilson, 1995), a metabolite produced by a wide range of fungi. Weathering of calcium and magnesium silicates by lichens can result in thallial precipitation of sparingly soluble oxalate salts such as whewellite (calcium oxalate monohydrate), glushinskite (magnesium-oxalate dihydrate) and oxalates of manganese, Fe(III), lead and copper (Wilson & Jones, 1984; Purvis, 1996; Sarret *et al.*, 1998). Edwards *et al.* (1993) demonstrate that calcium oxalate crystals can occur down to 20 mm beneath the surface of lichen-colonized rocks. Calcium oxalate precipitation may also serve as a detoxification response, caused by photo-oxidative stress from intense sunlight (Modenesi *et al.*, 1998). Ascaso and Wierzchos (1995) report calcium oxalate accumulation near the lichen–rock interface of endolithic lichens *Parmelia conspersa, Aspicilia intermutans* and *Lecidea auriculata*. Purvis (1984) and Chisholm *et al.* (1987) report vivid blue inclusions of copper oxalate 'moolooite' in *Acarospora rugulosa*, including one sample that contained 16% dry weight copper, equivalent to ~40% copper oxalate. Wilson *et al.* (1981) report significant accumulation of glushinskite on *Lecanora atra* growing on chrysotile serpentinite. Johnston and Vestal (1993) conducted a detailed study of oxalic acid production and mineralization in cryptoendolithic lichens from the dry Ross Valley of the Antarctic. The authors found abundant oxalic acid in the thallus just below an epilithic silica crust, and beneath the thallus in a weathering zone characterized by ferrihydrite accumulation. The authors found that visible deterioration of the sandstone substrate of cryptoendolithic colonies was caused primarily by leaching of iron, silicon, aluminium, phosphate and potassium via oxalate complexation. Beneath the lichen colony heterotrophic opportunistic bacteria were partially responsible for mineralization of oxalate to CO_2, resulting in release and precipitation of a ferric oxyhydroxide layer. Epilithic crusts enclosing lichens were attributed to oxalate-promoted leaching of iron and aluminium from sandstone cements, yielding a residue of amorphous silica and iron oxide. Photochemical oxidation of oxalate to CO_2 was also indicated, which may also be responsible for the formation of siliceous desert varnish above cryptogamic colonies in arid settings.

A group of organic acids produced uniquely by lichens, collectively referred to as 'lichen acids' have also been implicated in the promotion of metal complexation and silicate weathering (Williams & Rudolf, 1974). Most lichen acids are depsides and depsidones having carboxyl and hydroxyl groups that weakly dissociate in water, and are therefore sparingly soluble compounds in aqueous solution. Based on their low solubility these compounds would be expected to have a minimal effect on bulk silicate alteration or dissolution reactions, because acid dissociation and solubility is required for aqueous metal–organic complexation and for direct acid attack by proton production. However, Schatz (1963) demonstrates that lichen thallus fragments in contact with rock particles in water can result in solubilization of pigmented aqueous complexes within hours, caused by chelation of lichen acids with silicate cations. Schatz (1963) did not examine the composition of aqueous chelates produced, but did demonstrate that aqueous complexes absorbing in visible light were eluted most strongly from granite and mica in contact with lichens containing atranorin, baeomycesic, fumarprotocetraric, salazinic, usnic, gyrophoric and lecanoric acids. Experiments also showed that elution of pigmented complexes from granite and mica occurs rapidly in the presence of purified physodic and lobaric acids, both monocarboxylic depsidones. Similar results were obtained from granite in the presence of fumarprotocetraric acid from *Parmelia conspersa* (Syers, 1969). Iskandar and Syers (1972) and Ascaso *et al.* (1976) report experimental studies showing that metal:lichen-acid chelation can increase bulk mineral dissolution rates. Williams and Rudolf (1974) show that squamatic acid derived from *Cladonia squamosa* was effective in chelating iron from sandstone. The limited solubility of most lichen acids at circumneutral pH does not appear to preclude their mobility in soil solutions, as shown by Dawson *et al.* (1984), who report significant advective transport of lichen acids, including atranorin, usnic, rangiformic and psoromic acids, in soil profiles below *Cladonia mitis*. The authors concluded that lichen acids were significant in soil formation and mineral weathering processes in podzols.

At present it is difficult to compare the effects of different lichen taxa, or different lichen acids, on bulk silicate mineral weathering rates under field conditions. An abundant literature depicts the physical and chemical distribution of minerals, alteration products and organic components during lichen growth, mostly provided through SEM, EDS, XRD and

TEM studies. However, less information is available depicting the governing chemical relations involving minerals and lichen acids. Takani *et al.* (2002) conducted spectroscopic investigations of Cu^{2+} and Pd^{2+} binding to usnic acid, and provide data that describe both the structures and stabilities of metal–organic complexes involving this lichen acid. However, most of the experimental literature describing reactions of lichen acids with minerals does not quantitatively constrain stoichiometry, equilibrium mass action parameters, or kinetic rate constants. To understand and predict the effects of lichens on mineral weathering and soil formation processes on a large scale, or through long intervals of time, it is necessary to evaluate these parameters. Such information would greatly facilitate the quantification of lichen, and lichen acid, influences on mineral and soil geochemistry.

Effects of metal chelation and silicate weathering reactions

Organic acids appear to play a governing role in the stimulation of weathering rates by lichens. Metal–acid chelation results in elution of nutrient elements and micronutrients, which become available for biological uptake. Fungal penetration of hyphae into the substrate not only delivers organic acids to mineral surfaces, but also provides a conducive microenvironment for sustained weathering reactions. Hyphae and extracellular mucopolysaccharides coating fungal and algal cells help to maintain localized wetness during periods of sustained ambient desiccation, which increases the contact time between aqueous solution and minerals, thus increasing the overall bulk reaction rate. Alternating wet–dry periods, along with physical infiltration of hyphae, promote physical disaggregation of the substrate, which increases mineral reactive surface area, further increasing reaction rates. Enclosure of grains, and filling of void spaces, by lichen biomass and exudates is also likely to promote the development of local microenvironments where the activity of organic acids far exceeds their apparent concentration in the bulk soil solution.

Ligand-promoted dissolution (LPD) is the governing mechanism responsible for mineral weathering in the presence of organic acids. Unlike strong acids, which fully dissociate in solution, most organic acids do not yield stoichiometric concentrations of H^+, and therefore do not promote direct acid dissolution of minerals by H^+ exchange. Ligand-promoted dissolution occurs as a result of equilibrium competitive complexation reactions among cations in a crystal lattice, ligands bonded with cations inside a crystal lattice and aqueous ligands capable of chelating the cations. A mineral is a structured lattice wherein each cation is linked through (usually) covalent-ionic chemical bonds with a number of surrounding coordinated ligands. In silicates the ligand is an oxide. A metal cation can be coordinated by a maximum number of ligands (the *coordination number*) depending on cation charge and ionic radius. In a ligand exchange reaction one or more ligands dissociate from the inner coordination sphere of a cation, and are replaced by one or more ligands that collectively occupy an equal number of coordination positions around the cation. Ligand-promoted dissolution occurs when exchange involves replacement of oxygens (in silicate minerals) in the cationic coordination sphere with dissolved organic ligands. Multiprotic organic acids are generally more effective in LPD than are monoprotic organic acids or inorganic acids, because multifunctional organic molecules can occupy more coordination positions around a cation. Thus, oxalic acid, a diprotic carboxylic acid, is more effective in promoting metal oxide dissolution than are monoprotic organic acids such as acetic or formic (Stone, 1997), and polyprotic organic acids such as EDTA (ethylenediaminetetraacetic acid) and most siderophore compounds are generally more effective than diprotic acids.

Experimental studies of mineral weathering rates in the presence of oxalic acid demonstrate the importance of LPD. For example, in the presence of ~1 mM oxalic acid, rates of silica elution from feldspar can increase up to 15-fold at circumneutral pH, while Al elution rates can increase by two orders of magnitude (Barker *et al.*, 1997). Similar results are reported for quartz and olivine (Grandstaff, 1986; Bennett *et al.*, 1988), and indicate that oxalate leaching of aluminium, calcium, magnesium and other cations from primary silicate minerals can yield a silica-rich residue similar to that found in association with endolithic lichens (Johnston & Vestal, 1993; Lee & Parsons, 1999).

The rate of LPD of a metal oxide phase is directly dependent on the stability of the resulting metal–organic aqueous complex. Ludwig *et al.* (1995) and Casey and Ludwig (1995) demonstrate that experimentally determined dissolution rates of bunsenite (NiO) in the presence of dissolved organic acids strongly correlate with the thermodynamic formation stability constants of nickel–organic acid 1:1 aqueous complexes. Faster LPD reactions are facilitated where there is a strong thermodynamic driving force promoting cation-aqueous ligand coordination. Mineral dissolution rates also correlate with the number of coordinated ligand atoms occupying positions in the inner coordination sphere of the cation. Both these results demonstrate that multifunctional ligands and strongly chelating ligands result in faster mineral dissolution. These correlations are complicated by the fact that different organic acids dissociate to different extents as a function of pH, and ligand charge partially governs the extent to which an organic ligand adsorbs onto a mineral surface. Furthermore, different deprotonated acid species complex with surface and dissolved cations according to different mass law and mass action parameters. Understanding the acid–base chemistry and surface complexation relations occurring in a mineral + organic acid system is crucial in predicting the effects an organic acid will have on mineral stability. Oxalic acid and lichen acids appear to be effective agents in promoting specific mineral dissolution beneath a lichen thallus, but the relative importance of different lichen acids, and thus the relative importance of different lichen species in mineral weathering and soil formation, is difficult to assess due to a lack of quantitative chemical data describing metal–ligand interactions. Future work defining the chemical properties of lichen acids, and their complexation equilibria with mineral surfaces and with dissolved cations, is a necessary initial step in building a quantitative model of lichen biogeochemistry and pedogenesis.

Experimental studies of metal uptake by lichens

Understanding quantitatively the chemical mechanisms by which lichen and fungal biomass adsorb and retain metal cations requires obtaining specific chemical and thermodynamic parameters that may be used to develop a generalized and mechanistically sound framework for predicting the cation-coordinative properties of lichens under a wide range of environmental conditions. There is a significant body of experimental literature (e.g. see reviews by Nieboer *et al.*, 1978; Tyler, 1989; Brown, 1991; Richardson, 1995) describing metal sorption by lichens and fungi, and the localization of metal uptake within lichen thalli. Lichen biomass strongly accumulates metal cations from aqueous solutions, primarily through non-metabolic electrostatic and chemical coordination reactions at the cell–aqueous interface. Metals retained as surface complexes on lichen cell walls and extracellular polysaccharides may be initially precipitated as insoluble metal–organic biominerals, or may be converted to these forms by subsequent production of organic compounds by the lichen. It is possible that some organic acid production is directed specifically toward removal and sequestration of potentially toxic metals (Purvis, 1996).

Early studies of metal uptake by lichens presented metal sorption as an ion exchange process, and reported metal accumulation in terms of relative uptake capacity, or the capacity of one metal to displace another as an adsorbed complex. Competitive uptake studies by Puckett *et al.* (1973)

determined that anionic sites on the lichen surface prefer to coordinate metal ions in the relative order $Fe(III) \gg Pb > Cu \gg Ni, Zn > Co$. A similar competitive assessment of lichen metal affinity was presented by Richardson and Nieboer (1981), in which the authors demonstrated that during metal uptake by *Umbilicaria* and *Cladonia* thalli, divalent metals were displaced by monovalent metals in a consistent ratio of 1:2, indicating that sorption occurred by specific coordination with anionic functionalities at the cell–aqueous interface, and that two monovalent cations were required to substitute for one divalent cation in lichen surface complexation reactions. Puckett *et al.* (1973) and Richardson and Nieboer (1981) demonstrated that metal sorption by lichens is rapid, reaching effective equilibrium within minutes to hours. Using SEM and EDS methods, many authors have examined the localization of sorbed metals within thalli (e.g. Garty & Galun, 1979; Goyal & Seaward, 1981; 1982; Jones *et al.*, 1982; Richardson & Nieboer, 1983; Garty & Theiss, 1990; Garty & Delarea, 1991; Haas *et al.*, 1998). These studies show that metal localization varies among lichen taxa, but that aqueous cations tend to sorb most strongly to cortical structures such as rhizines, tomentum and apothecia. In contrast, less metal accumulation tends to occur within medullary tissue and the algal zone. Metal biosorption by melanins may be responsible for high uranium, iron and copper concentrations observed in the outer parts of melanized apothecia (Purvis *et al.*, 2004). Electron probe microanalysis of the lichen *Peltigera membranacea* following uranium uptake experiments show that uranium correlates strongly with phosphorus in the thallus (Haas *et al.*, 1998). Transmission electron microscopy analyses further revealed acicular nanocrystals of uranyl phosphate along cell walls, within the extracellular mucopolysaccharide gel and in intracellular concentric bodies (proteinaceous organelles) (Suzuki & Banfield, 1999). Paul *et al.* (2003) report the formation of intracellular manganese polyphosphate granules in the lower cortex of *Hypogymnia physodes*. Experimental observations are generally consistent with an expectation that environmentally exposed biomass, including upper and lower cortical structures, would tend to exhibit strong metal-binding affinities compared with interior medullary tissue.

Mechanisms of metal sorption by lichens

Metal sorption appears to concentrate primarily in the mycobiont. Thus, understanding the metal uptake process in free-living fungi provides crucial data that substantially inform lichen studies. Gadd (1993) provides a thorough review of studies investigating metal uptake by fungi, and only some aspects of this voluminous literature will be discussed here. The cell walls and extracellular exudates of fungi exhibit a variety of Lewis-acid functional groups, including carboxyl, phosphoryl, amine and hydroxyl moieties (Galun *et al.*, 1983; Tsezos, 1985; Muraleedharan & Venkobachar, 1994; Guibal *et al.*, 1998), most of which have a strong binding affinity for cationic aqueous species. Chief structural components of fungal cell walls are chitin and chitosan, comprised of polymerized N-acetyl-D-glucosamine and D-glucosamine, respectively, along with a variety of extracellular polymers including α- and β-glucans, cellulose, and a range of mannoproteins and glycoproteins. Extracellular polymeric and polysaccharide substances excreted by lichenized fungi include α- and β-glucans, galactomannans, D-galactosamine polymers, melanins and lipids (Peberdy, 1990), along with organic acids such as citric, oxalic and the family of endemic lichen acids. A few complex heteroglycans have been described in lichen fungal biomass, but less than 100 lichen species have been investigated for polysaccharide content (Olafsdottir & Ingolfsdottir, 2001).

Anionic functionalities on fungal cell wall components appear to adsorb cations in a manner similar to that observed for bacterial cell membranes, algal cells and mineral surfaces (Davis & Kent, 1990; Fein *et al.*, 1997). Acidic functional groups may also act as nucleation sites for authigenic mineral formation (Urrutia & Beveridge, 1994; Barker & Banfield, 1996; Fortin *et al.*, 1997). At environmental solution pH conditions most carboxyl, phosphoryl and some hydroxyl functionalities are negatively charged, and will tend to attract dissolved cations by a combination of electrostatic and chemical (i.e. covalent) forces. A surface containing charged functional groups, when exposed to an aqueous solution, will attract countervailing charges in the bulk solution by electrostatic attraction. Counterions that come in contact with the charged surface may bond with charged functional groups. By convention, electrostatically bonded surface complexes are referred to as *outer-sphere surface complexes*, in which the cation's hydration sphere separates it from the surface ligand. This depiction is consistent with spectroscopic investigations that directly interrogate surface complex structure (Brown *et al.*, 1995). By contrast, in *inner-sphere surface complexes* adsorbing cations bond directly with surface ligands. In inner-sphere complexes the bond is partially covalent in character, and tends therefore to be more stable over a wider range of pH and adsorbate concentration. So-called 'hard' Lewis bases tend to form bonds that are more ionic in character, favouring outer-sphere surface complexes. 'Hard' cations include the alkali (Li^+, Na^+, K^+) and alkaline earth metals (Mg^{2+}, Ca^{2+}, Sr^{2+}, Ba^{2+}), group III metals (Al^{3+}, Ga^{3+}), some transition elements (e.g. Sc^{3+}, Ti^{4+}, Zr^{4+}, Th^{4+}) and trivalent lanthanides (Stumm & Morgan, 1996). In contrast, 'soft' Lewis bases, including mainly transition elements such as Zn^{2+}, Cd^{2+}, Hg^{2+}, Pb^{2+}, Sn^{2+} and Tl^{3+}, and actinides such as UO_2^{2+} (Stumm & Morgan, 1996), tend to form inner-sphere surface complexes. 'Transitional' cations, such as Fe^{3+}, Fe^{2+}, Cr^{2+}, Cu^{2+}, Ni^{2+} and Co^{2+}, having electronegativities intermediate between hard and soft characterizations, have an equivalent tendency to form inner- and outer-sphere complexes. These divisions are not absolute, however, and in practice the formation of inner-versus outer-sphere complexes depends strongly on pH and cation aqueous speciation (e.g. hydrolysis, complexation with carbonate, chloride, etc.).

Inner-sphere complexes are relatively stable in comparison to outer-sphere complexes under equivalent solution conditions (i.e. pH, ionic strength), and in a competitive situation will tend to displace less stable adsorbates. This is a fundamental property of coordination reactions, and explains the observed trends in metal uptake 'preference' observed in lichen studies (Puckett *et al.*, 1973). Metal sorption results previously attributed to ion exchange reactions are more precisely described as resulting from competitive surface complexation reactions involving multiple cation types. Strictly speaking, each metal adsorption reaction can be described using a discrete mass law relation, such as

$$>L^{m-} + M^{n+} = >L - M^{n-m} \qquad (15.1)$$

where $>L$ represents a ligand functional group attached to the surface, M stands for an adsorbing metal cation, $>L - M$ stands for the resulting surface complex and n and m represent the cation and anion charges, respectively. The thermodynamic driving force for a reaction such as Eq. (15.1) is described through the conventional mass action expression

$$K_{ads} = [>L - M^{n-m}]/[>L^{m-}][M^{n+}] \qquad (15.2)$$

where square brackets indicate the concentration of each chemical species and K_{ads} denotes the apparent equilibrium constant for the adsorption reaction. In a situation where two cations compete for adsorption onto a surface composed of one ligand site type, the concentrations of the two metals at the surface will vary as a ratio of their adsorption equilibrium constants. For example, if one assumes the following mass law and mass action relations,

$$>L^{m-} + B^{n+} = >L - B^{n-m} \qquad log\ K_B = 3.0, \text{ and} \qquad (15.3)$$

$$>L^{m-} + C^{n+} = >L - C^{n-m} \quad log\ K_C = 7.0 \quad (15.4)$$

where B^{n+} and C^{n+} are different cations of equal charge, a combined competitive adsorption reaction involving both metals would take the form

$$>L - B^{n-m} + C^{n+} = >L - C^{n-m} + B^{n+}$$

$$log\ K_{exchange} = logK_C - logK_B \quad (15.5)$$

where at equilibrium the concentration of $>L - C^{n-m}$ would be 10^4 times greater ($log\ K_{exchange} = 10^4$) than that of $>L - B^{n-m}$. Using an equilibrium thermodynamic treatment such as that shown above obviates the need to tabulate individual cation replacement partition coefficients referenced to an arbitrary metal (e.g. Richardson & Nieboer, 1981), by instead formulating reversible metal uptake in conventional chemical terms.

Because surface complex formation is dependent on the extent of surface ligand dissociation, pH is a governing variable in this phenomenon. Numerous studies have demonstrated that metal uptake by lichens, algae and fungi is pH dependent (e.g. Tuominen, 1967; Strandberg et al., 1981; Tsezos & Volesky, 1982; Galun et al., 1987; Andres et al., 1993; Brady et al., 1994; Bengtsson et al., 1995; Sandau et al., 1996; Texier et al., 1997; Antonelli et al., 1998; Guibal et al., 1998; Haas et al., 1998). Similar results are obtained for metal adsorption onto bacterial surfaces (Fein et al., 1997, 1999; Daughney et al., 1998; Schlekat et al., 1998; Cox et al., 1999; Fowle & Fein, 1999, 2000; Small et al., 1999; Fein, 2000; Yee et al., 2000; Haas et al., 2001; Yee & Fein, 2002) and mineral surfaces (e.g. see reviews of Davis & Kent, 1990; Dzombak & Morel, 1990). At higher pH values, more Lewis-acid organic functional groups on cell surfaces are deprotonated, conferring a greater net negative charge to the whole surface.

Using surface titrations to understand lichen–solute interactions

Keeping track of the pH dependence of surface charge is facilitated through surface titration studies. In a surface titration the mineral or biomass is titrated as a function of pH using a strong acid or base (i.e. HCl or NaOH, respectively), and responsive changes to solution pH, resulting from H^+ exchange with the surface, are recorded. The resulting data provide a measurement of net solution H^+ versus pH. If the surface possesses no proton-exchange capacity net H^+ will track with pH, but for most hydrophilic surfaces proton uptake and release by the surface will buffer solution pH. Tuominen (1967) provides experimental titration data for the lichens *Cladonia alpestris* and *Umbilicaria deusta*, showing that both lichens distinctly buffer proton concentrations through the pH range 2–10. The titrations are reversible, indicating that proton exchange does not irreversibly damage or alter the lichen biomass, or elute chelating organic acids from the biomass surface. Fig. 15.4 shows *C. alpestris* titration data from Tuominen (1967), including both forward and reverse titration results. In comparison with a calculated titration of pure water (upper curve), *C. alpestris* evidences significant pH buffering capacity.

Titration data such as that reported by Tuominen (1967) may be modelled using a discrete multisite surface complexation approach, to estimate model-dependent values for the concentrations and acid dissociation constants of the predominant ligand types that undergo reversible H^+ exchange. Such calculations are facilitated by least-squares regression algorithms such as FITEQL (Westall, 1982), or geochemical speciation computer codes such as JCHESS (van der Lee & de Windt, 1999). Tuominen (1967) provides calculated pK_a (–log of the acid dissociation constant) values of 3.3, 4.4 and 8.3 for *C. alpestris*, which the author attributed to surface carboxyl (3.3), phosphoryl (4.4) and amine or hydroxyl (8.3)

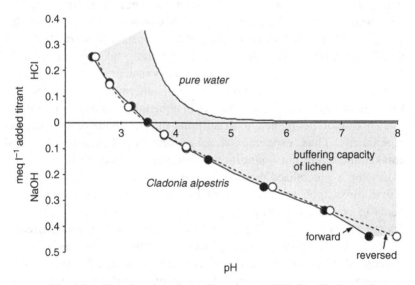

Fig. 15.4. Titration data from Tuominen (1967) for *Cladonia alpestris*, depicted as a function of pH versus concentration of added titrant. The closed circles represent forward titration data, while open circles stand for reversed titration data points. The upper curve is a calculated titration curve in pure water. The shaded area denotes the extent of pH buffering capacity exhibited by the lichen, relative to a non-buffering solution of pure water.

groups. These results are consistent with titration data for bacteria (e.g. Fein et al., 1997; Cox et al., 1999; Haas et al., 2001), which also show the predominance of cell-surface carboxyl, phosphoryl and hydroxyl functionalities, having similar pK_a values. These model-dependent titration results are verified by spectroscopic studies showing a dominant role for carboxyl and phosphoryl groups in reversible metal uptake by bacteria (e.g. Kelly et al., 2001).

Surface complexation models

The experiments of Tuominen (1967) pioneered the application of quantitative equilibrium chemical principles to the study of metal adsorption by lichens; however, his approach has not been widely adopted in mycology. Most experimental studies have sought to depict reversible metal sorption by lichens and fungi in terms of simplified operational models that do not accurately account for sorption under different conditions of pH, ionic strength and sorbent concentration. For example, Tsezos (1985) reviews data from previous experimental studies, depicting metal uptake by fungi and bacteria in terms of empirical adsorption isotherms. Bengtsson et al. (1995) use a Langmuir isotherm model to describe uranium uptake by the fungus Talaromyces emersonii. A similar approach is taken by Antonelli et al. (1998) for metal uptake by the lichen Evernia prunastri. Most lichen–metal studies have taken this approach. This is unfortunate, because empirical models such as distribution coefficients, Langmuir isotherms and Freundlich isotherms do not consider pH and ionic strength-dependent changes in aqueous and surface speciation, and so cannot be applied to conditions outside those of the original experiments. Thus, experimental data from most lichen–fungal metal uptake studies are not quantitatively comparable, making it impossible to develop a generalized model of metal adsorption by lichens or fungi. Worse, direct application of Langmuir isotherm data and partition coefficients to conditions outside those of the original experiment can yield dramatically erroneous results. Veith and Sposito (1977) and Koretsky (2000) review the shortcomings of using adsorption isotherms in lieu of surface complexation models (SCMs) in understanding adsorption data.

A more mechanistic and robust depiction of reversible metal adsorption is provided by SCMs that account explicitly for competitive speciation reactions using an equilibrium thermodynamic framework. Examples of SCMs in current use include the constant capacitance model (CCM), the diffuse double-layer model (DDLM), and the triple-layer model (TLM) (Stumm & Morgan, 1996; Koretsky, 2000). Each of these models envisages the surface–aqueous interface as comprising an array of variably charged surface sites, which interact electrostatically and chemically with solute ions in the surrounding bulk solution. The net electrostatic field exhibited by charged surface functional groups tends to attract countervailing ions from the bulk solution, which form a layer of compensating charge near the surface. This two-layer arrangement composed of fixed surface charges and countervailing solute ions is called the electrical double layer (EDL). Ion interactions with the bulk solution are partially shielded by the EDL, and this affects the equilibrium state of ion adsorption reactions. To account for non-ideal electrostatic effects, a correction to the equilibrium expression is made, yielding an intrinsic equilibrium constant (K_{int}) for a surface reaction, depicted using the relation

$$\log K_{int} = \log K[\exp(-F\psi/2.303RT)] \qquad (15.6)$$

where F stands for the molar electron charge (96 485 Coulombs mol^{-1}), R represents the gas constant, T depicts temperature in Kelvins and ψ stands for the electrical potential at the mineral surface (Stumm & Morgan, 1996). As a practical matter the value of the term ψ is determined based on a choice of assumptions regarding the field distribution of electrical potential as a function of distance from the surface. Varying depictions of the electrical potential near the surface constitute the functional differences among the major SCMs.

By applying SCMs to understanding lichen–metal adsorption data, it would be possible to establish thermodynamic values that are applicable to a wide range of conditions beyond those of the original experiments. Equilibrium constants derived using an SCM are intrinsic, surface site-specific reaction parameters that take account of pH and compositional differences, and may be employed to quantitatively compare metal uptake potentials among different metals and among different taxa. Ionic strength effects can also be accounted for using an SCM treatment. Ionic strength effects can be ignored in very dilute solutions, such as rainwater; however, in situations where periodic desiccation occurs, transient high ionic strength conditions can manifest at the lichen–aqueous interface, and an SCM approach could be used to quantify solute speciation effects under such conditions.

An SCM approach has been used in numerous studies examining metal–bacteria interactions, and these studies should guide future investigations of lichens and fungi. For example, Fein *et al.* (1997) modelled acid–base titration and metal adsorption data for the Gram-positive bacterium *Bacillus subtilis* using a CCM depiction. The authors

determined that *B. subtilis* surfaces could be approximated as an array
of three major types of surface sites; carboxyl, phosphoryl and hydroxyl.
The pK_a values and concentrations for these sites averaged to 4.82
(1.2×10^{-4} mol g^{-1} bacteria), 6.9 (4.4×10^{-5} mols g^{-1} bacteria), and 9.4
(6.2×10^{-5} mol g^{-1} bacteria), respectively. The authors used these estimated
parameters to quantitatively assess pH- and concentration-dependent metal
adsorption data in terms of discrete coordination equilibria, using a least-
squares regression procedure and the program FITEQL (Westall, 1982) to
fit the experimental data using the CCM. The results showed that reversible Cd^{2+}, Pb^{2+} and Cu^{2+} uptake in the pH range 2–11 could be modelled
by monodentate adsorption reactions onto surface carboxyl and phosphoryl groups. A similar approach was used by Daughney *et al.* (1998),
Fowle and Fein (1999), Yee *et al.* (2000) and Haas *et al.* (2001), among
many other studies (see review by Fein, 2000).

As an example of how an SCM approach can be used to help understand
metal adsorption data, Fig. 15.5 shows calculated curves for the extent of
Pb^{2+}, Cu^{2+} and Cd^{2+} adsorption onto the bacterium *B. subtilis* as a

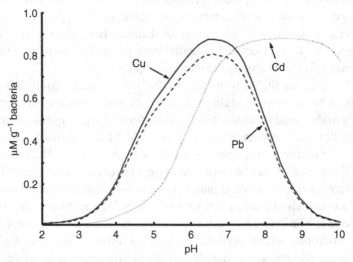

Fig. 15.5. Calculated metal sorption curves for Pb^{2+}, Cu^{2+} and Cd^{2+} onto
the bacterium *Bacillus subtilis*, shown as a function of pH versus
the concentration of sorbed metal. Curves are calculated based on
experimental metal sorption data of Fein *et al.* (1997), and were
computed using the geochemical speciation programme JCHESS. The
solution depicted contains 1 g l^{-1} bacteria dry wt (155 m^2 g^{-1} surface area,
8.0 C m^{-2} electrical double layer capacitance), 1 mM dissolved CaCO$_3$
and 1 μM dissolved lead, copper and cadmium. Adsorption was
calculated using a CCM treatment.

function of pH, using data from Fein *et al.* (1997), and calculated using a CCM depiction and the geochemical speciation programme JCHESS, along with thermodynamic parameters for all pertinent aqueous complexation reactions. Ion activity coefficients were calculated using the Davies equation (Stumm & Morgan, 1996). The solution depicted in Fig. 15.5 contains 1 gl^{-1} of bacteria, 1 mM dissolved $CaCO_3$ as a background electrolyte (representing surface and ground waters), and 1 μM of lead, copper and cadmium. Total ionic strength is 0.06 M. The curves in Fig. 15.5 illustrate the concentrations of each metal sorbed onto the bacteria. For clarity, aqueous complexes are not shown (e.g. $PbCO_3^0$, $CdOH^+$, etc.). Figure 15.5 demonstrates that the extent of metal adsorption varies dramatically with pH, as does the concentration ratio of each sorbed metal to the others. In particular, the concentration of adsorbed cadmium varies substantially with pH relative to lead and copper. This is due to strong aqueous complexation of lead and copper by dissolved carbonate at circumneutral pH, resulting in equilibrium desorption of these metals, and allowing for more extensive cadmium adsorption. Distribution coefficients or molar ion exchange ratios obtained at a single pH would be unable to predict the behaviour shown in Fig. 15.5, and in fact might substantially mislead investigators seeking to interpret observed patterns of metal retention by biomaterials in nature. For example, ion exchange data (for *B. subtilis*) measured at pH 4–5 would predict preferential uptake of lead and copper over cadmium. If high sorbed concentrations of cadmium were found on *B. subtilis* in nature, but low concentrations of lead or copper, one might interpret these data as indicating little to no lead or copper fall-out was occurring. In fact, at natural surface and ground water pH values of 7–8, substantial lead or copper fall-out might not be recorded due to the effects of pH-dependent aqueous complexation. A similar result would be expected for lichen or fungal biomass, which exhibit functional groups and metal uptake behaviour that are similar to bacteria (e.g. Tuominen, 1967).

A further advantage of an SCM approach is that thermodynamic parameters for adsorption reactions can be compared with the thermodynamic properties of other types of reactions, to assess correlative relations that may be used to predict the behaviour of chemical species for which experimental data are currently lacking. In Fein *et al.* (1997) the authors compared model results for equilibrium constants of metal–bacteria adsorption reactions, and found a strong correlative relationship between these values and corresponding equilibrium constants for metal-oxalate and metal-tiron (4,5-dihydroxy-m-benzenedisulphonic acid) aqueous complexation reactions. On this basis, the authors proposed predictive relations that could be used to estimate the adsorption equilibrium constants of many other metals onto *B. subtilis*. The use of linear free-energy correlations to predict thermodynamic properties for reactions where experimental data are incomplete is a widely used and successful technique in geochemical research (e.g. Helgeson, 1985; Shock *et al.*, 1992, 1997; Shock & Koretsky, 1993, 1995; Haas *et al.*, 1995; Prapaipong *et al.*, 1999; Prapaipong & Shock, 2001). If thermodynamic adsorption data were available for numerous lichen species, such correlative schemes could be developed for most, if not all, of the metals relevant to pollution-monitoring studies. In such a case, it would no longer be necessary to keep track of cation exchange and affinity ratios, or Langmuir isotherm parameters for each potential pH condition and each metal. Instead, an internally consistent set of thermodynamic properties for different metals could be used to assess quantitatively the cation adsorption capacity and exchange potential of a lichen for any desired combination of elements. The development of a universal database of thermodynamic parameters for lichen–metal interactions would provide a useful tool in biomonitoring studies, by facilitating interregion, interspecies and interelement comparisons and calibrations.

Conclusions

At present the state of knowledge regarding cation bioaccumulation by lichens is fragmentary and inconsistent. A wealth of largely descriptive information is available regarding the uptake and localization of metals by lichens, the occurrence of oxalate and lichen-acid biominerals within lichen thalli and the effects of lichens on bulk mineral weathering rates. However, the development of a generalized model of lichen–metal interactions, including complexation and solubility relations involving organic acids and extracellular polysaccharides, and the precise effects of lichen–metal coordination on ligand-assisted mineral weathering, is hampered at present because of a paucity of rigorous, quantitative data. Much work remains to be done. Potential studies that would most benefit and expand our understanding of lichen–metal and lichen–mineral interactions include:

- Comprehensive studies of metal adsorption by lichens and fungi as a function of pH, ionic strength and solute:biomass concentration, using lichens of different growth form, taxonomic affinity and character of organic acid production.

- Acid–base titration studies measuring the concentrations and acid dissociation properties of functional groups on lichen cell walls, extracellular polysaccharides and lichen acids.
- Quantitative modelling studies that constrain the properties of metal–lichen adsorption reactions in terms of thermodynamic mass law and mass action equations, using SCMs.
- Spectroscopic and bulk aqueous complexation studies of major lichen acids, including derivation of the structures and stabilities of metal–organic complexes and precipitates involving a wide range of elements.
- Spectroscopic investigations of the speciation of solid phases encountered as trapped particulate material in lichen biomass. Particular emphasis should be placed on understanding the dissolution of trapped particles and the precipitation of secondary solid phases on or within the thallus and their relationship with organic phases in the lichen. Fourier transform infrared spectroscopy (FTIR), X-ray photoelectron spectroscopy (XPS), X-ray absorption spectroscopy (XAS) in parallel with microscopy studies would be particularly useful in such investigations.
- Experimental studies quantifying LPD reaction equilibria and kinetics involving major rock-forming minerals and important lichen acids.

This is an incomplete list of potential research directions that would provide crucial information elucidating the biogeochemical influences of lichens in nature. The results of these and related investigations would not only promote the use and utility of lichens as pollution biomonitors, but also provide key information revealing the role of lichens in mineral weathering and soil formation processes, biogenic deterioration of monuments and the global biogeochemical cycling of carbon through geologic time.

Acknowledgements
We are grateful to Rosmarie Honegger for allowing us to adapt Fig. 15.1 in Honegger (1997), to Alan Orange for permission to include his drawing of *Physconia* rhizines, Takani Masako for the proposed Cu-norstictic acid diagram and Liz Bailey and Judith McLean for allowing us to include Figs 15.2a and b. Barbara Pawlik-Skowronska is thanked for stimulating discussions, Claudia Cabarcas Vergara for bibliographic assistance and Jim Chisholm, Gordon Cressey, Eva Fejer, John Francis and Gary Jones for assistance with XRD and IR mineral analysis.

370   *J. R. Haas and O. W. Purvis*

References

Adamo, P. & Violante, P. (2000). Weathering of rocks and neogenesis of minerals associated with lichen activity. *Applied Clay Science*, **16**, 229–56.

Adamo, P., Colombo, C. & Violante, P. (1997). Iron oxides and hydroxides in the weathering interface between *Stereocaulon vesuvianum* and volcanic rock. *Clay Minerals*, **32**, 453–61.

Andres, Y., MacCordick, J. J. & Hubert, J. C. (1993). Adsorption of several actinide (Th, U) and lanthanide (La, Eu, Yb) ions by *Mycobacterium smegmatis*. *Applied Microbiology and Biotechnology*, **39**, 413–17.

Antonelli, M. L., Ercole, P. & Campanella, L. (1998). Studies about the adsorption on lichen *Evernia prunastri* by enthalpimetric measurements. *Talanta*, **45**, 1039–47.

Ascaso, C. & Wierzchos, J. (1995). Study of the biodeterioration zone between the lichen thallus and the substrate. *Cryptogamic Botany*, **5**, 270–81.

Ascaso, C., Galvan, J. & Ortega, C. (1976). The pedogenic action of *Parmelia conspersa, Rhizocarpon geographicum* and *Umbilicaria pustulata*. *Lichenologist*, **8**, 151–71.

Asta, J., Orry, F., Toutain, F., Souchier, B. & Villemin, G. (2001). Micromorphological and ultrastructural investigations of the lichen-soil interface. *Soil Biology and Biochemistry*, **33**, 323–37.

Bargagli, R. & Mikhailova, I. (2002). Accumulation of inorganic contaminants. In *Monitoring with Lichens-Monitoring Lichens* Vol. 7, ed. P. L. Nimis, C. Scheidegger & P. A. Wolseley. Dordrecht: Kluwer Academic Publishers, pp. 65–84.

Barker, W. W. & Banfield, J. F. (1996). Biologically versus inorganically mediated weathering reactions: relationships between minerals and extracellular microbial polymers in lithobiontic communities. *Chemical Geology*, **132**, 55–69.

Barker, W. W., Welch, S. A. & Banfield, J. F. (1997). Biogeochemical weathering of silicate minerals. In *Geomicrobiology: Interactions between Microbes and Minerals, Reviews in Mineralogy*, Vol. 35, ed. J. F. Banfield & K. H. Nealson. Chelsea, Michigan: Mineralogical Society of America, pp. 391–428.

Beckett, P. J., Boileau, L. J. R., Padovan, D. & Richardson, D. H. S. (1982). Lichens and mosses as monitors of industrial activity associated with uranium mining in northern Ontario, Canada – Part 2: Distance dependent uranium and lead accumulation patterns. *Environmental Pollution (Series B)*, **4**, 91–107.

Bengtsson, L., Johansson, B., Hackett, T. J., McHale, L. & McHale, A. P. (1995). Studies on the biosorption of uranium by *Talaromyces emersonii* CBS 814.70 biomass. *Applications in Microbiology and Biotechnology*, **42**, 807–11.

Bennett, P. C., Melcer, M. E., Siegel, D. I. & Hassett, J. P. (1988). The dissolution of quartz in dilute aqueous solutions of organic acids at 25 °C. *Geochimica et Cosmochimica Acta*, **52**, 1521–30.

Berner, R. A. & Cochran, M. F. (1998). Plant-induced weathering of Hawaiian basalts. *Journal of Sedimentary Research*, **68**, 723–6.

Boileau, L. J. R., Beckett, P. J., Lavoie, P. & Richardson, D. H. S. (1982). Lichens and mosses as monitors of industrial activity associated with uranium mining in northern Ontario, Canada – Part 1: Field procedures, chemical analysis and interspecies comparisons. *Environmental Pollution (Series B)*, **4**, 69–84.

Boileau, L. J. R., Nieboer, E. & Richardson, D. H. S. (1985a). Uranium accumulation in the lichen *Cladonia rangiferina* (L.) Wigg. Part I. Uptake of cationic, neutral, and anionic forms of the uranyl ion. *Canadian Journal of Botany*, **63**, 384–9.

Boileau, L. J. R., Nieboer, E. & Richardson, D. H. S. (1985b). Uranium accumulation in the lichen *Cladonia rangiferina* (L.) Wigg. Part II. Toxic effects of cationic, neutral and anionic forms of the uranyl ion. *Canadian Journal of Botany*, **63**, 390–7.

Brady, D., Stoll, A. & Duncan, J. R. (1994). Biosorption of heavy metal cations by non-viable yeast biomass. *Environmental Technology*, **15**, 429–38.

Branquinho, C. (2001). Lichens. In *Metals in the Environment: Analysis by Biodiversity*, ed. M. N. V. Prasad. New York: Marcel Dekker, pp. 117–57.

Brightman, F. H. & Seaward, M. R. D. (1977). Lichens of man-made substrates. In *Lichen Ecology*, ed. M. R. D. Seaward. London: Academic Press, pp. 253–93.

Brown, D. H. (1991). Lichen mineral studies – Currently clarified or confused? *Symbiosis*, **11**, 207–23.

Brown, G. E., Jr, Parks, G. A. & O'Day, P. A. (1995). Sorption at mineral-water interfaces: macroscopic and microscopic perspectives. In *Mineral Surfaces, The Mineralogical Society Series*, Vol. 5, ed. D. J. Vaughan & R. A. D. Pattrick. Cambridge, UK: Chapman & Hall, pp. 129–84.

Casey, W. H. & Ludwig, C. (1995). Silicate mineral dissolution as a ligand-exchange reaction. In *Chemical Weathering Rates of Silicate Minerals, Reviews in Mineralogy*, Vol. 31, ed. A. E. White & S. L. Brantley. Chelsea, Michigan: Mineralogical Society of America, pp. 87–118.

Chen, J., Blume, H. P. & Beyer, L. (2000). Weathering of rocks induced by lichen colonization – a review. *Catena*, **39**, 121–46.

Chisholm, J. E., Jones, G. C. & Purvis, O. W. (1987). Hydrated copper oxalate, moolooite, in lichens. *Mineralogical Magazine*, **51**, 715–18.

Cicek, A. & Koparal, A. S. (2003). The assessment of air quality and identification of pollutant sources in the Eskisehir region Turkey using *Xanthoria parietina* (L.) Th.Fr. (1860). *Fresenius Environmental Bulletin*, **12**, 24–8.

Cox, J. S., Smith, D. S., Warren, L. A. & Ferris, F. G. (1999). Characterizing heterogeneous bacterial surface functional groups using discrete affinity spectra for proton binding. *Environmental Science and Technology*, **33**, 4514–21.

Czehura, S. J. (1977). A lichen indicator of copper mineralization, Lights Creek District, Plumas County, California. *Economic Geology*, **72**, 796–803.

Daughney, C. J., Fein, J. B. & Yee, N. (1998). A comparison of the thermodynamics of metal adsorption onto two common bacteria. *Chemical Geology*, **144**, 161–76.

Davis, J. A. & Kent, D. B. (1990). Surface complexation modelling in aqueous geochemistry. In *Mineral-Water Interface Geochemistry, Reviews in Mineralogy*, Vol. 23, ed. J. M. F. Hochella & A. F. White. Chelsea, Michigan: Mineralogical Society of America, pp. 177–260.

Dawson, H. J., Hrutfiord, B. F. & Ugolini, F. C. (1984). Mobility of lichen compounds from *Cladonia mitis* in arctic soils. *Soil Science*, **138**, 40–5.

Dzombak, D. A. & Morel, F. M. M. (1990). *Surface Complexation Modelling: Hydrous Ferric Oxide*. Place: Wiley Interscience.

Easton, E. M. (1994). Lichens and rocks – a review. *Geoscience Canada*, **21**, 59–76.

Edwards, H. G. H., Farwell, D. W. & Lewis, I. R. (1993). FT Raman microscopy and lichen biodeterioration. *Bruker Report*, **139**, 8–11.

Fahselt, D., Wu, T. W. & Mott, B. (1995). Trace element patterns in lichens following uranium mine closures. *The Bryologist*, **98**, 228–34.

Fein, J. B. (2000). Quantifying the effects of bacteria on adsorption reactions in water-rock systems. *Chemical Geology*, **169**, 265–80.

Fein, J. B., Daughney, C. J., Yee, N. & Davis, T. A. (1997). A chemical equilibrium model for metal adsorption onto bacterial surfaces. *Geochimica et Cosmochimica Acta*, **61**, 3319–28.

Fein, J. B., Boily, J. F., Guclu, K. & Kaulbach, E. (1999). Experimental study of humic acid adsorption onto bacteria and Al-oxide mineral surfaces. *Chemical Geology*, **162**, 33–45.

Fortin, D., Ferris, F. G. & Beveridge, T. J. (1997). Surface-mediated mineral development by bacteria. In *Geomicrobiology: Interactions between Microbes and Minerals, Reviews in Mineralogy*, Vol. 35, ed. J. F. Banfield & K. H. Nealson. Chelsea, Michigan: Mineralogical Society of America, pp. 161–80.

Fowle, D. A. & Fein, J. B. (1999). Competitive adsorption of metals onto bacterial surfaces. *Geochimica et Cosmochimica Acta*, 63, 3059–67.

Fowle, D. A. & Fein, J. B. (2000). Experimental measurements of the reversibility of metal-bacteria adsorption reactions. *Chemical Geology*, 168, 27–36.

Gadd, G. M. (1993). Interactions of fungi with toxic metals. *New Phytologist*, 124, 25–60.

Galun, M., Keller, P., Malki, D. *et al.* (1983). Removal of uranium (VI) from solution by fungal biomass and fungal wall-related biopolymers. *Science*, 219, 285–6.

Galun, M., Galun, E., Siegel, B. Z. *et al.* (1987). Removal of metal ions from aqueous solutions by *Penicillium* biomass: kinetic and uptake parameters. *Water, Air, and Soil Pollution*, 33, 359–71.

Garty, J. (2001). Biomonitoring atmospheric heavy metals with lichens: theory and application. *Critical Reviews in Plant Sciences*, 20, 309–71.

Garty, J. & Delarea, J. (1991). Localization of iron and other elements in the lichen *Nephroma arcticum* (L.) Torss. *Environmental and Experimental Botany*, 31, 367–75.

Garty, J. & Galun, M. (1979). Localization of heavy metals and other elements accumulated in the lichen thallus. *New Phytologist*, 82, 159–68.

Garty, J. & Theiss, H. B. (1990). The localization of lead in the lichen *Ramalina duriaei* (De Not.) Bagl. *Botanica Acta*, 103, 311–14.

Garty, J., Harel, Y. & Steinberger, Y. (1995). The role of lichens in the cycling of metals in the Negev desert. *Archives of Environmental Contamination and Toxicology*, 29, 247–53.

Gilbert, O. L. (1990). The lichen flora of urban wasteland. *Lichenologist*, 22, 87–101.

Goyal, R. & Seaward, M. R. D. (1981). Metal uptake in terricolous lichens I. Metal localization within the thallus. *New Phytologist*, 89, 631–45.

Goyal, R. & Seaward, M. R. D. (1982). Metal uptake in terricolous lichens III. Translocation in the thallus of *Peltigera canina*. *New Phytologist*, 90, 85–90.

Grandstaff, D. E. (1986). The dissolution rate of forsteritic olivine from Hawaiian beach sand. In *Rates of Chemical Weathering of Rocks and Minerals*, ed. S. M. Colman & D. P. Dethier. Orlando, Florida: Academic Press, pp. 41–59.

Guibal, E., Milot, C. & Tobin, J. M. (1998). Metal-anion sorption by chitosan beads: equilibrium and kinetic studies. *Industrial Engineering and Chemical Research*, 37, 1454–63.

Haas, J. R., Shock, E. L. & Sassani, D. C. (1995). Rare earth elements in hydrothermal systems: Estimates of standard partial molal thermodynamic properties of aqueous complexes of the rare earth elements at high pressures and temperatures. *Geochimica et Cosmochimica Acta*, 59, 4329–50.

Haas, J. R., Bailey, E. H. & Purvis, O. W. (1998). Bioaccumulation of metals by lichens: Uptake of aqueous uranium by *Peltigera membranacea* as a function of time and pH. *American Mineralogist*, 83, 1494–502.

Haas, J. R., DiChristina, T. J. & Wade, R. Jr. (2001). Thermodynamics of U(VI) sorption onto *Shewanella putrefaciens*. *Chemical Geology*, 180, 33–54.

Hallbauer, D. K. & Jahns, H. M. (1977). Attack of lichens on quartzitic rock surfaces. *Lichenologist*, 9, 119–22.

Helgeson, H. C. (1985). Some thermodynamic aspects of geochemistry. *Pure and Applied Chemistry*, 57, 31–44.

Hickmott, M. (1980). Lichens on lead. *Lichenologist*, 12, 404–6.

Honegger, R. (1997). Metabolic interactions at the mycobiont-photobiont interface in lichens. In *The Mycota*, Vol. 5A, *Plant Relationships. Part A*. ed. G. C. Carroll & P. Tudzynski. New York: Springer-Verlag, pp. 209–21.

Iskandar, I. K. & Syers, J. K. (1972). Metal-complex formation by lichen compounds. *Journal of Soil Science*, **23**, 255–65.

Jacquiot, L. & Daillant, O. (1997). Bio-accumulation des métaux lourds et d'autres eléments traces par les lichens. Revue bibliographique. *Bulletin de l'Observatoire Mycologique*, **12**, 2–31.

Jacquiot, L. & Daillant, O. (1999). Bio-accumulation des radioéléments par les lichens. Revue bibliographique. *Bulletin de l'Observatoire Mycologique*, **16**, 2–23.

Johnston, C. G. & Vestal, J. R. (1993). Biogeochemistry of oxalate in the Antarctic cryptoendolithic lichen-dominated community. *Microbial Ecology*, **25**, 305–19.

Jones, D. (1988). Lichens and pedogenesis. In *CRC Handbook of Lichenology*, vol. III, ed. M. Galun: Boca Raton: CRC Press, pp. 109–24.

Jones, D., Wilson, M. J. & Laundon, J. R. (1982). Observations on the location and form of lead in *Stereocaulon vesuvianum*. *Lichenologist*, **14**, 281–6.

Kasama, T., Murakami, T., Ohnuki, T. & Purvis, O. W. (2001). Effects of lichens on uranium migration. In *Scientific Basis for Nuclear Waste Management* XXIV, ed. K. P. Hart & G. R. Lumpkin. Pittsburgh, PA: Materials Research Society, pp. 683–90.

Kelly, S. D., Boyanov, M. I., Bunker, B. A. *et al.* (2001). XAFS determination of the bacterial cell wall functional groups responsible for complexation of Cd and U as a function of pH. *Journal of Synchrotron Radiation*, **8**, 946–8.

Knops J. M. H., III, Nash, T. H. N., Boucher, V. L. & Schlesinger, W. H. (1991). Mineral cycling and epiphytic lichens: implications at the ecosystem level. *Lichenologist*, **23**, 309–21.

Koretsky, C. (2000). The significance of surface complexation reactions in hydrologic systems: a geochemist's perspective. *Journal of Hydrology*, **230**, 127–71.

Lee, M. R. & Parsons, I. (1999). Biomechanical and biochemical weathering of lichen-encrusted granite: textural controls on organic-mineral interactions and deposition of silica-rich layers. *Chemical Geology*, **161**, 385–97.

Loppi, S. & Corsini, A. (2003). Diversity of epiphytic lichens and metal contents of *Parmelia caperata* thalli as monitors of air pollution in the town of Pistoia, Italy. *Environmental Monitoring and Assessment*, **86**, 289–301.

Loppi, S. & Pirintsos, S. A. (2003). Epiphytic lichens as sentinels for heavy metal pollution at forest ecosystems (central Italy). *Environmental Pollution*, **121**, 327–32.

Loppi, S., Frati, L., Paoli, L. *et al.* (2004). Biodiversity of epiphytic lichens and heavy metal contents of *Flavoparmelia caperata* thalli as indicators of temporal variations of air pollution in the town of Montecatini Terme (central Italy). *Science of the Total Environment*, **326**, 113–22.

Ludwig, C., Casey, W. H. & Rock, P. A. (1995). Prediction of ligand-promoted dissolution rates from the reactivities of aqueous complexes. *Nature*, **375**, 44–7.

McCarroll, D. & Viles, H. (1995). Rock-weathering by the lichen *Lecidea auriculata* in an Arctic alpine environment. *Earth Surface Processes and Landforms*, **20**, 199–206.

McLean, J., Purvis, O. W., Williamson, B. J. & Bailey, E. H. (1998). Role for lichen melanins in uranium remediation. *Nature*, **391**, 649–50.

Modenesi, P., Piana, M. & Pinna, D. (1998). Surface features in *Parmelia sulcata* (Lichenes) thalli growing in shaded or exposed habitats. *Nova Hedwigia*, **66**, 535–47.

Muraleedharan, T. R. & Venkobachar, L. I. (1994). Further insight into the mechanism of biosorption of heavy metals by *Ganoderma lucidum*. *Environmental Technology*, **15**, 1015–27.

Nash, T. H. (1996). Nutrients, elemental accumulation and mineral cycling. In *Lichen Biology*, ed. T. H. Nash. Cambridge, UK: Cambridge University Press, pp. 136–53.

Nash, T. H., Gries, C., Zschau, T. et al. (2003). Historical patterns of metal atmospheric deposition to the epilithic lichen *Xanthoparmelia* in Maricopa County, Arizona, USA. *Journal de Physique IV*, **107**, 921–4.

Nieboer, E. & Richardson, D. H. S. (1981). Lichens as monitors of atmospheric deposition. In *Atmospheric Pollutants in Natural Waters*, ed. S. J. Eisenreich. Ann Arbor: Ann Arbor Science, pp. 339–88.

Nieboer, E., Richardson, D. H. S. & Tomassini, F. D. (1978). Mineral uptake and release by lichens: an overview. *The Bryologist*, **81**, 226–46.

Nimis, P. L. & Purvis, O. W. (2002). Monitoring lichens as indicators of pollution. An introduction. In *Monitoring with Lichens – Monitoring Lichens*, ed. P. L. Nimis, C. Scheidegger & P. A. Wolseley. Dordrecht: Kluwer Academic Publishers, pp. 1–4.

Olafsdottir, E. S. & Ingolfsdottir, K. (2001). Polysaccharides from lichens: structural characteristics and biological activity. *Planta Medica*, **67**, 199–208.

Paul, A., Hauck, M. & Fritz, E. (2003). Effects of manganese on element distribution and structure in thalli of the epiphytic lichens *Hypogymnia physodes* and *Lecanora conizaeoides*. *Environmental and Experimental Botany*, **50**, 113–24.

Pawlik-Skowronska, B., di Toppi, L. S., Favali, M. A. et al. (2002). Lichens respond to heavy metals by phytochelatin synthesis. *New Phytologist*, **156**, 95–102.

Peberdy, J. F. (1990). Fungal cell walls – a review. In *Biochemistry of Cell Walls and Membranes in Fungi*, ed. P. J. Kuhn, A. P. J. Trinci, M. J. Jung, M. W. Goosey & L. E. Copping. Berlin; New York: Springer-Verlag, pp. 5–30.

Poelt, J. & Huneck, S. (1968). *Lecanora vinetorum* nova spec., ihre Vergesellschaftung, ihre Ökologie und ihre Chemie. *Österreichische botanische Zeitschrift*, **115**, 411–22.

Prapaipong, P. & Shock, E. L. (2001). Estimation of standard-state entropies of association for aqueous metal-organic complexes and chelates at 25 °C and 1 bar. *Geochimica et Cosmochimica Acta*, **62**, 3931–53.

Prapaipong, P., Shock, E. L. & Koretsky, C. M. (1999). Metal-organic complexes in geochemical processes: temperature dependence of the standard thermodynamic properties of aqueous complexes between metal cations and dicarboxylate ligands. *Geochimica et Cosmochimica Acta*, **63**, 2547–77.

Puckett, K. J., Nieboer, E., Gorzynski, M. J. & Richardson, D. H. S. (1973). The uptake of metal ions by lichens: a modified ion-exchange process. *New Phytologist*, **72**, 329–42.

Purvis, O. W. (1984). The occurrence of copper oxalate in lichens growing on copper sulphide-bearing rocks in Scandinavia. *Lichenologist*, **16**, 197–204.

Purvis, O. W. (1996). Interactions of lichens with metals. *Science Progress*, **79**, 283–309.

Purvis, O. W. (2000). Lichens. *The Natural History Museum*.

Purvis, O. W. & Halls, C. (1996). A review of lichens in metal-enriched environments. *Lichenologist*, **28**, 571–601.

Purvis, O. W. & James, P. W. (1985). Lichens of the Coniston copper mines. *Lichenologist*, **17**, 221–37.

Purvis, O. W., Gilbert, O. L. & James, P. W. (1985). The influence of copper mineralization on *Acarospora smaragdula*. *Lichenologist*, **17**, 111–14.

Purvis, O. W., Elix, J. A., Broomhead, J. A. & Jones G. C. (1987). The occurrence of copper-norstictic acid in lichens from cupriferous substrata. *Lichenologist*, **19**, 193–203.

Purvis, O. W., Elix, J. A. & Gaul, K. L. (1990). The occurrence of copper-psoromic acid from cupriferous substrata. *Lichenologist*, **22**, 345–54.

Purvis, O. W., Williamson, B., Bartok, K. & Zoltani, N. (2000). Bioaccumulation of lead by the lichen *Acarospora smaragdula* from smelter emissions. *New Phytologist*, **147**, 591–9.

Purvis, O. W., Bailey, E. H., McLean, J., Kasama, T. & Williamson, B. J. (2004). Uranium biosorption by the lichen *Trapelia involuta* at a uranium mine. *Geomicrobiology Journal*, **21**, 159–67.

Richardson, D. H. S. (1995). Metal uptake in lichens. *Symbiosis*, **18**, 119–27.

Richardson, D. H. S. & Nieboer, E. (1981). Lichens and pollution monitoring. *Endeavour*, **5**, 127–33.

Richardson, D. H. S. & Nieboer, E. (1983). The uptake of nickel ions by lichen thalli of the genera *Umbilicaria* and *Peltigera*. *Lichenologist*, **15**, 81–8.

Richardson, D. H. S., Kiang, S., Ahmadjian, V. & Nieboer, E. (1985). Lead and uranium uptake in lichens. In *Lichen Physiology and Cell Biology*, ed. D. H. Brown. New York: Plenum Press, pp. 227–46.

Saeki, M., Kunii, K., Seki, T. *et al.* (1977). Metal burdens in urban lichens. *Environmental Research*, **13**, 256–66.

Sandau, E., Sandau, P., Pulz, O. & Zimmermann, M. (1996). Heavy metal sorption by marine algae and algal by-products. *Acta Biotechnology*, **16**, 103–19.

Sarret, G., Manceau, A., Cuny, D. *et al.* (1998). Mechanisms of lichen resistance to metallic pollution. *Environmental Science and Technology*, **32**, 3325–30.

Schatz, A. (1963). Soil microorganisms and soil chelation. The pedogenic action of lichens and lichen acids. *Agricultural and Food Chemistry*, **11**, pp. 112–18.

Schlekat, C. E., Decho, A. W. & Chandler, G. T. (1998). Sorption of cadmium to bacterial extracellular polymeric sediment coatings under estuarine conditions. *Environmental Toxicology and Chemistry*, **17**, 1867–74.

Sensen, M. & Richardson, D. H. S. (2002). Mercury levels in lichens from different host trees around a chlor-alkali plant in New Brunswick, Canada. *Science of the Total Environment*, **293**, 31–45.

Shock, E. L. & Koretsky, C. M. (1993). Metal-organic complexes in geochemical processes: calculation of standard partial molal thermodynamic properties of aqueous acetate complexes at high pressures and temperatures. *Geochimica et Cosmochimica Acta*, **57**, 4899–922.

Shock, E. L. & Koretsky, C. M. (1995). Metal-organic complexes in geochemical processes: estimation of standard partial molal thermodynamic properties of aqueous complexes between metal cations and monovalent organic acid ligands at high temperatures and pressures. *Geochimica et Cosmochimica Acta*, **59**, 1497–532.

Shock, E. L., Oelkers, E. H., Johnson, J. W., Sverjensky, D. A. & Helgeson, H. C. (1992). Calculation of the thermodynamic behavior of aqueous species at high pressures and temperatures: effective electrostatic radii, dissociation constants, and standard partial molal properties to 1000 °C and 5 kb. *Journal of the Chemical Society (London) Faraday Transactions*, **88**, 803–26.

Shock, E. L., Sassani, D. C., Willis, M. & Sverjensky, D. A. (1997). Inorganic species in geologic fluids: correlations among standard molal thermodynamic properties of aqueous ions and hydroxide complexes. *Geochimica et Cosmochimica Acta*, **61**, 907–50.

Small, T. D., Warren, L. A., Roden, E. E. & Ferris, F. G. (1999). Sorption of strontium by bacteria, Fe(III) oxide, and bacteria–Fe(III) oxide composites. *Environmental Science and Technology*, **33**, 4465–70.

Stone, A. T. (1997). Reactions of extracellular organic ligands with dissolved metal ions and mineral surfaces. In *Geomicrobiology: Interactions between Microbes and Minerals, Reviews in Mineralogy*, Vol. 35, ed. J. F. Banfield & K. H. Nealson. Chelsea, Michigan: Mineralogical Society of America, pp. 309–44.

Strandberg, G. W., Shumate, S. E. & Parrott, J. R. (1981). Microbial cells as biosorbents for heavy metals: accumulation of uranium by *Saccharomyces cerevisiae* and *Pseudomonas aeruginosa*. *Applied and Environmental Microbiology*, **41**, 237–45.

Stumm, W. & Morgan, J. J. (1996). *Aquatic Chemistry*. New York: John Wiley & Sons.

Suzuki, Y. & Banfield, J. F. (1999). *Geomicrobiology of uranium*. In *Uranium: Mineralogy, Geochemistry and the Environment, Reviews in Mineralogy*, Vol. 38, ed. P. C. Burns & R. Finch. Chelsea, Michigan: Mineralogical Society of America, pp. 393–432.

Syers, J. K. (1969). Chelating ability of fumarprotocetraric acid and *Parmelia conspersa*. *Plant and Soil*, **31**, 205–8.

Takani, M., Yajima, T., Masuda, H. & Yamauchi, O. (2002). Spectroscopic and structural characterization of copper(II) and palladium(II) complexes of a lichen substance usnic acid and its derivatives. Possible forms of environmental metals retained in lichens. *Journal of Inorganic Biochemistry*, **91**, 139–50.

Texier, A. C., Andres, Y. & Cloirec, P. L. (1997). Selective biosorption of lanthanide (La, Eu) ions by *Mycobacterium smegmatis*. *Environmental Toxicology*, **18**, 835–41.

Tsezos, M. (1985). The selective extraction of metals from solution by micro-organisms. A brief review. *Canadian Metallurgical Quarterly*, **24**, 141–4.

Tsezos, M. & Volesky, B. (1982). The mechanism of thorium biosorption by *Rhizopus arrhizus*. *Biotechnology and Bioengineering*, **24**, 955–69.

Tuominen, Y. (1967). Studies of the strontium uptake of the *Cladonia alpestris* thallus. *Annales Botanici Fennici*, **4**, 1–28.

Tyler, G. (1989). Uptake, retention and toxicity of heavy metals in lichens. *Water, Air, and Soil Pollution*, **47**, 321–33.

Urrutia, M. M. & Beveridge, T. J. (1994). Formation of fine-grained metal and silicate precipitates on a bacterial surface (*Bacillus subtilis*). *Chemical Geology*, **116**, 261–80.

van der Lee, J. & de Windt, L. D. (1999). *CHESS Tutorial and Cookbook*. Fontainebleau, France: CIG-Ecole des Mines de Paris.

Veith, J. A. & Sposito, G. (1977). On the use of the Langmuir equation in the interpretation of "adsorption" phenomena. *Soil Science Society of America Journal*, **41**, 697–702.

Wessels, D. C. J. & Schoeman, P. (1988). Mechanism and rate of weathering of Clarens sandstone by an endolithic lichen. *South African Journal of Science*, **84**, 274–7.

Westall, J. C. (1982). *FITEQL, A computer program for determination of chemical equilibrium constants from experimental data. Version 2.0*. Department of Chemistry, Oregon State University.

Wild, H. (1968). Geobotanical anomalies in Rhodesia. 1. The vegetation of copper-bearing rocks. *Kirkia*, **7**, 1–72.

Williams, M. E. & Rudolf, E. D. (1974). The role of lichens and associated fungi in the chemical weathering of rock. *Mycologia*, **66**, 648–60.

Wilson, M. J. (1995). Interactions between lichens and rocks: a review. *Cryptogamic Botany*, **5**, 299–305.

Wilson, M. J. & Jones, D. (1984). The occurrence and significance of manganese oxalate in *Pertusaria corallina* (lichenes). *Pedobiologia*, **26**, 373–9.

Wilson, M. J., Jones, D. & McHardy, W. J. (1981). The weathering of serpentinite by *Lecanora atra*. *Lichenologist*, **13**, 167–76.

Yee, N. & Fein, J. B. (2002). Does metal adsorption onto bacterial surfaces inhibit or enhance aqueous metal transport? Column and batch reactor experiments on Cd-*Bacillus subtilis*-quartz systems. *Chemical Geology*, **185**, 303–19.

Yee, N., Fein, J. B. & Daughney, C. J. (2000). Experimental study of the pH, ionic strength and reversibility behavior of bacteria-mineral adsorption. *Geochimica et Cosmochimica Acta*, **64**, 609–17.

Yliruokanen, I. (1975). Uranium, thorium, lead, lanthanoids and yttrium in some plants growing on granitic and radioactive rocks. *Bulletin of the Geologic Society of Finland*, **47**, 71–8.

16
Fungi in subterranean environments

JOACHIM REITNER, GABRIELA SCHUMANN
AND KARSTEN PEDERSEN

Introduction

Exploration of the microbial world got off to a slow start some 350 years ago, when Leeuwenhoek and his contemporaries focused their microscopes on very small life forms. It was not until about 20 years ago, however, that exploration of the world of intra-terrestrial microbes gathered momentum. Until then, it was generally assumed that life could not persist deep underground, out of reach of the sun and a photosynthetic ecosystem base. In the mid 1980s, the drilling of deep holes for scientific research started. Holes up to thousands of metres deep were drilled in hard as well as sedimentary rock, and up came microbes in numbers equivalent to what could be found in many surface ecosystems (Pedersen, 1993). The deep subterranean biosphere had been discovered.

Defining the boundary between the ground-surface biosphere and the subterranean biosphere is problematic: various scientists define it differently, and there is no general consensus. For our purposes the main criterion is that the subterranean biosphere begins where contact with the surface biosphere is lost. This lies beneath soil and root zones, beneath the ground-water table, and beneath sediment and crust surfaces. A long time should have elapsed since last surface contact, 'long time' in this respect being at least several decades, preferably hundreds of years or more. In our view it is not depth per se that defines a subterranean ecosystem; rather, it is the duration of isolation from the surface.

A continuum of subterranean environments exists, ranging from very hard rocks such as granites and basalts, through sedimentary rocks and sandstones, to as yet unconsolidated, fairly soft sediments. Two main subterranean rock environments can be defined: *hard rocks* – those too hard and impermeable to allow microbes to pass, except via fractures, and

Fungi in Biogeochemical Cycles, ed. G. M. Gadd. Published by Cambridge University Press. © British Mycological Society 2006.

sedimentary rocks – those porous enough to allow microbes to penetrate. Hard rocks generally experienced temperatures high above the limit for life when they were formed and are therefore sterile after formation. With time, however, they fracture and can be colonized. Sedimentary rocks are formed slowly, mostly on the sea floor, and over geological time scales may appear on land as rocks. Sedimentary rocks that did not exceed the temperature limit for life (approximately 113 °C) when formed can harbour microbes that entered during the sedimentation process; microorganisms can, of course, also move in later.

Exploration of the subterranean biosphere poses different technical challenges when carried out at sea or on land. At sea, drilling must be performed from ships often through several kilometres of water. In contrast, drilling on land is relatively less technologically challenging. Furthermore, there are also pronounced geological differences between the sub-sea-floor and continental environments. The sea floor mostly consists of sediments underlain by magmatic hard rocks, while the continents are built of a range of hard, sedimentary and soft rocks of various origins. Finally, the sea floor is younger than many continental rock environments, which can be up to several billion years old. The oldest sea floor environment is about 170 Ma old, and is found under the Pacific Ocean, near Japan.

Comparing a surface environment, such as the sea, with most subterranean environments gives contradictory results. Spatially, the sea varies little within a small area, while a subterranean environment can vary greatly over short distances. However, when it comes to temporal variation the situation is reversed: the sea has daily and seasonal cycles and is strongly affected by weather, while subterranean environments vary little over time. This means that many more samples (from boreholes) need to be taken to get a statistically significant spatial overview of a subterranean than of a surface aquatic environment. Therefore these subterranean samples can be gathered over relatively long periods due to little temporal variability. However, the high cost of drilling greatly restricts the amount of such drilling. Therefore, our knowledge of the subterranean biosphere is still limited and thorough exploration has barely started.

There are two main routes for exploring the biogeochemistry of the subterranean biosphere: one is to sample and analyse the diversity, activity and distribution of active microbial life, while the other is to analyse traces of this biosphere. We can analyse various biosignatures such as anomalies in the stable isotope composition or the presence of organic molecules typical of life. Fossils are obvious evidence of subterranean life, provided it can be proven that the organisms of origin had been living in the explored subterranean environment. In this chapter, we will review the literature on the occurrence of fungi in subterranean environments and briefly speculate as to their possible roles in deep subterranean biogeochemical processes.

Present records: viable fungi

The search for viable fungi in deep ground water and sediments started concomitantly with the beginning of exploration of the subterranean biosphere in the early 1980s. Fungi have been isolated from various subsurface environments, such as ground water, rock and subsurface sediments, to learn more about their physiological potential. Deep sub-sea-floor sediments and continental sedimentary and hard rocks have been thoroughly scanned for the presence of viable yeasts and filamentous fungi, but the results have been relatively sketchy (Madsen & Ghiorse, 1993; Fredrickson & Onstott, 1996; Palumbo *et al.*, 1996; Ludvigsen *et al.*, 1999; Ekendahl *et al.*, 2003).

Sub-sea floor environments

Over two-thirds of the earth's surface is covered by oceans, and over 50% of these are from 3000 to 6000 m in depth, with an average depth of approximately 3200 m (Gage & Tyler, 1991). The sub-sea-floor can be considered to be an extreme environment for several reasons. It has a high hydrostatic pressure and a temperature that is low close to the sea floor, but that increases with depth and is very high around hydrothermal vents. It generally has a low nutrient concentration and is totally dark. Sediment and crust layers deep below the continental shelf and ocean floors, previously thought to be too nutrient depleted to sustain life, have now been found to harbour numerous bacteria (Parkes *et al.*, 1990, 1994, 2000; Fredrickson & Onstott, 1996; Wellsbury *et al.*, 1997; Torsvik *et al.*, 1998; McKinley *et al.*, 2000; Thorseth *et al.*, 2001; Wellsbury *et al.*, 2002; Newberry *et al.*, 2004). In contrast, fungi are rarely reported in deep-sea habitats (Kohlmeyer, 1969a, 1977; Raghukumar & Raghukumar, 1998; Soltwedel & Schewe, 1998; Zande, 1999).

Gage and Tyler (1991) stated that fungi, as prominent heterotrophic organisms in the oceans, are transported passively from the surface to the deep sea by rapidly sinking water masses in the Arctic and Antarctic regions and/or by attachment to sinking particulate substrate (Van Uden & Fell, 1968; Kohlmeyer & Kohlmeyer, 1979). Several substrates, such as wood, particulate organic matter, or chitin from the exoskeleton of marine crustaceans, have been shown to be degraded by autochthonous fungi at even greater ocean depths (Kohlmeyer, 1969b; Kohlmeyer & Kohlmeyer, 1979).

The deepest sediment ever sampled was from the Mariana Trench at 10 897 m below sea level. Thousands of microorganisms were isolated from this mud, and yeasts and other fungi were observed (Takami *et al.*, 1997). Further characterization resulted in the proposed new species *Penicillium lagena* and *Rhodotorula mucilaginosa* (Takami, 1999). Molecular analysis of those deep-mud samples did not return eukaryotic DNA, suggesting low numbers of yeasts (Kato, 1999). In other investigations 13 yeast strains were isolated from 4500–6500-m-deep sediments from the Japan Trench (Abe *et al.*, 2001), and a novel species of *Cryptococcus* was isolated from sediments from Surgua Bay, Japan (Nagahama *et al.*, 2003).

The available literature clearly shows that fungi are present in even the deepest sediments of the sea, and therefore could potentially be buried with sediments. Little is known about marine filamentous fungi and yeasts, and almost nothing about their life-cycles or metabolism under deep-sea and sub-sea-floor conditions. If they survive burial, they will eventually become part of the sub-sea-floor biosphere. No direct data describe how long fungi can survive in subsurface environments. Data from dried soil specimens indicate that fungi survive fewer than 50–100 years separated from their autochthonous surface environment (Sneath, 1962). However, sub-sea-floor conditions may very well be more favourable for preservation than are soil conditions.

Continental sedimentary rocks

There are many reports of fungi in shallow ground water and in unsaturated ground environments. For instance, sandy aquifers in the Segeberger Forest, Germany, were found to contain a diversity of fungi when analysed (Hirsch & Rades-Rohkohl, 1983; Hirsch *et al.*, 1992). A more recent investigation reported a great diversity of aquatic hyphomycetes in metal-contaminated ground water (Krauss *et al.*, 2003). A total of 13 boreholes 6–50 m deep were analysed, and up to 20 species were detected in a single borehole. All these boreholes, however, contained significant levels of oxygen. As depth increased and investigators approach the subterranean biosphere, defined by its isolation from the surface biosphere and its anaerobic and reduced conditions, observations of fungi grow rarer (Madsen & Ghiorse, 1993).

Contamination is a threat to all microbiological sampling and is particularly difficult to control when investigating the microbiology of deep subterranean environments. Thus it must be determined whether the organisms isolated or otherwise detected were present in the sample before drilling. Contamination of sediment samples with organisms from the drill water is usually controlled by adding dissolved and particulate tracers (e.g. Russell *et al.*, 1992; Smith *et al.*, 2000). After taking such precautions, fungi have been concluded to exist in sediments of the Atlantic Plain of North America down to a depth of 236 m, albeit in very low numbers (Fliermans, 1989; Sinclair & Ghiorse, 1989). Extraction and analysis of rRNA have occasionally revealed eukaryotic rRNA at depth. Sediments from depths of 173–217 m have been found to contain up to approximately 25% eukaryotic rRNA, suggesting the presence of protozoa and/or fungi (Ogram *et al.*, 1995). In the early 1980s, a paucity of eukaryotic life was concluded to be characteristic of deep, oligotrophic sediment formations (Ghiorse & Balkwill, 1983), and this conclusion still seems to be valid.

Continental hard rocks

It is more difficult to control contamination during drilling in hard rock than in sedimentary rock. The contaminated outer parts of a sedimentary core can be peeled off if needed to reach cells in the unconta-minated inner part of the core. The hard rock core, however, does not itself have microbes inside; instead, life dwells in fractures in the rock and these are very exposed to the drill water. A successful option is to flush the borehole after drilling and leave it flushing or pumping for weeks or even months. This will remove groundwater that is contaminated with drill water, and pristine groundwater will return to the fractures around the borehole. Using this flushing method combined with cultivation and rRNA sequencing methods, it was demonstrated that although contami-nants were found in the groundwater directly after drilling, they were absent about six months later (Pedersen *et al.*, 1997b).

A first indication that fungi can exist at depth in crystalline rock came in 1987 (Pedersen, 1987) when moulds were found at depths ranging between 807 and 1232 m. Plate counts showed between 130 and 630 colony-forming units (CFU) ml^{-1}, approximately similar to the plate counts obtained for bacteria. The next suggestion came when 16S rRNA genes obtained from deep ground water were scanned; yeast sequences repeatedly appeared in samples from boreholes representing a depth range of 40–626 m (Pedersen *et al.*, 1996, 1997b). Those observations eventually triggered a methodo-logically focused investigation that specifically scanned for fungi in bore-holes and open fractures along the Äspö Hard Rock Laboratory (HRL) tunnel between 201 and 448 m below sea level (Ekendahl *et al.*, 2003).

Three of the eight boreholes and all four fractures sampled were found to harbour yeasts and/or filamentous fungi.

To extend the knowledge of eukaryotic life in crystalline rock fractures, five yeasts, three yeast-like fungi and 17 filamentous fungal strains were isolated from Äspö HRL groundwater (Ekendahl *et al.*, 2003). Phenotypic testing and phylogenetic analysis of 18S rDNA sequences of the five yeast isolates revealed their relationships to *Rhodotorula minuta* and *Cryptococcus* spp. Scanning and transmission electron microscopy demonstrated that the strains possessed morphological characteristics typical of yeasts, although they were relatively small, having an average length of 3 μm. Enumeration through direct counting and most probable number methods revealed low numbers of fungi, between 0.01 and 1 CFU ml^{-1}, at some sites. Five of the strains were characterized physiologically to determine whether they were adapted to life in the deep biosphere. These studies revealed that the strains could grow at up to pH 10. They grew well at low temperatures down to 4 °C, but would not grow above 25–30 °C. All strains preferred less than 1% (w/v) NaCl, although they could all grow at up to 7% (w/v) NaCl concentration. These growth parameters suggest a degree of adaptation to the ground water at Äspö HRL. The pH of the sampled Äspö ground water was about 8, the temperature was 10 °C at 200 m and 15 °C at 450 m, and the salinity approximately 1% (w/v). Despite the fact that such eukaryotic microorganisms may be transient members of the deep biosphere microbial community, these observations suggested that they are capable of growing in this subterranean environment.

Reality or experimental bias?

Despite the range of painstaking approaches taken to discover viable fungal life in subterranean environments, negative or very low numbers are commonly reported. Perhaps the largest problem facing subsurface fungi is the absence of oxygen in most deep, subterranean environments. Most of the dissolved oxygen in ground water is consumed at very shallow depths (Banwart *et al.*, 1996). However, there are many types of facultative anaerobic yeasts, and strictly anaerobic fungi are known from the organotrophic environment of the rumen, so anaerobic conditions do not pose an absolute obstacle. Aerobic subterranean fungi could possibly produce spores under oxygen-limiting conditions, an occurrence that should be favoured under stressful environmental conditions. Therefore, though oxygen limitation does not rule out the existence of yeasts in the deep subsurface, it may curb their ability to grow.

The infrequency of fungal observations in subterranean environments may very well be due to low actual numbers. But many investigations of subterranean environments do not include specific methods for detecting fungi, so a lack of detection cannot be held as proving the absence of fungi. Further, a large discrepancy exists between the number and diversity of terrestrial versus subterranean sites investigated for fungal diversity. Far more terrestrial than subterranean environments have been searched for fungi. As viable subterranean fungi seem very difficult to detect due to low numbers, obvious contamination possibilities, unknown cultivation requirements, etc., it may be more profitable to search for *remnants* of fungi such as fossils.

Records from the past: fossil fungi

Fossil remains of a deep subsurface biosphere are unequivocal traces of former deep life, as contamination via drilling or mining procedures can be excluded. However, traces of fossilized eukaryotic life in deep subterranean environments are extremely rare (Schumann *et al.*, 2004). There are fossil representatives of all known domains of life, and while fossilized fungi do exist, they are exceptionally rare. The standard paleobotanical literature refers to the lack of fossil records of fungi (Pia, 1927; Gothan & Weyland, 1964; Pirozynski, 1976).

The genuine lack of definite fossil evidence is attributable to several factors. (1) The vegetative states of fungi are the same in the fossil record and in recent nature, i.e. they lack characteristic features required for accurate taxonomic determination. (2) The sexual states required for reliable taxonomic identification are so small and short-lived that it is very difficult to find them. (3) Fungi can migrate into deep rock environments a long time after rock formation and spores of recent fungi are ubiquitous in air, water and terrestrial environments, thus creating difficulties in distinguishing recent contaminants from genuine fossils (Stubblefield & Taylor, 1988).

Since there is limited information available as to how found fossil organisms lived or how they reproduced in the past, their true affinities may never be known. In addition, most fungi are not very well preserved in the fossil record, so it has been difficult to interpret the fossil record of fungi, leaving open the possibility of an earlier, unrecorded history (Gray & Shear, 1992; Taylor & Taylor, 1993; Taylor *et al.*, 1994; Hibbett *et al.*, 1995).

According to estimates deduced from a molecular clock calibrated on the basis of the fossil record, fungi have existed for over 900 Ma

(Blackwell, 2000; Berbee & Taylor, 2001; Heckman *et al.*, 2001). Genetic analyses of a gene coding for the ribosomal subunit of modern fungi have placed their origin at just 600 million years ago (mya). Attempts to match molecular data concerning fungal phylogeny to the geological record generally agree, but also indicate conflict between the two types of data (Berbee & Taylor, 1993).

History of fossil fungi

The fossil record of fungi was summarized by Tiffney and Barghoorn (1974) and Pirozynski (1976), followed by updates given by Stubblefield and Taylor (1988), Waggoner (1994) and Brown (2004). Based on the available fossil record and estimations of molecular data, fungi are presumed to have been present in the late Proterozoic (900–570 mya). However, there is no good fossil record of fungi from the late Pre-Cambrian, nor from the beginning of the Phanerozoic. The good fossil record starts in the early Devonian, possibly in the Ordovician. A fossilized cyanolichen (*Winfrenatia reticulata*) was found by Taylor *et al.* (1997) in the early Devonian (408–360 mya) Rhynie chert of Aberdeenshire, Scotland. Additionally, fossil hyphae in association with wood decay and several chytridiomycetes and various other unknown fungi and arbuscular mycorrhizal representatives (Glomeromycota) associated with plants originate nearly exclusively from this single site (Kidston & Lang, 1921; Pirozynski & Dalpé, 1989; Hass & Remy, 1992; Simon *et al.*, 1993; Hass *et al.*, 1994; Remy *et al.*, 1994; Taylor *et al.*, 1992, 1994, 1995; Redecker *et al.*, 2000). The tiny, predominantly aquatic chytrids are the most common form of fungi found in the Rhynie chert. Nevertheless, the earliest record of an arbuscular mycorrhizal fungus, in association with vascular plants, was found in the Guttenberg Formation, mid-Ordovician (460 and 455 mya) dolomite of Wisconsin, USA, which was presumably deposited in a shallow marine setting (Redecker *et al.*, 2002). Arbuscular mycorrhizal fungi are thought to be the oldest group of asexual multi-cellular organisms (Smith & Read, 1997). Fungal fossil diversity increased throughout the Paleozoic Era (Taylor & Taylor, 1993, 1997) with all modern classes being reported in the Pennsylvanian (320–286 mya). More recently, the earliest ascomycetes, evidenced by perfectly preserved perithecia and asci of *Pyrenomycetes*, have been described from the Lower Devonian (400 mya) Rhynie chert by Taylor *et al.* (1999). Possibly earlier ascomycetes were reported from the Silurian of Sweden (Sherwood-Pike & Gray, 1985). Fossilized slime moulds occurring in Baltic amber (Tertiary, 58–36 mya) were described by Domke (1952) and by Dörfelt *et al.* (2003).

Some younger myxomycete spores from the Oligocene (36–24 mya) and Pleistocene (1.8–0.01 mya) were found by Graham (1971) and Tiffney and Barghoorn (1974), and within Dominican amber (40–25 mya) by Waggoner and Poinar (1992). However, the oldest known fossilized fungal spores were found in amber dating back 225 Ma (Poinar *et al.*, 1993). As for the mushrooms, the first evidence of basidiomycetes in the fossil record dates from the middle Pennsylvanian (Dennis, 1970). The earliest unambiguous basidiocarps in the fossil record are of various agaricoid basidiomata in amber of the mid Cretaceous (Turonian) and Eocene (Poinar & Singer, 1990; Hibbett *et al.*, 1997; Hibbett *et al.*, 2003). Recently, a Cretaceous basidiocarp fragment was described by Smith *et al.* (2004); this provides an unequivocal earlier minimum age estimate of the Basidiomycota based on fruiting bodies and extends the known paleogeographical distribution of the phylum. In fact, all four major groups of modern fungi have been found in Devonian strata.

Using a combination of the fossil record and molecular phylogeny, Van der Auwera and De Wachter (1996) have selected the *Chytridiomycota* as the oldest and most basal group of fungi. Berbee and Taylor (2001), on the other hand, have indicated some fastidious parasitic fungi and commensals of the most common invertebrates (*Trichomycetes* and *Entomorphorales*) affiliated to the *Zygomycetes* as being the most basal group of fungi, together with some free-living flagellated chytrids as a sister taxon.

Fossil fungi of the sub-sea-floor biosphere

Fungi are, besides cyanobacteria, algae and sponges, well known as endolithic or boring microorganisms (Golubic *et al.*, 1975; Vogel *et al.*, 1987). Since the late nineteenth century boring activities have been noted on the part of fungi (Bornet & Flahaut, 1889) and of lichens (Mellor, 1923; Fry, 1927). Microborers attack carbonate by a process of biochemical dissolution (Hutchings, 1986; Ehrlich, 1990). Some promote rock weathering by mobilizing mineral constituents using excreted inorganic acids, organic acids, or ligands. An overview of the acid production of fungal species occurring on rocks is given by Sterflinger (2000). Other fungi promote rock weathering by redox attack of mineral constituents such as iron and manganese (Ehrlich, 1998). These activities result in an intricate network of fine, often elongated borings of typically 1–20 μm in diameter (Golubic *et al.*, 1975). This boring activity represents a major destructive process affecting sediment preservation. There is fossil evidence of a history of boring fungi extending as far back as to the Upper Devonian, approximately 370 mya (Kobluk & Risk, 1974). Hard substrates attacked by these microborers include wood, dentine, bone, fish scales, integument of arthropods, calcareous shells and limestone (Bromley, 1970). Zebrowski (1936) described fungal organisms in shell fragments of molluscs, foraminifera and ostracods, as well as in spicules of calcareous sponges, and dated these fossil fungi to the Cambrian period (580–500 mya). According to Miliman *et al.* (1972), filamentous fungal borings represent the most abundant endoliths found in the south-eastern Atlantic continental margin, being contained in sediments, some deposited during the Holocene, but the majority during the Pleistocene. Microboring fungal fossils are abundant in most carbonate-dominated sedimentary environments (Perkins & Halsey, 1971; May & Perkins, 1979; Radtke, 1993). Endolithic organisms displaying various fungal morphologies have occasionally been described from deep-sea locations as deep as 1450 m below sea level (Fremy, 1945; Cavaliere & Alberte, 1970; Zeff & Perkins, 1979). Budd and Perkins (1980) investigated the bathymetric zonation of fossilized microboring organisms in carbonate sediments of the Puerto Rican shelf and slope at depths ranging from intertidal to 530 m. In their study they showed that endolithic fungi are dominant in the aphotic zone where algae are excluded. As a result, they identified distinct endolithic communities dominated by fungi at greater depths (85–530 m). Even at a depth of 450 m, Edwards and Perkins (1974) found that endolithic fungi actively penetrate carbonate substrates such as molluscan, coral and echinoid fragments. Microboring assemblages from tropical environments (e.g. Perkins & Tsentas, 1976; May *et al.*, 1982; Günther, 1990) are distinct from those recognized in more temperate settings, which are typically dominated by fungal borers (Perkins & Halsey, 1971), and from those characterizing non-marine or brackish environments, where fungi and other heterotrophic species dominate (e.g. Radtke, 1993). In tropical environments infestation occurs rapidly, normally within days (Perkins & Tsentas, 1976; Kobluk & Risk, 1974), and microboring is thus considered to be a key taphonomic process.

   In modern reef and reef-related environments, microborers have an important role as agents of sediment destruction (Swinchatt, 1965; Tudhope & Risk, 1985; Vogel, 1993) as well as in the formation of destructive micrite envelopes (Bathurst, 1966). As a result, endolithic fungi are believed to have been playing a significant role in the production and destruction of carbonate sediments over long periods of geological time.

   Besides carbonate, fungi are also easily able to degrade aluminosilicates and silicates, and as in the case of carbonate, the most important mechanisms are the production of organic and inorganic acids, alkalis and complexing agents (Rossi & Ehrlich, 1990). Oxalate in particular, a strong solubilizing agent of silicate minerals, is mainly produced by fungi. In laboratory experiments, Mehta *et al.* (1979) were able to simulate the degradation of basalt with *Penicillium simplicissimum.*

Recently, the first fossil fungi have been found in the submarine, deep basaltic earth crust (Schumann *et al.*, 2004). The fossils were detected in thin sections obtained from drilled basaltic cores collected during Ocean Drilling Program, Leg 200, in the North Pacific Ocean (Stephen *et al.*, 2003). Unique filamentous fossilized fungi were observed in the carbonate-filled vesicles of a massive tholeiitic lava flow from the upper oceanic crust at a depth of 51 m below sea floor, beneath sediment and magmatic rocks, and under an overlying water column of about 5000 m (Fig. 16.1). These

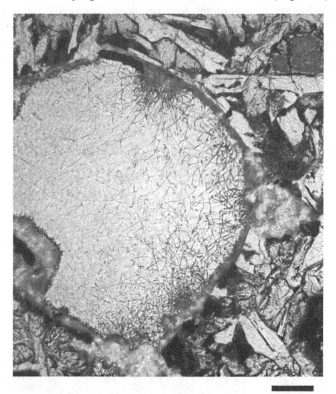

Fig. 16.1. Photomicrograph of a petrographic thin section from North Pacific tholeiitic lava obtained at Leg 200 of the Ocean Drilling Program (ODP) from a depth of 51 m below the sea floor (approximately 21 m below the sediment–basement contact layer) showing a $CaCO_3$-filled vesicle in massive tholeiitic lava flow. The vesicle is filled with mineralized filamentous fungal hyphae. Scale bar represents 100 µm.

particular basaltic rocks never reached subaerial environments, but have continuously been confined to the deep sea. Both light and emission scanning electron microscopy detected the presence of characteristic fungal structures, i.e. septa and central pores that suggest affiliation with higher fungi (Schumann *et al.*, 2004). The vesicles within the basalt are 0.3–3 mm in diameter, and the filling matrix is calcite and/or aragonite. They probably developed as gas pipes during magma cooling, resulting in the building of a complex, three-dimensional pore channel system. The basalt was determined to be 46 Ma old by means of marine magnetic anomalies (Stephen *et al.*, 2003) and ^{40}Ar/^{39}Ar analysis (Acton, personal communication). Based on calcareous nannofossil data (Stephen *et al.*, 2003), most or all of the approximately 30 m of clay-rich sediment overlying the basement is thought to have been deposited in the Eocene (Lutetian) when the site was still relatively near the ridge axis. The fungal thallus consists of hyphae that branch in all three dimensions with a cross-sectional dimension of about 5–10 µm (Fig. 16.2). As the preservation of the mycelia is extraordinary, the cell septa of the hyphae and even central

Fig. 16.2. The net of fungal hyphae extends from the basalt–carbonate boundary into the inner space (confer Fig. 16.1). The cross-sectional dimension of these hyphae is about 5–10 µm. Scale bar represents 20 µm.

pores in the septa are clearly visible. The mycelium completely fills the whole pore space, passing through several carbonate crystals originating from different individual seed crystals, evidencing the endolithic character of these fungi. Within some pores more than one hyphal morphology could be detected, differing in terms of size of hyphal cross-section. The additional presence of pyrite inclusions up to several hundred micrometres in size in the carbonate filling indicates anaerobic conditions during final mineralization in this habitat.

A completely different situation was found in fractures and veins of the same basaltic samples, where fungi have also been found (Stephen *et al.*, 2003) and the hyphae were again detected within a carbonate matrix. Due to encrustation with iron hydroxide, caused by the contact with seawater indicating anoxic to dysaerobic conditions, the morphological features of the hyphae are not so well preserved (Fig. 16.3).

As soon as it was known what to look for, sub-fossilized fungi also began to be found in the Holocene oceanic basement from the Kolbeinsey Ridge (north of Iceland) from a water depth of nearly 1500 m. This basalt was also riddled with pores, but most were open, some filled with water and some partly filled with sediment or authigenic minerals (organominerals or microbialites). Most of these microbialites are iron hydroxide

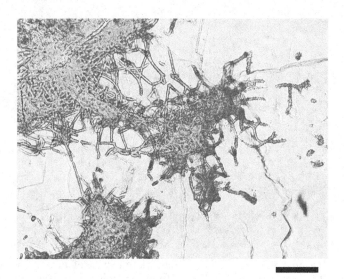

Fig. 16.3. Photomicrograph showing fossilized fungal hyphae in a carbonate-filled vein in the tholeiitic basalt. Note the encrustation of the hyphae with iron hydroxide indicating dysaerobic conditions. Scale bar represents 100 μm.

390   *J. Reitner* et al.

encrustations, manganese oxides, framboid pyrite and microlaminated and filamentous structures. These filamentous structures, which are encrusted mainly by iron minerals and amorphous silica, could also be affiliated with fungi, as shown in Figs. 16.4 & 16.5 (Reitner & Schumann unpublished data). These fungal structures are common in pores partly filled with sediment, where they probably metabolized organic sedimentary remains.

Deep subsurface fossil fungal remains in veins and fractures of continental rocks

Triberg granite and the Krunkelbach Uranium Mine in the Black Forest, south-western Germany In 1985 and 1986 the central Black Forest area of south-western Germany was intensively geologically surveyed to find good locations for very deep boreholes (Gehlen *et al.*, 1986). This survey was part of the German Deep Continental Drilling Programme (KTB: Kontinentales Tiefbohrprogramm). The area in the vicinity of

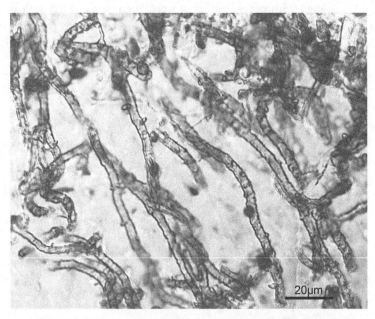

Fig. 16.4. Kolbeinsey Ridge. Network of hyphae of an unknown marine fungus in a pillow basalt pore from the Holocene oceanic crust of the Kolbeinsey Ridge (north of Iceland) from a water depth of nearly 1500 m. The old hyphae are partly covered with iron and manganese oxides. Fungi are very common in pillow basalts and play a central role during rock alteration and the formation of certain microbialites.

Windischeschenbach in the Fichtelgebirge Mountains (northern Bavaria) was finally selected as the site for drilling two deep boreholes into crystalline rocks. Several shallow boreholes to a depth of 300 m were drilled in the Black Forest area and 100% cored. From one borehole drilled to a depth of 250 m in the late-Varscian porphyritic Triberg granite near Moosengrund a detailed vein and fracture survey was carried out by one of the authors (JR) to find fossil remains of a possible deep biosphere. This investigation, based on thin sections, was subject to geochemical analysis via laser ablation coupled to ion charged plasma mass spectroscopy (ICP-MS) for rare earth elements (REE), electron microprobe for main and trace-element distribution, field emission scanning electron microscopy (SEM) for crystal structures, and linked electron dispersive X-ray (EDX) analysis for light elements such as carbon.

The aim was to investigate deep chert-haematite-bearing veins and fractures. The fractures below 100 m are nearly completely cemented (shallower ones are filled with deep ground water and were not investigated in this study). It is possible to distinguish two types of fractures: wide shear fractures several centimetres in diameter filled with breccia of the surrounding host rock cemented by fine-grained haematite and disseminated chert, and fractures 1 mm–1 cm wide with a clear zonation of

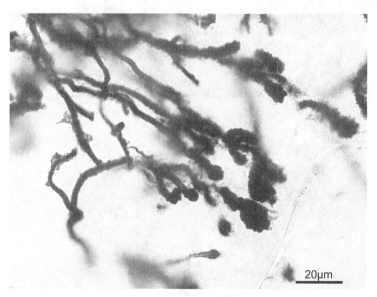

Fig. 16.5. Kolbeinsey Ridge. Terminal ends of an another type of unknown marine fungus from a pillow basalt of the Kolbeinsey Ridge. The thickened ends of the hyphae are probably conidia.

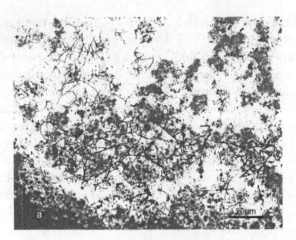

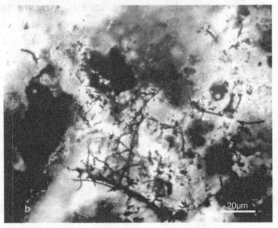

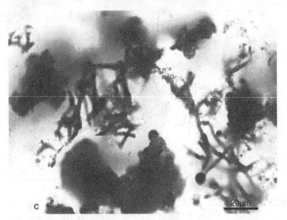

cements. The mineralization phase in the second type of fracture starts with small blue fluorites and is followed by granular quartz with laminated ghost structures and many fluid inclusions. The central areas of the fractures are filled with a very fine-grained chert, probably originating from an opaline silicate. Networks of haematite-covered fibres are present within these cherts. The overall mineral paragenesis originated in low-temperature hydrothermal fluids, which were rarely hotter than 100 °C and maybe cooler (Simon & Hoefs, 1985, 1987; Behr & Gerler, 1987). These particular fracture cements are post-Variscan in age, possibly Mesozoic and/or Tertiary, based on K/Ar, $^{40}Ar/^{39}Ar$, and Rb/Sr isotopic ages (Lippolt *et al.*, 1983; Mertz *et al.*, 1986; Hofmann, 1989). This means that the sedimentary cover of approximately 500 m, based on the post-Variscan sediment packages in the near vicinity, was higher than it is today, especially before the formation of the Tertiary tectonic Rhine-Graben structure. Therefore it is supposed that the formation of fracture fills and cements happened at depths of at least 800–1000 m.

The haematite-rich fractures often bear networks of filaments, which are covered with small haematite crystals. The filaments have diameters of 1–5 μm, and the entire networks extend for several millimetres (Fig. 16.6a, b). However, the filaments do not exhibit any septation, possibly because of diagenesis or various mineralization events. The branching mode of the filaments is very similar to that of fungal mycelia. In nearly all observed cases secondary hyphae branch from a primary hypha (Fig. 16.6b). At approximately 5 μm in diameter the main hyphae are much larger than the secondary ones, which are only approximately 1 μm in diameter. At the end of the filamentous hyphae oval bodies are often present, which are comparable with conidial cells or chlamydospores (Fig. 16.6c). The diameters of the filaments are small and the density of the hyphal bundles (synnemata) is low. It is supposed that the nutrient content of the environment was very low. It could be speculated that chemolithotrophy may have played a role in the growth of the fungi. The mycelia

---

Fig. 16.6. Triberg mycelia. (a) Haematite-covered hyphae of unknown fungal mycelia from a chert fracture of a granite core of approximately 200 m depth, Moosengrund borehole, Triberg Granite. Flocculent structures are fine-grained haematite which sometimes resembles *Metallogenium* colonies. The mycelia originate from the fracture margins and are embedded in a fine-grained chert/chalcedony matrix. The entire mycelia structure resembles the taxon Mucorales. (b) Detail of Fig. (a) exhibiting a thick strand (synnema). (c) The haematite thickening linked with the hyphae are probably remains of chlamydospores.

exhibit some similarities with the *Mucorales* based on thickenings of the primary hyphae, which likely are the large cells characteristic of the *Mucorales*. Generally, many round-shaped structures were observed together with filaments of uncertain origin. The haematite mineralization often causes spherical or oval-shaped structures. Krumbein (1981) described *Metallogenium* colonies, which are probably symbionts of the fungal community. These probable *Metallogenium* colonies were preserved as irregularly shaped spherical to star-shaped bodies with diameters of 20–50 μm, sometimes linked with fungal hyphae (Fig. 16.6a). Comparable structures have been described by Hofmann (1989) in a uranium mine near Krunkelbach (Menzenschwand, Black Forest, Germany). These *Metallogenium* colonies may help the fungi to survive in this extreme environment. Hofmann (1996) described filamentous microbial fossils as well. The filaments described are similar in terms of size and morphology to the fungal hyphae from the Triberg granite. Unfortunately, they do not exhibit any cellular structures. However, the hyphae were covered with uraninite and not with haematite. Uranium mineralization is closely related to these structures, and Hofmann (1989, 1996) supposed that the formation of the uranium mine was caused by the presence and activity of these organisms. The filaments are also embedded in fine crystalline chalcedony (chert) and exhibit a similar low-temperature hydrothermal environment, as seen in the fractures of the Triberg granite. The overall morphology of the Krunkelbach Mine hyphal structure is somewhat different from that of the Triberg granite. Though synnemata structures were not observed, knot-shaped concentrations of hyphal filaments are common. Therefore it is possible to distinguish between primary and secondary hyphae. Many of the hyphae bear small terminal knots comparable with conidia or chlamydospores, which also attests to their fungal origin.

Within the Triberg granite haematite-chert fractures, small spherical to oval bodies with a central opening are very common in some fractures. They are 10–15 μm in diameter and hundreds of them are often enriched in small pockets of the fractures (Fig. 16.7a, b). The cells are a transparent brownish colour and in a few cases exhibit budding structures (Fig. 16.7a). Electron dispersive X-ray analysis has shown that the brownish skin of the cells contains traces of carbon. Iron enrichment of the hyphal network is sometimes common and is probably restricted to mature cells. The size and overall morphological characteristics are typical of yeast cells. Single yeast cells are often fixed on the side walls of small fractures, forming a special type of biofilm (Fig. 16.7b). Beside the yeast cells, smaller rod- and coccoid-shaped cells are present, possibly bacteria (cf. Pedersen *et al.*, 1997a). Ekendahl *et al.*

(2003) isolated small yeast cells from deep ground-water samples obtained from a tunnel at the HRL in Äspö (Sweden) at depths of 201–448 m below sea level. (This observation corresponds well with fossil occurrences, which, however, probably come from even deeper environments.) The yeast cells from the Triberg granite fractures grew under anaerobic conditions, as did the yeasts from the Äspö ground waters. The observed remains of bacterial biofilms may explain the nutrient support for the yeast cells in this probably extremely oligotrophic environment. Fungi prefer mono- or polysaccharides as carbon and energy sources, and the bacterial biofilms would provide such compounds. The main question is the oxygen source. Perhaps the observed yeasts have a strictly anaerobic metabolism, as found in rumen habitats, or are facultative anaerobic yeasts, possibly using oxygen-rich groundwater which would sometimes influence the pore water fluids.

Fossil fungi from iron-stromatolites from deep karst environments near Warstein, north-western Germany The most intriguing fossil fungi remains are known from limestone quarries in the Rhenish Massif in the vicinity of Warstein, north-western Germany (Kretzschmar, 1982). The fungi and occasional bacteria are organized in yellow-brownish stromatolitic layers and form decimetre-thick occurrences. The fungal stromatolites are cemented by fine-grained cherts, and the hyphae are covered with iron hydroxide (goethite). The stromatolites are restricted to deep limestone caves formed in the Mesozoic. The fungal stromatolites are probably of

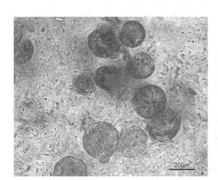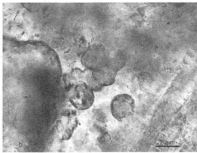

Fig. 16.7. Triberg yeasts. (a) Colony of round and oval-shaped yeast cells enriched in a small pocket in a granite fracture in the vicinity of the haematite-covered mycelia. The yeast cells are preserved in a yellow to orange colour. Electron dispersive X-ray analyses of the cellular wall found an enrichment of carbon. Many yeast cells are completely covered with small haematite crystals and therefore exhibit an opaque optical character. (b) In a few tight fractures the yeast cells are part of microbial biofilms. In such fractures mycelia networks are not present.

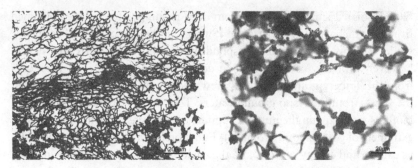

Fig. 16.8. Warstein mycelia. (a) Fossil fungal mycelium in a chert/ferrugenous stromatolite from a Tertiary deep karst environment from a limestone quarry of Warstein (Rhenish Massif, NW Germany). The stromatolite structure is caused by dense portions of hyphae filaments possibly resulting from seasonal ecological variation. The entire mycelium structure exhibits similarities with the fungal taxon *Papulaspora*. (b) Detail of the *Papulaspora* mycelium with conidia or chlamydospores.

Oligocene to Miocene origin, as determined by microthermometry (Kretzschmar, 1982). The chert mineralization and the supply of iron was controlled by low-temperature hydrothermal fluids. In the immediate vicinity of the fungal stromatolites, haematitic veins are present and support the assumption of hydrothermal influence.

Stromatolitic layers are formed by thick fungal mycelia with main and secondary hyphae (Fig. 16.8a). The vegetatively formed hyphae often form survival structures, such as sclerotia and chlamydospores, which are very common in the stromatolites. The main hyphae often bear thickened terminal ends, which are probably large cells similar to those found in the *Mucorales*. Synnematid hyphal bundles are common and comprise the architectonic base of the stromatolitic structure. Fructification structures may also be present; however, it is often impossible to decide whether they are conidia or chlamydospores (Fig. 16.8b). Kretzschmar (1982) has identified various fungal taxa, i.e. *Papulospora*, *Preussia* and *Chaetomium*, based on overall morphological characteristics.

The fungal remains in the deep karst environment of Warstein exhibit some similarities with the Black Forest occurrences. However, the nutrient supply in the karst environment was much better than it was in the very deep fractures of the Black Forest granites.

Conclusions

Geomycology is an intriguing new geoscientific field. The activity of fungi in rocks, minerals, and in deep fractures of continental and marine igneous rocks and in sediments is as yet poorly understood. If the subterranean biosphere is real, active and widespread, how important is the presence of deep cryptic life to biogeochemical processes? Only a few papers summarize the current results of this new field (Sterflinger, 2000; Burford *et al.*, 2003).

The most favourable environment for fungi is probably aerobic, rich in organic carbon, liquid or very moist, and of acidic or occasionally alkaline pH. The anaerobic, oligotrophic and slightly alkaline conditions that characterize most subterranean environments diverge greatly from these ideal conditions. However, the environment is stable and allows sites for both attached and unattached growth as well as the opportunity for adaptation. Fungi may take many forms in the subterranean biosphere. Their role may be completely transitory, barely surviving or even forming parasitic or symbiotic relationships with bacteria. The importance of fungi in the deep biosphere may be minor, given their low numbers, unfavourable growing conditions and competition with prokaryotic microorganisms. Nevertheless, various types of fungi have found niches on rock surfaces. For example, some fungi participate in rock weathering and solubilization of silica and phosphate (e.g. Jongmans *et al.*, 1997). Boring into minerals and causing rock dissolution may be the two most obvious biogeochemical activities fungi can have in subterranean environments. Fungal activity within the deep biosphere is a little-understood geological force that will be an important focus of future research.

Acknowlededgements
We thank Professor Dr Helmut Keupp, Freie Universität Berlin, for supporting us with the samples from Warstein.

References

Abe, F., Miura, T., Nagahama, T. *et al.* (2001). Isolation of a highly copper-tolerant yeast, *Cryptococcus* sp. from the Japan Trench and the induction of superoxide dismutase activity by Cu^{2+}. *Biotechnology Letters*, **23**, 2027–34.

Banwart, S., Tullborg, E.-L., Pedersen, K. *et al.* (1996). Organic carbon oxidation induced by largescale shallow water intrusion into a vertical fracture zone at the Äspö Hard Rock Laboratory (Sweden). *Journal of Contaminant Hydrology*, **21**, 115–25.

Bathurst, R. G. C. (1966). Boring algae, micrite envelopes and lithification of molluscan biosparites. *Geological Journal*, **5**, 15–32.

Behr, H. J. & Gerler, J. (1987). Inclusions of sedimentary brines in post-Varscian mineralizations in the Federal Republic of Germany. A study by neutron activation analysis. *Chemical Geology*, **61**, 63–77.

Berbee, M. L. & Taylor, J. W. (1993). Dating the evolutionary radiations of the true fungi. *Canadian Journal of Botany*, **71**, 1114–27.

Berbee, M. L. & Taylor, J. W. (2001). Fungal molecular evolution: gene trees and geologic time. In *The Mycota*, Vol. VII. *Systematics and Evolution. Part B*, ed. D. J. McLaughlin, E. G. McLaughlin & P. A. Lemke. Berlin: Springer-Verlag, pp. 229–45.

Blackwell, M. (2000). Terrestrial life-fungal from the start? *Science*, **289**, 1884–5.

Bornet, E. & Flahaut, C. (1889). Sur quelques plantes vivantes dans le test calcaire des mollusques. *Bulletin de la Société Botanique de France*, **36**, 147–76.

Bromley, R. (1970). Borings as trace fossils and *Entobio cretacea* Portlock, as an example. In *Trace Fossils*, eds. T. P. Crimes & J. C. Harper. Liverpool, UK: Seel House Press, pp. 49–90.

Brown, O. L. M. (2004). Fossil fungi or paleomycology. *Interciencia*, **29**, 94–8.

Budd, D. A. & Perkins, R. D. (1980). Bathymetric zonation and paleoecological significance of microborings in Puerto Rican shelf and slope sediments. *Journal of Sedimentary Petrology*, **50**, 881–904.

Burford, E. P., Kierans, M. & Gadd, G. M. (2003). Geomycology: fungi in mineral substrata. *Mycologist*, **17**, 98–107.

Cavaliere, A. R. & Alberte, R. W. (1970). Fungi in animal shell fragments. *Journal of the Elisha Mitchell Scientific Society*, **86**, 203–6.

Dennis, R. L. (1970). A Middle Pennsylvanian basidiomycete mycelium with clamp connections. *Mycologia*, **62**, 578–84.

Dörfelt, H., Schmidt, A. R., Ullmann, P. & Wunderlich, J. (2003). The oldest fossil myxogastroid slime mould. *Mycological Research*, **107**, 123–6.

Domke, W. (1952). Der erste sichere Fund eines Myxomyceten im Baltischen Bernstein (*Stemonitis splendens* Rost. fa. *succini* fa. nov. foss.). *Mitteilungen aus dem Geologischen Staatsinstitut in Hamburg*, **21**, 154–61.

Edwards, B. D. & Perkins, R. D. (1974). Distribution of microborings within continental margin sediments of southeastern United States. *Journal of Sedimentary Petrology*, **44**, 1122–35.

Ehrlich, H. L. (1990). Microbial formation and degradation of carbonates. In *Geomicrobiology*, ed. H. L. Ehrlich. New York: Marcel Dekker, pp. 157–95.

Ehrlich, H. L. (1998). Geomicrobiology: its significance for geology. *Earth-Science Reviews*, **45**, 45–60.

Ekendahl, S., O'Neill, A. H., Thomsson, E. & Pedersen, K. (2003). Characterisation of yeasts isolated from deep igneous rock aquifers of the fennoscandian shield. *Microbial Ecology*, **46**, 416–28.

Fliermans, C. B. (1989). Microbial life in the terrestrial subsurface of southeastern coastal plain sediments. *Hazardous Waste and Hazardous Materials*, **6**, 155–72.

Fredrickson, J. K. & Onstott, T. C. (1996). Microbes deep inside the earth. *Scientific American*, **8**, 42–7.

Fremy, P. (1945). Contributions à la physiologie des thallophytes marines perforants et cariants les roches calcaires et les coquilles. *Annales de l'Institut Océanographique*, **22**, 107–44.

Fry, E. J. (1927). The mechanical action of crustaceous lichens on substrata of shale, schist, gneiss, limestone and obsidian. *Annals of Botany*, **41**, 437–60.

Gage, J. D. & Tyler, P. A. (1991). *Deep-Sea Biology: A Natural History of Organisms on the Deep Sea Floor*. Cambridge: Cambridge University Press.

Gehlen, K. V., Kleinschmidt, G., Stenger, R., Wilhelm, H. & Wimmenauer, W. (1986). Kontinentales Tiefbohrprogramm der Bundesrepublik Deutschalnd KTB. Ergebnisse der Vorerkundungsarbeiten Lokation Schwarzwald. 2. *KTB-Kolloquium Seeheim/Odenwald*, 1–160.

Ghiorse, W. C. & Balkwill, D. L. (1983). Enumeration and morphological characterization of bacteria indigenous to subsurface sediments. *Developments in Industrial Microbiology*, **24**, 213–24.

Golubic, S., Perkins, R. D. & Lukas, K. J. (1975). Boring micro-organisms and microborings in carbonate substrates. In *The Study of Trace Fossils*, ed. R. W. Frey. Berlin: Springer-Verlag, pp. 229–59.

Gothan, W. & Weyland, H. (1964). *Lehrbuch der Paläobotanik*, 2nd edn. Berlin: Akademie-Verlag.

Graham, A. (1971). The role of *Myxomyceta* spores in palynology (with a brief note on the morphology of certain algal zygospores). *Review of Palaeobotany and Palynology*, **11**, 89–99.

Gray, J. & Shear, W. (1992). Early life on land. *American Scientist*, **80**, 444–56.

Günther, A. (1990). Distribution and bathymetric zonation of shell-boring endoliths in recent reef and shelf environments: Cozumel, Yucatan (Mexico). *Facies*, **22**, 233–62.

Hass, H. & Remy, W. (1992). Devonian fungi. Interactions with the green algae *Palaeonitella*. *Mycologia*, **86**, 901–10.

Hass, H., Taylor, T. N. & Remy, W. (1994). Fungi from the Lower Devonian Rhynie chert: Mycoparasitism. *American Journal of Botany*, **81**, 29–37.

Heckman, D. S., Geiser, D. M., Eidell, B. R. *et al.* (2001). Molecular evidence for the early colonization of land by fungi and plants. *Science*, **293**, 1129–33.

Hibbett, D. S., Grimaldi, D. & Donoghue, M. (1995). Cretaceous mushrooms in amber. *Nature*, **377**, 487.

Hibbett, D. S., Donoghue, M. J. & Tomlinson, P. B. (1997). Is *Phellinites digiustoi* the oldest homobasidiomycete? *American Journal of Botany*, **84**, 1005–11.

Hibbett, D. S., Binder, M. & Wang, Z. (2003). Another fossil agaric from Dominican amber. *Mycologia*, **95**, 685–7.

Hirsch, P. & Rades-Rohkohl, E. (1983). Microbial diversity in a ground water aquifer in northern Germany. *Developments in Industrial Microbiology*, **24**, 183–200.

Hirsch, P., Rades-Rohkohl, E., Kölbel-Boelke, J. & Nehrkorn, A. (1992). Morphological and taxonomic diversity of ground water micro-organisms. In *Progress in Hydrogeochemistry*, ed. G. Matthess, F. H. Frimmel, P. Hirsch, H. D. Schulz & E. Usdowski. Berlin: Springer-Verlag, pp. 311–25.

Hofmann, B. (1989). Genese, Alteration und rezentes Fliess-System der Uranlagerstätte Krunkelbach (Menzenschwand, Schwarzwald). *NAGRA Technischer Bericht*, **88–30**, 1–195.

Hofmann, B. (1996). Earth science collections of the Natural History Museum Bern (NMBE) – a review. *Jahrbuch des Naturhistorischen Museums Bern*, **12**, 115–34.

Hutchings, P. A. (1986). Biological destruction of coral reefs: a review. *Coral Reefs*, **4**, 239–52.

Jongmans, A. G., van Breemen, N., Lundström, U. *et al.* (1997). Rock-eating fungi. *Nature*, **389**, 682–3.

Kato, C. (1999). Molecular analyses of the sediment and isolation of extreme barophiles from the deepest Mariana Trench. In *Extremophiles in Deep-sea Environments*, ed. K. Horikoshi & K. Tsujii. Tokyo: Springer-Verlag, pp. 27–37.

Kidston, R. & Lang, W. H. (1921). On Old Red Sandstone plants showing structure, from the Rhynie Chert bed, Aberdeenshire. Part V. The Thallophyta occurring in the peat-bed; the succession of the plants throughout a vertical section of the bed, and the conditions of accumulation and preservation of the deposit. *Transactions of the Royal Society of Edinburgh*, **52**, 855–902.

Kobluk, D. & Risk, M. (1974). Devonian boring algae or fungi associated with micrite tubules. *Canadian Journal of Earth Sciences*, **11**, 1606–10.

400   *J. Reitner* et al.

Kohlmeyer, J. (1969a). Deterioration of wood by marine fungi in the deep sea. In *Materials Performance and the Deep Sea. American Society for Testing and Materials, Special Technical Publication*, **445**, 20–9.

Kohlmeyer, J. (1969b). The role of marine fungi in the penetration of calcareous substances. *American Zoologist*, **9**, 741–6.

Kohlmeyer, J. (1977). New-genera and species of higher fungi from deep-sea. *Revue de Mycologie*, **41**, 189–206.

Kohlmeyer, J. & Kohlmeyer, E. (1979). *Marine Mycology: The Higher Fungi*. London: Academic Press.

Krauss, G., Sridhar, K. R., Jung, K. *et al.* (2003). Aquatic hyphomycetes in polluted groundwater habitats of central Germany. *Microbial Ecology*, **45**, 329–39.

Kretzschmar, M. (1982). Fossile Pilze in Eisen-Stromatolithen von Warstein (Rheinisches Schiefergebirge). *Facies*, **7**, 237–60.

Krumbein, W. E. (1981). Biogenic rock varnishes of the Negev Desert (Israel). An ecological study of iron and manganese transformation by cyanobacteria and fungi. *Oecologica*, **50**, 25–38.

Lippolt, H. J., Schleicher, H. & Raczek, I. (1983). Rb-Sr systematics of Permian volcanites in the Schwarzwald (SW Germany). Space of time between plutonism and late orogenic volcanism. *Contribution of Mineralogy and Petrology*, **84**, 272–80.

Ludvigsen, L., Albrechtsen, H. J., Ringelberg, D. B., Ekelund, F. & Christensen, T. H. (1999). Distribution and composition of microbial populations in landfill leachate contaminated aquifer (Grindsted, Denmark). *Microbial Ecology*, **37**, 197–207.

McKinley, J. P., Stevens, T. O. & Westall, F. (2000). Microfossils and paleoenvironments in deep subsurface basalt samples. *Geomicrobiology Journal*, **17**, 43–54.

Madsen, E. L. & Ghiorse, W. C. (1993). Groundwater microbiology: subsurface ecosystem processes. In *Aquatic Microbiology*, ed. T E. Ford. London: Blackwell Scientific Publications, pp. 167–213.

May, J. A. & Perkins, R. D. (1979). Endolithic infestation of carbonate substrates below the sediment–water interface. *Journal of Sedimentary Petrology*, **49**, 357–78.

May, J. A., Macintyre, I. G. & Perkins, R. D. (1982). Distribution of microborers within planted substrates along a barrier reef transect, Carrie Bow Cay, Belize. In *Smithsonian Contribution to the Marine Science*, Vol. 12. *The Atlantic Barrier Reef Ecosystem at Carrie Bow Cay, Belize I: Structure and Communities*, eds. K. Rützler & I. G. Macintyre. pp. 93–107.

Mehta, A. P., Torma, A. E. & Murr, L. E. (1979). Effect of environmental parameters on the effciency of biodegradation of basalt rock by fungi. *Biotechnology and Bioengineering*, **21**, 875–85.

Mellor, E. (1923). Lichens and their action on the glass and leadings of church windows. *Nature*, **112**, 299–300.

Mertz, D. F., Lippolt, H. J. & Huck, K.-H. (1986). K-Ar, Ar40/Ar39 and Rb-Sr investigations on the genesis of the Clara vein deposits/Central Schwarzwald. *46. Jahrestagung der Deutschen Geologischen Gesellschaft ILP Karlsruhe*, Abstracts, 235.

Miliman, J. D., Pilkey, O. H. & Ross, D. A. (1972). Sediments of the continental margin off the eastern United States. *Geological Society of America Bulletin*, **83**, 1315–34.

Nagahama, T., Hamamoto, M., Nakase, T., Takaki, Y. & Horikoshi, K. (2003). *Cryptococcus surugaensis* sp. nov., a novel yeast species from sediments collected on the deep-sea floor of Suruga Bay. *International Journal of Systematic and Evolutionary Microbiology*, **53**, 2095–8.

Newberry, C. J., Webster, G., Cragg, B. A. *et al.* (2004). Diversity of prokaryotes and methanogenesis in deep subsurface sediments from the Nankai Trough, Ocean Drilling Program Leg 190. *Environmental Microbiology*, **6**, 274–87.

Ogram, A., Sun, W., Brockman, F. J. & Fredrickson, J. K. (1995). Isolation and characterization of RNA from low-biomass deep-subsurface sediments. *Applied and Environmental Microbiology*, **61**, 763–8.

Palumbo, A. V., Zhang, C. L., Liu, S. *et al.* (1996). Influence of media on measurement of bacterial populations in the subsurface – Numbers and diversity. *Applied Biochemistry and Biotechnology*, **57**, 905–14.

Parkes, R. J., Cragg, B. A., Fry, J. C. *et al.* (1990). Bacterial biomass and activity in deep sediment layers from the Peru Margin. *Philosophical Transactions of the Royal Society of London Series A – Mathematical Physical and Engineering Sciences*, **331**, 139–53.

Parkes, R. J., Cragg, B. A., Bale, S. J. *et al.* (1994). Deep bacterial biosphere in Pacific Ocean sediments. *Nature*, **371**, 410–13.

Parkes, R. J., Cragg, B. A. & Wellsbury, P. (2000). Recent studies on bacterial populations and processes in subsea floor sediments: A review. *Hydrogeology Journal*, **8**, 11–28.

Pedersen, K. (1987). *Preliminary Investigations of Deep Groundwater Microbiology in Swedish Granitic Rock.* SKB Technical report 88–01. Stockholm: Swedish Nuclear Fuel and Waste Management Co., pp. 1–22.

Pedersen, K. (1993). The deep subterranean biosphere. *Earth-Science Reviews*, **34**, 243–60.

Pedersen, K., Arlinger, J., Ekendahl, S. & Hallbeck, L. (1996). 16S rRNA gene diversity of attached and unattached groundwater bacteria along the access tunnel to the Äspö Hard Rock Laboratory, Sweden. *FEMS Microbiology Ecology*, **19**, 249–62.

Pedersen, K., Ekendahl, S., Tullborg, E.-L. *et al.* (1997a). Evidence of ancient life at 207 m depth in a granitic aquifer. *Geology*, **25**, 827–30.

Pedersen, K., Hallbeck, L., Arlinger, J., Erlandson, A.-C. & Jahromi, N. (1997b). Investigation of the potential for microbial contamination of deep granitic aquifers during drilling using 16S rRNA gene sequencing and culturing methods. *Journal of Microbiological Methods*, **30**, 179–92.

Perkins, R. D. & Halsey, S. D. (1971). Geologic significance of microboring fungi and algae in Carolina shelf sediments. *Journal of Sedimentary Petrology*, **41**, 843–53.

Perkins, R. D. & Tsentas, C. I. (1976). Microbial infestation of carbonate substrates planted on the St. Croix shelf, West Indies. *Geological Society of America Bulletin*, **87**, 1615–28.

Pia, J. (1927). Fungi. In *Handbuch der Paläobotanik*, ed. M. Hirmer. München: R. Oldenbourg, pp. 112–130.

Pirozynski, K. A. (1976). Fossil fungi. *Annual Review of Phytopathology*, **14**, 237–46.

Pirozynski, K. A. & Dalpé, Y. (1989). Geological history of the Glomaceae with particular reference to mycorrhizal symbiosis. *Symbiosis*, **7**, 1–36.

Poinar, G. O. & Singer, R. (1990). Upper Eocene gilled mushroom from the Dominican Republic. *Science*, **248**, 1099–101.

Poinar, G. O., Waggoner, B. M. & Bauer, U. C. (1993). Terrestrial soft-bodied protists and other micro-organisms in Triassic amber. *Science*, **259**, 222–4.

Radtke, G. (1993). The distribution of microborings in molluscan shells from recent reef environments at the Stocking Island, Bahamas. *Facies*, **29**, 81–92.

Raghukumar, C. & Raghukumar, S. (1998). Barotolerance of fungi isolated from deep-sea sediments of the Indian Ocean. *Aquatic Microbial Ecology*, **15**, 153–63.

Redecker, D., Kodner, R. & Graham, L. E. (2000). Glomalean fungi from the Ordovician. *Science*, **289**, 1920–1.

Redecker, D., Kodner, R. & Graham, L. E. (2002). *Palaeoglonius grayi* from the Ordovician. *Mycotaxon*, **84**, 33–7.

Remy, W., Taylor, T. N., Hass, H. & Kerp, H. (1994). Four hundred-million-year-old vesicular arbuscular mycorrhizae. *Proceedings of the National Academy of Science of the United States of America*, **91**, 11841–3.

Rossi, G. & Ehrlich, H. L. (1990). Other bioleaching processes. In *Microbial Mineral Recovery*, eds. H. L. Ehrlich & C. L. Brierley. New York: McGraw-Hill, pp. 149–70.

Russell, B. F., Phelps, T. J., Griffin, W. T. & Sargent, K. A. (1992). Procedures for sampling deep subsurface microbial communities in unconsolidated sediments. *Ground Water Monitoring Review*, **12**, 96–104.

Schumann, G., Manz, W., Reitner, J. & Lustrino, M. (2004). Ancient fungal life in North Pacific Eocene oceanic crust. *Geomicrobiology Journal*, **21**, 241–6.

Sherwood-Pike, M. A. & Gray, J. (1985). Silurian fungal remains: probable records of Ascomycetes. *Lethaia*, **18**, 1–20.

Simon, K. & Hoefs, J. (1985). Geochemische Untersuchungen an hydrothermal überprägten Graniten und Gneisen des Südschwarzwaldes. *Forstschritte Mineralogie*, **63**, 253–61.

Simon, K. & Hoefs, J. (1987). Effects of meteoric water interaction on Hercynian granites from the Südschwarzwald, SW Germany. *Chemical Geology*, **61**, 253–61.

Simon, L., Bousquet, J., Lévesque, R. C. & Lalonde, M. (1993). Origin and diversification of endomycorrhizal fungi and coincidence with vascular land plants. *Nature*, **363**, 67–9.

Sinclair, J. L. & Ghiorse, W. C. (1989). Distribution of aerobic bacteria, protozoa, algae, and fungi in deep subsurface sediments. *Geomicrobiology Journal*, **7**, 15–31.

Smith, D. C., Spivack, A. J., Fisk, M. R. *et al.* (2000). Tracer-based estimates of drilling-induced microbial contamination of deep sea crust. *Geomicrobiology Journal*, **17**, 207–19.

Smith, S. E. & Read, D. J. (1997). *Mycorrhizal Symbiosis*, San Diego: Academic Press.

Smith, S. Y., Currah, R. S. & Stockey, R. A. (2004). Cretaceous and Eocene poroid hymenophores from Vancouver Island, British Columbia. *Mycologia*, **96**, 180–6.

Sneath, P. H. A. (1962). Longevity of micro-organisms. *Nature*, **195**, 643–6.

Soltwedel, T. & Schewe, I. (1998). Activity and biomass of the small benthic biota under permanent ice coverage in the central Arctic Ocean. *Polar Biology*, **19**, 52–62.

Stephen, R. A., Kasahara, J., Acton, G. D. & Shipboard Scientific Party. (2003). *Proceedings of the Ocean Drilling Program*, Initial Reports, 200 [CD-ROM]. Available from: Ocean Drilling Program, Texas A&M University, College Station TX 77845–9547, USA.

Sterflinger, K. (2000). Fungi as geologic agents. *Geomicrobiology Journal*, **17**, 97–124.

Stubblefield, S. P. & Taylor, T. N. (1988). Recent advances in Paleomycology. *New Phytologist*, **108**, 3–25.

Swinchatt, J. P. (1965). Significance of constituent composition, texture and skeletal breakdown in some bonate sediments. *Journal of Sedimentary Petrology*, **35**, 71–90.

Takami, H. (1999). Isolation and characterization of micro-organisms from deep-sea mud. In *Extremophiles in Deep-sea Environments*, ed. K. Horikoshi & K. Tsujii. Tokyo: Springer-Verlag, pp. 3–26.

Takami, H., Ioue, A., Fuji, F. & Horikoshi, K. (1997). Microbial flora in the deepest sea mud of the Mariana Trench. *FEMS Microbiology Letters*, **152**, 279–85.

Taylor, N. & Taylor, E. L. (1993). *The Biology and Evolution of Fossil Plants*. Englewood Cliffs, New Jersey: Prentice Hall.

Taylor, N. & Taylor, E. L. (1997). The distribution and interactions of some Paleozoic fungi. *Review of Palaeobotany and Palynology*, **95**, 83–94.

Taylor, T. N., Remy, W. & Hass, H. (1992). Fungi from the lower Devonian Rhynie Chert – Chytridiomycetes. *American Journal of Botany*, **79**, 1233–41.

Taylor, T. N., Remy, W. & Hass, H. (1994). Allomyces in the Devonian. *Nature*, **367**, 601.

Taylor, T. N., Remy, W., Hass, H. & Kerp, H. (1995). Fossil arbuscular mycorrhizae from the early Devonian. *Mycologia*, **87**, 560–73.

Taylor, T. N., Hass, H. & Kerp, H. (1997). A cyanolichen from the Lower Devonian Rhynie chert. *American Journal of Botany*, **84**, 992–1004.

Taylor, T. N., Hass, H. & Kerp, H. (1999). The oldest fossil ascomycetes. *Nature*, **399**, 648.

Thorseth, I. H., Torsvik, T., Torsvik, V., Daae, F. L. & Pedersen, R. B. (2001). Diversity of life in ocean floor basalt. *Earth and Planetary Science Letters*, **194**, 31–7.

Tiffney, B. H. & Barghoorn, E. S. (1974). The fossil record of fungi. *Occasional Papers of the Farlow Herbarium of Cryptogamic Botany Harvard University*, **7**, 1–42.

Torsvik, T., Furnes, H., Muehlenbachs, K., Thorseth, I. H. & Tumyr, O. (1998). Evidence for microbial activity at the glass-alteration interface in oceanic basalts. *Earth and Planetary Science Letters*, **162**, 165–76.

Tudhope, A. W. & Risk, M. J. (1985). Rate of dissolution of carbonate sediments by microboring organisms, Davies Reef, Australia. *Journal of Sedimentary Petrology*, **55**, 440–7.

Van der Auwera, G. & De Wachter, R. (1996). Large-subunit rRNA sequence of the chytridiomycete *Blastocladiella emersonii*, and implications for the evolution of zoosporic fungi. *Journal of Molecular Evolution*, **43**, 476–83.

Van Uden, N. & Fell, J. W. (1968). Marine yeasts. In *Advances in Microbiology of the Sea*, eds. M. Droop & E. J. F. Wood. London: Academic Press, pp. 167–201.

Vogel, K. (1993). Bioerosion in fossil reefs. *Facies*, **28**, 109–14.

Vogel, K., Golubic, S. & Breh, C. E. (1987). Endolithic associations and their relation to facies distribution in the Middle Devonian of New York State, U. S. A. *Lethaia*, **20**, 263–90.

Waggoner, B. M. (1994). Fossil micro-organisms from upper Cretaceous amber of Mississippi. *Review of Palaeobotany and Palynology*, **80**, 75–84.

Waggoner, B. M. & Poinar, G. O. (1992). A fossil myxomycete plasmodium from Eocene-Oligocene amber of the Dominican Republic. *Journal of Protozoology*, **39**, 639–42.

Wellsbury, P., Goodman, K., Barth, T. *et al.* (1997). Deep marine biosphere fuelled by increasing organic matter availability during burial and heating. *Nature*, **388**, 573–6.

Wellsbury, P., Mather, I. & Parkes, R. J. (2002). Geomicrobiology of deep, low organic carbon sediments in the Woodlark Basin, Pacific Ocean. *FEMS Microbiology Ecology*, **42**, 59–70.

Zande, J. M. (1999). An ascomycete commensal on the gills of *Bathynerita naticoidea*, the dominant gastropod at Gulf of Mexico hydrocarbon seeps. *Invertebrate Biology*, **118**, 57–62.

Zebrowski, G. (1936). New genera of Cladochytriaceae. *Annals of Missouri Botanical Garden*, **23**, 553–64.

Zeff, M. L. & Perkins, R. D. (1979). Microbial alteration of Bahamian deep-sea carbonates. *Sedimentology*, **26**, 175–201.

17

The role of fungi in carbon and nitrogen cycles in freshwater ecosystems

VLADISLAV GULIS, KEVIN KUEHN
AND KELLER SUBERKROPP

Introduction

Fungi are adapted to a diverse array of freshwater ecosystems. In streams and rivers, flowing water provides a mechanism for downstream dispersal of fungal propagules. The dominant group of fungi in these habitats, aquatic hyphomycetes, have conidia that are morphologically adapted (tetraradiate and sigmoid) for attachment to their substrates (leaf litter and woody debris from riparian vegetation) in flowing water (Webster, 1959; Webster & Davey, 1984). In freshwater wetlands and lake littoral zones, production of emergent aquatic macrophytes is often extremely high, resulting in an abundance of plant material that eventually enters the detrital pool. The dead shoot material of these macrophytes (leaf blades, leaf sheaths and culms) often remains standing for long periods of time before collapsing to the sediments or water. This plant matter is colonized by fungi that are adapted for surviving the harsh conditions that prevail in the standing-dead environment (Kuehn et al., 1998). There are a number of other freshwater ecosystems where fungi are present and exhibit interesting adaptations, e.g. aero-aquatic fungi in woodland ponds, zoosporic organisms (Chytridiomycota and Oomycota) in a variety of habitats including the pelagic zones of lakes, and Trichomycetes that inhabit the guts of a variety of aquatic insects. Despite the well-known occurrence of these fungal groups in aquatic habitats, virtually nothing is known concerning their roles in biogeochemical processes. Overall, the contributions of fungi to biogeo-chemical cycles have been understudied in most freshwater ecosystems.

Most studies examining fungal participation in biogeochemical cycles in freshwater ecosystems focused on the role of fungi in the decomposition of plant litter. Historically, the lack of appropriate methods to accurately quantify

Fungi in Biogeochemical Cycles, ed. G. M. Gadd. Published by Cambridge University Press. © British Mycological Society 2006.

fungal biomass and rates of biomass production was a major reason for the paucity of knowledge concerning the role of fungi in litter decomposition (Gessner *et al.*, 1997). However, a growing body of evidence has emerged over the last two decades on the usefulness of the fungal sterol, ergosterol, in the quantification of fungal biomass within decaying plant litter and the technique for measuring *in situ* instantaneous growth rates of fungi from rates of [^{14}C]acetate incorporation into ergosterol (Gessner & Newell, 2002 and references therein). Both of these methodological developments are increasingly used today. This has been particularly useful in allowing quantitative assessment of the magnitude of fungal contributions to the cycling of carbon and the flow of energy in freshwater ecosystems. Results of these studies have indicated that fungi are significant decomposers of particulate detritus and play key roles in detrital food webs of freshwater ecosystems.

During plant litter decomposition, fungi are involved in a variety of processes that result in the conversion of plant carbon into fungal biomass and also into CO_2 as a result of their respiratory activities (Gessner *et al.*, 1997). In addition, fungal decomposition of plant litter, as well as feeding activities of detritivore consumers (invertebrates) on this microbially colonized plant litter facilitates in the export of plant carbon as either fine particulate organic matter (FPOM) or dissolved organic matter (DOM). This chapter will examine the role of fungi in carbon and nitrogen cycles during plant litter decomposition in two freshwater ecosystems, streams and wetlands. The use of quantitative methods has led to a greater appreciation of the impact of these organisms on biogeochemical processes within both of these ecosystems.

Freshwater streams

Riparian vegetation shades woodland streams and limits the magnitude of primary production occurring within these ecosystems. This vegetation also contributes the bulk of organic matter inputs to the stream, primarily in the form of leaves and woody debris. Woodland streams have been shown to receive up to 99% of their organic carbon from the riparian vegetation (Fisher & Likens, 1972; Webster & Meyer, 1997). Since woodland streams are dependent on leaf litter as a major source of carbon and energy, considerable attention has focused on the decomposition of this detritus and its links to higher trophic levels (Webster & Benfield, 1986; Suberkropp, 1998b).

Carbon cycle

In temperate streams, leaf resources enter as a pulse during autumn leaf fall (Fig. 17.1a) as temperatures are declining. Once leaf litter enters a stream, it is rapidly colonized by aquatic fungi and bacteria.

Aquatic hyphomycetes are ubiquitous in streams and concentrations of their conidia in the water typically reach annual maxima shortly after peaks of leaf inputs in the autumn (Fig. 17.1b, Iqbal & Webster, 1973; Bärlocher, 2000; Gulis & Suberkropp, 2004). As aquatic hyphomycete hyphae penetrate and grow in leaf litter, they secrete an array of extracellular enzymes (e.g. cellulases, xylanases, pectinases) that digest leaf

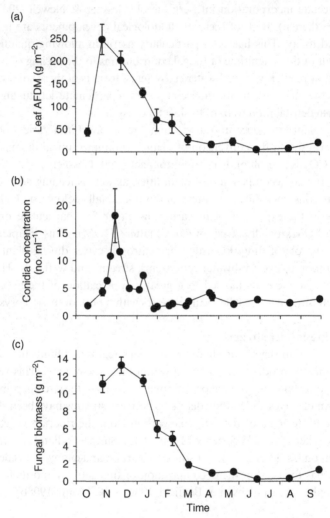

Fig. 17.1. Leaf litter standing crop as ash free dry mass (AFDM) per unit of stream bottom area (a), conidia concentration of aquatic hyphomycetes in water (b) and fungal biomass associated with leaves as dry mass (DM) per unit of stream bottom area (c) in a headwater southern Appalachian stream. Data from Suberkropp (1997). Symbols indicate means ±1 SE (n = 3–10).

polysaccharides, thereby allowing organic carbon to be assimilated by fungi (Suberkropp *et al.*, 1983; Chamier, 1985; Shearer, 1992). Fungal growth and enzymatic digestion also cause softening or maceration of leaf tissue (Suberkropp & Klug, 1980; Chamier & Dixon, 1982), contributing to the production and release of fine particulate organic carbon in streams.

Fungal biomass During the decomposition of leaves, fungal biomass typically increases to a maximum and then stabilizes or declines as conidia are released and hyphae senesce (Gessner & Chauvet, 1994). Fungal biomass can account for as much as 18%–23% of the total mass of detritus (Gessner & Chauvet, 1994; Methvin & Suberkropp, 2003). Fungal biomass associated with leaf detritus expressed on an areal basis reaches annual maximum values after leaf litter enters a stream and becomes colonized (Fig. 17.1c). Maximum fungal biomass of up to $27\,\mathrm{g\,m^{-2}}$ of stream bed has been found associated with naturally occurring leaves (Suberkropp, unpublished results). Fungal biomass associated with decaying leaves enclosed in litter bags typically reaches values that are one to two orders of magnitude greater than bacterial biomass found on the same leaves (Table 17.1). Likewise, microbial biomass inhabiting naturally occurring leaves in streams is also dominated by fungi (Table 17.1).

Table 17.1. *Fungal and bacterial biomass associated with decomposing leaves in streams. All values are maximum biomass estimates from litter bag decomposition studies except the study by Findlay* et al. *(2002b) where average microbial biomass from randomly collected leaves was estimated*

Leaf species	Biomass as percent of detritus		Reference
	Fungal	Bacterial	
Platanus hybrida	4.8	0.5	Baldy *et al.* (1995)
Populus nigra	9.9	0.3	Baldy *et al.* (1995)
Salix alba	7.8	0.3	Baldy *et al.* (1995)
Liriodendron tulipifera	14.0	0.08	Weyers & Suberkropp (1996)
Populus nigra	8.0	0.06	Baldy *et al.* (2002)
Alnus glutinosa	7.7	0.4	Hieber & Gessner (2002)
Salix fragilis	7.0	0.3	Hieber & Gessner (2002)
Acer rubrum	17.4	0.6	Gulis & Suberkropp (2003c)
Rhododendron maximum	10.8	0.2	Gulis & Suberkropp (2003c)
Leaf detritus from 9 streams	1.5	0.1	Findlay *et al.* (2002b)

Factors such as chemical characteristics of the leaf species (Gessner & Chauvet, 1994) and concentrations of nutrients (N and P) in the water (Grattan & Suberkropp, 2001; Niyogi *et al.*, 2003) can affect the amount of fungal biomass associated with leaves as they decompose. The strong correlation between exponential decay rates of different leaf species and measures of fungal biomass and activity (e.g. maximum ergosterol concentrations, net mycelial production and sporulation rates of aquatic hyphomycetes) indicate that fungi play a key role in regulating the decomposition of leaves (Gessner & Chauvet, 1994). Significant correlations of ergosterol concentrations with mass loss of wood veneers (Fig. 17.2) and decay rates of grass leaves (Niyogi *et al.*, 2003) suggest that this conclusion holds true for various types of plant detritus decomposing in streams.

After colonization and growth, one major fate of fungal biomass includes the production of conidia that are carried downstream. Conidia may provide an inoculum for leaves entering the stream, may be captured by filter-feeding invertebrates or may undergo decomposition by bacteria. Fungal biomass within leaf litter also serves as an important food resource for detritivores that consume leaf detritus. Microbially colonized detritus

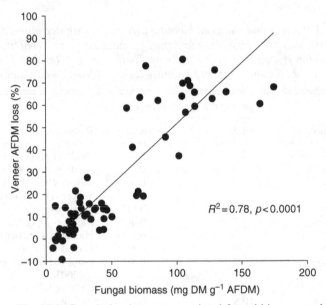

Fig. 17.2. Correlation between associated fungal biomass and mass loss of wood veneers decomposing in headwater streams. Data from Gulis *et al.* (2004).

is a more palatable and nutritious food source than dead leaves for invertebrate detritivores (Bärlocher, 1985; Suberkropp, 1992). Microbial growth and enzymatic decay of leaf polymers increases the nutritional quality of leaf detritus since microbial biomass is more digestible for detritivore consumers than recalcitrant leaf tissue (i.e. lignocellulose). In addition, microbial enzymes can partially degrade leaf polymers making them more digestible and microbial enzymes may remain active in the digestive tract of some detritivores (Bärlocher, 1985). Since the biomass of fungi associated with leaf detritus is much higher than that of bacteria, fungal biomass should provide a larger portion of the nutrition in the diets of invertebrate detritivores than bacterial biomass. Consequently, fungi occupy a central position in the trophic structure of stream food webs and mediate the cycling of carbon and flow of energy to higher trophic levels. However, fungal species that decompose leaf litter cannot be placed in a single black box within the carbon cycle, since species are not identical in their food quality for detritivore consumers. Fungal species are known to differ in their overall palatability (Bärlocher & Kendrick, 1973a; Suberkropp *et al.*, 1983) and nutritional value (Bärlocher & Kendrick, 1973b; Arsuffi & Suberkropp, 1986, Graça *et al.*, 1993) for these animals.

Fungal production Recently, production of fungi associated with decomposing plant material has been estimated by determining rates of incorporation of radiolabelled acetate into ergosterol. This method, introduced by Newell and Fallon (1991) for fungi decomposing *Spartina* litter in salt marshes, has been applied to the fungi decomposing leaves in freshwater streams (Suberkropp & Weyers, 1996; Gessner & Chauvet, 1997). Determination of fungal production allows estimation of the rate at which carbon from decomposing plant litter is converted into fungal biomass. This technique is particularly useful when losses in fungal biomass from leaf detritus are occurring (e.g. to sporulation, detritivore consumption, hyphal senescence and death) and biomass accumulation does not give a good indication of total fungal production.

Fungal production is typically higher than bacterial production associated with decomposing leaves. For example, maximum fungal production associated with decomposing *Liriodendron tulipifera* leaves was $7\,\text{mg}\,\text{g}^{-1}\,\text{d}^{-1}$ compared to $0.3\,\text{mg}\,\text{g}^{-1}\,\text{d}^{-1}$ for maximum bacterial production (Weyers & Suberkropp, 1996). Similarly, fungal production associated with *Populus nigra* leaves in a large river reached maximum values of $1.3–1.4\,\text{mg}\,\text{g}^{-1}\,\text{d}^{-1}$ whereas bacterial production achieved a maximum of $0.4 \, \text{mg g}^{-1} \text{d}^{-1}$ (Baldy *et al.*, 2002). Only when green, non-senescent *Alnus glutinosa* leaves were placed in a stream during the summer were bacterial rates of production ($1.2 \, \text{mg g}^{-1} \text{d}^{-1}$) slightly higher than those of fungi ($1 \, \text{mg g}^{-1} \text{d}^{-1}$) even though fungal biomass still accounted for 95%–99% of the total microbial biomass (Baldy & Gessner, 1997).

Although annual fungal production associated with naturally occurring leaves has only been estimated in a limited number of streams, it exhibits a wide range (Table 17.2). The amount of leaf litter present throughout an annual cycle appears to be an important factor controlling fungal production in a stream as annual fungal production is significantly correlated with the annual mean leaf standing crop (Fig. 17.3). The mean standing crop of leaf detritus is a function of both the input of leaf litter and the retentiveness of the stream. Most of the streams that have been examined exhibit relatively low retention of leaf litter, since winter storms (January to February) generally wash the bulk of autumn leaf litter from the stream (Suberkropp, 1997; Methvin & Suberkropp, 2003; Carter & Suberkropp, 2004). However, in streams that retain leaf detritus very efficiently (e.g. Coweeta 53, Table 17.2), the mean standing crop of leaf detritus is high and annual fungal production on areal basis is correspondingly high. Typically most fungal production per m^2 occurs in the autumn and winter when the greatest amount of leaf detritus is present in the stream even though temperatures are at annual minima (Suberkropp, 1997; Methvin & Suberkropp, 2003). During the summer, when temperatures are higher, there is generally little leaf detritus remaining in streams. This may account for the relatively long turnover times calculated for fungi based on annual production to biomass ratios (18–44 days, Table 17.2).

Rates of fungal production can be used to estimate the fraction of leaf detritus that is assimilated by fungi (for mycelial biomass, sporulation, respiration). Net production efficiencies (production/production + respiration) determined for two aquatic hyphomycete species growing on leaf litter in microcosms ranged from 24% to 46% (Suberkropp, 1991). A third species exhibited production efficiencies of 32% and 60% depending on the nutrient concentrations in the water, and production efficiencies tended to decrease when bacteria were present (Gulis & Suberkropp, 2003b). Using the lowest and highest values together with estimates of leaf litter inputs to the streams, the percentage of leaf input to streams that is assimilated by fungi can be calculated (Table 17.2). In most cases the percentage of leaf inputs assimilated by fungi is significant. For streams in which leaf detritus is washed out by January–February (the first five streams in Table 17.2), fungi are estimated to assimilate only 5%–40%

Table 17.2. *Annual production, production to biomass (P/B) ratios and turnover times of fungi with annual leaf litter inputs and the percentage of the leaf input that was assimilated by fungi in different streams*

Stream	Annual fungal production ($g\,m^{-2}$)	Annual P/B	Turnover time (d)	Annual leaf input ($g\,m^{-2}$)	% of leaf input assimilated	Reference
Payne Creek	16 ± 6	11.5	32	492 ± 19	5–14	Carter & Suberkropp (2004)
Hendrick Mill Branch	27 ± 9	20.8	18	422 ± 23	11–27	Methvin & Suberkropp (2003)
Basin Creek	32 ± 11	13.3	27	379 ± 26	14–35	Methvin & Suberkropp (2003)
Walker Branch	37 ± 8	8.2	44	460 (estimate)	13–33	Suberkropp (1997)
Lindsay Spring Branch	46 ± 25	13.0	28	478 ± 37	16–40	Carter & Suberkropp (2004)
Coweeta 53	193 ± 54	15.6	23	617–826	39–97	Suberkropp et al. (unpublished)

of leaf litter inputs. In contrast, in streams that retain leaf detritus more efficiently (e.g. Coweeta 53), fungi may assimilate as much as 39% to 97% of leaf litter inputs.

Respiration In studies examining leaf decomposition using litter bags or leaf packs, 17%–56% of the carbon loss from leaves has been found to be released as CO_2 by the microbial assemblages (Elwood *et al.*, 1981; Baldy & Gessner, 1997; Gulis & Suberkropp, 2003c). Both the whole-stream nutrient enrichment with nitrogen and/or phosphorus (Gulis & Suberkropp, 2003c; Ramírez *et al.*, 2003) and microcosm nutrient additions (Gulis & Suberkropp, 2003a, b) caused significant increases in the amount of leaf carbon lost through microbial respiration in comparison with detritus decomposing at ambient nutrient levels. In microcosms, fungi have been found to convert similar amounts of organic carbon into CO_2 to that observed for leaves colonized in streams, i.e. 14%–48% of the leaf carbon that is lost (Suberkropp, 1991; Gulis & Suberkropp, 2003b). Lower estimates have been reported for naturally decomposing leaves in two Alabama streams (Carter & Suberkropp 2004) where respiration by

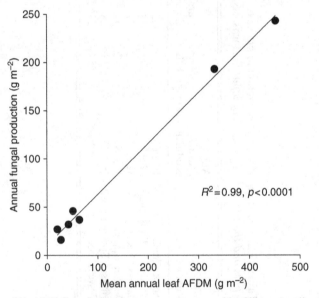

Fig. 17.3. Correlation between mean annual leaf litter standing stock and mean annual fungal production in streams. Based on data from Suberkropp (1997), Methvin and Suberkropp (2003), Carter and Suberkropp (2004) and Suberkropp *et al.* (unpublished).

microbial communities accounted for 7%–13% of the leaf litter input. However, both of these streams were not retentive and most of the leaf litter was thought to have been washed downstream before it decomposed.

Microbial respiration also accounts for a considerable mass loss of decomposing wood in streams (7%–44%, Collier & Smith, 2003). Respiration rates reported from submerged wood are generally lower than those from leaf litter if calculated per unit of mass (or volume) of detritus (Tank *et al.*, 1993; Fuss & Smock, 1996) because of lower surface to volume ratio of sticks versus leaves and fungal activity restricted mostly to the outer layers of wood. However, if expressed on a surface area basis, respiration rates from submerged wood are higher than those from leaves (Tank *et al.*, 1993; Fuss & Smock, 1996). Respiration rates from submerged decomposing thin wood veneers are comparable to those from leaf litter (Simon & Benfield, 2001; Stelzer *et al.*, 2003) and together with relatively high fungal biomass estimates (Simon & Benfield, 2001; Stelzer *et al.*, 2003; Gulis *et al.*, 2004) suggest that fungi are important participants in the decomposition of submerged wood.

Nitrogen cycle
Fungi have lower C/N ratios than the substrates they colonize; consequently they should either retain substrate nitrogen more efficiently than carbon or acquire nitrogen from exogenous sources. Aquatic fungi are capable of utilizing nitrogen from both organic substrates and the overlying water (Suberkropp, 1995). These fungi are often nitrogen (and phosphorus) limited due to relatively low nitrogen concentrations of submerged substrates (leaf litter and especially wood) and also of water. The relative importance of each nitrogen source depends on organic substrate qualities and water chemistry.

Nitrogen in water Concentrations of dissolved inorganic nutrients (e.g. N and P) vary dramatically across aquatic ecosystems and, along with other factors, determine the level of fungal activity. The concentration of dissolved inorganic nitrogen in stream water (ammonium, nitrite and nitrate) depends on watershed characteristics, such as bedrock and soil chemical properties, land use, stream hydrology and biotic activity. Nitrate predominates in well-oxygenated waters while ammonium (and nitrite) concentrations can be high in anoxic waters, often as a result of human activities. Most aquatic hyphomycetes (and presumably ascomycetes) can use both organic and inorganic nitrogen (Thornton, 1963, 1965), while some chytrids and oomycetes are unable to utilize nitrate. Ammonium can be

assimilated directly while all other nitrogen sources should be first transformed into ammonium either with nitrate and nitrite reductases or through
deamination.

There is a wealth of evidence that aquatic fungi obtain substantial
amounts of their nitrogen from the water column. Fungal biomass accrual
and sporulation of aquatic hyphomycetes are stimulated by dissolved
nitrogen in both laboratory (Suberkropp, 1998a; Sridhar & Bärlocher,
2000; Bärlocher & Corkum, 2003) and field experiments (Fig. 17.4;
Suberkropp & Chauvet, 1995; Tank & Dodds, 2003). Leaf litter nitrate
uptake rates are correlated with decomposition rates, leaf toughness and
respiration (Quinn *et al.*, 2000), which are functions of microbial activity.
Increases in absolute nitrogen content of leaf litter have been reported in
many studies (Hynes & Kaushik, 1969; Kaushik & Hynes, 1971;
Suberkropp *et al.*, 1976; Triska & Sedell, 1976; Chauvet, 1987; Molinero
et al., 1996), suggesting that nitrogen is actively taken up from the water
column by litter-inhabiting microorganisms. Selective antibiotic experiments (Kaushik & Hynes, 1971) and laboratory studies with pure cultures
growing on submerged wood (Gunasekera *et al.*, 1983) indicate that fungi
are responsible for the majority of this uptake. Enrichment of stream water

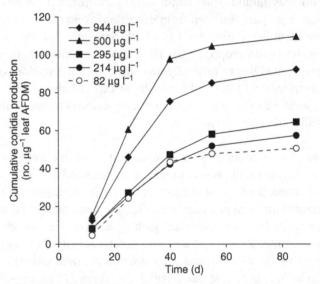

Fig. 17.4. The effect of whole-stream nitrate addition on sporulation of
aquatic hyphomycetes on oak leaves. Average stream water NO_3-N
concentrations at leaf bag stations during the study period are given
(Ferreira, Gulis & Graça unpublished).

with [^{15}N]ammonium and subsequent comparison of δ^{15}N values for bulk detritus and microbial nitrogen associated with decaying leaves and wood indicate microbial nitrogen immobilization (Mulholland *et al.*, 2000; Tank *et al.*, 2000; Sanzone *et al.*, 2001). Since fungi contribute 95% to over 99% of total microbial (fungi plus bacteria) biomass associated with decomposing leaf litter in streams (Baldy *et al.*, 1995; Hieber & Gessner, 2002; Gulis & Suberkropp, 2003c) it is likely that fungi, not bacteria, are largely responsible for the observed nitrogen immobilization.

On an ecosystem scale, Qualls (1984) estimated that 25% of dissolved inorganic nitrogen inflow can be immobilized by leaf litter in a blackwater swamp stream within a 1 km reach. Hamilton *et al.* (2001) found that 29% of stream water ammonium was taken up by heterotrophs associated with organic detritus. Estimates of the actively cycling fraction of nitrogen in leaves and small wood during ^{15}N enrichments are almost identical to the measured microbial nitrogen fraction (Tank *et al.*, 2000; Hamilton *et al.*, 2001). Because of fungal dominance on coarse particulate organic matter in streams, these nitrogen transformations were likely to have been controlled by fungal activities.

Ample supply of nitrogen from stream water often leads to increases in microbial respiration (Stelzer *et al.*, 2003) and leaf litter decomposition (Fig. 17.5; Suberkropp & Chauvet, 1995; Huryn *et al.*, 2002) that generally correlates with measures of fungal activity (Fig. 17.2, Gessner & Chauvet, 1994; Niyogi *et al.*, 2003). This stimulation of microbial carbon utilization demonstrates a tight coupling of carbon and nitrogen cycles in freshwater ecosystems. Consequently, eutrophication due to human activity may alter both nitrogen and carbon cycling within these habitats.

Nitrogen in organic substrates Initial nitrogen content was first suggested to affect microbially mediated leaf litter decomposition in water (Kaushik & Hynes, 1971). Among leaves of common riparian trees, alder leaves have a high nitrogen concentration because of nitrogen fixation by tree symbionts, decompose quickly and typically support high fungal biomass and conidia production of aquatic hyphomycetes soon after submergence (Gessner & Chauvet, 1994). However, it is apparent that leaf breakdown is also affected by fibre content, chemical inhibitors (phenolics) and physical barriers (Webster & Benfield, 1986). Gessner and Chauvet (1994) found that leaf litter decomposition rate, maximum fungal biomass, mycelial production and sporulation rate of aquatic hyphomycetes were negatively correlated with leaf litter initial lignin content but were not affected by nitrogen concentration. It appears that breakdown and fungal activity depend on complex interplay of factors, e.g. lignin/nitrogen ratio as suggested by Melillo *et al.* (1982) for leaf litter and also external factors such as stream water nutrient concentrations (Gessner & Suberkropp, unpublished). A similar relationship was suggested for wood decomposing in streams (Melillo *et al.*, 1983). Decomposition rate of wood was also linked to the activity of enzymes involved in nutrient sequestration (Sinsabaugh *et al.*, 1993), which depends on nitrogen and phosphorus availability.

The nitrogen concentrations of autumn-shed leaves of deciduous trees vary between 0.5%–3% averaging around 1% (e.g. Gessner & Chauvet, 1994; Ostrofsky, 1997) while the nitrogen content of fungal mycelium is about 3%–10% (e.g. Paul & Clark, 1989; Högberg & Högberg, 2002). Since fungi can contribute up to 18%–23% of total detrital mass of submerged decaying leaves (Suberkropp, 1995; Methvin & Suberkropp, 2003), it is not surprising that leaf litter nitrogen concentration typically increases during decomposition and frequently doubles (e.g. Kaushik & Hynes, 1971; Suberkropp *et al.*, 1976). Increases in leaf litter nitrogen

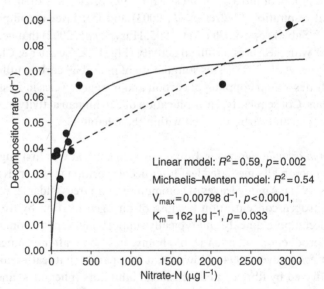

Fig. 17.5. Decomposition rates of alder leaf litter in streams differed in nitrate concentration in water. Both linear and Michaelis–Menten model $V = (V_{max} \cdot [S])/(K_m + [S])$, where (here) V_{max} is the maximum decomposition rate, K_m is the nitrate concentration at which half rate of decomposition is achieved, [S] is nitrate concentration, gave fairly good fit (Gulis, Ferreira & Graça, unpublished).

content correlate well with fungal biomass accrual (Fig. 17.6). Rier *et al.* (2002) reported that fungal biomass was negatively correlated with detrital C/N ratio (i.e. inverse of %N) in a study comparing decomposition of leaf litter grown under ambient and CO_2-enriched atmosphere and hence having different initial C/N ratios. Increases in nitrogen content during decomposition in streams also occur in wood (Sinsabaugh *et al.*, 1993; Gulis *et al.*, 2004). Overall, nitrogen increases are thought to coincide with increases in protein concentration due to fungal growth, thereby rendering plant litter a more palatable and nutritious food source for invertebrates (Kaushik & Hynes, 1971; Bärlocher, 1985; Suberkropp, 1992) and facilitating transfer of nutrients to higher trophic levels.

Dissimilatory nitrogen transformations Nitrification and denitrification are important processes of the global nitrogen cycle. Traditionally, bacteria are thought to be responsible for these transformations (including in aquatic ecosystems; Allan, 1995). However, nitrification has been reported for a wide range of terrestrial fungi (e.g. Falih & Wainwright, 1995) and fungi were confirmed to be capable of denitrification, reducing nitrate or nitrite to nitrous oxide or ultimately N_2 (e.g. Shoun *et al.*, 1992). Furthermore, fungi were found to dominate both microbial nitrification

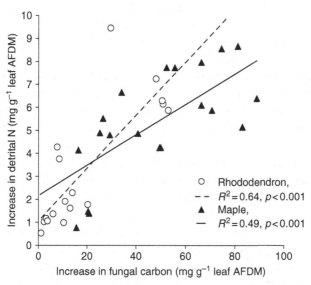

Fig. 17.6. The relationship between increases in fungal biomass and leaf litter nitrogen content of two species of leaves decomposing in a headwater stream. Based on data from Gulis and Suberkropp (2003c).

and denitrification in soil (Laughlin & Stevens, 2002). The potential for nitrification or denitrification has not been demonstrated specifically in aquatic fungi, but it is reasonable to expect that some species may be capable of denitrification, especially since it has been reported for *Fusarium*, *Cylindrocarpon* and *Nectria* (Hypocreales, Ascomycota) (Shoun *et al.*, 1992) and these genera are commonly found in aquatic environments. Low oxygen concentrations in benthic sediments are a prerequisite for denitrification by aquatic fungi. Recently, Guest and Smith (2002) proposed using fungi for biological nitrogen reduction of wastewater, because of their greater denitrification rate and other advantages over bacteria.

Stoichiometric perspective Even though carbon and nitrogen cycles in aquatic ecosystems are coupled, different stoichiometric ratios of water, detritus and consumers lead to different incorporation ratios of carbon and nitrogen (Frost *et al.*, 2002). Initial atomic C/N ratios were estimated to be around 20–96 for leaves entering aquatic ecosystems (calculated from %N in Melillo *et al.* (1982) and Gessner & Chauvet (1994); assumed carbon content of 50%) and 140–1100 for wood (Melillo *et al.*, 1984; Stelzer *et al.*, 2003). Carbon/nitrogen ratios have been estimated at 4.5–22 for terrestrial fungi (Paul & Clark, 1989; Högberg & Högberg, 2002; Wallander *et al.*, 2003, 2004) and 8–13 for stream fungi associated with leaf litter in two temperate streams (Sanzone *et al.*, 2001). Despite a certain plasticity of C/N ratios in microbes, considerable disparity between the elemental composition of the substrate and fungal mycelium would be likely to affect the growth rate of fungi. Thus, substrates with similar elemental composition to fungal biomass (i.e. lower C/N ratio or higher %N, see above) would support higher fungal activity. Elemental imbalance between the substrate and fungi can also be compensated by nitrogen uptake from the water column. Consequently, it is not surprising that the response of fungi (biomass, respiration and hence decomposition rate) to experimental nitrogen and phosphorus whole-stream enrichments was much higher for wood having an extremely high C/N ratio rather than for leaf litter (Stelzer *et al.*, 2003; Gulis *et al.*, 2004).

Spiralling In stream ecology, nutrient cycling is best described through the nutrient spiralling concept (Newbold *et al.*, 1981) that takes into account the unidirectional downstream movement of water in streams and rivers. It implies that an atom (e.g. nitrogen) is transported some distance downstream in an inorganic form before being incorporated into biota (which can also be transported downstream) and eventually released in inorganic form. With respect to aquatic fungi, nitrogen assimilated at a given stream location can be transported downstream as mycelia growing within a substrate or as fungal propagules where fungal nitrogen can be transferred to higher trophic levels through invertebrate feeding, remain within the fungal compartment through mycelial growth or released as inorganic nitrogen. Notably, aquatic hyphomycetes can convert up to 80% of fungal production (or 8%–12% of leaf mass loss) into conidia (Suberkropp, 1991) that are transported downstream. Aquatic hyphomycetes also contribute to maceration of leaf litter (Suberkropp & Klug, 1980) and generation of dissolved and fine particulate organic matter and, consequently, facilitate in the transport of nutrients downstream. On the other hand, fungal growth on high C/N substrates and nitrogen immobilization from the overlying water column may affect dissolved nitrogen uptake length and nitrogen recycling.

Freshwater Wetlands

Wetlands have been described as among the most biologically diverse and productive ecosystems on earth. These ecosystems represent a unique transition between the terrestrial and aquatic environment, have extensive food webs and provide a variety of habitats for a wide diversity of flora and fauna (Mitsch & Gosselink, 2000). A distinguishing feature of many freshwater wetlands and lake littoral zones is the presence of emergent vascular plants, such as *Phragmites, Typha* and *Juncus*. These plants frequently constitute a major fraction of the organic matter produced in wetlands, with estimates of annual net above-ground primary production often exceeding $2 \, kg \, m^{-2} \, yr^{-1}$ (Wetzel & Howe, 1999; Mitsch & Gosselink, 2000).

Very little of the living plant biomass is consumed by herbivores during the growing season (Dvorák & Imhof, 1998), since much of the carbon within macrophyte tissues resides in the structural material of cell walls (i.e. lignocellulose), which is not readily digested and assimilated by animal consumers (Mann, 1988; Kreeger & Newell, 2000). As a consequence, most of the plant matter produced eventually enters the detrital pool following plant senescence and death, where it is transformed by microbial decomposers (bacteria and fungi) and detritivore consumers (invertebrates). Microbial assemblages, particularly fungi, associated with decaying wetland plant litter are recognized as an important food source for detritus feeding consumers (Newell & Bärlocher, 1993; Bärlocher & Newell, 1994; Graça *et al.*, 2000; Newell & Porter, 2000; Silliman & Newell, 2003), thereby serving as important intermediaries in the flow of both carbon and nutrients to higher trophic levels. Similar to the stream systems described above, many characteristics related to nutrient cycling and energy flow in wetland ecosystems are regulated by the metabolic activities of microbial assemblages associated with this decaying plant matter.

Despite the overwhelming importance of plant detritus in wetlands, plant decomposition and the role of associated microorganisms, particularly fungi, is a seldom appreciated or investigated component of wetland processes (Newell, 1993). Previous studies of emergent plant decomposition in wetlands have often focused on litter that had been prematurely harvested and placed at or buried within the surface sediments (e.g. Moran *et al.*, 1989; Thormann *et al.*, 2001). As a result, litter decomposition was thought to be (1) primarily restricted to the sediment–water interface and (2) predominantly mediated by sediment-associated bacterial assemblages (Moran *et al.*, 1988). However, in many emergent macrophytes the collapse of above-ground plant matter and incorporation of litter into the sediments does not typically occur following senescence and death of the plant shoot (Newell, 1993). Much of the dead plant litter remains attached to the parent plant in an aerial standing-dead position for extended periods of time, where it undergoes initial microbial decay prior to its entry into the aquatic environment (Newell, 1993; Newell *et al.*, 1995; Bärlocher & Biddiscombe, 1996; Kuehn & Suberkropp, 1998a; Kuehn *et al.*, 1999; Gessner, 2001; Findlay *et al.*, 2002a; Welsch & Yavitt, 2003).

Fungi extensively colonize and reproduce on and within standing and submerged litter of emergent macrophytes. For example, over 600 species of fungi have now been recorded from *Phragmites australis* (e.g. Gessner & van Ryckegem, 2002). However, despite the wealth of qualitative evidence showing extensive fungal colonization, few investigators have attempted to quantify the role of fungi in litter decay in freshwater wetlands (Findlay *et al.*, 1990, 2002a; Newell *et al.*, 1995; Kuehn & Suberkropp, 1998a; Kuehn *et al.*, 1999, 2000; Gessner, 2001; Newell, 2003; Welsch & Yavitt, 2003) or their overall contribution to total ecosystem metabolism (Kuehn & Suberkropp, 1998b; Kuehn *et al.*, 2004).

Carbon cycle

Fungal biomass and production Recent studies conducted in subtropical and temperate freshwater wetlands provide compelling evidence that fungal participation in the aerial standing-dead litter phase can contribute significantly to overall carbon and nutrient cycling in wetlands (Kuehn & Suberkropp, 1998a, b; Kuehn *et al.*, 1999). Kuehn and Suberkropp (1998a) reported that living biomass of fungi associated with standing leaf litter of the rush, *Juncus effusus*, accounts for *c.* 5% of the total detrital mass. Once established, a relatively constant level of living fungal biomass was maintained for over 800 days and over 30 species of fungi were identified from standing *J. effusus* litter. Seasonal estimates of above-ground detrital mass of *J. effusus* at the wetland study site (Wetzel & Howe, 1999) ranged from 0.6 to 2.3 kg AFDM m^{-2} (Fig. 17.7a). When integrated on an areal basis, fungal biomass associated with standing *J. effusus* litter ranged from 24 to 116 g m^{-2} (Fig. 17.7b) emphasizing the quantitative significance of fungal decomposers in this ecosystem. Likewise, when viewed on a whole ecosystem scale, mean annual living fungal biomass within this relatively small wetland (15 ha, *J. effusus* 64.8% coverage) equals 2–11 tons.

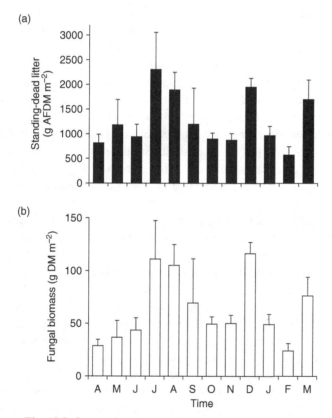

Fig. 17.7. Seasonal estimates of standing-dead litter of *J. effusus* (a) and litter-associated fungal biomass (b) at the Talladega Wetland Ecosystem, Alabama. Data from Wetzel and Howe (1999) and Kuehn and Suberkropp (1998a). Vertical lines indicate +1 SE (n = 6).

In addition to accumulating large quantities of biomass in decaying standing litter, fungal decomposers also exhibit high rates of biomass production. Newell *et al.* (1995) reported that microbial biomass and production associated with naturally standing and fallen litter of the fresh-water sedge, *Carex walteriana*, were dominated by fungal decomposers, with bacterial biomass and production often accounting for less than 0.5% that of fungi. Rates of fungal biomass production associated with stand-ing-dead litter of *J. effusus* accounted for >94% of the total microbial production, averaging $42 \mu g \, C \, g^{-1}$ AFDM h^{-1} (Kuehn *et al.*, 2000). Microbial biomass and rates of biomass production were also dominated by fungi (>99% of the total microbial production) in both standing and submerged litter of *Typha angustifolia* and *P. australis* in a tidal freshwater wetland of the Hudson River (Findlay *et al.*, 2002a).

The collapse of emergent plant litter into the water often leads to fungal succession and distinct changes in the biomass and activity of associated fungi. Kuehn *et al.* (2000) reported that fungal and bacterial biomass and production decrease rapidly following submergence of *J. effusus* litter, suggesting that the resident microbiota associated with decaying standing litter could not adapt to or survive the abrupt changes in condi-tions from an aerial to an aquatic environment. The initial decline is typically followed by an increase in fungal biomass and production during later stages of submerged decomposition. Similar changes in the activities of microbial assemblages were observed following submergence of standing-dead *P. australis* litter in a temperate lake littoral wetland (Komínková *et al.*, 2000).

After submergence of *J. effusus* litter and decreases in litter-associated microbial biomass, fungal decomposers continued to comprise the major microbial assemblage on/within decaying litter (Kuehn *et al.*, 2000). Estimates of fungal biomass and production greatly exceeded correspond-ing estimates of bacterial biomass and production throughout submerged litter decay. The comparison of the contribution of fungi and bacteria to carbon loss of *J. effusus* litter under submerged conditions (Table 17.3) indicates that fungi could explain a substantial portion (68%) of the litter mass loss observed. Similar findings of fungal dominance in other fresh-water wetlands (Findlay *et al.*, 1990, 2002a; Newell *et al.*, 1995; Sinsabaugh & Findlay, 1995) suggest that fungi are an important microbial assemblage involved in submerged macrophyte decay. These findings contrast sharply with previous studies that have reported a more predominant role of bacteria in wetland carbon cycling (Benner *et al.*, 1986; Moran *et al.*, 1988; Buesing, 2002).

Table 17.3. *Net production, production to biomass (P/B) ratio, turnover time and contribution of fungal and bacterial decomposers to submerged* J. effusus *litter decay (from Kuehn et al., 2000)*

Parameter	Fungi	Bacteria
Total net production (mg C g^{-1} initial leaf C)	44	3
Mean biomass (mg C g^{-1} initial leaf C)	16	0.2
P/B ratio	2.8	15
Turnover (d)	68	13
% of initial leaf C assimilated	13	2
% contribution to overall leaf C loss	68	11

Respiration Fungi associated with standing and submerged plant litter convert a considerable portion of the plant carbon into CO_2 as a result of their respiration. The microbiota, particularly fungi, associated with standing-dead wetland plant litter are well adapted to the fluctuating environmental conditions of the aerial habitat (Kuehn & Suberkropp, 1998b; Kuehn *et al.*, 1998, 1999; Kuehn *et al.*, 2004). Rates of microbial respiration (CO_2 evolution) from standing-dead litter exhibit pronounced diel periodicity (Fig. 17.8) and are positively correlated with night-time increases in plant litter water potentials (Kuehn & Suberkropp, 1998b; Kuehn *et al.*, 2004). Temperature-driven increases in relative humidity and subsequent dew formation is the primary mechanism underlying night-time increases in water availability and microbial activities within standing-dead litter. In contrast, respiratory activities virtually cease during the day as a result of increased desiccation stress.

Large differences can occur among plant litter types (species and organ) in terms of microbial colonization and metabolic response of micro-organisms to water availability. Microbial respiration rates associated with different *P. australis* litter fractions vary considerably (Kuehn *et al.*, 2004). Maximum respiration rates from leaf blades were higher (24%–42%) than those from sheath litter under the same conditions (Fig. 17.8a) while maximum respiration rates from culm litter were consistently an order of magnitude lower than rates from both leaf and sheath litter. The observed differences in rates of respiration among *P. australis* litter fractions were consistent with differences in litter water absorption patterns, known structural characteristics among litter fractions (e.g. lignocellulose) and degree of fungal colonization (Fig. 17.9, Kuehn *et al.*, 2004). Microbial respiration correlates well with litter-associated fungal biomass (ergosterol) (Fig. 17.9), providing convincing evidence that ergosterol is a good indicator of living

424   *V. Gulis* et al.

fungal mass (but see Mille-Lindblom *et al.*, 2004) and that fungi are likely to be responsible for most of the respiratory carbon release from standing litter in wetland habitats.

A rough budget also suggests that a considerable portion of plant carbon is likely to be converted into CO_2 under standing-dead conditions (Table 17.4). Only a small portion of the CO_2 flux from standing litter is due to wetting via precipitation, with most being accounted for by recurring night-time dew formation. In contrast to *P. australis* leaf and sheath litter, very little culm material is degraded under standing-dead conditions, perhaps because culms of *Phragmites* contain more recalcitrant and water-repellent tissues (Rodewald-Rudescu, 1974) and have low concentrations of fungal biomass. Hence, culm material appears to undergo more extensive microbial decay once shoots have collapsed to the sediments or water.

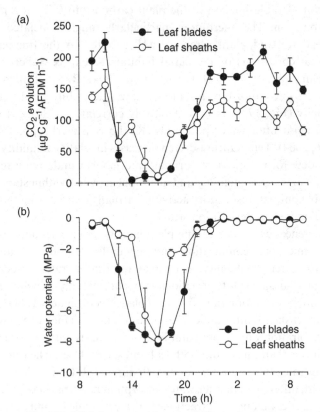

Fig. 17.8. Diel changes in (a) rates of CO_2 evolution from standing-dead leaf and sheath litter of *P. australis* and (b) plant litter water potential during field studies conducted in a littoral reed stand of Lake Hallwil, Switzerland. Data from Kuehn *et al.* (2004). Symbols indicate means ±1 SE (n = 3).

How common is the diel pattern in fungal respiration and what are the potential implications for wetland carbon cycling on an ecosystem scale? Remarkably similar diel patterns in respiration have been reported from standing-dead *J. effusus* litter in Alabama, USA (Kuehn & Suberkropp, 1998b) and *P. australis* litter from a temperate littoral reed stand in Switzerland (Kuehn *et al.*, 2004), indicating that this phenomenon is not restricted to subtropical regions, but is common even at northern latitudes where most wetlands occur on a global scale. When integrated on an areal and temporal basis, diel fluctuations in microbial respiration rates are a potentially significant source of CO_2 from wetlands and may represent a pathway of carbon flow that has gone largely unnoticed in prior chamber-based estimates and models of total wetland CO_2 flux (Yavitt, 1997; Updegraff *et al.*, 2001; Chimner *et al.*, 2002; Larmola *et al.*, 2003). Kuehn and Suberkropp (1998b) estimated daily fluxes of 1.37 to 3.35 $g\,C\,m^{-2}\,d^{-1}$ from microbial (fungal) assemblages inhabiting standing *J. effusus*, which were equal to or exceeded sediment CO_2 flux rates from the same wetland site (0.12 to 2.43 $g\,C\,m^{-2}\,d^{-1}$; Roden & Wetzel, 1996). Carbon dioxide flux rates reported for microbial assemblages inhabiting standing *P. australis* litter (Kuehn *et al.*, 2004) were lower

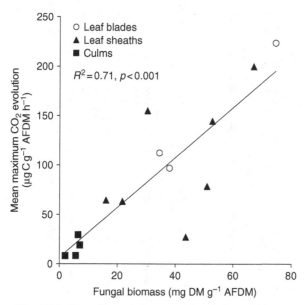

Fig. 17.9. The effect of litter-associated fungal biomass on the mean maximum rate of CO_2 evolution from standing-dead *P. australis* leaf litter. Data from Kuehn *et al.* (2004).

Table 17.4. *Estimate of the annual above-ground production of J. effusus and P. australis that is lost as CO_2 due to microbial, primarily fungal, respiration while in the standing-dead litter phase. Modified from Kuehn et al. (2004). Data for J. effusus were taken from Kuehn and Suberkropp (1998b)*

Site	Plant	Plant organ	Plant production (g C m^{-2} yr^{-1})	Microbial respiration (g C m^{-2} yr^{-1})	Lost due to microbial respiration (%)	
					Total	During rainfall
Talladega, Alabama	J. effusus	Leaves	2086*	579	27.7	0.8
Neuchatel, Switzerland	P. australis	Leaves	272	23	8.5	1.0
		Sheaths	191	57	29.8	5.7
		Culms	631	13	2.0	0.4
Hallwil, Switzerland	P. australis	Leaves	193	15	7.8	1.2
		Sheaths	88	25	28.0	3.9
		Culms	308	11	3.6	0.6

*Estimates of standing-dead biomass for *J. effusus* were taken from Wetzel and Howe (1999).

$(0.05-0.57\,\mathrm{g\,C\,m^{-2}\,d^{-1}})$ but still within the range of CO_2 flux estimates reported for wetland sediments in northern temperate climates (e.g. Scanlon & Moore, 2000).

Under submerged conditions, rates of microbial respiration associated with macrophyte litter are similar to maximum rates reported above for standing-dead litter under water-saturating conditions. Komínková *et al.* (2000) reported respiration rates from submerged *P. australis* leaves ranging from 29 to $127\,\mu\mathrm{g}\,CO_2\text{-}C\,\mathrm{g}^{-1}$ AFDM h^{-1}, with fluctuations in rates following changes in lake water temperature. In addition, as observed in standing litter, significant differences in fungal biomass and rates of respiration were also noted between *P. australis* leaf and culm litter, with submerged culm litter having considerably lower fungal biomass and rates of respiration than corresponding leaf litter. These comparisons with respiration rates from standing litter reveal that microbial respiration associated with submerged emergent macrophyte litter can also be a sizeable contribution. A partial decomposition budget of submerged *P. australis* leaves (Komínková *et al.*, 2000), indicated that *c.* 65% of detrital carbon loss could be accounted for by microbial respiration. It is likely that a sizeable portion of this respiratory carbon loss was due to the metabolic activities of fungal decomposers, since total litter-associated microbial biomass was predominantly fungal (>90%).

Nitrogen cycle

Research examining the quantitative contribution of microorganisms to total detrital nitrogen pools in freshwater wetlands is limited. Previous studies have suggested that litter-inhabiting microbial assemblages (bacteria and fungi) account for only a minor fraction of the total nitrogen content in decaying plant matter (Mann, 1988). However, these conclusions were often based on methods that underestimated microbial biomass within decaying plant litter, particularly fungal. Studies by Newell (1993 and references therein), using improved methods for estimating fungal biomass (i.e. immunoassay and ergosterol), were among the first to recognize that fungi could account for a substantial portion of the total plant litter nitrogen. Fungal assemblages (mainly ascomycetes) associated with standing-dead litter of the saltmarsh cordgrass, *Spartina alterniflora*, immobilized nearly all (99%) of the available litter nitrogen (assuming 4% N in fungal biomass, see Newell & Statzell-Tallman, 1982). Similar studies in freshwater wetlands reveal that fungal contributions to total detrital nitrogen can also be significant (Newell *et al.*, 1995; Kuehn & Suberkropp, 1998a; Kuehn *et al.*, 2000; Findlay *et al.*, 2002a); however, estimates are considerably lower than those obtained from saltmarsh systems.

Newell *et al.* (1995) reported that living fungal biomass accounted for 6%–35% of the total nitrogen in decaying standing and fallen litter of the freshwater sedge *C. walteriana*. Similar findings were reported for standing and submerged litter of *J. effusus* (Kuehn & Suberkropp, 1998a; Kuehn *et al.*, 2000), *T. angustifolia* and *P. australis* (Findlay *et al.*, 2002a), with fungal nitrogen accounting for as much as 40% of the total litter nitrogen content. However, note that in these studies, fungal contribution to total detrital nitrogen reflected only estimates of nitrogen within living fungal biomass and thus may be an underestimate. Residual non-living fungal biomass within decaying litter (e.g. N-acetylglucosamine) may account for a substantial portion of the total detrital nitrogen. Total fungal biomass (living plus dead hyphae) may be two-fold greater than living fungal biomass alone (Newell & Porter, 2000).

Conclusions

In freshwater ecosystems where fungi have been studied, particularly in plant litter decomposition subsystems, they have been found to play key roles both in the cycling of carbon and nitrogen and in mediating energy flow to higher trophic levels. Fungi undoubtedly have a number of other roles in these ecosystems, as pathogens, symbionts and decomposers of other types of organic matter. As methods are developed, adapted and applied to other subsystems of aquatic ecosystems and particularly when fungal activities become amenable to quantification, the roles played by fungi in aquatic ecosystems will become better understood.

Acknowledgements

Financial support of our research on aquatic fungal ecology from the US National Science Foundation and through the EU 5th Framework Programme, project 'RivFunction' is gratefully acknowledged.

References

Allan, J. D. (1995). *Stream ecology: Structure and Function of Running Waters*. London: Chapman & Hall.
Arsuffi, T. L. & Suberkropp, K. (1986). Growth of two stream caddisflies (Trichoptera) on leaves colonized by different fungal species. *Journal of the North American Benthological Society*, **5**, 297–305.

Baldy, V. & Gessner, M. O. (1997). Towards a budget of leaf litter decomposition in a first-order woodland stream. *Comptes Rendus de l'Academie des Sciences – Series III – Sciences de la Vie*, **320**, 747–58.

Baldy, V., Gessner, M. O. & Chauvet, E. (1995). Bacteria, fungi and the breakdown of leaf litter in a large river. *Oikos*, **74**, 93–102.

Baldy, V., Chauvet, E., Charcosset, J. Y. & Gessner, M. O. (2002). Microbial dynamics associated with leaves decomposing in the mainstem and floodplain pond of a large river. *Aquatic Microbial Ecology*, **28**, 25–36.

Bärlocher, F. (1985). The role of fungi in the nutrition of stream invertebrates. *Botanical Journal of the Linnaean Society*, **91**, 83–94.

Bärlocher, F. (2000). Water-borne conidia of aquatic hyphomycetes: seasonal and yearly patterns in Catamaran Brook, New Brunswick, Canada. *Canadian Journal of Botany*, **78**, 157–67.

Bärlocher, F. & Biddiscombe, N. R. (1996). Geratology and decomposition of *Typha latifolia* and *Lythrum salicaria* in a freshwater marsh. *Archiv für Hydrobiologie*, **136**, 309–25.

Bärlocher, F. & Corkum, M. (2003). Nutrient enrichment overwhelms diversity effects in leaf decomposition by stream fungi. *Oikos*, **101**, 247–52.

Bärlocher, F. & Kendrick, B. (1973a). Fungi and food preferences of *Gammarus pseudolimnaeus*. *Archiv für Hydrobiologie*, **72**, 501–16.

Bärlocher, F. & Kendrick, B. (1973b). Fungi in the diet of *Gammarus pseudolimnaeus* (Amphipoda). *Oikos*, **24**, 295–300.

Bärlocher, F. & Newell, S. Y. (1994). Growth of the saltmarsh periwinkle *Littoraria irrorata* on fungal and cordgrass diets. *Marine Biology*, **118**, 109–14.

Benner, R., Moran, M. A. & Hodson, R. E. (1986). Biochemical cycling of lignocellulosic carbon in marine and freshwater ecosystems: relative contributions of prokaryotes and eukaryotes. *Limnology and Oceanography*, **21**, 89–100.

Buesing, N. (2002). Microbial productivity and organic matter flow in a littoral reed stand. Unpublished Ph.D. thesis, Swiss Federal Institute of Technology, Zurich, Switzerland.

Carter, M. D. & Suberkropp, K. (2004). Respiration and annual fungal production associated with decomposing leaf litter in two streams. *Freshwater Biology*, **49**, 1112–22.

Chamier, A. C. (1985). Cell-wall-degrading enzymes of aquatic hyphomycetes: a review. *Botanical Journal of the Linnaean Society*, **91**, 67–81.

Chamier, A. C. & Dixon, P. A. (1982). Pectinases in leaf degradation by aquatic hyphomycetes: the enzymes and leaf maceration. *Journal of General Microbiology*, **128**, 2469–83.

Chauvet, E. (1987). Changes in the chemical composition of alder, poplar and willow leaves during decomposition in a river. *Hydrobiologia*, **148**, 35–44.

Chimner, R. A., Cooper, D. J. & Parton, W. J. (2002). Modelling carbon accumulation in Rocky Mountain fens. *Wetlands*, **22**, 100–10.

Collier, K. J. & Smith, B. J. (2003). Corrigendum to 'Role of wood in pumice-bed streams II: breakdown and colonization' [For. Ecol. Manage. 177 (2003) 261–276]. *Forest Ecology and Management*, **181**, 375–90.

Dvorák, J. & Imhof, G. (1998). The role of animals and animal communities in wetlands. In *The Production Ecology of Wetlands*, ed. D. F. Westlake, J. Kvêt & A. Szczepanski. Cambridge: Cambridge University Press, pp. 211–318.

Elwood, J. W., Newbold, J. D., Trimble, A. F. & Stark, R. W. (1981). The limiting role of phosphorus in a woodland stream ecosystem: effects of P enrichment on leaf decomposition and primary producers. *Ecology*, **62**, 146–58.

Falih, A. M. K. & Wainwright, M. (1995). Nitrification in-vitro by a range of filamentous fungi and yeasts. *Letters in Applied Microbiology*, **21**, 18–19.

Findlay, S., Howe, K. & Austin, H. K. (1990). Comparison of detritus dynamics in two tidal freshwater wetlands. *Ecology*, **71**, 288–95.

Findlay, S. E. G., Dye, S. & Kuehn, K. A. (2002a). Microbial growth and nitrogen retention in litter of *Phragmites australis* compared to *Typha angustifolia*. *Wetlands*, **22**, 616–25.

Findlay, S., Tank, J., Dye, S. *et al.* (2002b). A cross-system comparison of bacterial and fungal biomass in detritus pools of headwater streams. *Microbial Ecology*, **43**, 55–66.

Fisher, S. G. & Likens, G. E. (1972). Stream ecosystem: organic energy budget. *Bioscience*, **22**, 33–5.

Frost, P. C., Stelzer, R. S., Lamberti, G. A. & Elser, J. J. (2002). Ecological stoichiometry of trophic interactions in the benthos: understanding the role of C:N:P ratios in lentic and lotic habitats. *Journal of the North American Benthological Society*, **21**, 515–28.

Fuss, C. L. & Smock, L. A. (1996). Spatial and temporal variation of microbial respiration rates in a blackwater stream. *Freshwater Biology*, **36**, 339–49.

Gessner, M. O. (2001). Mass loss, fungal colonisation and nutrient dynamics of *Phragmites australis* leaves during senescence and early decay in a standing position. *Aquatic Botany*, **69**, 325–39.

Gessner, M. O. & Chauvet, E. (1994). Importance of stream microfungi in controlling breakdown rates of leaf litter. *Ecology*, **75**, 1807–17.

Gessner, M. O. & Chauvet, E. (1997). Growth and production of aquatic hyphomycetes in decomposing leaf litter. *Limnology and Oceanography*, **42**, 496–595.

Gessner, M. O. & Newell, S. Y. (2002). Biomass, growth rate, and production of filamentous fungi in plant litter. In *Manual of Environmental Microbiology*, 2nd edn, ed. C. J. Hurst, R. L. Crawford, G. R. Knudsen, M. J. McInerney & L. D. Stetzenbach. Washington, DC: ASM Press, pp. 390–408.

Gessner, M. O. & van Ryckegem, G. (2002). Water fungi as decomposers in freshwater ecosystems. In *Encyclopedia of Environmental Microbiology*, ed. G. Bitton. New York: Wiley & Sons, pp. 3353–64.

Gessner, M. O., Suberkropp, K. & Chauvet, E. (1997). Decomposition of plant litter by fungi in marine and freshwater ecosystems. In *The Mycota*, Vol. IV. *Environmental and Microbial Relationships*, ed. D. T. Wicklow & B. E. Söderström. Berlin: Springer-Verlag, pp. 303–22.

Graça, M. A. S., Maltby, L. & Calow, P. (1993). Importance of fungi in the diet of *Gammarus pulex* and *Asellus aquaticus* II. Effects on growth, reproduction and physiology. *Oecologia*, **96**, 304–9.

Graça, M. A., Newell, S. Y. & Kneib, R. T. (2000). Grazing rates of organic matter and living fungal biomass of decaying *Spartina alterniflora* by three species of salt-marsh invertebrates. *Marine Biology*, **136**, 281–9.

Grattan, R. M. & Suberkropp, K. (2001). Effects of nutrient enrichment on yellow poplar leaf decomposition and fungal activity in streams. *Journal of the North American Benthological Society*, **20**, 33–43.

Guest, R. K. & Smith, D. W. (2002). A potential new role for fungi in wastewater MBR biological nitrogen reduction system. *Journal of Environmental Engineering and Science*, **1**, 433–7.

Gulis, V. & Suberkropp, K. (2003a). Effect of inorganic nutrients on relative contributions of fungi and bacteria to carbon flow from submerged decomposing leaf litter. *Microbial Ecology*, **45**, 11–19.

Gulis, V. & Suberkropp, K. (2003b). Interactions between stream fungi and bacteria associated with decomposing leaf litter at different levels of nutrient availability. *Aquatic Microbial Ecology*, **30** 149–57.

Gulis, V. & Suberkropp, K. (2003c). Leaf litter decomposition and microbial activity in nutrient-enriched and unaltered reaches of a headwater stream. *Freshwater Biology*, **48**, 123–34.

Gulis, V. & Suberkropp, K. (2004). Effects of whole-stream nutrient enrichment on the concentration and abundance of aquatic hyphomycete conidia in transport. *Mycologia*, **96**, 57–65.

Gulis, V., Rosemond, A. D., Suberkropp, K., Weyers, H. S. & Benstead, J. P. (2004). Effects of nutrient enrichment on the decomposition of wood and associated microbial activity in streams. *Freshwater Biology*, **49**, 1437–47.

Gunasekera, S. A., Webster, J. & Legg, C. J. (1983). Effect of nitrate and phosphate on weight losses of pine and oak wood caused by aquatic and aero-aquatic hyphomycetes. *Transactions of the British Mycological Society*, **80**, 507–14.

Hamilton, S. K., Tank, J. L., Raikow, D. F. *et al.* (2001). Nitrogen uptake and transformation in a midwestern US stream: a stable isotope enrichment study. *Biogeochemistry*, **54**, 297–340.

Hieber, M. & Gessner, M. O. (2002). Contribution of stream detritivores, fungi, and bacteria to leaf breakdown based on biomass estimates. *Ecology*, **83**, 1026–38.

Högberg, M. N. & Högberg, P. (2002). Extramatrical ectomycorrhizal mycelium contributes one-third of microbial biomass and produces, together with associated roots, half the dissolved organic carbon in a forest soil. *New Phytologist*, **154**, 791–5.

Huryn, A. D., Butz Huryn, V. M., Arbuckle, C. J. & Tsomides, L. (2002). Catchment land-use, macroinvertebrates and detritus processing in headwater streams: taxonomic richness versus function. *Freshwater Biology*, **47**, 401–15.

Hynes, H. B. N. & Kaushik, N. K. (1969). The relationship between dissolved nutrient salts and protein production in submerged autumnal leaves. *Verhandlungen der Internationale Vereinigung für Theoretische und Angewandte Limnologie*, **17**, 95–103.

Iqbal, S. H. & Webster, J. (1973). Aquatic hyphomycete spora of the River Exe and its tributaries. *Transactions of the British Mycological Society*, **61**, 331–46.

Kaushik, N. K. & Hynes, H. B. N. (1971). The fate of dead leaves that fall into streams. *Archiv für Hydrobiologie*, **68**, 465–515.

Komínková, D., Kuehn, K. A., Büsing, N., Steiner, D. & Gessner, M. O. (2000). Microbial biomass, growth, and respiration associated with submerged litter of *Phragmites australis* decomposing in a littoral reed stand of a large lake. *Aquatic Microbial Ecology*, **22**, 271–82.

Kreeger, D. A. & Newell, R. I. E. (2000). Trophic complexity between producers and invertebrate consumers in salt marshes. In *Concepts and Controversies in Tidal Marsh Ecology*, ed. P. Weinstein & D. A. Kreeger. Dordrecht: Kluwer Academic Publishers, pp. 187–220.

Kuehn, K. A. & Suberkropp, K. (1998a). Decomposition of standing litter of the freshwater macrophyte *Juncus effusus* L. *Freshwater Biology*, **40**, 717–27.

Kuehn, K. A. & Suberkropp, K. (1998b). Diel fluctuations in microbial activity associated with standing-dead litter of the freshwater emergent macrophyte *Juncus effusus*. *Aquatic Microbial Ecology*, **14**, 171–82.

Kuehn, K. A., Churchill, P. F. & Suberkropp, K. (1998). Osmoregulatory strategies of fungal populations inhabiting standing dead litter of the emergent macrophyte *Juncus effusus*. *Applied and Environmental Microbiology*, **64**, 607–12.

Kuehn, K. A., Gessner, M. O., Wetzel, R. G. & Suberkropp, K. (1999). Standing litter decomposition of the emergent macrophyte *Erianthus giganteus*. *Microbial Ecology*, **38**, 50–7.

Kuehn, K. A., Lemke, M. J., Suberkropp, K. & Wetzel, R. G. (2000). Microbial biomass and production associated with decaying leaf litter of the emergent macrophyte *Juncus effusus*. *Limnology and Oceanography*, **45**, 862–70.

Kuehn, K. A., Steiner, D. & Gessner, M. O. (2004). Diel mineralization patterns of standing-dead plant litter: implications for CO_2 flux from wetlands. *Ecology*, **85**, 2504–18.

432   V. Gulis et al.

Larmola, T., Alm, J., Juutinen, S., Martikainen, P. J. & Silvola, J. (2003). Ecosystem CO_2 exchange and plant biomass in the littoral zone of a boreal eutrophic lake. *Freshwater Biology*, **48**, 1295–310.

Laughlin, R. J. & Stevens, R. J. (2002). Evidence for fungal dominance of denitrification and codenitrification in a grassland soil. *Soil Science Society of America Journal*, **66**, 1540–8.

Mann, K. H. (1988). Production and use of detritus in various freshwater, estuarine and coastal marine ecosystems. *Limnology and Oceanography*, **33**, 910–30.

Melillo, J. M., Aber, J. D. & Muratore, J. F. (1982). Nitrogen and lignin control of hardwood leaf litter decomposition dynamics. *Ecology*, **63**, 621–6.

Melillo, J. M., Naiman, R. J., Aber, J. D. & Eshleman, K. N. (1983). The influence of substrate quality and stream size on wood decomposition dynamics. *Oecologia*, **58**, 281–5.

Melillo, J. M., Naiman, R. J., Aber, J. D. & Linkins, A. E. (1984). Factors controlling mass-loss and nitrogen dynamics of plant litter decaying in northern streams. *Bulletin of Marine Science*, **35**, 341–56.

Methvin, B. R. & Suberkropp, K. (2003). Annual production of leaf-decaying fungi in 2 streams. *Journal of the North American Benthological Society*, **22**, 554–64.

Mille-Lindblom, C., von Wachenfeldt, E. & Tranvik, L. J. (2004). Ergosterol as a measure of living fungal biomass: persistence in environmental samples after fungal death. *Journal of Microbiological Methods*, **59**, 253–62.

Mitsch, W. J. & Gosselink, J. G. (2000). *Wetlands*, 3rd edn. New York: Wiley & Sons.

Molinero, J., Pozo, J. & Gonzalez, E. (1996). Litter breakdown in streams of the Agüera catchment: Influence of dissolved nutrients and land use. *Freshwater Biology*, **36**, 745–56.

Moran, M. A., Legovic, T., Benner, R. & Hodson, R. E. (1988). Carbon flow from lignocellulose: a simulation analysis of a detritus-based ecosystem. *Ecology*, **69**, 1525–36.

Moran, M. A., Benner, R. & Hodson, R. E. (1989). Kinetics of microbial degradation of vascular plant material in two wetland ecosystems. *Oecologia*, **79**, 158–67.

Mulholland, P. J., Tank, J. L., Sanzone, D. M. *et al.* (2000). Nitrogen cycling in a forest stream determined by a N-15 tracer addition. *Ecological Monographs*, **70**, 471–93.

Newbold, J. D., Elwood, J. W., O'Neill, R. V. & van Winkle, W. (1981). Measuring nutrient spiralling in streams. *Canadian Journal of Fisheries and Aquatic Sciences*, **38**, 860–3.

Newell, S. Y. (1993). Decomposition of shoots of a saltmarsh grass: methodology and dynamics of microbial assemblages. *Advances in Microbial Ecology*, **13**, 301–26.

Newell, S. Y. (2003). Fungal content and activities in standing-decaying leaf blades of plants of the Georgia Coastal Ecosystem research area. *Aquatic Microbial Ecology*, **32**, 95–103.

Newell, S. Y. & Bärlocher, F. (1993). Removal of fungal and total organic matter from decaying cordgrass leaves by shredder snails. *Journal of Experimental Marine Biology and Ecology*, **171**, 39–49.

Newell, S. Y. & Fallon, R. D. (1991). Toward a method for measuring instantaneous fungal growth rates in field samples. *Ecology*, **72**, 1547–59.

Newell, S. Y. & Porter, D. (2000). Microbial secondary production from salt marsh-grass shoots, and its known potential fates. In *Concepts and Controversies in Tidal Marsh Ecology*, ed. M. P. Weinstein & D. A. Kreeger. Dordrecht: Kluwer Academic Publishers, pp. 159–86.

Newell, S. Y. & Statzell-Tallman, A. (1982). Factors for conversion of fungal biovolume values to biomass, carbon and nitrogen: variation with mycelial ages, growth conditions, and strains of fungi from a salt marsh. *Oikos*, **39**, 261–8.

Newell, S. Y., Moran, M. A., Wicks, R. & Hodson, R. E. (1995). Productivities of microbial decomposers during early stages of decomposition of leaves of a freshwater sedge. *Freshwater Biology*, **34**, 135–48.

Niyogi, D. K., Simon, K. S. & Townsend, C. R. (2003). Breakdown of tussock grass in streams along a gradient of agricultural development in New Zealand. *Freshwater Biology*, **48**, 1698–708.

Ostrofsky, M. L. (1997). Relationship between chemical characteristics of autumn-shed leaves and aquatic processing rates. *Journal of the North American Benthological Society*, **16**, 750–9.

Paul, E. A. & Clark, F. E. (1989). *Soil Microbiology and Biochemistry*. San Diego: Academic Press.

Qualls, R. G. (1984). The role of leaf litter nitrogen immobilization in the nitrogen budget of a swamp stream. *Journal of Environmental Quality*, **13**, 640–4.

Quinn, J. M., Burrell, G. P. & Parkyn, S. M. (2000). Influences of leaf toughness and nitrogen content on in-stream processing and nutrient uptake by litter in a Waikato, New Zealand, pasture stream and streamside channels. *New Zealand Journal of Marine and Freshwater Research*, **34**, 253–71.

Ramírez, A., Pringle, C. M. & Molina, L. (2003). Effects of stream phosphorus levels on microbial respiration. *Freshwater Biology*, **48**, 88–97.

Rier, S. T., Tuchman, N. C., Wetzel, R. G. & Teeri, J. A. (2002). Elevated-CO_2-induced changes in the chemistry of quaking aspen (*Populus tremuloides* Michaux) leaf litter: subsequent mass loss and microbial response in a stream ecosystem. *Journal of the North American Benthological Society*, **21**, 16–27.

Roden, E. E. & Wetzel, R. G. (1996). Organic carbon oxidation and suppression of methane production by microbial Fe(III) oxide reduction in vegetated and unvegetated freshwater wetland sediments. *Limnology and Oceanography*, **41**, 1733–48.

Rodewald-Rudescu, L. (1974). Das Schilfrohr. (Die Binnengewässer, vol. 27). Stuttgart: Schweizerbart.

Sanzone, D. M., Tank, J. L., Meyer, J. L., Mulholland, P. J. & Findlay, S. E. G. (2001). Microbial incorporation of nitrogen in stream detritus. *Hydrobiologia*, **464**, 27–35.

Scanlon, D. & Moore, T. (2000). Carbon dioxide production from peatland soil profiles: the influence of temperature, oxic/anoxic conditions and substrate. *Soil Science*, **165**, 153–60.

Shearer, C. A. (1992). The role of woody debris. In *The Ecology of Aquatic Hyphomycetes*, ed. F. Bärlocher. Berlin: Springer-Verlag, pp. 77–98.

Shoun, H., Kim, D. H., Uchiyama, H. & Sugiyama, J. (1992). Denitrification by fungi. *FEMS Microbiology Letters*, **94**, 277–81.

Silliman, B. R. & Newell, S. Y. (2003). Fungal farming in a snail. *Proceedings of the National Academy of Sciences of the United States of America*, **100**, 15643–8.

Simon, K. S. & Benfield, E. F. (2001). Leaf and wood breakdown in cave streams. *Journal of the North American Benthological Society*, **20**, 550–63.

Sinsabaugh, R. L. & Findlay, S. (1995). Microbial production, enzyme activity, and carbon turnover in surface sediments of the Hudson River estuary. *Microbial Ecology*, **30**, 127–41.

Sinsabaugh, R. L., Antibus, R. K., Linkins, A. E. *et al.* (1993). Wood decomposition: nitrogen and phosphorus dynamics in relation to extracellular enzyme activity. *Ecology*, **74**, 1586–93.

Sridhar, K. R. & Bärlocher, F. (2000). Initial colonization, nutrient supply, and fungal activity on leaves decaying in streams. *Applied and Environmental Microbiology*, **66**, 1114–19.

Stelzer, R. S., Heffernan, J. & Likens, G. E. (2003). The influence of dissolved nutrients and particulate organic matter quality on microbial respiration and biomass in a forest stream. *Freshwater Biology*, **48**, 1925–37.

Suberkropp, K. (1991). Relationships between growth and sporulation of aquatic hyphomycetes on decomposing leaf litter. *Mycological Research*, **95**, 843–50.

Suberkropp, K. (1992). Interactions with invertebrates. In *The Ecology of Aquatic Hyphomycetes*, ed. F. Bärlocher. Berlin: Springer-Verlag, pp. 118–34.

Suberkropp, K. (1995). The influence of nutrients on fungal growth, productivity, and sporulation during leaf breakdown in streams. *Canadian Journal of Botany*, **73** (Suppl. 1), S1361–9.

Suberkropp, K. (1997). Annual production of leaf-decaying fungi in a woodland stream. *Freshwater Biology*, **38**, 169–78.

Suberkropp, K. (1998a). Effect of dissolved nutrients on two aquatic hyphomycetes growing on leaf litter. *Mycological Research*, **102**, 998–1002.

Suberkropp, K. (1998b). Microorganisms and organic matter processing. In *River Ecology and Management: Lessons from the Pacific Coastal Ecoregion*, eds. R. J. Naiman, & R. E. Bilby. New York: Springer-Verlag, pp. 120–43.

Suberkropp, K. & Chauvet, E. (1995). Regulation of leaf breakdown by fungi in streams: Influences of water chemistry. *Ecology*, **76**, 1433–45.

Suberkropp, K. & Klug, M. J. (1980). The maceration of deciduous leaf litter by aquatic hyphomycetes. *Canadian Journal of Botany*, **58**, 1025–31.

Suberkropp, K. & Weyers, H. (1996). Application of fungal and bacterial production methodologies to decomposing leaves in streams. *Applied and Environmental Microbiology*, **62**, 1610–15.

Suberkropp, K., Godshalk, G. L. & Klug, M. J. (1976). Changes in the chemical composition of leaves during processing in a woodland stream. *Ecology*, **57**, 720–77.

Suberkropp, K., Arsuffi, T. L. & Anderson, J. P. (1983). Comparison of degradative ability, enzymatic activity, and palatability of aquatic hyphomycetes grown on leaf litter. *Applied and Environmental Microbiology*, **46**, 237–44.

Tank, J. L. & Dodds, W. K. (2003). Nutrient limitation of epilithic and epixylic biofilms in ten North American streams. *Freshwater Biology*, **48**, 1031–49.

Tank, J. L., Webster, J. R. & Benfield, E. F. (1993). Microbial respiration on decaying leaves and sticks in a southern Appalachian stream. *Journal of the North American Benthological Society*, **12**, 394–405.

Tank, J. L., Meyer, J. L., Sanzone, D. M. *et al.* (2000). Analysis of nitrogen cycling in a forest stream during autumn using a N-15-tracer addition. *Limnology and Oceanography*, **45**, 1013–29.

Thormann, M. N., Bayley, S. E. & Currah, R. S. (2001). Comparison of decomposition of belowground and aboveground plant litters in peatlands of boreal Alberta, Canada. *Canadian Journal of Botany*, **79**, 9–22.

Thornton, D. R. (1963). The physiology and nutrition of some aquatic hyphomycetes. *Journal of General Microbiology*, **33**, 23–31.

Thornton, D. R.(1965). Amino acid analysis of fresh leaf litter and the nitrogen nutrition of some aquatic hyphomycetes. *Canadian Journal of Microbiology*, **11**, 657–62.

Triska, F. J. & Sedell, J. R. (1976). Decomposition of four species of leaf litter in response to nitrate manipulation. *Ecology*, **57**, 783–92.

Updegraff, K., Bridgham, S. D., Pastor, J., Weishampel, P. & Harth, C. (2001). Response of CO_2 and CH_4 emissions from peatlands to warming and water table manipulation. *Ecological Applications*, **11**, 311–26.

Wallander, H., Nilsson, L. O., Hagerberg, D. & Rosengren, U. (2003). Direct estimates of C:N ratios of ectomycorrhizal mycelia collected from Norway spruce forest soils. *Soil Biology & Biochemistry*, **35**, 997–9.

Wallander, H., Goransson, H. & Rosengren, U. (2004). Production, standing biomass and natural abundance of N-15 and C-13 in ectomycorrhizal mycelia collected at different soil depths in two forest types. *Oecologia*, **139**, 89–97.

Webster, J. (1959). Experiments with spores of aquatic hyphomycetes I. Sedimentation, and impaction on smooth surfaces. *Annals of Botany New Series*, **23**, 595–611.

Webster, J. R. & Benfield, E. F. (1986). Vascular plant breakdown in freshwater ecosystems. *Annual Review of Ecology and Systematics*, **17**, 567–94.

Webster, J. & Davey, R. A. (1984). Sigmoid conidial shape in aquatic fungi. *Transactions of the British Mycological Society*, **83**, 43–52.

Webster, J. R. & Meyer, J. L. (eds.) (1997). Stream organic matter budgets. *Journal of the North American Benthological Society*, **16**, 3–161.

Welsch, M. & Yavitt, J. B. (2003). Early stages of decay of *Lythrum salicaria* L. and *Typha latifolia* L. in a standing-dead position. *Aquatic Botany*, **75**, 45–57.

Wetzel, R. G. & Howe, M. J. (1999). High production in a herbaceous perennial plant achieved by continuous growth and synchronized population dynamics. *Aquatic Botany*, **64**, 111–29.

Weyers, H. S. & Suberkropp, K. (1996). Fungal and bacterial production during the breakdown of yellow poplar leaves in two streams. *Journal of the North American Benthological Society*, **15**, 408–20.

Yavitt, J. B. (1997). Methane and carbon dioxide dynamics in *Typha latifolia* (L.) wetlands in central New York State. *Wetlands*, **17**, 394–406.

18
Biogeochemical roles of fungi in marine and estuarine habitats

NICHOLAS CLIPSON, MARINUS OTTE
AND ELEANOR LANDY

Introduction

Oceans cover around 70% of the global surface area, yet remain one of the least explored regions for fungal diversity; consequently knowledge of the fungal contribution to ecosystem processes in these marine environments is extremely limited. For the purposes of this review, marine habitats are defined as those influenced in some way by seawater, generally from existing saline water bodies. In some cases, saline habitats have resulted from salt accumulation in soils originating from ancient seas. Broadly, marine ecosystems divide between those influenced in some way by terrestrial environments, generally situated close to coastal regions, and those associated with the open ocean. Broad boundaries within marine environments are detailed in Table 18.1, where coastal and open ocean, and the effect of depth within open oceans, is differentiated. Marine environments tend also to be strongly linked, representing movement between different regions of seas and oceans, as summarized in Fig. 18.1. In coastal regions, numerous types of marine environment develop, including saline wetlands and lagoons on low-energy coasts, estuarine systems where there is freshwater influx, and a range of beach and splash communities on high-energy coasts. Such ecosystems are reviewed in more detail by Packham and Willis (1997). Adjacent to coastal regions, and where continental shelves are shallow, coastal sea communities form, including coral reefs, which are found in both tropical and cold seas. A number of inland saline environments also exist, such as salt pans and salt deserts.

Environmental parameters within marine environments

In terms of biogeochemical conditions and elemental composition, the open oceans are amongst the more stable systems on Earth.

Fungi in Biogeochemical Cycles, ed. G. M. Gadd. Published by Cambridge University Press. © British Mycological Society 2006.

Table 18.1. *Systems within oceanic and coastal environments, each characterized with more or less distinct biogeochemical cycles from an organismal perspective. See also Fig. 18.1*

Type	Characteristics
Open Ocean (No direct interactions with terrestrial environments)	
Near surface (Pelagic)	Depth 0–200 m, photic, water–atmosphere interface, turbulent, wind-driven, variable temperature, above thermocline, no interactions with bottom
Deep layer (Mesopelagic/ Abyssopelagic)	Depth >200 m, no light, high pressures, below thermocline, cold, no interactions with atmosphere or bottom, mostly lateral currents
Near bottom (Benthic)	Depth >200 m, no light, sediment–water interface, no interactions with atmosphere, mostly lateral currents
Deep-sea hydrothermal vents	Depths variable, but typically >1000 m, no light, high pressures, warm–hot
Coastal (Strong, direct interactions with terrestrial environments)	
Shallow seas	Depth 0–200 m, photic
Coral reefs	Depth 0–200 m, photic
Rocky shores	Highly turbulent
Sandy shores/beaches	Moderately–highly turbulent
Estuaries	Freshwater/seawater mixing, tidal, strongly affected by watershed of river
Salt marshes	Can be estuarine or lagoon-type, tidal

Defining to all marine systems is a high concentration of NaCl. In the open ocean, NaCl concentrations are relatively stable, generally remaining around 500 mM, although these can be different in confined seas. For example, the Black and Baltic Seas have narrow tidal ranges, substantial inputs of fresh water and restricted connection to the open ocean resulting in seawater NaCl concentrations in the range 50–200 mM. Typical constituents of seawater from the open ocean are given in Table 18.2. Nearshore marine systems on the other hand are highly variable. Salt concentrations are more variable and dynamic, and affected strongly by terrestrial and climatological influences. Saline wetlands such as salt marshes or mangrove swamps are tidally inundated, but also subject to strong evapotranspirational forces together with freshwater inputs from rainfall and streams (Jefferies *et al.*, 1979), are waterlogged and consequently tend to have highly negative redox potentials (Armstrong *et al.*, 1985). Obviously, such environments represent transitional zones between terrestrial and marine environments, but for practical purposes, all environments with salinities in excess of 0.5 ppt are generally considered marine. Coastal

Table 18.2. *Typical ionic composition of seawater from the open ocean (adapted from Martin, 1970)*

Ionic species	Concentration	
	g kg^{-1}	mM
Chloride	19.35	548
Sodium	10.76	470
Sulphate	2.71	28
Magnesium	1.29	54
Calcium	0.413	10
Potassium	0.387	10
Bicarbonate	0.142	2
Bromide	0.067	0.8
Strontium	0.008	1.5
Boron	0.0045	0.4
Fluoride	0.001	0.07

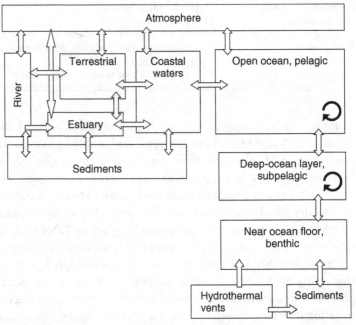

Fig. 18.1. Interactions between and within marine ecosystems. While most interactions are two directional, interactions between rivers and estuaries and between hydrothermal vents and their surroundings are unidirectional. Circular arrows within oceanic compartments indicate cycles that are largely confined within those compartments (after Wangersky, 1980; Salomons & Förstner, 1984).

Table 18.3. *Known elemental concentrations for marine environments*

Type

Open Ocean (no direct interactions with terrestrial environments)
Near surface (Pelagic)
Water (μg l^{-1}) after Salomons and Förstner (1984)

Cu	North Pacific – 0.003
	North Atlantic – 0.006
Zn	North Pacific – 0.007
	North Atlantic – 0.007

Water (mg l^{-1}) after Zehnder and Zinder (1980)

| S | as dimethylsulphide – 1.2×10^{-5} |
| | as carbon disulphide – 5.2×10^{-7} |

Deep layer (Mesopelagic/Abyssopelagic)
Water (μg l^{-1}) after Salomons and Förstner (1984)

Cu	North Pacific – 0.18
	North Atlantic – 0.09
Zn	North Pacific – 0.52
	North Atlantic – 0.13

Near bottom (Benthic)
Deep-sea sediments (μg g^{-1}) after Salomons and Förstner (1984)

Cu	30–394
Zn	35–170
Fe	9000–65 000

Coastal (strong, direct interactions with terrestrial environments)
Shallow seas

Mediterranean Sea water (μg l^{-1}) after Salomons and Förstner (1984)

| Cu | 0.2–36.2 |
| Zn | 1–365 |

Stagnant Bay water (mg l^{-1}) after Zehnder and Zinder (1980)

| S | as carbon disulphide 5.4×10^{-6} |

Typical nearshore mud sediments (μg g^{-1}) after Salomons and Förstner (1984)

Cu	48
Zn	95
Fe	70 000

systems also tend to have much higher elemental concentrations than open ocean, with some comparative data for selected elements given in Table 18.3.

It is beyond the scope of this article to detail the complete individual elemental cycles for these systems. For detailed discussions on biogeochemical cycling of elements the reader is referred to Hutzinger (1980) and Salomons and Förstner (1984). Typically, the open oceans contain lower concentrations of elements than nearshore systems, such as shallow seas and estuaries (Table 18.2). Transport from terrestrial systems to nearshore systems via rivers is relatively fast, while transport to deeper ocean systems is slow (see also Fig. 18.1). Many elements are immobilized in sediments and these therefore act as sinks from which remobilization occurs slowly. This is particularly true of phosphorus and most metals, for which the atmospheric component of the cycle is generally considered to be negligible (Elmsley, 1980; Salomons & Förstner, 1984). Elements for which the atmospheric component is relatively important include carbon, sulphur, nitrogen and some metals and metalloids that can be methylated, such as mercury, arsenic and selenium. The latter group of elements is important in the context of this paper, as several fungi have been shown to methylate metalloids, such as arsenic (Craig, 1980).

Highly saline environments are not only directly associated with present seas and oceans, but also with former seas which have led to salt deposition. These are generally hypersaline environments and may include salt lakes such as the Dead Sea, where salt concentrations may reach 4–5 M NaCl (Buchalo *et al.*, 1998), together with salt pans and flats. In many cases, these are dominated by other ions such as potassium, magnesium, calcium, sulphate, carbonate and bicarbonate, as well as sodium and chloride. Flowers *et al.* (1986) estimated that about 10% of global land area was occupied by soils too saline for the growth of non-halophiles.

Most of the ocean floor appears to be an area of low primary productivity and low biomass density (Nybakken, 1997). The exceptions to this are the hydrothermal vent and cold seep communities discovered by deep-sea submersibles (Ballard, 1977). Hydrothermal vents have temperatures in the range 8–16 °C (as opposed to the normal 2 °C of most of the deep-sea), are associated with the edge of tectonic plate systems and are very rich in reduced sulphur compounds such as H_2S. Primary productivity is driven by prokaryal chemolithoautotrophs, which drive relatively rich food webs with high productivity (Nybakken, 1997). Cold seeps are often associated with hypersaline brines or hydrocarbon seeps. All these ecosystems are characterized by high pressures exerted by the overlying water column. Only one record to date of fungal occurrence has been reported.

Fungal diversity and distribution

That fungi are important components of marine ecosystems has been increasingly recognized over the last 50 years or so, with the first major report by Barghoorn and Linder (1944), coining the term 'marine fungi' to describe fungi isolated from marine environments (see also

Johnson & Sparrow, 1961). Nevertheless, with the vast geographical extent of the oceans, this remains a very under-researched group, with a Science Citation Index search using the terms 'marine' and 'fungus' only yielding 738 hits from 27.2 million records.

There has been some discussion as to how marine fungi should be defined, with Kohlmeyer and Kohlmeyer (1979) advancing three ecological groupings of fungi occurring in marine ecosystems; obligate marine fungi, facultative marine fungi and terrestrial fungi (see also Kohlmeyer & Volkmann-Kohlmeyer, 2003). They defined obligate marine fungi by their inability to complete their life-cycle outside the marine environment. In order to accommodate the possibility that terrestrial species might also be active in the seas, facultative marine fungi were defined as those originating from freshwater or terrestrial environments but also capable of growth and sporulation in the sea. Terrestrial fungi do not have the ability to complete life-cycles in marine environments, and although commonly isolated may just exist in a dormant state until more favourable conditions prevail for spore germination (Kohlmeyer & Kohlmeyer, 1979). These divisions relate largely to ecological conditions, but the physiological basis of these groupings has not been established although may relate to ability to sporulate or attach. There are examples of terrestrial fungi being physiologically active in marine environments; for example, the terrestrial species *Aspergillus sydowii* and *Aspergillus fumigatus* cause mortality of the Caribbean sea fan *Gorgonia ventalina*, with disease-causing populations probably maintained by continual influx of vegetative material from neighbouring terrestrial environments (Smith *et al.*, 1996; Geiser *et al.*, 1998).

Fungi have been isolated from a very diverse range of marine habitats including salt marshes, mangroves, deep-sea vents, inland salt seas and salterns (Table 18.4). It is difficult to estimate precisely how many species have been isolated to date from marine ecosystems, although Jones and Mitchell's estimate of around 1000 species either isolated and described, or newly discovered (Jones & Mitchell, 1996), including around 170 Thraustochytrids and Lower Fungi, is probably reasonable (see Table 18.5). Such figures are rather unreliable as estimates of total global fungal diversity because there is differential geographical effort in species discovery with some regions better described than others (e.g. Volkmann-Kohlmeyer & Kohlmeyer, 1993). Table 18.5 also details marine fungal diversity divided into broad phylogenetic groups. Apart from marine lichens, marine fungi are dominated by the Ascomycotina, particularly by the Halosphaeriales. Many of the genera found have common

Table 18.4. *List of the main marine environments from which fungi have been isolated including information on their biogeography*

Marine Environment	Fungal Group/Isolate	Geographical Location	Reference
Wood blocks submerged in marine ecosystems	*Digitatispora marina* or *Nia vibrissae* with some hetero-basidiomycetous yeasts	Worldwide	Jones (1982)
Shoreline	Higher and lower marine fungi	European Atlantic Coast and North Sea	Rees *et al.* (1979); Wilson (1954, 1965)
Sand dunes	Higher and lower marine fungi	European Atlantic Coast and North Sea	Farrant *et al.* (1985); Lutley & Wilson (1972)
Coastal grasslands	Higher and lower marine fungi	European Atlantic Coast and North Sea	Lucas & Webster (1967); Apinis & Chesters (1964); Jones (1962); Rees *et al.* (1979); Cotton (1907–1911); Dickinson (1965)
Harbour area	Higher and lower marine fungi	Langstone Harbour, UK; Tenerife, Europe	Jones (1962); Kohlmeyer (1967); Peterson & Koch (1996)
Seawater	Ascomycetes and Yeasts	Atlantic Coast, Ireland; Pacific Coast, USA	Duffy *et al.* (1991); Meyers *et al.* (1967); Shinano (1962)
Deep-sea water	Pink basidiomycetous yeasts	—	Lorenz & Molitoris (1997); Nagahama *et al.* (2001)
Mangroves with loose sediment, submerged mangrove roots, trunks and branches	Higher and lower marine fungi	—	Kathiresan & Bingham (2001); Raghukumar *et al.* (1994); Venkateswara-Sarma *et al.* (2001)
Marine sediment	*Beauveria bassiana*	—	Suresh & Chandra-Sekaran (1998)

Bathyal region	Higher and lower marine fungi	Western Coral Sea	Alongi (1987)
Salt marshes	Higher and lower marine fungi	North Sea and Atlantic Coasts, Europe; Atlantic Coast, USA	Goodman et al. (1959); Jones (1962); Pugh (1962); Apinis & Chesters (1964); Dickinson (1965); Dickinson & Pugh (1965); Dickinson & Morgan-Jones (1966); Sivanesan & Manners (1970); Dickinson & Kent (1972); Boyle (1976); Gessner & Kohlmeyer (1976); Burge & Perkins (1977); Gessner (1977); Kohlmeyer & Kohlmeyer (1979); Newell & Fell (1982); Hendrarto & Dickinson (1984); Newell (1991); Bergbauer & Newell (1992); Cooke et al. (1993); Newell (1996); Newell et al. (1996a); Barato et al. (1997); Ellis & Ellis (1997); Raybould et al. (1998)
Miscellaneous	Higher and lower marine fungi	Pohnpei Japan; Antarctic	Kobayashi et al. (2003); Pugh & Jones (1986)

Table 18.5. *Marine fungal groups and estimated numbers of species (from Jones & Mitchell (1996), with permission). Figures with asterisks represent numbers of incompletely described or new species*

Fungal group	Estimated number of species
Lower fungi	100 + 32*
Thraustochytrids	40
Hemi-ascomycetes	50
Eu-ascomycetes	305 + 143*
Deuteromycotina	79 + 200*
Basidiomycotina	7 + 7*
Trichomycetes	23
Marine lichens	18 + 410*

characteristics (reproduction in aquatic habitats; thin-walled, unitunicate, deliquescing asci; central pseudoparenchyma in immature asci; hyaline, bicelled ascospores with polar or equatorial appendages), which Jones (1995) considers as adaptations to aquatic habitats.

Biogeographically, marine fungi appear to be found worldwide including polar regions (Pugh & Jones, 1986; Grasso *et al.*, 1997), although probably the highest levels of diversity so far have been found in the Tropics associated with coastal ecosystems, especially mangroves. The true extent of global distribution is also difficult to quantify because the intensity of investigation has been highly localized and large areas remain almost totally underinvestigated. For example, Clipson *et al.* (2001) examined the extent of fungal diversity within the European Union (EU), identifying 317 described marine fungal species, mostly associated with the Atlantic coastline, which probably coincides with the area of most intense study. Given that there are around 1000 species of marine fungi so far isolated or described, and that tropical regions tend to be more diverse than temperate areas, the extent of European fungal diversity simply reflects that EU coasts are better explored than other regions.

Another issue to be considered when discussing marine fungal diversity is that of culturability. Most diversity estimates have involved isolation into culture media, or identification from fruiting bodies growing on natural substrates. It is likely that many marine fungi are unculturable and their presence and taxon will only be demonstrated by molecular techniques. There have been very few fungal molecular studies of environmental samples from marine systems (e.g. Buchan *et al.*, 2003), and until such approaches are more widely used, our view of marine fungal diversity will remain limited.

Fungal survival in marine environments

Fungi living in marine environments face a number of potentially detrimental environmental factors for which adaptation is presumably required. Salinity is a feature of all marine environments by definition but, depending on which marine ecosystem is being considered, a number of other factors may be influential including pressure, redox and temperature in abyssal systems; temperature in high latitudes; and redox combined with general environmental fluctuation in coastal zones. Adaptation to salinity is by far the best understood of fungal adaptations to marine ecosystems and has been extensively reviewed (Clipson & Jennings, 1992; Blomberg & Adler, 1993; Hooley *et al.*, 2003). Much of our understanding of the mechanisms by which marine fungi adapt to salt comes from the physiological studies on the marine hyphomycete *Dendryphiella salina* by Jennings and co-workers in the 1980s and 1990s, and by subsequent genetic studies on this and other non-marine fungi (e.g. Clement *et al.*, 1999; and see Bohnert *et al.*, 2001; Hooley *et al.*, 2003).

Central to fungal survival in marine ecosystems is maintenance of the osmotic gradient across the fungal cell membrane to maintain inwardly directed water transport (Hooley *et al.*, 2003). The walled nature of fungal cells means that maintenance of turgor is essential for growth, turgor being the driving force for apical growth and expansion (Money, 1997). Seawater has a water potential of around $-2.3\,\mathrm{MPa}$, meaning that cellular water potential has to be kept more negative by the accumulation or synthesis of osmotically active solutes. Ionic solutes, particularly sodium and chloride, are obviously widely available to marine organisms, but are not particularly compatible with cytoplasmic metabolism. Studies of the effect of NaCl on activity of the cytoplasmic enzymes malate dehydrogenase and NAD-linked glycerol dehydrogenase indicated substantial inhibition in vitro at quite low NaCl (100 mM) concentration (Paton & Jennings, 1988). This may indicate that the fungal cytoplasm is salt-sensitive, as found for other biological groups such as plants, although some ion effects may be mitigated in vivo by up-regulated protein production or effects of the counteranion environment (Nilsson & Adler, 1990; Hooley *et al.*, 2003).

In fungi, the principal organic solutes used to generate osmotic potential are the polyols, including glycerol, mannitol, arabinitol and erythritol (Blomberg & Adler, 1993). These solutes are compatible with cellular metabolism in that they have little or no detrimental effect on cellular metabolism or enzyme stability. In many fungi, mannitol is the primary osmoresponsive polyol of growing fungal cells (Beever & Laracy, 1986). For species such as *D. salina*, *Debaryomyces hansenii* and *Zygosaccharomyces rouxii*, where polyol concentrations have been measured under saline conditions in vitro, polyols generated between 36% and 75% of cellular osmotic potential (Hooley *et al.*, 2003). Other compatible solutes may also be important including proline and trehalose (Jennings & Burke, 1990; Davis *et al.*, 2000).

Ions also have a role in generation of cellular osmotic potentials, despite their cytotoxic properties. In a study on *D. salina* growing at near seawater concentrations (*c.* 500 mM NaCl) using X-ray microanalysis, cytoplasmic sodium concentration was estimated at 173 mM, which is probably still compatible with cytosolic enzyme activity, albeit with some inhibition of more sensitive enzymes (Clipson *et al.*, 1990).There was no evidence that there was preferential ion accumulation in vacuoles, although these did not occupy large volumes of these cells. Vacuolar localization of ions may be more important as vacuoles are largely inert at the metabolic level and may be particularly significant in fungal species where a greater proportion of the cell is vacuolar.

Critical to cellular osmotolerance in marine fungi is regulation of osmotic processes through gene expression and the regulation of gene activity. There have been few genetic studies of osmotolerance in marine fungi, but the processes occuring in yeasts are now fairly well understood and are summarized in Yale and Bohnert (2001) and Hooley *et al.* (2003). A number of genes involved in stress responses by the marine yeast *Debaryomyces hansenii* have been characterized including the cloning of the genes for superoxide dismutase (Hernandez-Saavedra & Romero-Geraldo, 2001), a homologue of the high osmolarity glycerol response gene *HOG1* encoding for a MAPkinase (Bansal & Mondal, 2000), and the identification of several ion transport-related genes including *DhENA1* and *DhENA2* for sodium ATPases (Almagro *et al.*, 2001). Much less is known about the genetic control of osmotolerance in filamentous marine fungi.

Critical to life in the deep sea is tolerance of high pressures. Raghukumar and Raghukumar (1998) demonstrated that deep-sea filamentous fungal isolates grown under high pressure were able to undergo hyphal extension and conidiate, suggesting barotolerance. Nevertheless, the mechanisms conferring barotolerance in fungi in the deep sea are yet to be elucidated.

Marine fungal involvement in biogeochemical cycles

Conceptual cycles

'Biogeochemical cycling' here refers to all processes and pathways involved in the cycling and turnover of elements. This therefore does not just include chemical transformations, but also physical processes, such as adsorption/desorption, and biological transfer within (compartmentation) and between organisms. Which processes and compartments predominate in a particular cycle depends on the type of environment. From a global perspective all compartments are connected, but this is not so from an organismal (fungal) perspective (Hutzinger, 1980). Open ocean near-surface ecosystems and their biogeochemical cycles for example will be dominated by water–atmosphere interactions and will be hardly affected by what is happening at the ocean floor. Likewise, the effects of atmospheric processes on ecosystems associated with deep-ocean hydrothermal vents ('smokers') are negligible. Within the marine environment therefore a number of more or less distinct systems can be identified that each have their own biogeochemical cycles (Table 18.1).

In the next section, what is known or what can be suggested about the involvement of marine fungi in the cycling of carbon, nitrogen, phosphorus, sulphur and metals will be discussed, followed by a section assessing fungal involvement in elemental cycling in some of the marine compartments identified in Fig. 18.1. Firstly, however, it is important to note that the biogeochemical cycles of carbon, nitrogen, phosphorus, sulphur and the metals are intrinsically linked (see also Fomina *et al.*, Chapter 10 this volume). For example, the cycles of nitrogen, phosphorus and sulphur in particular include organic compounds such as proteins, amino acids, ATP and DNA, while reduction reactions involving metals carried out by microorganisms are typically linked to organic matter oxidation reactions.

Carbon

Fungal distribution in marine systems, as in any other ecosystem, is very much determined by the quality and availability of substrates based upon organic carbon. Secretion of enzymes is central to the fungal lifestyle, with fungi particularly suited to the degradation and recycling of organic polymers. Although most attention has been given to marine saprotrophs, availability of hosts for parasitic or mutualistic symbionts will also be important to those fungal species exhibiting that type of life-cycle. A wide range of fungal species causing epizootics and epiphytotics in the marine environment has been reported (see Porter, 1986, for overview).

Distribution of marine fungal species will follow substrate availability, particularly for more recalcitrant materials, which are less suitable for recycling by bacteria and other degraders, and for which there is less competition from other saprotrophic groups. Although the water column of the open ocean can at times have relatively high levels of organic matter when planktonic blooms occur, most marine waters are low in available organic material and have very low fungal biomass. Additionally in these regions, fungal communities are largely yeast-dominated rather than comprising filamentous species (Sieburth, 1979). In a study of the Pacific, van Uden and Fell (1968) found viable yeast cell numbers varying between 13 and 274 cells per litre. In general, yeast cell densities decline with increased depth and distance from land (Gadanho *et al.*, 2003), presumably reflecting substrate availability and favourable environmental conditions. In a study based in the Atlantic south of Portugal, Gadanho *et al.* (2003) identified 31 yeast taxa using micro-satellite primed PCR. How yeast cells derive nutrition in these environments is yet to be determined.

In coastal areas, biomass availability and complexity is much higher, particularly through the presence of highly productive plant-dominated ecosystems such as salt marshes and mangroves. These provide polymer-rich substrates, particularly lignocellulosic materials. Critical to the utilization of these substrates is exoenzyme production. There have been a number of studies investigating the ability of marine filamentous fungi to produce enzymes (Torzilli, 1982; Molitoris & Schaumann, 1986; Schaumann *et al.*, 1986; Torzilli & Andrykovitch, 1986; Lorenz & Molitoris, 1997; Rohrmann *et al.*, 1992), and it is clear that at least for those marine fungal species tested, a very wide range of exoenzymes can be produced. These are briefly summarized in Table 18.6.

Nitrogen

Nitrogen is an essential nutrient for all organisms and it is therefore no surprise that growth and activity of fungi, including those associated with marine environments, is affected by nitrogen supply (Sguros & Simms, 1963; Jennings, 1989; Feeney *et al.*, 1992; Edwards *et al.*, 1998; Ahammed & Prema, 2002). Both organic and inorganic nitrogen sources can be utilized depending on the species and their environment.

Although not many reports exist in relation to marine nitrogen cycling and fungi, it appears that, like their terrestrial counterparts, marine fungi have a wide range of substrate requirements. Nitrogen cyclers are either generalists or specialists, e.g. Kimura *et al.* (2002) investigated the trichomycete fungus *Entromyces callianassae*, which occurs exclusively in the foregut lining of the callianassid shrimp *Nihonotrypaea harmandi* and appears to be involved in the hydrolysation of certain nitrogen-containing compounds. Furthermore, Kirchner (1995) found that the yeast-like fungus *Aureobasidium pullulans* was involved in the degradation of chitin in the moults and carcasses of the marine copepod *Tisbe holothuriae*.

Table 18.6. *Production of exoenzymes for the breakdown of substrate polymers demonstrated in isolates of filamentous marine fungi (from Torzilli, 1982; Schaumann et al., 1986; Rohrmann et al., 1992)*

Marine fungal species	Exoenzyme produced
Lulworthia spp.	Carrageenase
Halocyphina villosa, Lulworthia spp.	Laminarinase
Halocyphina villosa	Gelatinase
Halocyphina villosa,	Caseinase
Lulworthia spp. *Corollospora maritima,*	
Digitatispora marina,	
Cirrenalia pygmea,	
Varicosporina ramulosa,	
Lulworthia spp.	Alginase
Pleospora pelagica, Pleospora vagans,	Pectinase
Phaeosphaeria typharum,	
Halocyphina villosa,	Xylanase
Pleospora pelagica,	
Pleospora vagans,	
Phaeosphaeria typharum,	
Lulworthia spp.	Agarase
Lulworthia spp.	Lipase
Lulworthia spp.	Amylase
Lulworthia spp.	Chitinase
Corollospora maritima, Cirrenalia pygmea,	Cellulase complex
Varicosporina ramulosa,	
Pleospora pelagica, Pleospora vagans,	
Phaeosphaeria typharum,	
Lulworthia spp.	Laccase
Lulworthia spp.	Tyrosinase

Like their terrestrial counterparts, some marine fungi have been found to degrade nitrogen-rich materials like chitin that are not easily utilized by other organisms and therefore play an important role in biogeochemical cycling of nitrogen. Quantification of the contribution of marine fungi to the biogeochemical cycle of nitrogen is extremely difficult and Newell and co-workers working on coastal salt marshes have found that the role of fungi has been grossly underestimated. Fungal biomass can account for virtually all of the nitrogen present in decaying standing plant biomass in salt marshes (Newell, 1996). In the more oligotrophic open ocean, availability of nitrogen may be an important factor limiting fungal growth.

Phosphorus

As for all other organisms, phosphorus is as important to fungi as nitrogen (Beever & Burns, 1980; Jennings, 1989, 1995). The involvement of fungi in phosphorus uptake and metabolism in mycorrhizal associations is particularly well researched (Smith & Read, 1997). However, very few reports exist concerning marine fungi in relation to phosphorus. While working on aspects of the habitat, seasonality and species distribution of thraustochytrids and marine fungoid protists, Bongiorni and Dini (2002) found that among the nutrients assessed in the water column, only total phosphorus was related to variations of thraustochytrid densities. Sengupta and Chaudhuri (2002) reported on mycorrhizal associations in mangroves and their involvement in phosphorus acquisition. In salt marshes, mycorrhizal associations do occur, although these are often restricted to a few species (Carvalho *et al.*, 2001). In general, the most common plant families found on salt marshes (Chenopodiaceae and Graminae) are not strongly mycorrhizal. In contrast, halophytic members of the Asteraceae do tend to be functionally mycorrhizal (Rozema *et al.*, 1986; Carvalho *et al.*, 2001). In the open ocean, a much more oligotrophic environment than coastal waters, phosphorous availablity is known to limit bacterial productivity (Van Wambeke *et al.*, 2002), although effects on yeast populations are not known.

Sulphur

While sulphur plays an important role in global biogeochemical cycling in general, and in marine environments in particular, few reports concern involvement of fungi in these processes. Seawater is rich in sulphate, and sulphide is present in high concentrations in anaerobic marine environments. Sulphur is also an essential element to all organisms, including fungi (Slaughter, 1989; Jennings, 1995). Fungi are known to be able to oxidize inorganic sulphur (Wainwright, 1989) and to produce compounds like dimethyl sulphide (Slaughter, 1989), which is an important compound in the cycling of sulphur in marine environments. Fungi can therefore be expected to be involved in biogeochemical cycling of sulphur in marine sediments. A few studies report on the ability of fungi, not necessarily marine, to degrade both organic and inorganic sulphur compounds. Phae and Shoda (1991) described a fungus that is able to degrade hydrogen sulphide, methane ethiol, dimethyl sulphide and dimethyl disulphide, while Faison *et al.* (1991) reported on *Paecilomyces* sp., a coal-solubilizing fungus, which is able to degrade organic sulphur compounds, such as ethyl phenyl sulphide, diphenyl sulphide and dibenzyl sulphide. A few reports exist on the involvement of fungi associated with marine organisms in relation to sulphur. Zande (1999) describes an ascomycete on the gills of the gastropod *Bathynerita naticoidea*, which the author suggests could be involved in detoxification of the abundant sulphide compounds in its habitat. Also associated with sulphide-rich marine environments is the fungus *Fusarium lateritium*, which is able to degrade dimethylsulphoniopropionate derived from algae and the saltmarsh grass *Spartina alterniflora* (Bacic & Yoch, 1998).

Metals

Most of the elements in the periodic table are metals, yet the vast majority of research on the biogeochemistry of elements has focused on non-metals: carbon, nitrogen, phosphorus and sulphur. Despite the fact that the biogeochemistry of some metals, such as iron and manganese, has in recent decades received more attention, particularly in relation to biogeochemistry of marine sediments, most metals have been under-researched in this respect. Still some studies provide scattered information about the possible involvement of marine fungi in the biogeochemistry of metals. One very interesting type of habitat concerns deep-sea hydrothermal vents. Both Duhig *et al.* (1992) and Burgath and Vonstackelberg (1995) speculate that filamentous growth structures associated with iron and manganese pyritic deposits and incrustations are of bacterial or fungal origin. Additionally, Schumann *et al.* (2004) attributed branching fossil structures showing septa and central pores and associated with pyrite deposits in deep-ocean basaltic rocks to fungi. This is of course no proof that marine fungi are actively involved in the biogeochemistry of metals. More convincing, however, though not from a marine environment, is that manganese-oxidizing fungi were isolated from the streambed of a Japanese river by Tani *et al.* (2003). While some marine fungi might oxidize iron and manganese as part of a detoxification mechanism in environments where these metals are in abundance, other species may occupy nutrient-poor open ocean habitats. In such habitats the ability to recruit metals might be an advantage and from this perspective it is of great interest that siderophore-producing facultative marine fungi were identified by Vala *et al.* (2000).

Importance of fungi in biogeochemistry of marine environments

Due to the paucity of reports, we can only speculate on the importance of marine fungi in biogeochemical cycling. Also, because of the dependence of fungi on carbon, it can be expected that their abundance will be determined mostly by the availability of carbon, which increases from the open oceans to coastal seas, to coastal and estuarine ecosystems, such as salt marshes. No doubt partly due to their relative accessibility, salt marshes are the most studied marine ecosystems in relation to fungi.

Coastal and estuarine systems: salt marshes

At the terrestrial–marine interface, e.g. estuaries and salt marshes, high productivity results from nutrient recycling, riverine external loading, tidally mediated net export of organic matter, balancing of nutrient inputs and outputs as well as aerobic and anaerobic microbial activity (Adam, 1993). Much of the work to date on the role of fungi in such a marine system has taken place in tidal salt marshes along the Atlantic and Gulf coast of the USA (Gessner & Kohlmeyer, 1976; Gessner, 1977; Newell *et al.*, 2000). The saltmarsh cordgrass *S. alterniflora* Loisel. is an abundant and highly productive plant species in this system and has frequently been used as the model for understanding fungal–plant interactions in this unusual environment (Newell, 1993a, b). Organic matter originating from *S. alterniflora* has been reported as a key contributor to saltmarsh productivity via input to detritus food webs where microorganisms (via detrital formation and enrichment) are involved in herbivore nutrition (Gessner, 1977).

Fungal mycelia and their hyphal sheaths are complexly intertwined with plant tissues (>70% lignocellulose) in the *S. alterniflora* decomposition process (Newell *et al.*, 1996a, b). According to Newell (1996), there are four main strategies used by marine microbes to capture metabolizable organic carbon: (1) maximized use of surface area coupled with high substrate affinity, and suitable enzymes that are capable of diffusing into solid particles; (2) penetration by tunnelling or surface erosion; (3) penetration by absorptive ectoplasmic nets or rhizoids, and (4) pervasion via networks of self-extending tubular flowing cytoplasm within rigid microfibrillar tubes of chitin laminaran or cellulose laminaran. A key role of aerobic filamentous fungi is lignocellulose degradation (decomposition of plant matter where fungi efficiently break down the lignin polymer into smaller, low-molecular-weight fragments available either for conversion to fungal biomass or further decomposition by bacteria). Lyons *et al.* (2003) reported 15 distinct laccase sequence types from a group of fungi isolated from Newell's saltmarsh decay system, demonstrating high sequence diversity of this functional gene in a natural fungal community.

The life-cycle of *S. alterniflora* involves shoot growth from the perennial rhizomes from April to July, maturation and seed production from July to October, frost-mediated death in October, and an ensuing decomposition period aided by weathering, shearing and tidal action. Some of the fungi reported to be involved in this process were *Dreschslera halodes, Halosphaeria hamata, Phaeosphaeria typharum, Stagonospora* sp., *Buergenerula spartinae, Alternaria alternata, Epicoccum nigrum, Claviceps purpurea, Leptosphaeria obiones, Leptosphaeria pelagica, Pleospora pelagica, Pleospora vegans* and *Lulworthia* sp. (Gessner, 1977). However, more recently ascomycetous fungi have been reported as the major secondary producers within standing decaying leaves of smooth cordgrass *S. alterniflora* L. throughout the geographical range of this saltmarsh grass (Newell *et al.*, 2000; Newell, 2001a, b), further supported by the molecular data of Buchan *et al.* (2003). The four ascomycetes that appear to be most involved in this system are: *Phaeosphaeria spartinicola* Leuchtmann, *Mycosphaerella* sp. 2 (of Kohlmeyer & Kohlmeyer, 1979), *Phaeosphaeria halima* Johnson and *Buergenerula spartinae* Kohlm. et Gessner (Newell, 2001a, b).

The role of fungi in nutrient cycling has been further elucidated by data (also generated by Newell's group) from a salt marsh at Sapelo Island, Georgia, where significant differences in ergosterol (the biochemical indicator of living fungal biomass) content of decaying leaves were found to indicate differences in nitrogen (but not phosphorus) availability (Newell, 2001a, b). Incidentally, nitrogen is thought to be a limiting nutrient for both smooth cordgrass (Mendelssohn & Morris, 2000) and its associated fungal populations (Newell *et al.*, 1996b). Newell's group found that the ratio between fungal productivity and average ergosterol contents was significantly higher in winter/spring than summer/autumn, indicating that there may have been greater access by mycophagous invertebrates and/or bacterial competitors due to higher tides and lower elevation of leaves in those periods where fungi did not predominate.

Other marine ecosystems: coastal seas and oceans

Depending on currents, some areas of the open ocean are particularly low in certain nutrients, such as iron (e.g. Boyd, 2002; Turner *et al.*, 2004), which limits the growth of planktonic algal species and hence leads to low levels of primary productivity. This means that the open oceans tend to be generally poor in biomass, leading to low levels of decomposers. Although biomasses do tend to be low, it is not clear at present how important the fungal components of these systems are. It might be expected that at some depth, particularly below the photic zone, microorganisms including fungi are the most important organisms (relative to total biomass), growing on the detritus precipitating down from the ocean surface layers. At great depths, a surprising abundance of life can be found near 'smokers', deep-ocean hydrothermal vents, and it is very likely that fungi form a significant component of these systems (Duhig *et al.*, 1992; Burgath & Vonstackelberg, 1995; Nagahama *et al.*, 2001), although it is unclear to what extent fungi would be limited by low oxygen levels.

Conclusions

In comparison to terrestrial environments, knowledge concerning fungally mediated processes in marine environments is very limited. Although much effort has been expended in the isolation and identification of fungi from marine environments, as well as trying to understand their mechanisms of adaptation, the challenge of elucidating how fungi contribute to marine food webs and processes largely remains. It must be remembered that the global marine ecosystem consists of probably as many distinct ecosystems as found terrestrially, each requiring separate study and analysis. This has been carried out in a few notable cases, such as for salt marshes and some mangrove systems, but remains very sketchy for any open ocean system. Of course, such systems are difficult to study requiring use of research boat time and, in the case of deep-sea systems, the availability of submersibles. Nevertheless, these studies are carried out for other biological groupings such as marine bacteria and protozoa; a fungal component could be built into such research projects with relative ease. Marine mycologists have also been slow to incorporate molecular methods into their studies. Still, most estimates of diversity are isolate based, and there has been little attempt to incorporate functional gene studies into marine mycology programmes. It might be argued that without these approaches, combined with greater exploration of open sea systems, our view of how fungi contribute to the majority of marine food webs will remain extremely restricted. This should form the challenge for the current and next generation of marine mycologists.

References

Adam, P. (1993). *Saltmarsh Ecology*. Cambridge: Cambridge University Press.

Ahammed, S. & Prema, P. (2002). Influence of media on synthesis of lignin peroxidase from *Aspergillus* sp. *Applied Biochemistry and Biotechnology*, **102**, 327–36.

Almagro, A., Prista, C., Benito, B., Loureiro-Dias, M. C. & Ramos, J. (2001). Cloning and expression of two genes coding for sodium pumps in the salt tolerant yeast *Debaryomyces hansenii*. *Journal of Bacteriology*, **183**, 3251–5.

Alongi, D. M. (1987). The distribution and composition of deep sea microbenthos in a bathyal region of the western coral sea. *Deep Sea Research*, **34**, 1245–54.

Apinis, A. E. & Chesters, C. G. C. (1964). Ascomycetes of some saltmarshes and sand dunes. *Transactions of the British Mycological Society*, **47**, 419–35.

Armstrong, W., Wright, E. J., Lythe, S. & Gaynard, T. J. (1985). Plant zonation and the effects of the spring-neap tide cycle on soil aeration in a Humber saltmarsh. *Journal of Ecology*, **73**, 323–39.

Bacic, M. K. & Yoch, D. C. (1998). *In vivo* characterization of dimethylsulfoniopropionate lyase in the fungus *Fusarium laterilium*. *Applied and Environmental Microbiology*, **64**, 106–11.

Ballard, R. D. (1977). Notes on a major oceanographic find. *Oceanus*, **20**, 35–44.
Bansal, P. K. & Mondal, A. K. (2000). Isolation and sequence of the HOG1 homologue from *Debaryomyces hansenii* by complementation of the *hog1Delta* strain of *Saccharomyces cerevisiae*. *Yeast*, **16**, 81–8.
Barato, M., Basilio, M. C. & Baptista-Ferreira, J. L. (1997). *Nia globospora*, a new marine gasteromycete on baits of *Spartina maritima* in Portugal. *Mycological Research*, **101**, 687–90.
Barghoorn, E. S. & Linder, D. H. (1944) Marine fungi: their taxonomy and biology. *Farlowia*, **I**, 395–467.
Beever, R. E. & Burns, D. J. W. (1980). Phosphorus uptake, storage and utilization by fungi. *Advances in Botanical Research*, **8**, 127–219.
Beever, R. E. & Laracy, E. P. (1986). Osmotic adjustment in the filamentous fungus *Aspergillus nidulans*. *Journal of Bacteriology*, **168**, 1358–65.
Bergbauer, M. & Newell, S. Y. (1992). Contribution to lignocellulose degradation and DOC formation from a saltmarsh macrophyte by the ascomycete *Phaeosphaeria spartinicola*. *FEMS Microbiology Ecology*, **86**, 341–8.
Blomberg, A. & Adler, L. (1993). Tolerance of fungi to NaCl. In *Stress Tolerance of Fungi*, ed. D. H. Jennings. New York: Marcel Dekker, pp. 233–56.
Bohnert, H. J., Ayoubi, P., Borchert, C. *et al.* (2001). A genomics approach towards salt stress tolerance. *Plant Physiology and Biochemistry*, **39**, 295–311.
Bongiorni, L. & Dini, F. (2002). Distribution and abundance of thraustochytrids in different Mediterranean coastal habitats. *Aquatic Microbial Ecology*, **30**, 49–56.
Boyd, P. W. (2002). The role of iron in the biogeochemistry of the Southern Ocean and equatorial Pacific: a comparison of *in situ* iron enrichments. *Deep Sea Research Part II – Topical Studies in Oceanograph*, **49**, 1803–21.
Boyle, P. J. (1976). Ergot epiphytic on *Spartina* sp. in Ireland. *Irish Journal of Agricultural Research*, **15**, 419–24.
Buchalo, A. S., Nevo, E., Wasser, S. P., Oren, A. & Molitoris, H. P. (1998). Fungal life in the extremely hypersaline Dead Sea: first records. *Proceedings of the Royal Society, London Series B*, **265**, 1461–5.
Buchan, A., Newell, S. Y., Butler, M. *et al.* (2003). Dynamics of bacterial and fungal communities on decaying saltmarsh grass. *Applied and Environmental Microbiology*, **69**, 6676–87.
Burgath, K. P. & Vonstackelberg, U. (1995). Sulfide-impregnated volcanics and ferromanganese incrustations from the southern Lau basin (Southwest Pacific). *Marine Georesources and Geotechnology*, **13**, 263–308.
Burge, M. N. & Perkins, E. J. (1977). *Studies in the Distribution and Biological Impact of the Effluent Released by Albright and Wilson Ltd., Whitehaven*. Cumbria Sea Fisheries Committee Scientific Report 77/3.
Carvalho, L. M., Cacador, I. & Martins-Loucao, M. A. (2001). Temporal and spatial variation of arbuscular mycorrhizas in salt marsh plants of the Tagus estuary (Portugal). *Mycorrhiza*, **11**, 303–9.
Clement, D. J., Stanley, M. S., O'Neil, J. *et al.* (1999). Complementation cloning of salt tolerance determinants from the marine hyphomycete *Dendryphiella salina* in *Aspergillus nidulans*. *Mycological Research*, **103**, 1252–8.
Clipson, N. J. W. & Jennings, D. H. (1992). *Dendryphiella salina* and *Debaryomyces hansenii*: Models for ecophysiological adaptation to salinity by fungi that grow in the sea. *Canadian Journal of Botany.* **70**, 2097–105.
Clipson, N. J. W., Hajibagheri, H. A. & Jennings, D. H. (1990). X-ray microanalysis of the marine fungus *Dendryphiella salina* at different salinities. *Journal of Experimental Botany*, **41**, 199–202.
Clipson, N. J. W. Landy, E. T. & Otte, M. L. (2001). Fungi. In *European Register of Marine Species: A Checklist of the Marine Species in Europe and a Bibliography of*

Identification Guides, ed. M. Costello, C. Emblow & R. J. White, *Patriomoines Naturels*, **50**, 15–19. Paris: Publications Scientifiques du MNHN.

Cooke, J. C., Butler, R. H. & Madolo, G. (1993). Some observations on the vertical distribution of vesicular arbuscular mycorrhizae in roots of salt marsh grasses growing in saturated soils. *Mycologia*, **85**, 547–50.

Cotton, A. D. (1907–1911). Notes on marine pyrenomycetes. *Transactions of the British Mycological Society*, **3**, 94.

Craig, P. J. (1980). Metal cycles and biological methylation. In *The Handbook of Environmental Chemistry*, Vol. 1, Part A. *The Natural Environment and the Biogeochemical Cycles*, ed. O. Hutzinger. Berlin: Springer-Verlag, pp. 170–227.

Davis, D. J., Burlak, C. & Money, N. P. (2000). Osmotic pressure of fungal compatible osmolytes. *Mycological Research*, **104**, 800–4.

Dickinson, C. H. (1965). The mycoflora associated with *Halimione portulacoides* III. Fungi on green and moribund leaves. *Transactions of the British Mycological Society*, **48**, 603–10.

Dickinson, C. H. & Kent, J. W. (1972). Critical analysis of fungi in two sand dune soils. *Transactions of the British Mycological Society*, **58**, 269–80.

Dickinson, C. H. & Morgan-Jones, G. O. (1966). The mycoflora associated with *Halimione portulacoides* IV. Observations on some species of Sphaeropsidales. *Transactions of the British Mycological Society*, **49**, 43–55.

Dickinson, C. H. & Pugh, G. J. F. (1965). The mycoflora associated with *Halimione portulacoides*. I. The establishment of the root surface flora of mature plants. *Transactions of the British Mycological Society*, **48**, 381–90.

Duffy, A. P., Curran, P. M. T. & Muircheartaigh, I. M. O. (1991). Effect of temperature and nutrients on spore germination in marine and non-marine fungi. *Cryptogamic Botany*, **23**, 125–9.

Duhig, N. C., Davidson, G. J. & Stolz, J. (1992). Microbial involvement in the formation of Cambrian sea-floor silica iron-oxide deposits, Australia. *Geology*, **20**, 511–14.

Edwards, J., Chamberlain, D., Brosnan, G. *et al.* (1998). A comparative physiological study of *Dendryphiella salina* and *D. arenaria* in relation to adaptation to life in the sea. *Mycological Research*, **102**, 1198–202.

Ellis, M. B. & Ellis, J. P. (1997). *Microfungi on Land Plants – An Identification Handbook*. London: The Richmond Publishing Company Limited.

Elmsley, J. (1980). The phosphorus cycle. In *The Handbook of Environmental Chemistry*, Vol. 1, Part A. *The Natural Environment and the Biogeochemical Cycles*, ed. O. Hutzinger. Berlin: Springer-Verlag, pp. 147–67.

Faison, B. D., Clark, T. M., Lewis, S. N. *et al.* (1991). Degradation of organic sulfur compounds by a coal-solubilizing fungus. *Applied Biochemistry and Biotechnology*, **28/29**, 237–51.

Farrant, C., Hyde, K. D. & Jones, E. B. G. (1985). Further studies on lignicolous marine fungi from Danish sand dunes. *Transactions of the British Mycological Society*, **85**, 164–7.

Feeney, N., Curran, P. M. T. & O'Muircheartaigh, I. G. (1992). Biodeterioration of woods by marine fungi and *Chaetomium globosum* in response to an external nitrogen source. *International Biodeterioration and Biodegradation*, **29**, 123–33.

Flowers, T. J., Hajibagheri, M. A. & Clipson, N. J. W. (1986). Halophytes. *The Quarterly Review of Biology*, **6**, 313–37.

Gadanho, M., Almeida, J. & Sampaio, J. P. (2003). Assessment of yeast diversity in a marine environment in the south of Portugal by microsatellite-primed PCR. *Antonie van Leeuwenhoek*, **84**, 217–27.

Geiser, D. M., Taylor, J. W., Ritchie, K. B. & Smith, G. W. (1998). Cause of sea fan death in the West Indies. *Nature*, **394**, 137–8.

Gessner, R. V. (1977). Seasonal occurrence and distribution of fungi associated with *Spartina alterniflora* from a Rhode Island estuary. *Mycologia*, **69**, 477–91.

Gessner, R. V. & Kohlmeyer, J. (1976). Geographical distribution and taxonomy of fungi from salt marsh *Spartina*. *Canadian Journal of Botany*, **54**, 2023–37.

Goodman, P. J., Braybrooks, E. M. & Lambert, J. M. (1959). Investigations into die-back of *Spartina townsendii* agg. I. The present status of *Spartina townsendii* in Britain. *Journal of Ecology*, **47**, 651–77.

Grasso, S., Bruni, V. & Maio, G. (1997). Marine fungi in Terra Nova Bay. *New Microbiology*, **20**, 371–6.

Hendrarto, I. B. & Dickinson, C. H. (1984). Soil and root micro-organisms in four salt marsh communities. *Transactions of the British Mycological Society*, **83**, 615–20.

Hernandez-Saavedra, N. Y. & Romero-Geraldo, R. (2001). Cloning and sequencing the genomic encoding region of copper-zinc superoxide dismutase enzyme from several marine strains of the genus *Debaryomyces (Lodder and Kreger-van Rij)*. *Yeast*, **18**, 1227–38.

Hooley, P., Fincham, D. A., Whitehead, M. P. & Clipson, N. J. W. (2003). Fungal osmotolerance. *Advances in Applied Microbiology*, **53**, 177–211.

Hutzinger, O. (Ed.) (1980). *The Handbook of Environmental Chemistry*, Vol. 1, Part A. *The Natural Environment and the Biogeochemical Cycles*. Berlin: Springer-Verlag.

Jefferies, R. L., Davy, A. J. & Rudmik, T. (1979). The growth strategies of coastal halophytes. In *Ecological Processes in Coastal Environments*, ed. R. L. Jefferies & A. J. Davy. Oxford: Blackwell, pp. 243–68.

Jennings, D. H. (1989). Some perspectives on nitrogen and phosphorus metabolism in fungi. In *Nitrogen, Phosphorus and Sulphur Utilization by Fungi*, ed. L. Boddy, R. Marchant & D. J. Read. Cambridge: Cambridge University Press, pp. 1–32.

Jennings, D. H. (1995). *The Physiology of Fungal Nutrition*. Cambridge: Cambridge University Press.

Jennings, D. H. & Burke, R. M. (1990). Compatible solutes – the mycological dimension and their role as physiological buffering agents. *New Phytologist*, **116**, 277–83.

Johnson, T. W. & Sparrow, F. K. (1961). *Fungi in Oceans and Estuaries*. Weinhein: Cramer.

Jones, E. B. G. (1962). Marine fungi. *Transactions of the British Mycological Society*, **45**, 93–114.

Jones, E. B. G. (1982). Decomposition by basidiomycetes in aquatic environments. In *Decomposer Basidiomycetes: Their Biology and Ecology*, ed. J. C. Frankland, J. N. Hedger & M. J. Swift. Cambridge: Cambridge University Press, pp. 192–212.

Jones, E. B. G. (1995). Ultrastructure and taxonomy of the aquatic ascomycetous order Halosphaeriales. *Canadian Journal of Botany*, **73** (Suppl. 1), S790–S801.

Jones, E. B. G. & Mitchell, J. I. (1996). Biodiversity of marine fungi. In *Biodiversity, International Biodiversity Seminar*, ed. A. Climerman & N. Gunde-Climerman. Ljublijana: National Institute of Chemistry and Slovenia National Commission for UNESCO, pp. 31–42.

Kathiresan, K. & Bingham, B. J. (2001). Biology of mangroves and mangrove ecosystems. *Advances in Marine Biology*, **40**, 81–251.

Kimura, H., Harada, K., Hara, K. & Tamaki, A. (2002). Enzymatic approach to fungal association with arthropod guts: a case for the crustacean host, *Nihonotrypaea harmandi*, and its foregut fungus, *Enteromyces callianassae*. *Marine Ecology – Pubblicazioni della Stazione Zoologica di Napoli*, I **23**, 157–83.

Kirchner, M. (1995). Microbial colonization of copepod body surfaces and chitin degradation in the sea. *Helgolander Meeresuntersuchingen*, **49**, 201–212.

Kobayashi, H., Maguro, S., Yoshimoto, T. & Namikoshi, M. (2003). Absolute structure, biosynthesis and anti-microtubule activity of phomopsidin, isolated from a marine derived fungus *Phomopsis* sp. *Tetrahedron*, **59**, 455–9.

458   *N. Clipson* et al.

Kohlmeyer, J. (1967). Intertidal and phycophilous fungi from Tenerife (Canary Islands). *Transactions of the British Mycological Society*, **50**, 137–47.

Kohlmeyer, J. & Kohlmeyer, E. (1979). *Marine Mycology: The Higher Fungi*. New York: Academic Press.

Kohlmeyer, J. & Volkmann-Kohlmeyer, B. (2003). Fungi from coral reefs: a commentary. *Mycological Research*, **107**, 386–7.

Lorenz, R. & Molitoris, H.-P. (1997). Combined influence of salinity and temperature (*Phoma*-pattern) on growth of marine fungi. *Canadian Journal of Botany*, **70**, 2111–15.

Lucas, M. T. & Webster, J. (1967). Conidial states of British species of Leptosphaeria. *Transactions of the British Mycological Society*, **50**, 85–121.

Lutley, M. & Wilson, I. M. (1972). Development and fine structure of ascospores in the marine fungus *Ceriosporopsis halima*. *Transactions of the British Mycological Society*, **58**, 393–402.

Lyons, J. I., Newell, S. Y., Buchan, A. & Moran, M. A. (2003). Diversity of ascomycete laccase gene sequences in a southeastern US saltmarsh. *Microbial Ecology*, **45**, 270–81.

Martin, D. F. (1970). *Marine Chemistry*. New York: Marcel Dekker.

Mendelssohn, I. A. & Morris, J. T. (2000). Eco-physiological controls on the primary productivity of *Spartina alterniflora*. In *Concepts and Controversies in Tidal Marsh Ecology*, ed. M. P. Weinstein & D. A. Kreeger. Dortrecht: Kluwer Academic Publishers, pp. 59–80.

Meyers, S. P., Ahearn, D. G., Gunkel, W. & Roth, P. J., Jr. (1967). Yeasts from the North Sea. *Marine Biology*, **1**, 118–23.

Molitoris, H. -P. & Schaumann, K. (1986). Physiology of marine fungi: a screening programme for growth and enzyme production. In *The Biology Of Marine Fungi*. ed. S. T. Moss. Cambridge: Cambridge University Press, pp. 35–48.

Money, N. P. (1997). Wishful thinking of turgor revisited: The mechanics of fungal growth. *Fungal Genetics and Biology*, **22**, 173–87.

Nagahama, T., Hamamoto, M., Nakase, T. Takami, H. & Horikoshi, K. (2001). Distribution and identification of red yeasts in deep-sea environments around the northwest Pacific Ocean. *Antonie van Leeuwenhoek*, **80**, 101–10.

Newell, S. Y. (1991). *Phaeosphaeria spartinicola*, a new species on *Spartina*. *Mycotaxon*, **41**, 1–7.

Newell, S. Y. (1993a). Decomposition of shoots of a saltmarsh grass – methods and dynamics of microbial assemblages. *Advances in Microbial Ecology*, **13**, 301–26.

Newell, S. Y. (1993b). Membrane containing fungal mass and fungal specific growth rate in natural samples. In *Handbook of Methods in Aquatic Microbial Ecology*, ed. P. F. Kemp, B. F. Sherr, E. B. Sherr, & J. J. Cole. Boca Raton: Lewis publishers, pp. 579–86.

Newell, S. Y. (1996). Established and potential impacts of eukaryotic mycelial decomposers in marine-terrestrial ecotones. *Journal of Experimental Marine Biology and Ecology*, **200**, 187–206.

Newell, S. Y. (2001a). Multiyear patterns of fungal biomass dynamics and productivity within naturally decaying smooth cordgrass shoots. *Limnology and Oceanography*, **46**, 573–83.

Newell, S. Y. (2001b). Spore-expulsion rates and extents of blade occupation by ascomycetes of the smooth cordgrass standing decay system. *Botanica Marina*, **44**, 277–85.

Newell, S. Y. & Fell, J. W. (1982). Surface sterilization and the active mycoflora of leaves of a seagrass. *Botanica Marina*, **25**, 339–46.

Newell, S. Y., Arsuffi, T. L. & Palm, L. A. (1996a). Misting and nitrogen fertilization of shoots of a saltmarsh grass: effects upon fungal decay of leaf blades. *Oecologia*, **108**, 495–502.

Newell, S. Y., Porter, D. & Lingle, W. L. (1996b). Lignocellulosis by ascomycetes (fungi) on a saltmarsh grass (smooth cordgrass). *Microscopy Research and Technique*, **33**, 32–46.

Newell, S. Y., Blum, L. K., Crawford, R. E., Dai, T. & Dionne, M. (2000). Autumnal biomass and potential productivity of saltmarsh fungi from 29° to 43° North latitude along the United States Atlantic coast. *Applied and Environmental Microbiology*, **66**, 180–185.

Nilsson, A. & Adler, L. (1990). Purification and characterisation of glycerol-3-phosphate dehydrogenase (NAD⁺) in the salt tolerant yeast *Debaryomyces hansenii*. *Biochimica et Biophysica Acta*, **1034**, 180–185.

Nybakken, J. W. (1997). *Marine Biology – An Ecological Approach*. Menlo Park: Addison–Wesley–Longman.

Packham, J. R. & Willis, A. J. (1997). *Ecology of Dunes, Saltmarsh, and Shingle*. Cambridge: Cambridge University Press.

Paton, F. M. & Jennings, D. H. (1988). Effect of sodium and potassium chloride and polyols on malate and glucose 6-phosphate dehydrogenase from the marine fungus *Dendryphiella salina*. *Transactions of the British Mycological Society*, **91**, 205–15.

Peterson, K. R. L. & Koch, J. (1996). *Ansiostigma rotundatum* gen et sp. nov.: a lignicolous marine ascomycete from Svanemollen harbour, Denmark. *Mycological Research*, **100**, 209–12.

Phae, C. G. & Shoda, M. (1991). A new fungus which degrades hydrogen sulfide, methanethiol, dimethyl sulfide and dimethyl disulfide. *Biotechnology Letters*, **13**, 375–80.

Porter, D. (1986). Mycoses of marine organisms: an overview of pathogenic fungi. In *The Biology of Marine Fungi*, ed. S. T. Moss. Cambridge: Cambridge University Press, pp. 141–54.

Pugh, G. J. F. (1962). Studies on fungi in coastal soils II. Fungal ecology in a developing salt marsh. *Transactions of the British Mycological Society*, **45**, 560–6.

Pugh, G. J. F. & Jones, E. B. G. (1986). Antarctic marine fungi: a preliminary account. In *The Biology of Marine Fungi*, ed. S. T. Moss. Cambridge: Cambridge University Press, pp. 323–30.

Raghukumar, C. & Raghukumar, S. (1998). Barotolerance of fungi isolated from deep-sea sediments of the Indian Ocean. *Aquatic Microbial Ecology*, **15**, 153–63.

Raghukumar, S., Sarma, S., Raghukumar, C., Sathe-Pathak, V. & Chandramohan, D. (1994). Thraustochytrid and fungal component of marine detritus IV. Laboratory studies on decomposition of leaves of the mangrove *Rhizophora apiculata* Blume. *Journal of Experimental Marine Biology and Ecology*, **183**, 113–31.

Raybould, A. F., Gray, A. J. & Clarke, R. T. (1998). The long-term epidemic of *Claviceps purpurea* on *Spartina anglica* in Poole Harbour: pattern of infection, effects on seed production, and the role of fusarium heterosporum. *New Phytologist*, **138**, 497–505.

Rees, G., Johnson, R. G. & Jones, E. B. G. (1979). Lignincolous marine fungi from Danish sand dunes. *Transactions of the British Mycological Society*, **72**, 99–106.

Rohrmann, S., Lorenz, R. & Molitoris, H. P. (1992). Use of natural and artificial seawater for the investigation of growth, fruit body production, and enzyme activities in marine fungi. *Canadian Journal of Botany*, **70**, 2106–10.

Rozema, J., Arp, W., Van Diggelen, J. *et al.* (1986). Occurrence and ecological significance of vesicular-arbuscular mycorrhiza in the salt marsh environment. *Acta Botanica Neerlandica*, **35**, 457–67.

Salomons, W. & Förstner, U. (1984). *Metals in the Hydrocycle*. Berlin: Springer-Verlag.

Schaumann, K., Mulach, W. & Molitoris, H.-P. (1986). Comparative studies on growth and exoenzyme production of different *Lulworthia* isolates. In *The Biology of Marine Fungi*, ed. S. T. Moss. Cambridge: Cambridge University Press, pp. 49–60.

Schumann, G., Manz, W., Reitner, J. & Lustrino, M. (2004). Ancient fungal life in North Pacific eocene oceanic crust. *Geomicrobiology Journal*, **21**, 241–6.

Sengupta, A. & Chaudhuri, S. (2002). Arbuscular mycorrhizal relations of mangrove plant community at the Ganges river estuary in India. *Mycorrhiza*, **12**, 169–74.

Sguros, P. L. & Simms, J. (1963). Role of marine fungi in biochemistry of oceans 2. Effect of glucose, inorganic nitrogen, and tris (hydroxymethyl) amino methane on growth and pH changes in synthetic media. *Mycologia*, **55**, 728–41.

Shinano, H. (1962). Studies on yeasts isolated from various areas in the north Pacific. *Bulletin of the Japanese Society of Scientific Fisheries*, **28**, 1113–22.

Sieburth, J. M. (1979). *Sea Microbes*. Oxford: Oxford University Press.

Sivanesan, A. & Manners, J. G. (1970). Fungi associated with *Spartina townsendii* in healthy and die back sites. *Transactions of the British Mycological Society*, **55**, 191–204.

Slaughter, J. C. (1989). Sulphur compounds in fungi. In *Nitrogen, Phosphorus and Sulphur Utilization by Fungi*, ed. L. Boddy, R. Marchant & D. J. Read. Cambridge: Cambridge University Press, pp. 91–106.

Smith, G. W., Ives, L. D., Nagelkerken, I. A. & Ritchie, K. B. (1996). Caribbean sea-fan mortalities. *Nature*, **383**, 487.

Smith, S. E. & Read, D. J. (1997). *Mycorrhizal Symbiosis*. 2nd Edn. San Diego: Academic Press.

Suresh, P. V. & Chandra-Sekaran, M. (1998). Utilization of prawn waste for chitinase production by the marine fungus *Beauveria bassiana* by solid state fermentation. *World Journal of Microbiology and Biotechnology*, **14**, 655–60.

Tani, Y., Miyata, N., Iwahori, K. *et al.* (2003). Biogeochemistry of manganese oxide coatings on pebble surfaces in the Kikukawa river system, Shizuoka, Japan. *Applied Geochemistry*, **18**, 1541–54.

Torzilli, A. P. (1982). Polysaccharidase production and cell wall degradation by several salt marsh fungi. *Mycologia*, **74**, 297–302.

Torzilli, A. P. & Andrykovitch, G. (1986). Degradation of *Spartina* lignocellulose by individual and mixed cultures of salt-marsh fungi. *Canadian Journal of Botany*, **64**, 2211–15.

Turner, S. M., Harvey, M. J., Law, C. S., Nightingale, P. D. & Liss, P. S. (2004). Iron-induced changes in oceanic sulfur biogeochemistry. *Geophysical Research Letters*, **31**, Art. No. L14307.

Vala, A. K., Vaidya, S. Y. & Dube H. C. (2000). Siderophore production by facultative marine fungi. *Indian Journal of Marine Sciences*, **29**, 339–40.

van Uden, N. & Fell, J. W. (1968). Marine yeasts. In *Advances in the Microbiology of the Sea*, Vol. 1., ed. M. R. Droop & E. J. F. Wood. London: Academic Press, pp. 167–202.

Van Wambeke, F., Christaki, U., Giannakourou, A., Moutin, T. & Souvemerzoglou, K. (2002). Longitudinal and vertical trends of bacterial limitation by phosphorus and carbon in the Mediterranean Sea. *Microbial Ecology*, **43**, 119–33.

Venkateswara-Sarma, V., Hyde, K. D. & Vittal, B. P. R. (2001). Frequency of occurrence of mangrove fungi from the east coast of India. *Hydrobiologia*, **455**, 41–53.

Volkmann-Kohlmeyer, J. & Kohlmeyer, J. (1993). Biogeographic observations on pacific marine fungi. *Mycologia*, **85**, 337–46.

Wainwright, M. (1989). Inorganic sulphur oxidation by fungi. In *Nitrogen, Phosphorus and Sulphur Utilization by Fungi*, ed. L. Boddy, R. Marchant, & D. J. Read. Cambridge: Cambridge University Press, pp. 73–90.

Wangersky, P. J. (1980). Chemical oceanography. In *The Handbook of Environmental Chemistry*, Vol. 1, Part A. *The Natural Environment and the Biogeochemical Cycles*, ed. O. Hutzinger. Berlin: Springer-Verlag, pp. 51–68.

Wilson, I. M. (1954). *Ceriosporopsis halima* Linder and *C. cambrensis* sp. Nov.: two marine pyrenomycetes on wood. *Transactions of the British Mycological Society*, **37**, 272–85.

Wilson, I. M. (1965). Development of the perithecium and ascospores of *Ceriosporosis halima*. *Transactions of the British Mycological Society*, **48**, 19–33.

Yale, J. & Bohnert, H. J. (2001). Transcript expression in *Saccharomyces cerevisiae* at high salinity. *Journal of Biological Chemistry*, **276**, 15 996–6007.

Zande, J. M. (1999). An ascomycete commensal on the gills of *Bathynerita naticoidea*, the dominant gastropod at Gulf of Mexico hydrocarbon seeps. *Invertebrate Biology*, **118**, 57–62.

Zehnder, A. J. B. & Zinder, S. H. (1980). The sulphur cycle. In *The Handbook of Environmental Chemistry*, Vol. 1, Part A. *The Natural Environment and the Biogeochemical Cycles*, ed. O. Hutzinger. Berlin: Springer-Verlag, pp. 105–46.

Index

Printed in the United States
By Bookmasters